Grzimek's Animal Life Encyclopedia focuses on the diversity of animal life on Earth. Each entry in the *Encyclopedia* discusses an animal group's evolution and systematics, and describes the taxonomic relationships both within the group and between it and other animal groups. Publication of this volume, dedicated to evolution, is a natural extension of the *Grzimek's* series, providing in-depth information on evolutionary processes and mechanisms and summarizing the latest research and discoveries. It brings up-to-date the Evolution volume that accompanied the original *Grzimek's* set.

Grzimek's Animal Life Encyclopedia
Evolution

• • • •

Grzimek's Animal Life Encyclopedia
Evolution

● ● ● ●

Michael Hutchins, Series Editor
Valerius Geist and Eric R. Pianka, Advisory Editors

● ● ● ●

In Association with The Wildlife Society

GALE
CENGAGE Learning™

Detroit • New York • San Francisco • New Haven, Conn • Waterville, Maine • London

Grzimek's Animal Life Encyclopedia

Evolution

Project Editor: Deirdre S. Blanchfield

Editorial: Jason Everett and Melissa McDade

Rights Acquisition and Management: Robyn
Young

Composition: Evi Abou-El-Seoud

Manufacturing: Wendy Blurton

Imaging: John Watkins

Product Design: Kristine Julien and Jennifer
Wahi

For product information and technology assistance, contact us at
Gale Customer Support, 1-800-877-4253.
For permission to use material from this text or product,
submit all requests online at **www.cengage.com/permissions.**
Further permissions questions can be emailed to
permissionrequest@cengage.com

Cover photograph reproduced by permission of Christian Ziegler/Minden
Pictures/National Geographic Stock (picture of six species of butterfly and
two species of moth).

LIBRARY OF CONGRESS CATALOGING-IN-PUBLICATION DATA

Grzimek, Bernhard. [Animal life encyclopedia evolution] Grzimek's
animal life encyclopedia evolution / Michael Hutchins, series editor;
Valerius Geist and Eric Pianka, advisory editors; in association with The
Wildlife Society.
    p. cm.
    Includes bibliographical references and index.
    ISBN 978-1-4144-8669-7 (hardcover) – ISBN 978-1-4144-8670-3 (e-book)
    1. Evolution (Biology) – Encyclopedias. 2. Evolution (Biology) – Study and
teaching. I. Hutchins, Michael. II. Geist, Valerius. III. Pianka, Eric R. IV.
Wildlife Society. V. Title. VI. Title: Animal life encyclopedia evolution.

QH360.2.G65 2011
576.8—dc22   2010038259

Gale
27500 Drake Rd.
Farmington Hills, MI, 48331-3535

ISBN-13: 978-1-4144-8669-7   ISBN-10: 1-4144-8669-3

This title is also available as an e-book.
ISBN-13: 978-1-4144-8670-3   ISBN-10: 1-4144-8670-7
Contact your Gale, a part of Cengage Learning sales representative for
ordering information.

Printed in China by China Translation & Printing Serivices Limited
1 2 3 4 5 6 7 15 14 13 12 11

· · · · ·

Editorial board

Contents

· · · · ·

Preface

The idea that species were divinely created and immutable persisted for centuries. Then, a studious young man decided to leave his comfortable home in England and take a long, difficult voyage around the world. Charles Darwin's now-famous trip on H.M.S. *Beagle* from 1831–1836 would ultimately alter our view of the origin of species.

Of course others—such as Darwin's grandfather, Erasmus Darwin, and the well-known French naturalist, Jean-Baptiste Lamarck—had speculated that species might change over time, but they could not identify the mechanism. Darwin's keen observations of nature and his landmark 1859 book—*The Origin of Species by Means of Natural Selection, or the Preservation of Favoured Races in the Struggle for Life*—would provide that answer and rock the scientific world. His writings and those of the many scientists that followed would forever alter our perceptions of ourselves and the other living creatures that share this planet.

Darwin's most significant contribution to the biological sciences was his theory of evolution through natural selection. The basic concept is simple and elegant: Individual animals possessing heritable traits that allow them to survive and reproduce will pass those traits on to the next generation, while those that do not survive and reproduce will not pass along their traits. The result is a gradual adaptation of a species to its environment, and the evolution of traits that promote survival and reproduction.

Of course, environments do not remain static, a factor also seen as fueling evolutionary change and ultimately leading to the creation of new species. Amazingly, this concept was "discovered" independently by another avid naturalist, Alfred Russell Wallace, a contemporary of Darwin's, who also spent many years in the field observing and recording details about the natural world.

Darwin and Wallace based their theories on direct observations of nature. However, it was Darwin who amassed the largest amount of evidence. This, and his prolific writing on the topic, explains why his name, not Wallace's, has become most closely associated with the discovery of evolution through natural selection.

Many of Darwin's most significant and perceptive observations occurred during his visit to the Galápagos Islands off the western coast of South America. He recorded several different species of giant tortoises, each inhabiting different islands in the archipelago, which suggested that they all had a common ancestor but had somehow changed over time. Darwin observed a new and unique species of iguana, which, unlike its mainland ancestor, swam underwater and grazed on algae. He documented a new species of cormorant that possessed under-developed wings and had completely lost the ability to fly. Furthermore, Darwin studied the many species of finches that, although they looked very similar, had differently shaped beaks that allowed them to exploit different food sources more efficiently. One even used cactus spines as a tool to pry insects from cracks and crevices.

I had the privilege of visiting the Galápagos Islands in 1979 and of observing these natural wonders firsthand. For biologists like me, a visit to the Galápagos is analogous to a pilgrimage, but unlike Darwin, I knew what to expect based on a long line of scientists who had gone before me. While sailing through these unique desert islands, I often wondered how Darwin felt during his first encounters with these unusual natural phenomena. I assume he was thrilled by the discoveries, but I also suspect that it challenged his preconceived views, and began his remarkable intellectual journey. It also shaped his philosophy of science and life. Once a deeply religious man who accepted what he was told purely on the basis of faith, Darwin gradually became more skeptical, listening to what the natural world "told" him through direct observation and experimentation. He once said: "Great is the power of steady misrepresentation, but the history of science shows how, fortunately, this power does not long endure."

It is also remarkable that neither Darwin nor Wallace had any idea of how physical (phenotypic) traits were passed from one generation to the next. That had to wait until the discipline of genetics developed, which began with work of an obscure Austrian monk, Gregor Mendel, who studied patterns of inheritance in garden peas. The development of this discipline eventually led, in the late1930s and early1940s, to another great synthesis between modern genetics and evolutionary theory. Led by such notables as Theodosius Dobzhansky and Ernst Mayer, this period produced the Hardy-Weinberg Equation, which provided a greater under-standing of how species change as the result of population-level changes in gene frequencies that code for particular traits. These concepts were further confirmed by the discovery of DNA—the building block of all life—by James Watson, Francis Crick and Rosalind Franklin in 1953.

Since then, the field of evolutionary biology and all of its sister disciplines—including geology, paleontology, genetics, ecology, animal behavior, zoology, botany and others—have accumulated vast stores of knowledge about the natural world. The power of evolutionary theory has been demonstrated time and time again through both observation and predictive experimental studies.

Does this mean that Darwinian natural selection can explain everything about the natural world? No. There are many nuances to science, and we admittedly still have much to learn. As an example: Niles Eldredge and Stephen J. Gould introduced the concept of "punctuated equilibrium" in 1972, a comparatively rapid form of evolution that can occur in response to environmental change and lead to speciation. Though it did not change the basic concept behind natural selection, it did challenge the idea that all evolution is uniform and gradual.

This volume is an attempt to summarize our current knowledge about the evolutionary process and the evidence that has accumulated over the past century and a half since Darwin's writings. With the discovery of genetics and DNA, which is common to all living organisms, evolution through natural selection has become the unifying theory in all of biology. As such, it is a powerful tool with which to understand the origin of many of nature's wonders, whether it be fish that change sex, flowers that resemble bees, or the colorful plumage and bizarre courtship behavior of New Guinea's birds of paradise.

Darwin's revelation challenged prevailing non-scientific explanations for the origin of life on Earth, including humans. His theory generates public controversy even to this day. Some U.S. states have attempted to ban the teaching of evolution in elementary schools, harkening back to the heady days of the Scopes Monkey Trial and the spirited debates of William Jennings Bryant and Clarence Darrow. However, the courts have consistently upheld the differences between science-based knowledge and religion, which is not testable via the scientific method. That being said, controversy persists, and education is key, making the dissemination of information about evolution vitally important. Indeed, evolution is not just about dinosaurs and other extinct species. On the contrary, it is highly relevant to today's society and we ignore it at our peril. Our understanding of modern medicine, agriculture, and conservation—just to name a few—are dependent on our knowledge of biological evolution. How else can we combat disease organisms that quickly adapt to antibiotics, ensure the future of domestic animals and plants, and conserve some semblance of nature for future generations in a world dominated by human influences?

Evolutionary theory is often misinterpreted. In the past, some people have used it to promote and justify social dominance of one group over another, so it is important that we remain vigilant. However, it is also important to note the positive message of evolution: that all living organisms are related and dependent on many of the same things for their survival. This should be an even more powerful argument that humans should show greater respect for nature and not drive other species toward irreversible extinction. Given the interdependencies that exist in nature, such attitudes may be critical for the future of *Homo sapiens*.

Grzimek's Animal Life Encyclopedia is a 17-volume set, covering the entire animal kingdom, from protozoa to mammals. Originally edited by the famous Frankfurt Zoo director and conservationist, Bernhard Grzimek, and translated into English in the early 1970s, the first English edition, consisting of 11 volumes, was published by Van Nostrand Reinhold in both England and the United States. That version was very popular, but after 30 years the series was in urgent need of updating. During that time, thousands of biologists, working in both the field and laboratory, had generated a tremendous amount of new information covering a wide range of species. Recognizing an urgent need to include the new information, the entire collection was updated, reformatted, and completely rewritten by English-speaking experts from 2001 to 2003 under the direction of new owner and publisher, Gale. I had the privilege of serving as the primary Series Editor for that project. However, the project was not complete, as the original series also included three additional topical volumes: Ethology (animal behavior), Ecology, and Evolution.

With the publication of *Grzimek's Evolution, Second Edition*, we have now updated, completely rewritten, and reformatted one of these topical volumes. Yet despite the many scientific advances detailed in this book, the basic goals and perspective remain the same as those expressed by the editor of the 1976 first edition, Herbert Wendt, who said: "In the individual articles here the reader will find material he has not encountered in other works. However, science never comes to a standstill; it knows no final conclusion, and each new work brings forth new information compiled from the latest research in that field. . . . The reader can become acquainted with current theoretical thinking about evolution in reading this volume." This volume is the next iteration in the continuing accumulation of knowledge about the origins and inner workings of our natural world.

I'd like to express my deep appreciation to publisher Gale; to The Wildlife Society Council for allowing my participation in this project; to my topic editors Valerius Geist and Eric Pianka; and to the many expert authors who contributed essays that make up this new volume.

Bethesda, Maryland, 2010
Michael Hutchins
Series Editor

• • • • •

The relevance of evolution to contemporary society

A lot of time and energy has been consumed debating the question of whether evolution has happened or not, but to people who are neither evolutionary biologists nor creationists, a better question might be "why does it matter?" In many branches of science, including evolutionary biology, the impacts of a particular theoretical advance or new bit of data can seem obscure, or even irrelevant, to many people. To illustrate this, consider the following titles from the prestigious journal *Nature* on the topic of evolution from the first half of 2010: "Allelic Variation in a Fatty-Acyl Reductase Gene Causes Divergence in Moth Sex Pheromones"; "Post-copulatory Sexual Selection and Sexual Conflict in the Evolution of Male Pregnancy"; "Co-option of the Hormone-Signalling Module Dafachronic Acid–DAF-12 in Nematode Evolution"; "The *Ectocarpus* Genome and the Independent Evolution of Multicellularity in Brown Algae"; and "Evolution of Self-Compatibility in *Arabidopsis* by a Mutation in the Male Specificity Gene." A simple search of any recent scientific journal will likely reveal equally dense titles.

These topics are certainly interesting to scientists who work on similar questions, and they may even be inherently interesting to people with a desire to understand the way the world works, but it can be hard to see how these discoveries will have a material impact on people's day-to-day lives. If there is no impact—if the study of evolution is merely interesting but not important—then why is the topic subject to such impassioned debate? Why do many individuals devote their entire careers to studying the evolutionary process? Why do taxpayers invest hundreds of thousands of dollars funding evolutionary research?

The answer is that the study of evolution has important implications for human welfare. The human species (and every other species on the planet) is the outcome of millions of years of evolution; every aspect of our bodies and minds has been honed by natural selection. Understanding human evolutionary biology allows us to care for our bodies and minds more effectively. Learning about the evolutionary process has also enabled us to harness this force for our own ends. Evolutionary principles have unwittingly been in use for thousands of years in the practice of agriculture; more recently, an explicit understanding of evolution has allowed us to advance agricultural science. Pathogens are constantly evolving to thwart our efforts to control them, but evolutionary studies of pathogens have enabled scientists to identify new emerging diseases, to determine how diseases are transmitted, and to design effective programs to halt epidemics. Several fields of study outside of biology have benefited from the insight that self-replicating systems evolve and have actually harnessed the evolutionary process to improve the design of, for example, computer programs. Finally, evolutionary biology may become a key tool to confront the challenge of a rapidly changing climate.

This entry offers examples of some of the benefits that evolutionary biology has bestowed on humanity. Many such examples exist, but only a limited number have been highlighted here, including some of the major contributions evolutionary biology has made to contemporary life. Some of the ways in which evolutionary biology is expected to assist humankind in the future are also discussed.

Topic overviewsFirst, however, it is critical to define the term *evolution*. Evolution simply refers to a pattern of change in the phenotype of members of a population (e.g., any characteristic of its behavior or anatomy) through time. Evolution refers only to changes in traits that are inherited from one generation to the next, usually because such changes are the result of genetic differences, though cultures can also evolve by processes analogous to biological evolution (see the entry "Evolution of language"). Several processes can lead to evolution. Genetic drift, the change in the genetic composition of a population due to random variation in birth and death, is one such process. Mutation, genetic changes due to errors during DNA replication, is another. However, the process that most evolutionary biologists credit with producing broad evolutionary patterns is natural selection. First elucidated by Charles Darwin in 1859 in his book *On the Origin of Species*, natural selection is the process by which individuals with advantageous variations (for example, animals that are faster or stronger) succeed in producing more offspring, while individuals with harmful variations fail to survive or reproduce. For natural selection to lead to evolution, individual variation must be passed from parent to offspring; in other words, it must be heritable. Over time, many small advantageous variations can accumulate, leading to larger patterns of change within the population. Evolutionary biologists propose that these processes acting together—natural selection, genetic drift, and mutation—can explain the diversity and complexity of all life on Earth.

A variety of methods are used to study the evolutionary process. Paleontologists study evolution using fossils to observe some of the different life-forms that have existed through history, including the ancestors of plants and animals alive today (see the entries "The fossil record: A window to the past" and "What does the fossil record tell us about evolution?). Phylogeneticists use similarities and differences between organisms to arrange species into phylogenetic trees that reflect the historical relationships among groups. Evolutionary ecologists and experimental evolutionary biologists measure selection within or among natural or experimental populations in real time (see the entry "Evolutionary ecology"). Evolutionary geneticists use molecular methods to trace the genetic changes that allow an organism to take on a new appearance (see the entry "Genetics: The blueprint of life"). Finally, some evolutionary biologists use computer simulations to predict what the outcome of selection might be under given conditions. Together, these methods have been used to advance our understanding of the process of evolution. This body of knowledge has been put to use for human benefit in a variety of ways.

Evolutionary agriculture

Thousands of years before Darwin articulated the theory of evolution by natural selection, humans were intuitively using evolutionary principles to shape domestic plants and animals. In fact, Darwin's knowledge of animal husbandry (including his own extensive work breeding pigeons) was integral to the development of his theory. Breeding experiments at home helped him to appreciate the variation among individuals of a species and the potential to harness that variation for a variety of purposes. By keeping and breeding livestock with especially useful characteristics, early humans (and Darwin) intentionally or unintentionally preserved desirable variations and eliminated undesirable ones. Likewise, by saving only the largest and most flavorful seeds to plant in the coming year, early humans were sure to gradually improve the quality of their crops. The steady accumulation of variation in a particular direction ultimately produced plants and animals so different from their wild counterparts that they could scarcely be recognized as relatives. Different varieties of the same species became as different as distinct species, or even genera.

Although agriculture is no longer the primary occupation of citizens in many developed countries, agricultural products are still one of the foundations on which civilization is built. Consider corn, one of the most successful and widespread agricultural products in history. About 800 million tons of corn are produced annually; corn is consumed directly, modified to make a variety of food additives, used as feed for domestic animals, and fermented to make ethanol for fuel. Yet the wild progenitor of corn, teosinte grass, is a humble plant with little nutritional or economic value. The transformation of corn began thousands of years ago when Central American people began to harvest and plant the seeds of teosinte. In its wild form, teosinte has only five to ten small kernels, and each kernel is encased in a hard shell that makes it difficult and time consuming to extract. A relatively small number of changes resulted in modern corn, including an increase in the size and number of kernels per ear, a decrease in the height and number of branches of the plant, and a reduction in the strength of the protective coat surrounding each kernel.

Research conducted since the early twentieth century has helped to identify the genes that are responsible for these changes. For example, it appears that a single amino acid change in the gene *teosinte glume architechture1* (*tga1*) is responsible for the reduction of the thickness of the seed coat in corn. Changes in the gene *Teosinte branched1* are responsible for a reduction in the height and number of branches in domestic corn versus wild corn. Genes have also been identified that control kernel color, starch content, inflorescence structure, and the circadian rhythms of domestic corn.

These studies have increased our understanding of how domestic corn differs from teosinte; they may also help scientists accelerate the development of new varieties. For example, plant breeding companies are working to identify the genetic changes that are responsible for their best varieties—several of these changes could then be combined in a single variety. They are also screening wild populations of teosinte (and other wild progenitors of domestic crops) for new gene variants that may prove useful. In some cases evolutionary geneticists have been able to increase the nutritional value of a crop by manipulating gene expression. For example, scientists have been able to increase the provitamin A content of several varieties of rice using a variant known from wild populations.

The transformations that humans have wrought in animals are in some cases far more dramatic than the changes observed in plants. For example, all breeds of the domestic dog are believed to be the result of a single domestication of the gray wolf (see the entry "Canid evolution: From wolf to dog").

In some cases, the domestication of wild animals has resulted in genetic changes to both animals and humans. For example, very few human cultures have begun using dairy products from cattle to supplement their nutritional needs. Dairy products such as milk and cheese contain a unique sugar, lactose, than can be digested only with the enzyme lactase. All people have the gene for lactase and produce copious quantities of the enzyme as infants when milk comprises the majority of their diet. The adult human diet, however, has not historically contained milk or milk products, and in most people the lactase gene is shut off later in life. For adults who are not producing lactase (who are lactose intolerant), the consumption of dairy products can result in severe gastrointestinal distress.

In the cultures in which dairy cattle are used, adults would clearly benefit from being able to consume lactose-containing milk as adults. This ability, also called lactase persistence, has in fact evolved independently in several regions in which dairy products are a part of the diet. Genetic mutations in the regulatory region of the gene have resulted in lactase expression in adulthood, allowing comfortable digestion of lactose. The genetic changes that cause lactase persistence are currently under positive selection, as the consumption of dairy has spread worldwide. Consequently, these genetic changes are becoming more common in the human population over time.

In addition to corn and cattle, evolutionary biologists are undertaking studies of many other domestic species of plants and animals. For example, microbiologists are evolving microbes that can be used to clean up nuclear waste and other pollutants. Biological sources of renewable energy, such as biofuels produced by algae, are being developed in part by human selection. The potential to use evolution to solve the problems facing the human population has never been greater.

Not all anthropogenic evolution currently underway is intentional, however. Human actions have altered the action of selection on many wild species, and they are evolving in response to these human actions. For example, fishermen tend to collect the largest fish they catch and throw back the smaller ones. A fish that reproduces early and at a small size will have an advantage under these conditions, first because it is likely to reproduce before it is caught and second because it is more likely to be thrown back if it is caught. This appears to be leading to the evolution of smaller body size in many economically important species of fish. In cities, high levels of ambient noise have made it more difficult for birds to hear one another's songs. As a result, bird songs are evolving to be louder and to stand out against the sounds of traffic. Birds in cities may also switch to another sensory mode to communi- cate—for example, through visual displays—and stop singing altogether.

Studies of the unintentional evolutionary consequences of human behavior may be as important as intentional human selection in shaping the future of many species. Evolutionary biology can illuminate our effect on plant and animal evolution, allowing us some measure of control over the impacts we have on evolution in the wild.

Evolution and epidemics

Evolutionary biology is also changing the way we understand and combat epidemic disease. The pathogens that cause disease are able to evolve substantially more quickly than humans can because of their large populations and rapid rate of reproduction. This means that they can evolve circles around us, while we struggle to evolve a response. Scientists, however, are learning the tricks that pathogens use to bypass human immune systems. Understanding how pathogens evolve in response to the actions of individuals will allow us to manipulate our own behavior to stop disease from spreading, to ameliorate the affects of infection on each individual, and to actually direct the evolution of a pathogen toward reduced virulence.

The potential for evolutionary biology to inform medicine has become especially clear in the decades following the outbreak of human immunodeficiency virus (HIV) around the world. Like a typical virus, HIV consists of a minimal bundle of genetic material coated in proteins that slips unnoticed into a host cell, inserts itself into the host genome, and hijacks the cellular machinery to produce more virions, which then infect

other cells within the same individual or are passed to another person. The host cells on which HIV specializes are part of the human immune system; when the immune system is compromised, many other pathogens are able to invade the body more easily. This is why acquired immunodeficiency syndrome (AIDS), the clinical manifestation of HIV infection, can take so many forms.

Unlike many viruses, HIV is a retrovirus, meaning that it encodes its genetic information in RNA rather than DNA as it moves from host cell to host cell. RNA is translated to DNA by the error-prone enzyme reverse transcriptase, leading to frequent mutations. As a consequence, retroviruses evolve especially quickly. Within a single individual, several lineages of virus can evolve and diverge over a short period of time. This rapid evolution makes HIV difficult to treat, but it also makes the virus especially amenable to evolutionary studies.

The evolution of the HIV virus happens quickly enough that phylogenetic trees can be used to trace transmission from one person to another. Some of the unique mutations that are present in an infected individual will be passed on when the virus is transmitted to another person, resulting in a greater similarity between the viruses in these two people than in the larger population. The newly infected person, however, will likely acquire only one or two of the strains of HIV that have evolved in the other person; their strain of the virus will be nested within the tree of viruses from the person who infected them.

A phylogenetic tree of HIV was used in court in 2002 to prosecute a gastroenterologist accused of attempting to murder his mistress by forcibly injecting blood drawn from an HIV-positive patient. The mistress, who had tested negative for HIV infection on several occasions when she donated blood, tested positive in 1994 after being stabbed with a needle by the doctor during an argument. Subsequent investigation revealed that the doctor had drawn blood in a suspicious manner from a patient who was HIV positive. The patient and the mistress had no known contact, but a phylogenetic analysis of the HIV strain in the victim clearly indicated that the source of the infection was the patient. The doctor was convicted of attempted murder. Subsequently, phylogenetic analyses have been used in other criminal investigations, including cases of rape and assault.

Evolutionary analyses have also helped scientists devise better treatment regimens for patients infected with HIV. The earliest treatments used to combat the virus included administration of azidothymidine (AZT), a chemical that interferes with the function of reverse transcriptase and therefore blocks the ability of the virus to incorporate itself into the host genome. However, the rapid pace of evolution of HIV allowed the virus to quickly evolve AZT resistance. Other drugs with similar modes of action likewise helped in the short term but failed in the long term because of viral evolution. Given the rapid rate of viral replication and the huge numbers of virions found within any particular patient, the evolution of resistance was a virtual certainty, even when the probability of a particular mutation for resistance was relatively small. While the probability of resistance evolving against a single drug is high, the probability of *several* rare mutations occurring simultaneously within a single virion is much, much lower. To take advantage of this, treatment has shifted so that patients receive several drugs at the same time, and the combination of drugs in use changes periodically. Although the virus will still eventually evolve resistance to the combination drug therapy, viral numbers will be reduced for a longer period of time than compared with treatment with a single drug. Also, because HIV evolves quickly, it quickly loses resistance to a drug when it is no longer being administered. These drugs can then be reintroduced to the patient when drug resistance has been lost. This treatment regimen has added many years to the life expectancy of a person diagnosed with AIDS.

In the long term, the impact of HIV on the population can be alleviated by manipulating the evolution of virulence, the ability of a pathogen to produce disease. The evolution of virulence is a classic theoretical application of evolutionary theory. A pathogen will benefit from being highly virulent only under some circumstances. In general, infections that are highly virulent are the result of pathogens that are reproducing at a high rate. The advantage to the virus of this accelerated reproductive rate is that it is able to produce many viral particles, and these make it more likely that an infected person will pass the virus on to others. The disadvantage to a virus is that it kills its host too quickly—before the host can pass the virus along—and therefore dies with its host. A virulent strain of virus will have an advantage when it has many, frequent opportunities to spread to other hosts, but the same strain will be at a disadvantage if opportunities to spread are few and far between.

Evolutionary theory therefore predicts that evolution will favor increased virulence in pathogens that spread quickly and will favor reduced virulence in pathogens that spread slowly. As a consequence, simple alterations to human behavior can dramatically alter the course of a disease in infected individuals. For example, the virulence of the common cold can be reduced by regular hand washing and by encouraging sick individuals to stay at home. These public health measures are cheap and easy—and hugely effective.

With regard to HIV, simple measures such as increasing the use of condoms, reducing the average number of sexual partners of each person, screening donated blood, and distributing clean needles to intravenous drug users can all work together to direct the evolution of HIV. These measures will not only reduce the number of new cases of HIV infection, they will also reduce the virulence—and hence the lethality—of the disease in infected individuals.

The same sort of thinking has been applied to the use of pesticides in agriculture in an attempt to forestall the evolution of pest resistance. These ideas have led to planting schemes that would not have been suggested otherwise, with big economic benefits. For example, large tracts of land planted with a single, genetically homogenous population of a crop are highly conducive to the rapid spread of a pathogen and are unlikely to contain resistance genes. Planting fields with a combination of different plants, arranging the landscape into a patchwork of different crops, and maintaining genetic diversity within crops can help limit the spread of pathogens and increase the

likelihood that resistance to a pathogen will evolve if an outbreak does occur.

Nonbiological evolution

Evolutionary theory has proven useful outside of biology as well. The conditions required for evolution by natural selection are quite simple: a self-replicating system in which variations in the ability to reproduce are passed from parents to offspring. It turns out that many systems meet these conditions and not all such systems are biological. For example, languages can be said to evolve, as can minerals, technology, and possibly even universes.

One area in which evolutionary thinking has led to major advances is computer programming. Using evolutionary thinking, computer programmers have designed systems in which different programs compete to accomplish a set task. At each iteration, variations are randomly introduced into program function in a manner analogous to genetic mutations. Variations that improve the performance of the program are preserved at the next stage, whereas less successful programs are eliminated. After many generations, an optimal programming solution to the given problem will be found. This algorithm gives programmers the ability to easily explore a vast swath of "program space" without writing each slight variant individually. Sometimes this method can produce solutions that are counterintuitive and that programmers would be unlikely to discover. Evolutionary programming has been used for such programming tasks as the design of electronics, quantum computing, and the development of competitive strategies.

Conclusion: The evolving present

"It may be said that natural selection is daily and hourly scrutinizing, throughout the world, every variation, even the slightest; rejecting that which is bad, preserving and adding up all that is good; silently and insensibly working, whenever and wherever opportunity offers, at the improvement of each organic being in relation to its organic and inorganic conditions of life."
—*Charles Darwin (1859, p.84)*

Now more than ever, it is important to understand how evolution works. Evolution is happening around us at every moment as living organisms face the unrelenting string of challenges that define the struggle to survive and reproduce. Increasingly, these challenges are the result of human activity in the world: hunting and gathering, habitat destruction and fragmentation, pollution, the introduction of invasive species, and the spread of novel pathogens, to name a few. Yet, of all the impacts humans have had on the planet so far, none compare to the possible impacts of the changes in climate that are already underway. Before the end of the twenty-first century, most species are likely to face dramatic shifts in average temperature, precipitation, and seasonality within their range. Early predictions suggest that many species will not survive this transformation. The loss of species diversity will deprive us of more than the fascination of biological diversity—in a less diverse world whole functions of Earth's ecosystems may fail. The planet-wide cycles of carbon, nitrogen, oxygen, water, and other critical nutrients may break down. Droughts, floods, hurricanes, and other catastrophes are likely to become more common. A reduction in biodiversity will also probably lead to more frequent and severe epidemics, both in humans and the plants and animals that feed us. These predictions might be considered alarmist by some; by others they are a conservative guess about the things that could go wrong in a system that is still very poorly understood.

In the face of such uncertainty, the future surely holds immense challenges for all living things. How will humankind confront ecological catastrophe when it occurs? Organisms can respond to a changing climate in three ways: They can adapt, they can move, or they can go extinct. There is already evidence of all three of these occurring. The question is: Which will predominate? The ability of organisms to move to keep up with habitat shifts has been compromised by the fragmentation and destruction of natural habitat. Adaptation has successfully overcome such challenges in the past, but contemporary climate change is happening more rapidly then in the past. Although it is now known that adaptation by natural selection can move more rapidly than was previously thought, it is unclear whether it can keep pace with climate change.

If we want to prevent species extinction, there are a few strategies suggested by evolutionary biology. First, we can facilitate range shifts by preserving natural habitat and increasing the connectivity among habitats. When this is not possible, we can aid adaptation in three ways. First, we can maintain large population sizes of species that are threatened. This will give natural selection more raw variation to act on and will help prevent the negative effects of inbreeding. Second, we can help spread helpful mutations when they arise. This might be accomplished in a couple of ways. We can move individuals with beneficial traits into other populations. Or we can identify the specific mutation that contributed to adaptation and genetically engineer it into other organisms. This strategy could be useful for such purposes as spreading genes for resistance to chytridiomycosis, a fungus that is causing substantial mortality in amphibians worldwide. Third, we can try to ameliorate climate change to give species time to adapt. Further evolutionary research may suggest additional measures to advance conservation (see the entry "Evolution and biodiversity conservation").

Evolution and the forces that drive it—natural selection, drift, and mutation—are ubiquitous in the world around us. Whether we choose to accept the reality of evolution or not, we are both shaping the course of evolution and in turn being shaped by it. We can, however, help to choose where evolution will take us by shaping our own behavior. In some of the areas of our lives in which evolution has the most direct impact—on the plants and animals used in agriculture, the ecosystem on which we depend, and the pathogens that infect us, to name a few—we have the power to direct the outcome of evolution for our own benefit and the benefit of the planet. The first step to harnessing this power is to make sure that the principles of evolution are widely taught and understood.

Resources

Books

Darwin, Charles. 1859. *On the Origin of Species by Means of Natural Selection.* London, UK: J. Murray.

Periodicals

Carroll, Sean P., Andrew P. Hendry, David N. Reznick, and Charles W. Fox. 2007. "Evolution on Ecological Time-Scales." *Functional Ecology* 21(3): 387–393.

Doebley, John F., Brandon S. Gaut, and Bruce D. Smith. 2006. "The Molecular Genetics of Crop Domestication." *Cell* 127 (7): 1309–1321.

Dunlop, Erin S., Katja Enberg, Christian Jørgensen, and Mikko Heino. 2009. "Toward Darwinian Fisheries Management." *Evolutionary Applications* 2(3): 245–259. Part of special issue of the same title.

Hillis, David M. 2005. "Health Applications of the Tree of Life." In *Evolutionary Science and Society: Educating a New Generation*, ed. Joel Cracraft and Rodger W. Bybee. Revised Proceedings of the BSCS, AIBS Symposium, Chicago, November 2004. Colorado Springs, CO: Biological Sciences Curriculum Study.

Ingram, Catherine J. E., Charlotte A. Mulcare, Yuval Itan, et al. 2009. "Lactose Digestion and the Evolutionary Genetics of Lactase Persistence." *Human Genetics* 124(6): 579–591.

Parmesan, Camille. 2006. "Ecological and Evolutionary Responses to Recent Climate Change." *Annual Review of Ecology, Evolution, and Systematics* 37: 637–669.

Alexis Harrison, Harvard University
Jonathan Losos, Harvard University

Contributing writers

David E. Alexander
Department of Ecology and
    Evolutionary Biology
University of Kansas

Bruce Anderson, Ph.D.
Department of Botany and Zoology
University of Stellenbosch

Elizabeth Ellwood, Ph.D.
Boston University

Steven A. Benner, Ph.D.
Created FFAME.org

Matthew C. Brandley, Ph.D.
Postdoctoral Fellow
University of Sydney

Guy L. Bush, Ph.D.
Emeritus Professor
Department of Zoology
Michigan State University

Angus Carroll
Manager, Darwin Census

Juliet Clutton-Brock, Ph.D, D.Sc.
Natural History Museum
London (retired)

Paul W. Ewald, Ph.D.
Evolutionary Medicine
University of Louisville

David Fitch, Ph.D.
Department of Biology
New York University

R. Allen Gardner, Ph.D.
Center for Advanced Study
University of Nevada

Michael A.D. Goodisman, Ph.D.
School of Biology
Georgia Institute of Technology

Lilach Hadany, Ph.D.
Department of Molecular Biology and
    Ecology of Plants
Faculty of Life Sciences
Tel Aviv University

Christian Hof, Ph.D.
Center for Macroecology, Evolution and
    Climate
University of Copenhagen

Ary Hoffmann, Ph.D.
The University of Melbourne

Thomas R. Holtz, Jr., Ph.D.
Department of Geology
University of Maryland

Lisa Horth, Ph.D.
Department of Biological Sciences
Old Dominion University

Michael Hutchins, Ph.D.
Executive Director/CEO
The Wildlife Society

Jon H. Kaas, Ph.D.
Department of Psychology
Vanderbilt University

Haagen D. Klaus, Ph.D.
Utah Valley University

Vladimír Kováč, Ph.D.
Comenius University
Faculty of Natural Sciences
Bratislava, Slovakia

Michael E. N. Majerus, Ph.D.
Department of Genetics

University of Cambridge

Katherine A. Mitchell, Ph.D.
The University of Melbourne

T. H. Oakley, Ph.D.
Ecology Evolution Marine Biology
University of CA-Santa Barbara

Kevin Padian, Ph.D.
Department of Integrative Biology and
    Museum of Paleontology
University of California

M. Sabrina Pankey
Ecology Evolution Marine Biology
University of CA-Santa Barbara

Kenneth Petren, Ph.D.
Department of Biological Sciences
University of Cincinnati

Eric R. Pianka, Ph.D., D.Sc.
Denton A. Cooley Centennial
    Professor of Zoology
Section of Integrative Biology
University of Texas at Austin

Richard B. Primack, Ph.D.
Biology Department
Boston University

Rob Roy Ramey II, Ph.D.
Wildlife Science International, Inc.

Derek A. Roff, Ph.D.
University of California

Michael Rowan-Robinson
Professor of Astrophysics
Blackett Laboratory
Imperial College London

Harry Roy, Ph.D.
Rensselaer Polytechnic Institute

Michael J. Ryan, Ph.D.
The University of Texas at Austin

Thomas W. Schoener, Ph.D.
Section of Evolution and Ecology
University of CA-Davis

James A. Serpell, Ph.D.
University of Pennsylvania
School of Veterinary Medicine

John J. Wiens, Ph.D.
Department of Ecology and Evolution
Stony Brook

Michael C. Wilson, Ph.D.
Department of Earth and
    Environmental Sciences
Douglas College

John van Wyhe
Departments of Biology and History,
    National University of Singapore
The Complete Work of Charles Darwin
    Online, Bye-Fellow Christ's College,
    Cambridge

Scientific methods and human knowledge

Human evolution

At the end of the Pleistocene (10,000 years before the present), only about 500 generations ago, humans were still hunter-gatherers, living off the land in small bands or tribes. Natural selection shaped these ancestral human populations, both physically and mentally, to cope with the natural world around them. Like all animals, humans dwell in a three-dimensional world that they perceive on a local spatial scale within a limited time horizon. Because our ancestors possessed very limited knowledge about how the world worked, they benefited by actively seeking evidence of connections among apparently unconnected events. As a result, humans today seem to be predisposed to look for causal connections by linking events whether or not they appear to be connected. We desperately want to understand and "know" things in order to exploit our environment for our own ends. Our brains, shaped by natural selection, allow us to do this fairly well for immediate events and simple local phenomena using our five senses and logic (Ornstein and Ehrlich 1989).

However, more complex, large-scale, time-delayed, and elusive things beyond the reach of manipulation, direct observation, or our senses are another matter. If people do not simply ignore them, they tend to resort to supernatural "explanations," invoking phenomena beyond those detected by sensory systems. Some have even suggested that the tendency to believe in such supernatural phenomena might have a genetic basis (Morrison 1999). The ability to believe and place faith in *super*natural (a misnomer, it should actually be called *sub*natural) phenomena has doubtless helped humans make apparent "sense" of otherwise puzzling or inexplicable events or phenomena. A predisposition for unquestioning belief in authority could spare each generation from repeating mistakes or having to rediscover or verify things that have already been discovered. It might also help reach consensus on "explanations" that cannot be verified. For example, many must have died in the process of finding out which plants and fungi were edible on a trial-and-error basis. Human ancestors no doubt sat around campfires telling one another stories, passing on such vital information from one generation to the next—this was the origin of human knowledge, culture, and the beginning of our domination of planet Earth.

Language

Development of verbal language allowed people to better exchange and expand ideas and concepts, no doubt facilitating control of their environment and thereby their survival and reproduction. Language is a double-edged sword, however: Words help to formulate concepts, but at the same time they limit the directions thought processes can take. The ways in which people envision the natural world around them are constrained by the words they develop, especially by the different meanings, attitudes, and emotions they can convey. Words, nouns in particular, can have very different referents between humans. For example, the word *mountain* means something quite different to someone raised in Colorado versus someone raised in Georgia. Precise definitions or universal agreement are needed to ensure accurate passage of understanding.

Frail and limited as humans are, they have struggled long and hard to comprehend the world around them. Bound by limited senses and life spans, people have nevertheless managed to begin to understand a fair bit about matter and nature. While people have difficulty imagining worlds with more than three dimensions or things without limits, it is a tribute to our intellect that there are words for concepts as elusive as hypervolumes, eternity, and infinity. Nevertheless, humans are much more comfortable with three dimensions and with things that begin and end.

Two different ways of "knowing"

Humans explain events and phenomena in two very different ways. One approach to knowing (sense 1, common sense) involves thinking and is objective, based on making repeatable observations that allow us to predict nature and future events; this rational, logical approach to knowing led to scientific methodology (Moore 1993). Another, very different, nonobjective approach to "knowing" (sense 2, faith-based) is based primarily on the invocation of "supernatural" explanations, bolstered by authorities who claim to have special access to "supernatural" sources. This nonscientific approach, championed by religions of all kinds, has helped many humans accept and cope with things they have no power to change or difficulty understanding rationally, such as unexpected deaths, other misfortunes, or seemingly random natural disasters.

Unfortunately, the power conferred on religious leaders has often led to serious abuses and resistance to accepting the rational understanding of the functioning of nature as demonstrated by new scientific discoveries. These two diametrically opposed ways people interpret and "know" (sense 1 versus sense 2) their environments have contributed to the regrettable past and modern-day conflicts between science and religion.

Human cognitive abilities

Human intelligence has also evolved so that people have remarkably good abilities to detect intentions of other humans in social interactions. This is their so-called intuition. People seem to have a propensity, however, for mysticism and a tendency to emphasize explanations that invoke intention over those based on sheer mechanism, situation, or circumstance. Indeed, humans may be predisposed to see intentions in their friends and enemies that may or may not exist. Similarly, they attribute conscious thought and intention to the actions of nonhuman animals (anthropomorphism). For example, predators "want" to kill us and prey "want" to escape from us. People even look for meaning and purpose in inanimate things such as the climate or the universe. Thus a destructive storm is interpreted as having occurred because people strayed from religious tradition or did something wrong and had to be "punished."

Everyone, religious or not, relies on objective rational thinking to handle problems encountered in everyday life. Thus, humans all know they must eat to stay alive, things fall down not up or sideways, they seek to avoid collisions when driving, balance their budgets, and so on. Remarkably, many people switch back and forth between rational knowing (sense 1) to faith-based "knowing" (sense 2) with ease. The brain may be organized in ways that promote such duality.

Adamant insistence on faith-based "knowing" coupled with careless use of words such as *believe* and *truth* have provided numerous opportunities to foment confusion and have allowed science to be deliberately maligned and misrepresented by those who stand to lose from changing sensibilities. Thus, religious leaders have often rejected new scientific evidence because it reduced the domain of processes over which religion could claim authority. As a result, scientific investigators have sometimes been vilified as Galileo Galilei (1564–1642) was during the Spanish Inquisition; scientists have even been tortured and executed because their views conflicted with mystical belief systems.

For various reasons, even many educated people still entertain faith-based systems of belief. They are comfortable

Scientific investigators have sometimes been vilified as Galileo Galilei (1564–1642) (on left) was during the Spanish Inquisition. © Bettmann/Corbis.

with "proofs" based on ancient mythology as the unchangeable "truth." People who "know" something or "believe" in supernatural "proof" are thus unable and/or unwilling to use logic and reason to comprehend reasoned alternatives—they cannot improve their understanding without substantial changes in their worldview (a paradigm shift) and thinking processes. Religious beliefs can be changed only when a believer or an authoritative leader has some new "revelation" or changes or rejects their current faith-based belief system. Sometimes, politically motivated charismatic leaders create their own new faiths.

Scientific methods, part 1

In contrast, scientific methods unrelentingly demand that one keeps an open mind, leading to a continually improved understanding of the natural world about them. Consider some definitions of science, such as: "Science is a set of cognitive and behavioral methods to describe and interpret observed or inferred phenomena, past or present, aimed at building a testable body of knowledge open to rejection or confirmation" (Shermer 2001, p. 98). In a 1987 U.S. Supreme Court case (*Edwards v. Aguillard*), an amicus curiae brief was tendered by a community of Nobel laureates, who defined science as follows: "Science is devoted to formulating and testing naturalistic explanations for natural phenomena. It is a process for systematically collecting and recording data about the physical world, then categorizing and studying the collected data in an effort to infer the principles of nature that best explain the observed phenomena." Their brief goes on to explain, somewhat apologetically, science's inability to explain putative "supernatural" events and phenomena:

> Science is not equipped to evaluate supernatural explanations for our observations; without passing judgment on the truth or falsity of supernatural explanations, science leaves their consideration to the domain of religious faith. Because the scope of scientific inquiry is consciously limited to the search for naturalistic principles, science remains free of religious dogma and is thus an appropriate subject for public-school instruction.

Scientists do not concern themselves with anything "supernatural"; they are interested only in naturally occurring phenomena. Motivated by curiosity about their surroundings, they assume that an organized reality exists in nature and that objective principles can be formulated, which will adequately reflect that natural order. This pivotal assumption that an external organized reality exists is *not* based on faith but is verified every day by observing predictable repeatable events such as day is followed by night. A scientist "believes" that an organized reality exists, but in a fundamentally different and much more rational way than a religious person "believes" in supernatural influences. Scientists go to great lengths to satisfy their desire to understand natural events and phenomena. They continually cross-check one another to verify currently accepted explanations. For a scientist, reason and logic always trump authority and faith as a way of knowing (Moore 1993; Pianka 2000). Some individuals have tried to characterize science as a form of religion, but the two forms of "knowing" are fundamentally different.

Ancient mariners believed the earth was flat and feared sailing off its edge. © Illustration Works/Alamy.

Superficial commonsense perceptions led human ancestors to invoke the longstanding notion of a "flat Earth" and "moving Sun." Under this now archaic geocentric view, Earth did not move but was at the center of the universe, with the Sun moving across the sky. Early mariners were actually afraid of falling off the flat Earth (strangely, people did not seem to be overly concerned about either what supported this flat Earth or about ocean waters spilling off and draining into some sort of bottomless abyss—why didn't the oceans drain dry?).

Over time, the understanding of the world has improved steadily as human knowledge has expanded. The quest for understanding has liberated and enlightened many. For example, in Europe during the Middle Ages, the origins of disease and other undesirable phenomena were thought to be the actions of demons—unseen creatures from Hell who wrecked havoc on the populace (Sagan 1995). Similarly, primitive peoples such as Australian, New Guinean, and African tribesmen attributed sickness to the influence of witches and spirits. It is now known that illnesses are frequently caused by microscopic organisms called germs and viruses, and at least this can give some level of comfort that lives are not controlled by unknown malevolent forces wishing to do harm. The ultimate result is that instead of continuing to burn witches at the stake, people have sought to create a medical profession.

Nevertheless, the capacity for ambiguity inherent in language has also provided a ready mechanism that has unfortunately permitted some to obfuscate, conflate, and misinterpret those same ideas and concepts. Communication is impaired when people use the same words in different ways, whether deliberately or not. Many commonly used words suffer because of just such a "failure to communicate." Problems arise especially when words convey divergent attitudes. In the context of scientific versus vernacular terminology, key words such as *fact*,

In Europe during the Middle Ages, the origins of disease and other undesirable phenomena were thought to be the actions of demons and witches were burned at the stake. © Bettmann/Corbis.

know, *truth*, *proof*, *faith*, *belief*, *design*, and *theory* are widely misconstrued because they convey different meanings to different audiences. Another term that is a source of considerable confusion is *random*.

What is a "fact?"

Take the term *fact*, for example. Most people consider a "fact" as "what really happened." Many "facts," however, are not so clean and simple—most involve varying levels of interpretation. Consider the apparently simple "fact" that the Sun rises each morning. Daily there is new evidence confirming this "fact." One can be quite confident that the Sun will rise again tomorrow. Under the now defunct concept of a "flat Earth" and "moving Sun," the Sun's movement across the sky was viewed from the perspective of a fixed nonmoving Earth at the center of the universe. Indeed, references to sunrise and sunset are based on this interpretation, which is supported by human superficial commonsense perceptions. Understandably, people think of themselves at the center of the universe and interpret other events and phenomena from such an anthropocentric frame of reference. But an understanding of cosmic events was greatly enhanced when instead of thinking of the Sun as moving, it is viewed as the center of a solar system, and Earth is interpreted as a rotating globe orbiting around a small star. The vocabulary has not caught up—clearly, one should refer to "sunrise" as "spinup" and "sunset" as "spindown" (Pianka 2000). In contrast, because the Moon does revolve around Earth, it is

appropriate to call its movements "moonrise" and "moonset." The German philosopher Friedrich Nietzsche (1844–1900) once wrote, "there are no facts, only interpretation." He somewhat overstated the case, because repeatable, observable, predictable events certainly qualify as "facts" even though they may often not be completely devoid of interpretation. People view the Sun's position change relative to the horizon every day, even though "sunrise" is a misinterpretation. Although it took a long while to become accepted, the heliocentric solar system perspective has now replaced the geocentric concept in the minds of most people.

Mythology

Some people seem to need to believe in a deity to make sense of their existence and the phenomena they perceive around them. Perhaps "knowing" that an omnipotent caring entity looks over them helps in confronting and coping with human weaknesses and limited life spans. People are expected to outgrow their belief in the Tooth Fairy, the Easter Bunny, and Santa Claus, but not the cherished myth of one or more omnipotent deities. Everybody wants to believe in a caring god and an afterlife, as comforting and irrational as that may be.

Darwin's dangerous idea

Darwin's ideas threatened to dethrone humans as such exalted creatures (Dennett 1995). Many people remain convinced that humans are fundamentally different from other life-forms—they find it odious even to contemplate that humans and great apes might have descended from a common ancestor—despite the fact that the vast majority of human genes are shared with chimpanzees and gorillas. Indeed, these apes have the same blood-group types as humans. All vertebrates share the same basic body plan. They are bilaterally symmetric, with a head, brain, nose, two eyes, paired forelimbs and hind limbs, stomach, intestine, heart, kidneys, liver, and assorted other internal organs. Even a tiny fish or lizard shares all these features with humans. Scientific evidence is overwhelming that all life on Earth arose from a single common ancestor. Human blood plasma approximates the salt concentrations of the oceans because life arose there. The genetic code is universal for all life-forms on Earth. Right now, genes that first evolved in bacteria billions of years ago in Earth's primeval seas operate respiratory metabolism within human bodies and keep humans alive from second to second. Green plants capture solar energy using genes perfected by ancient photosynthetic bacteria, providing the energetic foundation that supports all life on Earth. (Such microbes generated most of the oxygen that makes up Earth's current atmosphere, without which humans could not exist.) Such scientific evidence tells us that humans are simply one terminal branch of the vast tree of life. Microbes, fungi, and plants are our distant cousins. Some Eastern philosophies share the belief that we are one part of a huge river of life flowing through time. Hence we do have an afterlife, after all, in the form of the ongoing tree of life, especially our descendants.

Some religions, such as Buddhism, do not postulate a creator god, but nevertheless perceive nature as a creative process and therefore sacred to a high degree. Such a perspective essentially equates natural selection with god,

MR. BERGH TO THE RESCUE.

THE DEFRAUDED GORILLA. "That *Man* wants to claim my Pedigree. He says he is one of my Descendants."

Mr. BERGH. "Now, Mr. DARWIN, how could you insult him so?"

Mr. Bergh to the Rescue cartoon by Thomas Nast. Many people remain convinced that humans are fundamentally different from other life-forms, despite the fact that the vast majority of human genes are shared with chimpanzees and gorillas. © Corbis.

but is not reliant on the concept of a human-like, omniscient being that controls every action and event (Kauffman 2008).

Scientific methods, part 2

Science, especially at introductory levels, is too often taught as fact-based transmission of information, with inadequate attention paid to its process. People are taught that science is a body of answers, deliverable in absolutes ("learn this for your exam"), when they should be taught to think of science as a way of asking questions about the natural order of things. As a result, most people, including many who have taken several science courses, do not appreciate the scientific process of logical inquiry, especially the tentative and probabilistic nature of many scientific conclusions.

A widespread misconception is that science can explain everything—quite the contrary, science thrives on uncertainty because it always remains tentative (in science, nothing is ever known for certain). Scientists must always remain open minded, discarding weak explanations in favor of ones that better explain observed events. The strength of scientific methods is that, if these processes are adopted and followed rigorously, understanding and knowledge will improve steadily over time. Dogmatic faith-based belief systems impair such progress by vigorously defending and maintaining the status quo of archaic systems of "belief," preventing us from reaching our full human potential.

Despite such impediments, scientific methods have brought human understanding a long way.

Nevertheless, scientists, like all people, are fallible and subject to becoming dogmatic—not all scientists practice the methods of science correctly, nor honestly, nor use potentially confounding words properly. And, although plenty of zealots stand ready to pounce on any such mistake to discredit scientists, science, unlike religion, has powerful built-in self-correcting mechanisms. Hypotheses are tested and tests are replicated to double check on their accuracy.

Randomness

Another much-abused term with varied meanings is *random*. The dictionary definition is "without definite direction" or "lacking a definite plan, purpose, or pattern" or "equiprobable." Often the word is used to describe anything capricious or unpredictable, which usually means that something is simply so poorly understood that it appears indeterminate. Invoking "randomness" may often merely be a cover-up for our ignorance. Because the basic source of genetic change is "random" mutations, proponents of intelligent design mistakenly argue that "evolution is random." In actuality, natural selection favors highly nonrandom organisms whose adaptations to cope with their environments enhance their survival and reproductive success.

Scientific methods, part 3

Scientists begin an investigation by formulating hypothetical statements about how reality might work, called hypotheses (also known as models). All hypotheses make simplifying assumptions—some sacrifice precision for generality, whereas others sacrifice generality for precision (Levins 1966). Some hypotheses actually sacrifice certain aspects of realism itself! Hypotheses are mere "caricatures of nature . . . designed to convey the essence of nature with great economy of detail" (Horn 1979, p. 56). Many hypotheses are not "correct" or "true." Even a demonstrably false hypothesis can be useful in developing improved understanding. Any given hypothesis merely represents one particular attempt to explain reality. Hypotheses are like circus mirrors that do not reflect reality perfectly. Most hypotheses are to some extent incorrect, but scientists use them because nature is too complex to be investigated without employing simplifying assumptions. Hypotheses generate predictions that can be tested by confronting them with reality. Some types of hypotheses, such as historical ones involving evolutionary changes, are judged by their explanatory power, rather than their ability to predict. Evolutionary theory, however, does have strong predictive power. For example, hummingbird-pollinated flowers are often red: When an early naturalist exploring the New World tropics found a red flower with a long curved corolla in the jungle, he predicted that there must be a hummingbird with a long curved beak that had coevolved along with this plant to pollinate it. Years later, scientists discovered the hummingbird, as predicted.

For many years, some of the strongest evidence for continental drift was found in biological geography. One of the most conspicuous examples involves primitive freshwater lungfish, which occur in South America, Africa, and Australia.

The present-day geographic distributions of freshwater fish, such as the South American lungfish (*Lepidosiren paradoxa*), strongly suggest that the three great southern continents must have once been joined. © Tom McHugh/Photo Researchers, Inc.

Because these freshwater fish could hardly have crossed the oceans, their present-day geographic distributions strongly suggest that the three great southern continents must have once been joined. Distributions of presumably ancient flightless ratite birds (rheas, ostriches, and emus) suggest the same scenario, as do the distributions of some very old lineages of plants such as southern beech trees (*Nothofagus*), which are found on all the southern continents.

Most people do not realize that scientific knowledge usually does not advance because a scientist "proves" something, but rather science often proceeds because hypotheses are disproved. When the predictive or explanatory power of a hypothesis fails, it is either discarded or revised. Stated another way, if a hypothesis does not conform adequately to reality, it is replaced by another that reflects the real world more accurately. This iterative process of scientific inquiry is thus self-regulating; as time progresses, knowledge expands and is continually refined and improved to mirror external reality more and more accurately.

Observation and experiment play a vital role in science. They are used to test models and to refute inadequate hypotheses, and thus they help us formulate improved interpretations of natural phenomena. Some natural events cannot be manipulated. Thus, we cannot stop the Sun's fusion or Earth's rotation to test current ideas, but each daily observation of spinup or spindown nevertheless strengthens our confidence in the accepted interpretation of celestial events. Note, however, that such repeatability may be consistent with a hypothesis that is later shown to be incorrect. For example, predictions derived from a geocentric worldview survived tests for centuries until Nicolaus Copernicus (1473–1543) and Galileo Galilei (1564–1642) provided

convincing contrary evidence. This is why all scientific hypotheses and theories remain tentative and are always subject to being replaced when a superior explanation is discovered. Darwin's rational view of nature should now replace archaic myth-based creation stories.

Hypotheses become theories

In time, a well-substantiated hypothesis is elevated to become a robust scientific "theory" (nonscientists often comment "it's just a theory," invoking a much more speculative and demeaning attitude). Eventually, reliable scientific theories can even attain the status of "law," such as the laws of motion or the laws of thermodynamics. Darwin's mechanism of natural selection is truly a unifying theory of life, not even restricted to DNA-based life on Earth: It presumably would apply to any self-replicating entity (any life-form) anywhere in the entire megaverse (cosmos). Natural selection is as close to a "law" as we can get in biology. People's worldviews and personal philosophies would benefit greatly from embracing "Darwin's dangerous idea" (Dennett 1995), rather than naively rejecting it outright and refusing to examine the evidence for it. We are extremely fortunate to be able to learn from past genius and research effort. In a few hours of careful reading, anyone can now master material that required many lifetimes to acquire.

Darwin ended *On the Origin of Species* (1859) with the following passage:

> It is interesting to contemplate a tangled bank, clothed with many plants of many kinds, with birds singing on the bushes, with various insects flitting about, and with worms crawling through the damp earth, and to reflect that these elaborately constructed forms, so different from each other, and dependent upon each other in so complex a manner, have all been produced by laws acting around us. These laws, taken in the largest sense, being growth with reproduction; inheritance which is almost implied by reproduction; variability from the indirect and direct action of the conditions of life, and from use and disuse; a ratio of increase so high as to lead to a struggle for life, and as a consequence to natural selection, entailing divergence of character and the extinction of less improved forms. Thus, from the war of nature, from famine and death, the most exalted object which we are capable of conceiving, namely, the production of the higher animals, directly follows. There is grandeur in this view of life.... (pp. 373–374)

While an evolutionary perspective may dethrone humans as divine or special creations, it greatly enriches our understanding of the real world, much more so than one that invokes mythical or supernatural explanations such as some hypothetical "god" who intervenes at his whim.

Resources

Books

Darwin, Charles. 1994. *On the Origin of Species*. New York: Modern Library, Random House. (Orig. pub. 1859.)

Dawkins, Richard. 1995. *River out of Eden: A Darwinian View of Life*. New York: Basic Books.

Dennett, Daniel C. 1995. *Darwin's Dangerous Idea: Evolution and the Meanings of Life*. New York: Simon and Schuster.

Horn, Henry S. 1979. "Adaptation from the Perspective of Optimality." In *Topics in Plant Population Biology*, ed. Otto T. Solbrig, Subodh Jain, George B. Johnson, and Peter H. Raven. New York: Columbia University Press.

Kauffman, Stuart. 2008. *Reinventing the Sacred: A New View of Science, Reason, and Religion*. New York: Basic Books.

Moore, John A. 1993. *Science as a Way of Knowing: The Foundations of Modern Biology*. Cambridge, MA: Harvard University Press.

Morrison, Reg. 1999. *The Spirit in the Gene: Humanity's Proud Illusion and the Laws of Nature*. Ithaca, New York: Cornell University Press, Comstock Publishing.

Ornstein, Robert, and Paul Ehrlich. 1989. *New World New Mind: Moving toward Conscious Evolution*. New York: Doubleday.

Pianka, Eric R. 2000. *Evolutionary Ecology*. 6th edition. San Francisco: Benjamin-Cummings.

Sagan, Carl. 1995. *The Demon-Haunted World: Science as a Candle in the Dark*. New York: Random House.

Shermer, Michael. 2001. *The Borderlands of Science: Where Sense Meets Nonsense*. Oxford: Oxford University Press.

Periodicals

Levins, Richard. 1966. "The Strategy of Model Building in Population Biology." *American Scientist* 54(4): 421–431.

Other

Edwards v. Aguillard, 482 U.S. 578 (1987).

Holtz, Brian. 2005. "Human Knowledge: Foundations and Limits." Available from http://humanknowledge.net.

Eric R. Pianka

• • • • •

Creation stories: The human journey to understand the origins of life on Earth

Where did the earth and all its life come from? Why are we here? These questions have stirred human imagination since the dawning of civilization and probably for many millennia before. The human mind, forever curious, seeks answers to questions, both large and small. When humans first began to think abstractly and use language to convey their thoughts to others they likely also began to ask themselves probing questions about their world. The first "philosophers" among us began to tackle the most interesting questions of all—about our own origin and the origin of the stars, the oceans, the land, and the myriad of other creatures with which we share this planet. What resulted was a wide variety of explanations. These *creation stories*, as they are commonly called, filled a gaping void—a knowledge gap that was impossible to fill until the relatively recent emergence of science and the scientific method. Creation stories are as varied as the human experiences and cultures that spawned them and, as one would expect, closely reflect this cultural diversity and experience (Campbell 1990).

Creation stories are sometimes called *origin myths*, a term that presumes imaginary and unscientific roots. Paradoxically, *Merriam-Webster's Collegiate Dictionary*, Ninth Edition, defines the term *myth* as *both* a "traditional story of ostensibly historical events that serves to unfold part of the world view of a people" and "an unfounded or false notion." Not surprisingly, the use of the term often depends on one's perspective: One person's or culture's "myth" is another's historical "truth." This relativism is a major difference between faith-based "knowing" and the scientific method. In faith-based belief systems, the greatest importance is often attached to those aspects of myth that are the most fantastic and least probable. As Edmund R. Leach so aptly put it, "The non-rationality of myth is its very essence, for religion requires a demonstration of faith by the suspension of critical doubt" (1967, p. 1). When properly conducted, science is not a faith-based endeavor; in fact, it is just the opposite in that it requires verification through repeatable testing and observation. Thus, both reason and skepticism are central to a scientist's worldview. Theories or concepts that are unverifiable, or not supported by evidence, fall out of favor, whereas those with greater explanatory and predictive power emerge and gain wider acceptance. Creation stories are decidedly not science, but they are the beginnings of it, for the scientific process always begins with questions. As such, they are worthy of study and understanding, which is why they have been included in a volume on the scientific basis of and evidence for evolution.

Another legitimate reason for providing a broad sampling of the many thousands of creation stories that have emerged from human cultures over time is to combat intellectual, cultural, or theological arrogance—the arrogance of those who believe that they have a corner on "truth" and that their worldviews represent the one and *only* explanation for the origin of life on earth, including human life. This form of intolerance has led to xenophobia, religious wars, terrorism, and book burning, and has stifled human progress for centuries. Indeed, when proponents of creationism or its latest incarnation, so-called intelligent design, talk about giving "equal time" to theological *and* scientific explanations for human and nonhuman origins in our schools, they are referring to the Biblical version of creation. Yet, the Biblical (Hebrew) creation story is just one of many passed down from earlier generations and civilizations. It is no more or less unique or interesting, and certainly no more or less fantastic, than alternative worldviews. While there is wonder and beauty in all of these creation stories and in their very human attempts to seek explanation, they are not science-based and therefore, unlike evolutionary theory, cannot be tested through experimentation or observation, and therefore have no predictive power. That is precisely why such stories belong in courses on comparative religion or cultural anthropology, rather than in those focused on the biological sciences or related fields.

The following is intended to be a broad overview of some of the world's many creation stories broken down by geographical region. It would be impossible to provide a comprehensive overview of the vast number of stories that exist, to describe them in any detail, or to provide anything more than a cursory interpretation. Such stories often have many variations. The first five sections provide brief overviews of the creation stories of indigenous cultures from each geographical region of the world. The sixth section gives an overview of the creation stories of some of the world's contemporary pan-global (major) religions. Descriptions derive from a variety of sources, but primarily from the works of Sophia Lyon Fahs and Dorothy T. Spoerl (1958), Barbara C. Sproul (1979), and David Adams Leeming and Margaret Adams Leeming (1994).

Indigenous peoples: North, Central, and South America

People of Eurasian descent, also known as Native Americans or First Peoples, were the first to colonize the Western Hemisphere. Native American cultures were highly developed and diverse, perhaps reaching their pinnacle in the mound builders of North America and the Olmec, Toltec, Incan, Mayan, and Aztec cultures of Central and South America. The creation stories of the indigenous peoples of North, Central, and South America are diverse and reflect the diversity of their ecologies and lifestyles.

Aztec

The pre-Columbian Aztec civilization dominated large portions of what is now Central (Meso-) America from the fourteen to sixteen centuries C.E. In the Aztec worldview, two gods—the plumed serpent, Quetzalcoatl, and Tezcatlipoca—pulled the earth goddess, Coatlicue, from the heavens. The gods then turned into two large serpents and tore her body into two pieces, one of which became the earth and the other, the sky. In her earthly manifestation, Coatlicue was the source of all nature, with her hair becoming plants, her eyes and mouth, caves, and the source of all water, and other parts of her body, the mountains and valleys. Coatlicue was angry at what had been done to her, however, and this is why she demanded frequent human sacrifice, an integral and fearsome component of Aztec civilization.

Cherokee

The Cherokee people originally lived in the mountainous regions of the southeastern United States (in portions of present-day Alabama, Georgia, North and South Carolina, and Tennessee). When gold was discovered on their land, however, they were forced to move west to what is now the state of Oklahoma. According to Cherokee tradition, the world was once completely covered with water (a common theme among many cultures). At the time, all living things lived densely crowded into Galun'lati, the vault beyond the sky. Increasingly desperate to have more space, the animals chose Water Beetle, or Dayuni'si, to go out and explore. Water Beetle darted across the water in every conceivable direction but could find nowhere to rest. Next, he dived to the bottom of the waters and retrieved a piece of soft mud, which, when he reached the surface, expanded to become land. The Great Spirit fastened this "earth-island" to the sky with four rawhide cords that stretched between four sacred mountains, representing the four sacred directions. The earth was still muddy, however, and too soft to stand on. Buzzard was sent out to look for dry land, and he found a place that was suitable for habitation, a place that became Cherokee country. The powerful flapping of Grandfather Buzzard's great wings created the mountains and valleys that exist in the region today. When the land dried, all the animals descended from the vault to take up residence in their new home. But they were not happy because the land was dark. They decided to pull Sister Sun from behind the rainbow and give her a regular path to take across the sky. The Great Spirit then decided to make a man and a woman. When the man pushed a fish against the body of the woman, she became pregnant, and this happened every seven days, until the Great Spirit decided that women should give birth only once per year.

Eskimo

Eskimos (Inuit and Yupik) are indigenous peoples who have traditionally inhabited circumpolar regions. Their cultures are highly adapted to living in seasonally cold, harsh environments. Eskimo creation stories, like those of the Aborigines of Australia, tend to be more local than cosmic, and as a result there are many variations. Some groups tell of two men emerging from the earth following a great flood. The two men lived together as a couple and when one became pregnant, the other chanted magical words that resulted in the pregnant man's penis splitting into a vagina, thus making it possible for the first child to emerge. Another popular story features Raven, the trickster god. According to this version, the first man dropped to the ground fully formed from a pod growing on a vine. Raven was surprised as it was he himself who had created the vine. Raven then asked the man if he had

Coatlicue is represented as a woman wearing a skirt of writhing snakes and a necklace made of human hearts, hands, and skulls. Her face is formed by two facing serpents referring to the duality of her nature. © Travilepix/Alamy.

eaten anything, and the man said that he had drunk a fluid lying on the ground, which Raven explained was water. Raven brought berries for the man to eat and told him to plant some of the seeds. Many more men eventually grew on the pods. To sustain them, Raven made plants and animals, bringing them to life with the flapping of his wings. To keep men from greedily consuming everything, Raven then made the plants and animals hard to find and catch, and he also formed the dangerous bear out of clay, to frighten the men. Sensing that man was lonely and in need of companionship, Raven formed woman out of clay. Man was pleased and soon they had children, thus perpetuating humankind.

Guarayu-Guarani

According to the Guarayu-Guarani people of Bolivia, in the beginning, there were only water and bulrushes, and then came Mbir, the primordial worm, also known as Miracucha. Eventually, the worm transformed into a man-god, and in this form, he created the world and all living things out of chaos. Two other gods in Guarayu-Guarani legend are the sun, also known as Zaguagua or Zaguguayu, and another called

Haida artist Bill Reid's "The Raven and the First Men" carving. Museum of Anthropology University of British Columbia Vancouver, BC. © Keith Douglas/Alamy.

Abaangui. The sun god decorated himself so brightly that no one could bear to look at him until he descended from the sky in the evening. Abaangui tried very hard to become human and finally succeeded. But his nose was so large that he cut it off, whereupon it flew into the sky and became the moon. The first Guarayu-Guarani ancestor was named Tamoi, or Grandfather. Grandfather taught the people how to plant crops, brew beer, hunt, fish, and make fire for warmth and cooking. When his task was complete, Tamoi wandered off into the west, leaving his wife and child behind; they subsequently transformed into two sacred rocks. The souls of the dead also travel west, where they meet the Grandfather of Worms. If the person behaved badly in life, this worm becomes huge and blocks the soul's path. The people pray to Tamoi for good fortune and honor him in song and dance.

Haida

The Haida people live along the lush, rainy northwestern coast of North America, including the Queen Charlotte Islands of British Columbia, Canada, and the southern half of Prince of Wales Island in southeast Alaska. According to Haida tradition, there was a time when only the sea and the god, Raven, existed. Raven flew over the vast ocean and, when he saw a tiny island below him, he commanded it to become earth. As the land area expanded, Raven cut it into pieces, creating the Queen Charlotte Islands out of one small piece and the rest of the land's surface from a larger one. On his travels, Raven heard sounds coming from a large, white clamshell lying on a beach (Smelcer 1993). Looking into the shell, he saw a small face and soon others peered out. The creatures inside were afraid of Raven and huddled together in the farthest reaches of the shell. He eventually coaxed five little bodies out of the shell, and these were the first people. With their skinny arms and bare bodies, Raven thought these creatures were very odd indeed. Raven watched and played with the people for a long time, and taught them how to survive. Ever since that time, the Haida people have made their living from the sea and the coastline, and they have never returned to the clamshell where Raven discovered them.

Hopi

The Hopi people of southeastern Arizona traditionally were pueblo dwellers who lived in matrilineal clans. In one version of Hopi legend, only the earth goddess, Spider Woman, and the sun god, Tawa, existed in the beginning. While Tawa controlled the heavens, Spider Woman ruled over the earth. Though they were not husband and wife, the two lived together in Spider Woman's lair, Under-the-World. Both longed for companionship, so Tawa split himself into Tawa and the god of life energy, Muiyinwuh. Spider Woman became Spider Woman and Huzruiwuhti, goddess of life-forms. Tawa and Huzruiwuhti became a couple, and their union produced the Four Corners, the Up and the Down, and the Great Serpent. Tawa and Spider Woman created the world and placed it between the Up and Down and within the Four Corners. They also created all living creatures through their shared thoughts, which they expressed in song, and which Spider Woman formed in clay. Together they covered the new beings with a sacred blanket and sang a sacred song, which gave them life. But they were still not pleased with their work, so Tawa conceived of beings that were similar to himself and Spider Woman and who could rule the world they had created. So Spider Woman created man and woman from this new thought and held them in her arms until they stirred to life.

Maya

The great Mayan civilization, which reached its peak between 250 and 900 C.E., once extended from southern Mexico and the Yucatán Peninsula in the north to present-day El Salvador, Guatemala, Belize, and western Honduras in the south. The best known of the Maya were the Quiché people who lived in the highlands of Guatemala. Their history and traditions are preserved in a book known as the *Popol Vuh* or Book of the People.

According to Mayan belief, in the beginning, only the creator-gods—Tepu and the feathered serpent, Gucumatz—existed in the void. They possessed the power of the sun, and whatever sprang into their minds became a reality. When they said, "Let there be earth," the earth was created out of a mist; when they thought about mountains and trees, mountains and trees appeared. They created all of the animals, but when they called upon the animals to praise them for their good work, no praise came. Disappointed, Tepu and Gucumatz decided to make beings that were aware of them and their creations, and they formed humans out of clay. Unfortunately, their first attempt failed, as the clay fell apart when it became wet. They tried again using wood. The wood people did not fall apart when wet, but they proved to be too inflexible, mindless, and without inner light. In addition, they did not praise their creators and caused trouble, so Tepu sent a great flood to destroy them. The survivors were chased by the other animals into the forest and became monkeys.

The dawn of the world was approaching, however, and the creators had to hurry to develop a more acceptable being. They enlisted the aid of Coyote, Mountain Lion, Parrot, and Crow, who helped gather together food and drink for the people, primarily corn, beans, and water. Tepu and Gucumatz then created the four original ancestors of the Quiché people. These four men were wise and knew what to do to survive, and they also praised their creators appropriately. Tepu and Gucumatz were pleased, but they grew to fear their own creations, so they took some of man's powers away. At the same time they created wives for the original four men, and their pairings produced the Quiché people, as well as other tribes.

Sioux

Most Sioux were once a seminomadic people, inhabiting the Great Plains of what is now the north-central United States and subsisting on the vast herds of buffalo that roamed through this region. There are several Sioux tribes and also several variations of the creation story. According to the Lakota Sioux of North and South Dakota, the original man emerged like a blade of grass from the soil of the Great Plains. At first, only his head emerged and he looked at the nothingness that surrounded him. Gradually he pulled himself out and stood upon the soft earth. The sun eventually dried

and solidified the earth and gave strength to the man, who became the ancestor of the Lakota people.

Among the Brule Sioux, the creation story starts with a great flood. In this story, the first people were attacked by a huge Water Monster, who sent the waters to kill them. The people climbed a steep hill to escape, but the water immersed them and they drowned. The remaining pool of blood became a quarry from which the people made their sacred red stone pipes. These pipes had great power because their smoke represented the breath of the ancestors. After the flood, the Water Monster was turned into stone and became the Badlands. One person—a young girl—survived the flood, having been picked up by Wanblee, the eagle, and flown to his home in a tall tree. The girl became Wanblee's wife, and from their union came two pairs of twins, one set male and the other female. These were the parents of the Brule Sioux people, who take pride in being known as the eagle people.

Indigenous peoples: Africa

Bulu

European colonialism and exploitation heavily influenced the creation story of the Bulu people of the West African nation of Cameroon as reflected by the Bulu's negative self-image in the tale. According to Bulu tradition, in the beginning, there was only Membe'e or Mebe'e, the one who holds up the world. Membe'e's son, Zambe, created humans, gorillas, chimpanzees, and elephants, which he named after himself. He created two types of men—black and white—and gave them many useful things with which to enrich their lives, including water, gardening tools, fire, and books. The men stirred the fire to keep it going, but when smoke blew in the white man's eyes, he left with his book. Chimpanzee and gorilla also left the fire and went into the forest to gather fruit, while elephant just stood idly by. In contrast, the black man continued to stir the fire, not bothering to study the book. When Membe'e visited earth, he inquired what his creations had done with the gifts he had given them. When he heard what chimpanzee and gorilla had done, he decided that they would forever have hairy bodies and spend their lives in the forest eating fruit. Elephant was also sent away from the fire to live in the forest. Zambe then asked the black man what he had done with his book. He replied that he had not had time to read the book because he was busy tending the fire. Zambe said he would have to spend his days working for others because he had not obtained knowledge from books. He then asked the white man how he had used the gifts. The white man told him that he had only read the book, nothing else. Zambe told him that he would continue to learn many things, but that he would need the black man to take care of him because he would not know how to take care of himself. According to the story, this is why animals live in the forest, white men spend their time reading, and black men work hard but always have a good fire to keep them warm.

Bushmen

The so-called bushmen are indigenous hunter-gatherers who live across arid regions of southwestern Africa, encompassing portions of South Africa, Zimbabwe, Lesotho, Mozambique,

Swaziland, Botswana, Namibia, and Angola. According to Bushman tradition, their creator god was a praying mantis, known as Kaggen or Cagn. He created the world and, at one time, lived in harmony with people on earth. The foolishness of humans eventually drove him away, however, which is why life became hard and people became hungry. There are many stories told about this creator god, his wife, Coti, and their two sons who instructed the people how to find and prepare food. Kaggen had an ability to shift shapes and transform himself into any animal he wanted. His favorite, however, was the eland, a large, cowlike antelope. To this day, the bushmen believe that only the eland knows where Kaggen is.

Fon

Despite many variations, most Fon people of the West African nation of Benin (formerly known as Dahomey) believe that the moon goddess, Mawu, created the universe and earth and was the mother of the gods and people. She is often portrayed as two beings: Mawu (moon in the sky) and Mawu-Lisa (moon-sun), who is both male and female. In some versions, this mother-father-creator is known as Nana-Buluku. According to the story, Nana-Buluku gave birth to Mawu and Lisa, and then gave them power over all creation. Mawu was thought to live in the west and Lisa in the east, and an eclipse occurred when they were making love. In many versions of the story, Mawu created the world, riding from place to place on the back of Aido-Hwedo, the great rainbow serpent. When her creation was complete, Mawu asked the great serpent to coil himself around the earth to hold it steady. Mawu also created the oceans to help keep the great serpent cool as he strained to support the heavy world. Earthquakes or tidal waves occurred when the serpent shifted his weight. The Fon people say that one day Aido-Hwedo will swallow his own tail and that the earth will end by toppling into the sea.

Mande

The Mande people live in Mali and some other West African countries. According to their tradition, a supreme being, Mangala, created *Eleusine* seeds (a food plant that produces millet) and other seeds that formed the basis of creation. Some seeds produced the first people: two pairs of twins, one set male and the other female. All these seeds existed together in a large egg. One of the male twins, Pemba, believed he could assume power over all of creation, so he left the egg prematurely, his mother's placenta still attached. He drifted through space until part of the placenta became the earth and another part the sun. The earth was barren, which displeased Pemba, so he returned to the egg and stole some of the male seeds, which he subsequently planted in the earth created from his mother's placenta. Though the placenta's blood nourished the seeds, Pemba's act was considered incest, which rendered the earth impure. To atone for Pemba's bad behavior, Mangala sacrificed another male twin, called Faro, who earlier had assumed the form of a fish. His body was cut into sixty pieces, which become the earth's trees. Eventually, Mangala brought Faro back to life (trees are still considered a symbol of resurrection) and sent him to earth. With him Faro brought the original eight ancestors of the Mande people (Mande means "son of the fish"), as well as the first animals and plants, all of which possessed the male and female life

force in perfect balance. Thus, Faro created the world out of Mangala's seeds from the world egg. He taught the first people how to plant and grow crops.

Nandi

When the Nandi people first arrived in what is now Kenya in East Africa, they encountered the Dorobo people already living there. They assimilated into that culture and adopted the Dorobo people's creation mythology. According to Dorobo tradition, the world and God already existed, but at some point, God decided that the world needed more order. When God arrived on earth, he found that the Dorobo people, the elephant, and the thunder already existed. The elephant and thunder, however, were unhappy that people could turn over in their sleep without standing up—something they could not do. Thunder was so irritated at man's habits and so feared him because of his evil nature that he fled into the sky, where he resides to this day. Elephant taunted the thunder for being so afraid of such a small, insignificant creature. Man was also afraid of thunder, however, and happy that thunder had decided to retreat into the sky. Men then decided to kill the elephant with a poisoned arrow. So, elephant too became afraid of man and pleaded with thunder to transport him to the heavens. Thunder refused and chided elephant, saying "I thought you were not afraid of something so small." Man killed the elephant and continues to do so to this day. Eventually, people reproduced and assumed dominion over the entire world.

Yoruba

The Yoruba people of Nigeria call their creator god Olurun or Olodumare. They say that in the beginning there was only water. Then Olurun sent his assistant, the lesser god Obatala, down to earth to create land. Obatala took a shell containing some earth, some iron, and a bird (either a rooster or pigeon hen) with him. He placed the iron in the water, the earth on top of the iron, and the bird on top of the earth. The bird's scratching spread the earth about, thus creating land. Olurun named the earth Ife, which means "wide" in the Yoruba language. When the land was ready, some of the lesser gods descended to earth to live with Obatala. Chameleon was the first among them. Then Obatala created humans by sculpting them out of earth, and he bade Olurun to breathe life into them. One day, Obatala became drunk and, by mistake, started making crippled people, who even today are considered sacred to the Yoruba people. Ultimately it was Olurun who had the power to breathe life into the people Obatala made. Some believe that Obatala was jealous of Olurun's power and wanted to give people life himself. Others claim it is Obatala who was responsible for forming babies in their mother's wombs.

Zulu

The Zulu people, who live primarily in what is now South Africa, are the largest ethnic group in the country. Zulu tradition holds that their creator god, Unkulunkulu, or the Ancient One, is the source of all creation. Some say that Unkulunkulu originally emerged from reeds, because the Zulu word for reeds also means "the source." Unkulunkulu then broke off people from the reeds as well as cattle and all of the other animals and plants and earth's natural features, such as mountains and streams. He also taught the Zulu how to hunt, make fire, and grow crops. The people believe that Unkulunkulu is in everything and that he was the original and ancient source of all that exists.

Indigenous peoples: Europe and the Middle East

Egypt

Egyptian civilization is among the oldest and most complex of all human societies, with records dating back to around 3000 B.C.E., and a prehistoric era about which little is known. Egyptian creation stories are varied, depending on the historic period and religious center from which they originated. During prehistoric times, creation mythology seemed to be centered on a goddess called Nun, who was thought to be responsible for all creation. It is said that Atum, the male god who created the universe, arose from her or created himself by an act of will. Later remnants of this female creative power exist in the stories of Nut, Isis, and others, but by the time of the Great Pyramids and the patriarchal reign of the pharaohs (2780–2250 B.C.E.), male gods had achieved dominance. Many

changes took place in Egyptian religious thought over time, but certain aspects of the creation myth remained relatively constant.

According to early Egyptian tradition, for example, in the beginning, there was nothing but chaos and water (sometimes also depicted as a cosmic egg), which arose from the Great Mother Goddess (Nun). There was also the Eye or sun, which created the cosmos from the chaos of the surrounding waters. The sun—alternately known as Atum, Ra, Re, Khepera, Neb-er-tcher or Ptah in various incarnations—plays a particularly important role, and is often also associated with a primeval mound, which either symbolized the fertile hills of earth left after the Nile River floods and recedes, or the early morning sun coming over the horizon. The Great Pyramids represented such mounds. But by the time of the Pyramids, Atum had also been connected with the sun god Ra, and therefore also symbolized the emergence of light into chaotic darkness. Atum, often regarded as the bisexual god (the great He-She), also created other gods, which completed the cycle of creation.

The best-known event in the Egyptian creation story is the separation of the creator-gods Geb (earth) and Nut (sky). Their incestuous parents were Atum's brother/son and sister/daughter, Shu (air/life) and Tefnut (moisture/order). It is said

An Egyptian view of creation. Geb, the Earth God, lies on the ground with his sister Nut, the sky goddess, arched above him. Between them hovers the air god who is not involved. © Mary Evans Picture Library/Alamy.

that Atum (as Ra, Khepra or Neb-er-tcher) created his brother/son and sister/daughter out of nothing, either by a union with his own shadow, masturbation, or spitting them out, but some versions have them all being created simultaneously. Nut is typically depicted as arching over her brother Geb, who has an erect phallus. This symbolized the earth longing for the life-giving rain, which makes procreation possible. When Atum was reunited with his two created parts, Shu and Tefnut, he wept tears of joy, and according to some versions, from those tears emerged humankind. Veronica Ions (1973 [1965]) provides further details and variations of this complex story.

Greece

Greece was among the first great Western civilizations. Many creation myths originated in Greece, one of the first of which comes from the writings of Hesiod, who lived around 800 B.C.E. This story identified Earth and Heaven (Sky) as the primordial parents of all creation. However, as the Greek people moved from an agrarian to a more warlike, patriarchal culture of city-states, there was a transition from the worship of mother goddesses to a powerful, thunder-bearing male god, who eventually became Zeus. The creation story of this time is the best known, and it includes a panoply of gods that created and influenced the world as we know it.

According to this version, in the beginning there was complete chaos—a void. Out of nothing emerged Gaia, or Mother Earth, and Mount Olympus, the home of the gods. Gaia, through virgin birth, produced a son, Uranos or Ouranos, who ruled over the heaven and stars and became her husband. She also created the sea (Pontus) and earth's physical features of mountains, valleys, and hills. The union of Gaia with her husband-son, Uranos, produced the first gods, known as the Titans. Among the best known were Oceanus, the one-eyed Cyclops, Kronos, and the earth goddess Rhea. Most of the children of Gaia and Uranos hated their father, and he also hated them, which caused Gaia much pain and anguish. She plotted with her son Kronos to take his father's life. When Uranos came in lustful passion, bringing the night to Gaia, Kronos carried out his terrible mission with a sickle fashioned by Gaia herself. Kronos cut Uranos's body into many parts and flung them into the sea. Uranos's genitals floated upon the sea and their seed formed a thick foam. Out of this emerged Aphrodite, the goddess of love and desire.

With his father's demise, Kronos now presided as king over heaven and earth. Kronos's rape of his sister Rhea produced the first six of the family gods that would inhabit Mount Olympus: Demeter (earth goddess who would replace her grandmother, Gaia, and mother, Rhea), Hades (god of the underworld), Hera, Hestia (goddess of the hearth), Poseidon (god of the sea), and Zeus. Kronos had heard that he would be overthrown by one of his own children, and as each child was born, he ate it to prevent his own demise. Rhea was understandably horrified by Kronos's behavior, and when her last child, Zeus, was born, he was hidden away on the island of Crete to prevent his discovery. Rhea deceived Kronos by substituting a rock on his plate, which he subsequently consumed thinking it was his son Zeus.

Zeus eventually led a war against his father and the Titans, which he won, establishing himself as the undisputed ruler of Mount Olympus. He married his sister Hera; their union produced Hephaestus (the blacksmith god) and Ares (god of war). But Zeus was an unfaithful husband, and he had many children with both goddesses and mortal women. These offspring included Athena (goddess of wisdom), whom it is said emerged fully developed from Zeus's head; Hermes (the messenger god); Apollo (the god of prophesy and light); Artemis (goddess of the hunt); and Dionysus (god of wine, madness, and ecstasy). Some stories also have Zeus fathering Aphrodite. There are many Greek stories about the origin of humans. Some say that the Titan Prometheus, who fought on Zeus's side in the war against Kronos, and was later betrayed by him, created humans by sculpting them out of water and clay. Prometheus is also said to have given people fire, and therefore he is also sometimes known as the god of fire.

Hebrew

For many Jews and Christians, the Bible is considered the word of God. Genesis, the first book of the Old Testament, contains the Judeo-Christian creation story. In actuality two distinct, even occasionally contradictory, creation stories are contained in Genesis, apparently written by different authors

Zeus is the ruler of Mount Olympus and god of the sky and thunder in Greek mythology. © Peter Horree/Alamy.

(Leach 1967). One generally summarizes the world's origin, whereas the other provides a more detailed explanation of the origin of humans. Many scholars believe that these stories were heavily influenced by other, more ancient Near Eastern creation stories, and also by the historical climate at the time. According to Genesis 1, in the beginning there was nothing but darkness, water, and chaos. God moved to bring more order by creating heaven and earth over a seven-day period, purely by the force of his thoughts and words. During the first days, he created light to counter the darkness, separated the earth from the sky, and created time. On the fourth day, he created the sun and the moon to separate day from night as well as a calendar to track time. On the fifth day, God created animals and plants to populate the land and the oceans. On the sixth day, he created the first man and woman in his own image and told them to "go forth and multiply." On the seventh day, God rested, thereby reinforcing his power over all creation. Everything in creation was now deemed perfect and good.

In Genesis 2, as in many other creation stories in the Near East and elsewhere, God formed the first human (Adam) out of clay. Man essentially toiled for God, working in the garden and naming the various animals and plants, but he also had a special relationship with the deity, who created an earthly paradise specifically for him. God then created woman out of Adam's rib, which some have interpreted to mean that women are the inferior gender. In the original text, however, God concluded that man was not complete and that he needed companionship. He initially experimented by creating animals to fulfill this role, but this did not work out, thereby reinforcing a perception that man and nature, though close, may not be too close. Only when woman was created was humanity considered to be complete.

Genesis 3 tells the story of humankind's fall from grace, after a serpent tempted Eve to pick and eat fruit from the Tree of Knowledge. Conflict between the sexes arose because Adam blames Eve for tempting him to eat the fruit, an act that God had forbidden. God punished Adam and Eve with a loss of immortality, permanent banishment from the Garden of Eden, and a continuous struggle to survive. As a result, people would have to work for a living, rather than having everything given to them. Some authors, including David Adams Leeming and Margaret Adams Leeming (1994) believe that this story may be more about God giving humans freedom of choice and about human independence from God. They reason that had God not given humans that choice, their relationship with God would always have been one of complete dependence.

Norse

Germanic peoples of Iceland and Scandinavia, also known as Vikings or Norse people, have a complex creation story that comes originally from an old Icelandic text known as *The Younger Edda* or *The Prose Edda*. This text is based on a much older oral tradition passed on through generations. According to the story, a king named Gylfi once ruled the region now known as Sweden. A wise old woman had told the king about the Aesir, gods who dwelled in Asgard, also known as Valhalla. The king disguised himself and went to Valhalla, where he met the High One, who agreed to answer his questions about

the world and its beginnings. The High One told Gylfi that there were once two places: Muspell in the south was a place of fire and light, and Niflheim in the north was a dark, cold, and icy place. The two intersected in Ginnungagap, a place of great emptiness. It was here between two worlds that hot and cold mixed, creating mist, where life had its first beginnings.

The first living thing to form was the ice giant, Ymir. One of Ymir's legs mated with the other and produced a son, the first of a family of ogres. A giant cow, Auohumla or Audhumla, emerged from the melting ice in Ginnungagap, and her udder produced rivers of flowing milk, upon which Ymir and his family fed. As the giant cow herself fed upon the ice blocks that surrounded her, a man appeared. This was Buri, the Strong. His son, Bor, married Bestla, daughter of one of the ogres. From their union came the great god Odin and the lesser gods, Vili and Ve. These three sibling gods banded together and killed Ymir. Blood from his body created a massive flood that killed all the ogres. The gods transported Ymir's body to the center of Ginnungagap, where they transformed his skull into the sky, his brain into clouds, his hair into trees, his body into the earth, his blood into the oceans, his bones into mountains, and his teeth and jaws into rocks. From the southern lands of Muspell, the gods took hot embers and turned them into the sun, moon, and stars and positioned them over Ginnungagap. The three gods then made the first man (Ask) and woman (Embra) out of fallen ash and elm trees. Odin breathed life into them, Vili gave them the gift of intelligence, and Ve gave them vision and hearing.

Roman

The Roman Empire, centered in Italy and eventually extending from England to North Africa and the Middle East, had profound effects on European and world history and culture that are still felt today. The Roman creation story was heavily influenced by Greek mythology and was told by the great Roman poet Ovid (43 B.C.E.–17 C.E.) in his epic, *Metamorphoses*. According to Roman tradition, in the beginning there was nothing except chaos. At some point a god or strong natural force decided to bring order to this chaos. This force created the heavy, round earth, then the air and the seas. When all was prepared, the creator-god, whom many identified as Prometheus, decided to create people in his own image to witness his creation. The first man was fashioned out of rainwater and some of the heavier elements on earth. This heralded a Golden Age, when perfection reigned in heaven and on earth and when people were peaceful and happy and took advantage of all earth's gifts. But this Eden-like condition did not last. In the next era, known as the Silver Age, discord arose in heaven, resulting in Jupiter replacing Saturn as the king of the gods. Seasons developed on earth, agriculture emerged, and life became harder. The Bronze Age followed, a time when humans began waging war. Finally came the Iron Age, when goodness was overwhelmed by evil. This was a time when love and trust died among humans, resulting in frequent disharmony, war, greed, and disloyalty. Hardship was prevalent during these difficult times. Wars in heaven also raged between the gods in Mount Olympus and giants from ancient times. After settling their own affairs, the gods decided to eliminate humanity by sending a great flood. Only two individuals were fortunate enough to escape in a boat: Deucalion and Pyrra or Pyrrha, and their union signaled the beginning of a new human creation.

Indigenous peoples: Australia and the Pacific Islands

Australia

The indigenous peoples of Australia, the Aborigines, have lived on the continent for at least 40,000 years. When Europeans arrived to colonize the land, more than 250 languages were spoken and Aboriginal people lived in a variety of different habitats, ranging from lush rain forest to arid desert to temperate islands, such as Tasmania. Not surprisingly, many variations on Aboriginal creation stories exist. The one presented here comes from Cyril Havecker (1987), but others can be found in the works of A. W. Reed (1971), Leeming and Leeming (1994), and Barbara C. Sproel (1979).

In the beginning, the Great Spirit Baiame was deep in sleep when he began dreaming of life in the past, present, and future. His dream turned into a nightmare, however, which caused the universe to quiver, and which awakened his helpers from a previous time and dimension: Nungeena, Punjel, and Yhi. When Baiame awoke, his helpers told him that they knew of his desire to materialize his dream, and that they would help him, in a sort of creation by committee. Punjel suggested that it would take many millions of workers to form the edifice of life, and that this be done by utilizing portions of Baiame's Supreme Intelligence, which Punjel called yowies (souls). Punjel suggested that these yowies be given the power to mold forms, and that they also possess three inherent qualities or drives: a drive to seek nutrition to maintain their bodies for continued survival, a sex drive to ensure reproduction and continuation of their kind, and an ambition and desire to achieve. He also gave the yowies a spirit body known as a Dowie, which would allow them to recollect previous experiences. Given the awesome responsibility yowies possessed, memory was deemed important to bring order to creation, instead of chaos.

After these three drives and the power to mold form were bestowed upon each yowie, Punjel suggested that the units of intelligence be taken directly from Baiame's body and placed into a mixing bowl, where they would be stirred in a counterclockwise direction. The yowies could then be projected by thought to Tya (the earth) through a swirling motion. So the mass began to swirl inside Baiame's body, and the plan for creation began to materialize as Tya began to solidify and the yowies began molding their myriad of forms to recreate the Baiame's dream. Fog began to envelop Tya, and then Nungeena was called on to take the next step in creation: plant life. But there was no light, so the conditions for sustaining plants were not favorable. Nungeena went to Punjel and asked him to call on Yhi, the sun goddess, for help. Yhi responded by sending warming rays of light to Tya, which melted the ice that had formed after eons of time and covered the earth with water. Nungeena was displeased, however, because her plants were drowning.

The group discussed the situation and together they summoned Uluru, the great intelligent snake from higher planes of the universe. Suddenly, a colorful rainbow appeared, and from this rainbow materialized Uluru, the great Rainbow Serpent. Uluru burrowed into the earth, and the resulting holes filled with water to create the oceans, rivers, and lakes. The material he dug out became the hills, mountains, and valleys. But Uluru, in his hurry, had moved so much material that Tya became unbalanced and began to wobble in space. In his panicked attempt to correct the situation, the great serpent threw soil and rock into the air toward the sun goddess Yhi. Not wanting to be contaminated, Yhi demanded that this cease immediately. She then pondered what to do with the surplus material suspended in space and decided to leave it there, where it became Bahloo, the moon.

Other problems began to emerge. Nungeena had nurtured the plant life so well that it began to overgrow and spoil, and furthermore, various species were competing with one another for dominance. So, the creators had lengthy discussions to identify a solution. They decided it was time to introduce animals to the world to bring balance by helping to control the overgrowth of plants. Punjel called on the advanced yowies to inhabit animal forms, and the first creature formed was the jellyfish. Soon there were too many, and they were all uniform in appearance. Nungeena was concerned about the lack of diversity, so together they called on the rain spirits to bring rain so that the jellyfish would have to struggle to adapt and survive. Only the strongest of these aquatic animals would persist to change form and adapt to a new environment. Nungeena again was given the task of taking the next step in creation. She did so by creating the turtle, which emerged out of the water to lay its eggs on land. The yowies that emerged from these eggs eventually transformed to become lizards, goannas, and snakes. Then came the legendary bunyip, a large fearsome beast that dominated the land. But eventually Uluru brought cold to the land and many forms perished in the changing climate, but not all.

After eons of time had passed, life had become balanced on Tya. With the resulting harmony, Punjel thought it was time to attempt a special creation: humans. It was decided to create a man in the spirit form first. When the time came for the final molding of the human form, Baiame would become human himself and come to Tya where he would explain to the three existing kingdoms—mineral, plant, and animal—that they were about to be joined by another, more intelligent companion. Baiame subsequently descended to Tya with 300 men and women, who became members of the very first tribe. He taught them everything they needed to know to survive and also instructed them how to conduct their dances, ceremonies, and other tribal practices. Eventually, Baiame returned to the spirit world.

Fiji

According to Fijian legend, in the beginning there were only water and twilight. The gods lived on their own island, which floated around the edge of the world. The Fijian creator god was the serpent Ndengei. He was the original creator god that came before all others eventually brought by either the Polynesians or Europeans. He created the world out of nothing. People say that when Ndengei rolls over at night, his movement creates earthquakes. When he wakes up in the morning, it is daytime. But it was Ndengei's son, Rokomautu, who is credited with the creation of the Fiji Islands. It is said that Rokomautu gathered earth from the bottom of the sea and brought it to the surface to form the islands.

Hawaii

The Polynesian people that inhabit the Hawaiian Islands had developed a rich culture prior to the arrival of Europeans in the late 1700s. Native Hawaiians trace their ancestry to the early Marquesan and Tahitian people who colonized the islands by sea, possibly as early as 400 C.E. Hawaii is part of the Polynesian Triangle, a large region of the Pacific Ocean that is bounded by three island groups: Hawaii, Easter Island, and New Zealand. The many cultures that exist within the triangle share linguistic and cultural similarities, owing to their shared origin. All Polynesians are thought to have originally descended from Malayo-Polynesian peoples who migrated from Southeast Asia. The long and complex Hawaiian creation story is recounted in a 2,000-line poem, known as the "Kumulipo." The poem, which explains the genealogy of the original creator gods, was sung at the birth of royal children; it symbolized new beginnings and the connection between royalty and the plants and animals of the first creation.

According to Hawaiian tradition, in the beginning there was only darkness. Out of nothing, emerged the male Kumulipo and the female Po'ele, both of whom represented the essence of deep darkness. From their union came the children of darkness, which included shellfish from the depths of the sea and grubs that lived beneath the soil. As the world became lighter, the male Pouliuli (deep darkness) and his mate, Powehiwehi (darkness with a little light), were born. Their union produced all the fish in the sea. Then Po'el'ele (dark-night male) and his mate, Pohaha (night turning into dawn female), were born, and they became the parents of night-flying insects. An egg was born to them too, and out of that emerged the first bird, and then many more birds were produced. Next came Popanopano (male) and his mate, Polalowehi (female), whose union produced both land and sea animals. But it was still not light yet. Another couple, Po-kanokano (male) and Po-lalo-uli (female), were born at this stage, and their union produced Kamapua'a, the pig. Yet another couple, Po-hiolo (male) and Po-ne'a-aku (female), whose translated names mean "night ending," was born. Their union produced the rat, Pilo'i. The rat people damaged the land as a result of their incessant scratching and eating. Then came the birth of Po-ne'e-aku (male) and Poneiemai (female), whose names roughly translated mean "night leaving" and "night pregnant." Their union produced the dawn, the dog, and the wind. At this stage of creation, it was now almost light, but not quite day. Next were Po-kinikini and Po-he'enalu; their union produced the time in which men and women came into the world. Finally, La'ila'a (female) and Ki'i (male) were born; they were acquainted with Kane, the red-faced god. At this final stage, daytime had come to the world, and the world as we know it came into being.

New Guinea

The island of New Guinea—now split into the Indonesian province of Papua in the west (formerly known as Irian Jaya)

and the nation of Papua New Guinea in the east—was once connected to Australia, but was separated by flooding during the last interglacial period. Habitation of the island by Melanesian people dates back to at least 40,000 to 60,000 years ago. The degree of cultural diversity that developed on this heavily forested, tropical island is staggering; more than 1,000 different tribes exist, speaking as many distinct languages. Not surprisingly, creation stories of these people are highly varied as well. According to the Keraki people, the first humans emerged from a palm tree. Gainji, the Keraki's creator god, first heard the people speaking in their many languages; when they emerged from the tree, the people gathered in their own language groups and then went their separate ways. The Kiwai people identified Marunogere as their creator. He gave the people the highly prized pig, so central to Papuan culture, and instructed them how to conduct their sacred ceremonies.

New Zealand

The indigenous Polynesian people of New Zealand are called the Maori. According to Maori tradition, in the beginning there was only darkness and water. The Maori supreme being, Io, lived alone and did little. As he stirred into activity, however, Io spoke, calling upon the darkness to become light, which it subsequently did. He then called upon the light to become "dark-possessing light," upon which darkness returned and became the night. Io continued to speak, creating the world out of nothing with the power of his words. Io created the gods Rangi and Papa, also known as Sky Father and Earth Mother, respectively. From their union came many offspring, two of which, Rango (female) and Tane (male), created life. However, Tane, the god of life, separated his parents into Heaven and Earth to make more room for his creation. They were so sad at their separation that to this day Rangi's tears (rain) fall on Papa and Papa's sighs (mist) rise up to her husband. Some say that the god Tane sculpted the first Maori out of red clay. Others say the god Tiki created man in his image. Additional details can be found in the work of Antony Alpers (1964).

Tahiti

According to Tahitian tradition, their creator god was known as Taaroa or Ta'aroa. Taaroa existed alone at the beginning of time, when there was no earth, sky, sun, moon, stars, or living things, including humans. He created the universe from nothing and is the personification of the universe. The universe is said to be his shell. For a long time, Taaroa resided in his egglike shell until he broke out of it. He held up the upper dome of the shell to create the sky (Rumia). He fashioned the world out of his own body, with his spine becoming mountains; his ribs, hillsides; and his feathers, trees and other plants. He kept his head, however, for himself. Taaroa created the gods and all living things; he is a part of everything that exists. Just as Taaroa had a shell, so does everything else. The sky is a shell covering the earth, the earth is a shell for all that live on it, and, because all people are born of woman, her womb is a shell for all humankind. Taaroa created the first man and woman (Ti'i and Hina, respectively) out of the earth (out of himself). Ti'i was said to be bad and disliked humans, whereas Hina was good and protected them. Eventually wars arose between the gods in heaven and among the humans on earth. This angered Taaroa, and only the pleading of the good Hina prevented the total annihilation of the world.

Indigenous peoples: Asia

China

The sources of one Chinese creation story include both popular legend and a third-century text called the *San-wu li-chi* (Leeming and Leeming 1994, p. 47). However, many variations exist. According to ancient Chinese tradition, in the beginning, there was a huge cosmic egg. Inside the egg, it was chaotic, but also filled with the essence of yin and yang, the balancing opposites of male and female, hot and cold, dark and light, wet and dry, and passive and active. Also within the egg was Phan Ku, an amorphous giant, who, though yet unformed, broke forth from the egg to separate earth and sky and the other opposites. Once the earth was created, the animals and plants soon appeared, but no humans. For 18,000 years Phan Ku grew, pushing the earth and sky farther apart. When he did take shape, the giant Phan Ku was covered in hair and also had horns on his head and tusks coming out of his mouth. He created the mountains, valleys, rivers, and oceans by carving them out of solid rock with his great mallet and chisel. He also fashioned the sun, moon, and stars. When Phan Ku died, various portions of his body became the building blocks of creation; his skull became the uppermost portion of the sky, his blood the rivers, his breath and voice the wind and thunder, and so on. Some people also say that the fleas on his head became the first people. Other versions, however, have him sculpting humans out of moist clay (Eberhard 1965; Ferguson and Anesaki 1937). When they dried, they were imbued with the vital forces of yin and yang, and thus became human. After Phan Ku had fashioned numerous clay people, a huge rainstorm damaged some of his creations, thus explaining the existence of people who are lame, deaf, and crippled. Phan Ku represents everything that is yin–yang. Upon his death, a void was created from which suffering and sin emerged.

A much more philosophical version of creation comes from Chinese texts produced around 200 B.C.E. This version also begins with a universe in chaos. In the beginning, light turned into sky and darkness into the earth. Yin and yang were embodied in the contrast between darkness and light and pervaded everything that came after. As yin and yang merged into one, the five elements became separated and the first humans came into existence. As the first man watched the moon, the stars, and the sun, a golden being suddenly appeared before him. The Gold One taught the man (often referred to as the "Old Yellow One") everything he needed to know to survive. The Gold One also explained the origins of the earth and humankind. The life force that flows through all living things, he said, was created by heaven and earth, and he also explained the essence of yin and yang—that the yin principle gave and the yang received—and told the man how the sun and moon exchanged light, and how this results in the passage of time.

Taoism, developed by the great Chinese philosopher La-Tzu during the sixth century B.C.E., also contains elements of creation, as evidenced in this 1963 translation by D. C. Lau from the *Tao te Ching*:

> OThere is a thing confusedly formed,
> Born before heaven and earth.
> Silent and void
> It stands alone and does not change,
> Goes round and does not weary.
> It is capable of being the Mother of the World.
> I know not its name
> So I style it "the way."

Japan

Japanese creation stories are described in two primary sources: the *Kojiki* (Records of Ancient Matters), compiled by Futo no Yasumuro in 712 C.E., and the *Nihongi* (Chronicles of Japan), compiled anonymously in 720 C.E. Both of these texts are influenced by Chinese concepts of yin and yang and also by the Shinto religion, which worships the divine forces of nature. According to the *Kojiki*, in the beginning there was chaos, until Heaven and Earth separated. At this time, the Three High Deities created the first ancestors, a brother and sister: Izanagi (male, who invites) and Izanami (female, who invites), who were the basis of all creation. When he emerged into the light, Izanagi washed his eyes, and created the sun and moon. As he bathed in the sea, he released the gods of earth and sky.

In comparison, the *Nihongi*'s description of the creation is much more elaborate. In this version, in the beginning, Heaven and Earth were still unified. As in early Chinese cosmology, there was only an egg containing both chaos and the seeds of creation. Heaven raised itself first, and then Earth began to form. A huge plantlike form began to grow between Heaven and Earth, which eventually became a male god, and this was followed by the creation of two other gods, both male. The first ancestors, Izanami and Izanagi, wondered what lay below them, so they thrust their jeweled Spear of Heaven into the sea, spontaneously creating the first island, Onogoro-jima (spontaneously conceived island).

Wishing to marry, the brother and sister planned an elaborate courtship, wherein they would walk in opposite directions around the world. When they met, Izanami spoke first, but Izanagi objected, stating that as the man, he should have spoken first. So they repeated the process again. Upon their next meeting, Izanagi spoke first, and then he asked Izanami about her body. She told him about a part of her body that was empty and the essence of her femininity. Izanagi told her that there was a part of his body that was the source of his masculinity, and that perhaps if the two parts joined that procreation might be possible. Their union produced the eight-island chain of Japan and also the sun goddess, Ameratsu. Ameratsu shone so brilliantly that her parents sent her to the heavens to rule over the universe. Their next child was the moon god, and he became Ameratsu's mate. They produced children as well, some of them fearsome, such as the Impetuous One and the God of Fire. The first of these was exiled to the underworld (the land of Yomi), and the second burned his mother to death after she had given birth to Midzuhano-me (water goddess) and Haniyama-hime (earth goddess). The fire god took the earth goddess as his wife. The rest of this tragic story has the heartbroken Izanagi traveling to Yomi to find the dead Izanami. He finds her, but not before she had eaten the food of Yomi, thereby making it impossible for her to return to the world of the living. She pleaded with her husband not to look at her, but Izanagi lit a torch and saw her rotting body. Izanami was furious, and she and the furies (ugly females) chased Izanagi out of Yomi. Following his expulsion from the underworld, Izanagi's luck turned sour. After a cleansing bath in sacred waters, Izanagi sequestered himself on a distant island, and Izanami became queen of the land of the dead.

Malaysia (Negritos)

The Negritos people of the Malay Peninsula are among the first people to have inhabited the region, some 50,000 or more years ago. According to Negritos tradition, in the beginning there was only a divine couple, known as Pedn (male) and Manoid (female). The sun existed, but there was no earth. Dung Beetle created the earth by rolling up a ball of mud. Eventually Pedn and Manoid descended from the heavens to earth, where Manoid begged Pedn to have a child. Pedn agreed, but their children were to be produced by picking fruit from a tree. Their first child was a boy, named Kakuh-bird, who was formed when Pedn placed the fruit on a sacred cloth. The second child, Tortoise, was a girl. The siblings married and produced children. One of these shot an arrow into a rock, out of which flowed the first water.

Philippines (Babogo)

The Babogo people live on Mindanao, the second-largest and easternmost island in the Philippines. According to Babogo tradition, in the beginning only the creator god Melu existed. Melu resided in the heavens and was completely white in color and had gold teeth. He polished himself frequently and, with the dry skin that resulted, he created the earth and everything that exists. He also created two small people, both of whom lacked noses. Melu's brother, who was not very smart, offered to create the noses, and Melu reluctantly agreed. Unfortunately, the brother made the noses upside down, and when it rained, the people almost drowned. To save themselves, they stood on their heads under the protection of a tree. Melu came upon them and asked them what they were doing. He then noticed their misdirected noses and subsequently turned them around.

Pan-global religions

Buddhism

Buddhist tradition places little emphasis on creation, although an early Indian sect known as Theravada Buddhism includes a book of scripture, called the Pitaka, which includes writings that are attributed to the Buddha himself. In one section, these scriptures refer to the ending of our world and the beginning of a new creation. No creator is identified as such, but the passages suggest that when our world ends, only darkness and water will remain. There will be no sun, moon, or stars, and no living creatures, including humans, for a long period of time. Eventually, however, earth will form on the waters, and life, including humans, will emerge again.

Giant Buddha at Bodhgaya Bihar India. © Richard Robinson. Image from BigStockPhoto.com.

Christianity

See the "Hebrew" section above for a description of the best-known creation story from the Judeo-Christian tradition, in Genesis from the Old Testament. It is important to note, however, that another, more spiritual, version of creation can be found in the Gospel of John in the New Testament. It was never meant as an alternative to the Genesis account, but it does offer another perspective. In John, all creation comes about as the result of the Word, and the Word existed before anything else. In his prologue, John equates Jesus with the Word or Logos, which in Greek means reason or order (as opposed to chaos). Thus all life is seen as being alive with his light and being. Later Christians sometimes referred to Jesus as the "New Adam," with his resurrection symbolizing the possibility of a new beginning (or creation) based on Christian principles.

Hinduism

Creation is expressed in many different ways in Hinduism, now a pan-global religion centered in India. Hinduism was likely originally brought to that country by invaders from Iran. Hinduism is a complex polytheistic religion that has many gods, although all represent various aspects of a single principle called Brahman. According to the Hindu world-view, Brahman is everywhere and nowhere, everything and nothing. Creation in Hindu tradition can be seen arising

Trimurti, meaning "having three forms," is the term applied to the three main Hindu gods: Brahma, Vishnu, and Shiva. This Trimurti, or triad, represents all aspects of the Supreme Being. © Louise Batalla Duran/ Alamy.

from nothingness from Brahman's thoughts or as the result of the actions of the god Brahma, the male personification of Brahman. Hindu creation stories originate from the Hindu scriptures and epic poems, the oldest of which are found in the Rig Veda, written around 2000 B.C.E. These stories have many similarities to Egyptian and Greek mythology.

In the first and tenth books of the Rig Veda, the Phallus of Heaven, representing the vital male force, sought out his young daughter, the Earth. From this incestuous union, some seed was spilled onto the Earth, creating the Angirases, mediators between gods and humans, who are responsible for distributing the gifts of the Earth. In this story, Heaven is seen as the father of humankind and Earth as the mother. In yet another version, also from the tenth book, the gods sacrificed a 10,000-headed, thousand-footed man called Purusa to create the universe. Purusa's lower portion became the world; his head, the sky; his eye, the sun; his mind, the moon; his feet, the earth; his navel, the air; his breath, the wind; his arms, the warrior caste; his mouth, the god Indra and Brahman priests; his thighs, the common people; and his feet, the lower castes.

His sacrifice also produced animals, plants, sacred rituals and words, and the Rig Veda. Also in the tenth book of the Rig Veda, there is another hymn that, like the Chinese concept of yin and yang, emphasizes the importance of opposites. It asks questions about being and nonbeing, darkness and light, and life and death, and about the very essence and mystery of creation. One translated verse says:

> Then even nothingness was not, nor existence.
> There was no air then, nor the heavens beyond it.
> What covered it? Where was it? In whose keeping?
> Was there then cosmic water, in depths unfathomed?
> (Basham cited in Leeming and Leeming 1994)

Islam

The Holy book of Islam, the Koran, which was dictated by the angel Gabriel to the prophet Muhammad, makes several passing references to the creation of the earth, humans, and other living things. The Islamic creation story is one in which God creates our world from nothing with the power of his thoughts. In one passage the book asks people how they could not believe in a god powerful enough to create the world in two days, but in another passage says that the earth and heavens were created in six days. In yet another passage, the book says that Allah created heaven and earth by beckoning them "in obedience." In still another, Allah is said to have created the world "to set forth his truth," that man was created from a "moist germ," and that God also created cattle and other fruits of the earth. In a later passage, referring to Allah, says: "It is God who hath given you the earth as a sure foundation, and over it built up the Heaven, and formed you, and made your forms beautiful, and feedeth you with good things." And in another passage it says: "He it is who created you out of dust, then of the germ of life, then of thick blood and then brought you forth infants."

Conclusions

The world's creation stories are as diverse as the cultures that spawned them. Many recurrent themes exist, however, among both related and unrelated cultures and subgroups within cultures, which may reflect their members' shared humanity. Some of these include: creation by cosmic egg, sacrifice, thought, word, chaos, and incest. Clearly, these stories, many passed down from generation to generation through both oral and written tradition, were an attempt to seek answers to profound questions that could not be answered at the time, and to place them within a framework that people could comprehend. As the renowned mythologist Joseph Campbell once said: "The material of myth is the material of our life, the material of our body, and the material of our environment, and a living vital mythology deals with these in terms of what are appropriate to the nature of knowledge at the time" (1990, p. 1). As such, the gods that were thought to create our universe, world, and all living things, including humans, were imbued with human characteristics, including hate, love, lust, power, deceit, conceit, sadness, joy, creativity, and compassion (Smith 1952).

Unwavering belief in numerous creation stories persists to this day. And despite well-documented evidence to the contrary, many people still consider some creation stories to be literal descriptions of the origin of life on earth, including human life. Yet, that some stories have come to dominate others has less to do with their revealed truth than it does with historical trends in technological advancement, especially in the development of weaponry, access to vital resources, the spread of disease, and wars of conquest, with their resultant cultural hegemony and spread of dominant civilizations, as described in Jared Diamond's Pulitzer-prize-winning book *Guns, Germs, and Steel: The Fates of Human Societies* (1997).

Science and the scientific method have provided humankind with a viable alternative explanation for the origin and diversification of life on earth, one that unifies all of biology and has strong predictive power—that is, the process of evolution through natural selection. While some of the world's great religions have embraced evolutionary theory, still others continue to reject it outright, preferring instead to hold onto supernatural explanations from the past. Is evolutionary theory perfect? Do scientists know everything they would like to know about the evolutionary process? Absolutely not! Scientific knowledge is continually growing, and many gaps remain to be filled. For example, many questions still remain about the details of the evolutionary process, and many great mysteries regarding the origin of the ever-expanding universe persist. These great mysteries are at the very core of science and will challenge scientists for years to come. That being said, the scientific disciplines of astronomy (Barrow and Silk 1983), geology (Lutgens, Tarbuck, and Tasa 2009), and biology (Mayr 1970) have answered many basic questions about the origins of the universe, earth, nature, life, and humanity. As Charles Darwin himself once said: "There is grandeur in this view of life, with its several powers, having been originally breathed into a few forms or into one; and that, whilst this planet has gone cycling on according to the fixed law of gravity, from so simple a beginning endless forms most beautiful and most wonderful have been, and are being, evolved" (1859).

Resources

Books

Alpers, Antony. 1964. *Maori Myths and Tribal Legends*. Auckland, NZ: Longman Paul Limited.

Barrow, John D., and Joseph Silk. 1983. *The Left Hand of Creation: The Origin and Evolution of the Expanding Universe*. New York: Basic Books.

Campbell, Joseph. 1990. *Transformations of Myth through Time*. New York: Harper Perennial.

Darwin, Charles. 1859. *On the Origin of Species by Means of Natural Selection; or, The Preservation of Favoured Races in the Struggle for Life*. London: John Murray.

Diamond, Jared. 1997. *Guns, Germs, and Steel: The Fates of Human Societies*. New York: Norton.

Eberhard, Wolfram, ed. and trans. 1965. *Folktales of China*. Rev. edition. Chicago: University of Chicago Press.

Fahs, Sophia Lyon, and Dorothy T. Spoerl. 1958. *Beginnings: Earth, Sky, Life, Death*. Boston: Starr King Press.

Ferguson, John C., and Masaharu Anesaki. 1937. *The Mythology of All Races*, Vol. 8: *Chinese and Japanese*. Boston: Archaeological Institute of America and Marshall Jones Co.

Havecker, Cyril. 1987. *Understanding Aboriginal Culture*. Sydney, Australia: Cosmos.

Ions, Veronica. 1973 [1965]. *Egyptian Mythology*. London: Hamlyn.

Lao-Tzu. 1963. *Tao te Ching*, trans. D. C. Lau. Baltimore: Penguin.

Leach, Edmund R. 1967. "Genesis as Myth." In *Myth and Cosmos: Readings in Mythology and Symbolism*, ed. John Middleton. Austin: University of Texas Press.

Leeming, David Adams, and Margaret Adams Leeming. 1994. *A Dictionary of Creation Myths*. New York: Oxford University Press.

Lutgens, Frederick K., Edward J. Tarbuck, and Dennis Tasa. 2009. *Essentials of Geology*. 10th edition. Upper Saddle River, NJ: Pearson Prentice Hall.

Mayr, Ernst. 1970. *Populations, Species, and Evolution*. Cambridge, MA: Harvard University Press, Belknap Press.

Reed, A. W. 1971. *Myths and Legends of Australia*. Sydney, Australia: A. H. and A. W. Reed.

Smelcer, John E. 1993. *A Cycle of Myths: Native Legends from Southeast Alaska*. Anchorage, AK: Salmon Run Press.

Smith, Homer William. 1952. *Man and His Gods*. Boston: Little, Brown.

Sproul, Barbara C. 1979. *Primal Myths: Creation Myths Around the World*. San Francisco: Harper and Row.

Michael Hutchins

History of evolutionary thought

Anyone who thinks for a minute about biological diversity—the tremendous variety of organisms in the world, with all its colors, habits, ways of making a living, and so on—has to wonder about how it got that way. There are two possibilities: Either it has always been just as it is now, or it has changed through time. For centuries people thought that the world and its creatures have always been pretty much as they are now, because they saw little evidence of change. Many religious scriptures or accounts of creation still argue that this is the case. And yet since the 1850s the dominant explanation of life's diversity has relied on the centrality of evolution, the idea that all organisms have evolved from a common ancestry. One cannot understand how animals grow, reproduce, behave, and generally live without a strong foundation in evolutionary theory and the patterns of relationships it explains.

The Enlightenment

During the Enlightenment, which began in Europe in the seventeenth century, the view that the world was eternally unchanged since a divine creation began to lose universality of acceptance. Two main sources of evidence altered this view. The first came from rocks. Scholars began to learn more about how rocks were formed by looking at the geological formations and sediments that were then being laid down. They decided that things that were called "fossils" actually were the remains of ancient organisms, not just artifacts. They noticed that the organisms found as fossils were not the same as living organisms, and vice versa. By the early 1800s a rather good idea of the general outline of the geologic record, and all the fossils within it, was emerging. At the time there was no way to measure how old these things were, how long they had been buried, or how much older one bed of fossils was than another. But it became generally accepted that life had changed through the time represented by these rocks. And the farther back in the geologic record scholars went, the less the fossilized organisms looked like the contemporary organisms with which they were familiar. This cast considerable doubt on the idea that all organisms had been created at once and had survived unchanged to the present day.

The second line of evidence during the Enlightenment came from living organisms themselves. People began to realize that organisms, despite their differences, shared lots of similarities with each other. Comparative anatomy, which emerged as a science during the 1600s and 1700s, showed that all vertebrates, including humans, had a similar basic body plan, even though some vertebrates swam, whereas others ran on four legs, climbed, and flew. The bones of the skeleton, and increasingly of the skull, corresponded very well in particulars among the different vertebrate groups. Some animals even seemed to have transitional features between other distinct groups. Why would it make sense to build all these different adaptations in animals using the same kinds of bones? Why not wheels, pulleys, helicopter blades?

It had been common knowledge since the time of the ancient Greeks that organisms could be conveniently grouped together or classified based on their similarities: all the carnivores, all the hoofed animals, all the worms, and so on. Among the carnivores, there was a well-defined cat group, a dog group, another for bears, another for raccoons and similar things, and so on. Why were these organisms similar? Were they related? What did it mean to say that they were "related" if each species had been created independently? Or could it be that different cats, for example, had differentiated from a common cat ancestor?

During those early Enlightenment days, research was unsystematic, and traditional ideas persisted, as they usually do. Progress was not steady, and there was usually at least one step backward for every step forward. One could not advance ideas or even pretty clear-cut facts without one eye on what the church was going to say about it. Science and religion were not separate in Western Europe, and no finding that called itself science could contradict church doctrine without being called heresy.

But even apart from this intellectual atmosphere, there was a simple conceptual step that had to happen first: Ideas of evolution could not be seriously entertained until people understood something about how vast time really is. The whole extent of geologic time—what scientists call "deep time"—was unimaginable to people before the mid-nineteenth century. From the earliest humans right up through contemporary societies that build their lives around farming, hunting, and gathering, time had little meaning apart from the cycle of the seasons. Some advanced cultures had the luxury to keep detailed records of the passage of years, but few recorded these histories in years accurately. How far did even human time go back? There was no way to tell.

In the late 1700s these traditional ideas began to change. For centuries, people in England had collected fossil shells and bones, without necessarily realizing that they were actual remains of formerly living creatures. These collections, like those in Europe, became what were called natural history cabinets, and the nobility mostly owned them. But in the late 1700s in England, a practically educated civil engineer named William Smith (1769–1839) began to use fossils and the rocks that contained them in a new way. Smith supervised the digging of canals to bring water across large stretches between rivers. He noticed that different regions had different layers of rocks and, moreover, that the layers of rock lay atop one another in the same order from one place to another. He also observed that particular fossil shells were characteristic of each of these layers.

From this perceptive series of observations, Smith constructed the first geologic map of England, which was completed in 1801. His nephew John Phillips (1800–1874), who later became a professor at Oxford University, devised the first known description of the diversity of life through time, as based on the fossil record, in 1860. And it is remarkably similar to descriptions made 130 years later with much more information.

Meanwhile, in France, more progress was being made in understanding geological processes. In the early 1800s Alexandre Brongniart (1770–1847) and Georges Cuvier (1769–1832) were surveying the fossils of the cliffs and hills around Paris. They found a series of rocks that always occurred atop one another in the same order, with characteristic fossils. But they wondered particularly about why these different associations of fossil animals, which are called faunas, replaced each other in the way that they did. Cuvier, one of the greatest scientists of any age, noted that the animals found in these rocks did not currently inhabit those areas, or anywhere else for that matter. Many of them were marine animals, which also meant that seas had once covered the Paris area. Other faunas comprised terrestrial mammals. Moreover, the replacements of faunas from one rock layer to the next seemed to be very rapid. Cuvier concluded quite logically that the replacements had to be the result of environmental changes. The new animals that colonized the area after the climate stabilized again must have immigrated from elsewhere, so the catastrophes could not be worldwide and total. Cuvier called these *révolutions du globe* (a complex phrase that refers to geological change as well as to the "revolutions" of the Earth around the Sun). This idea was influential because it made people realize how detailed the testimony of the geologic and paleontological record was, and what it revealed about the unsuspected history of life on Earth.

But Cuvier had another important concept in mind. The animals in the fossil record that are so different from living forms are, he claimed, extinct. The concept of irrevocable and final disappearance shocked a lot of people, because it countered some interpretations of Biblical scripture. If God created everything, his creations must be perfect because he is perfect. Therefore extinction was inconceivable. And yet, there was no reasonable doubt about it.

Cuvier did not propose the idea of evolution. Nor did he support the idea of special creation. He thought it was most prudent for scientists to be conservative and stick to the facts, not devise elaborate systems of explanation for which there was

Jean-Baptiste de Lamarck with Musée National d'Histoire Naturelle in the background. Alex Bartel/Photo Researchers, Inc.

little evidence. This is one reason why he was so opposed to the ideas of his colleague Jean-Baptiste de Lamarck (1744–1829).

The center of scientific intellectual activity in natural history in Paris was the Jardin du Roi, later the Jardin des Plantes, which held all the specimen collections and the zoo. Its director in the mid-1700s was Georges-Louis Leclerc, comte de Buffon (1707–1788), who spent half the year at his estates in Burgundy, where he raised timber for ships and tried to develop the strongest wood for naval purposes. Buffon wrote a thirty-six-volume treatise on natural history that was very popular at the time. In it he proposed that species were not fixed entities, and that closely related species could have come from a common ancestor. But he had no mechanism to explain how such change could occur, nor any evidence to support it. This was the problem with many early ideas about evolution.

Lamarck and Darwin

Lamarck is probably the best-known name associated with the concept of evolution before Charles Darwin (1809–1882). Lamarck was a botanist and invertebrate zoologist who worked at the Jardin du Roi/Jardin des Plantes during the late 1700s and early 1800s. He was a colleague of Cuvier and other great scientists of the time, and he was the strongest proponent of evolution in France. He developed a detailed, complex theory that involved spontaneous generation, the migration of fluids through the body to produce new features, and the continual progress of simple organisms to more complex ones. Little of what Lamarck said was taken seriously during his lifetime, and none of it has survived to become part of current evolutionary theory. Lamarck accepted the inheritance of acquired characters, though he did not originate the concept. This was the idea that features acquired during an organism's lifetime could be passed to its descendants if advantageous. At the time, even basic reproductive biology, including genetics, was not understood, and so people did not realize that the sex cells in animals are produced when the animal is born, and that the genetic material is not affected by changes during the animal's lifetime.

Lamarck thought that the central process in evolution was progressive "improvement" of forms toward more complex forms. He recognized that adaptation took place, but he thought that adaptation was almost always a dead-end process: It created locally new forms changing in new directions, but it did not advance the organisms progressively along what might be called his "escalator of life" from simple to complex forms. This view was incomprehensible to Darwin, who saw no way to measure or assess "progress" and saw that adaptation was the principal way in which organisms evolved.

Lamarck also thought that organisms changed in response to what has been translated as "felt needs." The word in French is *besoins*, which simply means needs, such as the need to breathe, eat, and find warmth. There is nothing metaphysical about this; Lamarck's view was purely materialistic. One certainly can agree that organisms have to be opportunistic in taking advantage of environmental conditions that meet their needs. But the issue in which Lamarck is often

contrasted with Darwin has to do with those famous giraffes. In fact, Lamarck, like Darwin, devoted only one paragraph out of thousands of pages to the giraffe example. Some textbooks give the impression that they had a great debate over this, but their publications were separated by fifty years.

Lamarck suggested that giraffes, by stretching their necks to reach higher leaves, could acquire longer necks and pass these changes on to their offspring. Most scientists at the time, including Darwin for much of his career, thought that this inheritance of acquired characteristics was a reasonable explanation. Darwin's idea, though, was that in a species of giraffe there would be individuals of variably sized necks; the ones with longer necks, if this was a critical advantage, would be the ones more likely to survive to pass the characteristic on to their offspring.

A major problem with Darwin's explanation, according to many European biologists, was that he did not explain why some giraffes had longer necks than others, or where this variation came from. The mechanisms of heredity were not understood until the renaissance of interest in the work of the Austrian botanist Gregor Mendel (1822–1884) on inheritance in pea plants, which was overlooked until the end of the nineteenth century; but still, the origin of these variations remained mysterious.

Darwin's contributions

As a young man, Darwin was interested in natural history, but mostly in the context of hunting and collecting insects and other animals. He went to medical school at the University of Edinburgh, then the greatest center for medical studies in Britain, following his father's expectations. But his course of study there seemed only to reinforce his interest in natural history. He enrolled at the University of Cambridge and continued his studies in geology, botany, and zoology, intending to become a rural clergyman with time for extracurricular scientific pursuits. And then in 1831, when he was twenty-two, he was recommended by one of his professors for a very unusual position.

Captain Robert FitzRoy (1805–1865) of HMS *Beagle* was charged with the mission to sail around the world, checking the depths of waters around important ports, collecting plants and animals for scientific and potential economic uses, and fostering the possibility of expanded trade and colonization. FitzRoy was a first-rate scientist who had completely refitted his little ship for the voyage with extra decks that could store natural history specimens. He wanted to add to his crew a gentleman naturalist to converse with and contribute to the scientific mission of the expedition. Darwin and his Cambridge professors convinced his father that this would be a great opportunity for a gentleman of Darwin's interests. Over the next five years, Darwin collected plants, animals, and some of the first mammalian fossils from South America. The voyage profoundly changed Darwin's views of life. He saw plants and animals adapted to all kinds of surroundings. He noticed their geographic distributions and saw that some animals and plants were completely different from country to country and from habitat to habitat. His notebooks from the voyage of the *Beagle* reveal that he had begun to think deeply

Lamarck suggested that giraffes, by stretching their necks to reach higher leaves, could acquire longer necks and pass these changes on to their offspring. © Sean Nel Image from BigStockPhoto.com.

Illustration of the HMS *Beagle* carrying Charles Darwin's expedition in the Straits of Magellan, Mt. Sarmiento in the distance. © Bettmann/Corbis.

about distributions of plants and animals, and how organisms were adapted to their surroundings. But it was only after he returned to England in 1836 that he began to explore the reasons why. He opened his first notebook on what he called transmutation theory shortly thereafter.

An intellectual turning point for Darwin, as for many intellectuals of his time, came when he read Thomas Robert Malthus's *Essay on the Principle of Population* in 1838. Malthus (1766–1834) had argued forty years earlier that the increase in human population growth was greater than the increase in food supply, so there was likely to be a struggle for existence that explained why certain individuals survived instead of others. Building on Malthus's ideas, Darwin advanced several circumstances that supported natural selection:

1. More offspring are produced than can survive;

2. natural variation is relatively copious and unconstrained, though slight;

3. some variations give a better advantage to the organism's way of life;

4. these variations are heritable;

5. and those individuals most fit for their environments, by virtue of these advantageous variations, would on balance leave more offspring for the next generation.

How did Darwin formulate this idea? He devoted the first chapters of *On the Origin of Species by Means of Natural Selection* (1859) to how farmers and breeders try to improve their crops and livestock. Darwin saw, for example, how breeders could select for slight variations in pigeon color and feather structure. In only a few thousand years, Darwin noted, human breeders had produced all the varieties of plants and animals that comprise humankind's crops, livestock, and domestic animals. Given the eons of time available to nature, Darwin reasoned, what must natural selection have been able to accomplish? He concluded that all the diversity seen in the world today and in the past result from the natural actions of processes such as natural selection.

Alfred Russel Wallace (1823–1913), a collector of natural history specimens and a longtime correspondent of Darwin, devised the idea of natural selection simultaneously to Darwin. Wallace was not an independently wealthy gentleman, like Darwin; he did a great deal of exploring in southeastern Asia, Australia, and other lands. From his tent in Malaysia, where he was collecting specimens and studying biogeography, Wallace wrote to Darwin, sending him a manuscript and asking if Darwin could have it reviewed for publication.

Darwin was in a difficult spot, even though he had spent two decades gathering evidence and formulating his arguments. Wallace's letter made Darwin realize that he had to publish his theory. He had been discussing his ideas with a considerable circle of influential scientists, who knew very well how much work he had been putting into his theory. Darwin put the matter in their hands, and with Wallace's eager consent, the two men published their idea jointly in a publication of the Linnean Society of London in 1858. The paper actually had little effect when it was read at the Linnean Society's meeting; neither author was present. But when *Origin of Species* was published in late 1859, it sold out all its copies instantly and created an enduring scandal.

Origin of Species

Darwin's book was mainly about natural selection. His argument was that this natural mechanism, which did not require divine intervention, was able to explain how much of the diversity of life evolved. Darwin did not present experimental evidence that demonstrated natural selection, though he was a great experimenter in natural processes. And he did not focus much on how new adaptations might arise. He devised what he called "one long argument" for the idea that natural selection, a process on a far grander scale than artificial selection, could have accounted for the great range of biological diversity seen today and through time. His view was that a horse that was able to outrun and escape predators was more likely to have offspring (with its same traits) than a horse that could not run as fast, if running fast was a critical characteristic for survival. The slight variations in populations could be selected upon in this way, and so adaptations would gradually evolve in species and eventually result in the great diversity of life.

Darwin did not use the fossil record very much in his book; in fact, he bemoaned its incompleteness in two chapters (although he also defended its ability to address certain questions), and antievolutionists have continued to make much of his dismay until this day, even though scientists know a lot more about fossils than they did in Darwin's time. But Darwin's theory was really not about macroevolution; he thought that this would take care of itself if he could show how evolution worked among organisms in a species. It is virtually impossible to measure individual interactions and the effects of natural selection in fossil organisms, and Darwin knew this. That is why he did not fervently defend the fossil record as a strong source of evidence for his mechanism. He knew the fossil record was valuable, but it was valuable for something different than what he was arguing.

Many authors have stressed the importance to Darwin's theory of "gradual" change—the idea that organisms have evolved by very small steps, not by sudden large jumps. This was an important part of Darwin's argument, because he thought it was easy to defend. Thomas Henry Huxley (1825–1895) urged him against it, saying that he had unnecessarily restricted himself, and that he should allow that evolution could sometimes make substantial leaps. But Darwin wanted to be persuasive, so he emphasized small differences.

It is important to note, however, that the term *gradual* did not mean then what it means in the early twenty-first century. For example, when Darwin was on the *Beagle* he stopped in Chile and witnessed a large earthquake in Concepción that leveled dozens of houses and injured or killed hundreds of people. The next day he went down to the seacoast and saw that the cliff face had risen several meters out of the water as a result of the sudden earthquake. And he saw a whole series of terraces that had been lifted up through time in this way. Darwin described this as a "gradual" change because the term means "steplike," from the Latin *gradus*, or step. It does not necessarily mean "slow and steady" in an imperceptible way, although Darwin described this tempo of change as well.

Darwin wrote at least nine books, plus four monographs, on the taxonomy of fossil and living barnacles. His second most important evolutionary work is about sexual selection, which he rightly recognized as a very different evolutionary process than natural selection. Simply stated, sexual selection is the process by which one sex selects mates from the other sex, because one sex has a feature that is considered desirable in mates. These characteristics often reflect environmental fitness only indirectly or not at all. As one can readily see, in some cases the features that will make individuals attractive to potential mates may be deleterious to their own survival. A stock example is the peacock whose tail is so large that it outcompetes other males for mates; but the same tail impedes it from getting away from predators.

What were the immediate effects of Darwin's *Origin of Species* in 1859? There were the usual reactions: praise on one side, vilification on the other. Darwin's circle of colleagues, scientists in Cambridge and London, knew of his ideas and were glad to see them in print. In contrast, people who had clearly not read the book denounced it, burned it, and excoriated Darwin in the vilest terms. But some people who read it took issue with it as well, and these people included some of Darwin's friends, colleagues, and former teachers. Many objected to the idea that all creatures could be descended from common ancestors; some needed to see their ancestry as separate from those of apes and monkeys; and some thought that natural selection was not powerful enough to have shaped biological diversity in all of its myriad forms.

Historians of science have noted that even if Darwin may not have convinced most people immediately of the importance of natural selection, for many he provided the first convincing mechanism of how life could evolve. For many others he provided evidence that allowed them to accept that evolution had actually happened, whatever its mechanisms were, rather than special creation of each form of life. And it was this idea that changed the world.

Evolution after Darwin

Research into evolution as a process and a pattern flourished with the publication of Darwin's book. The fossil record became better known and was studied with the intent of discovering ancestors of living and other fossil forms. Travelers studied native faunas and floras with an eye toward their evolution and their relationships to other faunas and floras elsewhere. Classifications of species took on the added assumption that these were not just "arrangements" of groups that showed "affinities" to each other, but real relationships that were underpinned by common ancestry. Comparative anatomy, especially comparative embryology, began to take on strong evolutionary overtones. Embryologists noted that organisms that were more closely related had more similar embryological stages. Systematists began to assemble broad groups of organisms into trees of life, a process that still continues with revisions today.

In that late-nineteenth-century research activity, however, natural selection was either assumed or ignored, but seldom measured or tested. In fact, "neo-Lamarckian" ideas of inheritance of acquired characteristics held sway among a great many evolutionists, notably in the United States. Many paleontologists accepted the idea of orthogenesis, which is defined as "straight-line evolution," and it has to do with trends in macroevolution, many of which were well known in the late 1800s. Orthogenesis is associated with the idea that a trend may not be reversed once it gets going. Sometimes the idea that a lineage is "heading for a goal" is entailed here, but that is actually a very different idea called teleology.

Mutationism was another school of thought that developed in the early 1900s. Mendel's work on genetics was rediscovered at that time, and the English biologist William Bateson in 1900 coined the term *genetics* to describe this new study of heredity. Geneticists tried to investigate the hereditary material and see how it affected morphological expression. This was easiest to observe when the mutations were relatively strong in their effects. Most of these large mutations appeared to affect how animals and plants develop, and this led scientists such as Bateson to suspect that changes in development were of primary importance in evolution. Because most biologists at the time doubted that natural selection was a strong enough force to direct evolution primarily, they concluded logically that small mutations would be of little use in evolution and would tend to be swamped by blending of other characters by random mating. Mutation was sufficient for new species to form; therefore, natural selection was viewed as relatively unimportant.

Thus, the early decades of the 1900s witnessed an almost complete abandonment of Darwin's views of the importance of natural selection. It was not until the 1930s that scientists revived natural selection as an important evolutionary force and showed that mutations worked together with natural selection to produce evolutionary change. In the process, they also showed that most substantial mutations had deleterious effects and that there was little evidence they actually had contributed to substantial evolutionary change. As a consequence, by the 1930s small mutations—what would be called continuous variation as opposed to the discrete variation of larger mutations—were accepted as the genetic basis of nearly all evolution. What emerged to unify this chaos and disagreement in the years around 1930 was one of the most important advances in evolution of the twentieth century. It was called the modern synthesis of evolution. And like any great advance, it brought both benefits and problems.

The modern synthesis

A new vision of evolutionary biology began with the emergence of a new field of biology: theoretical population genetics. This field uses mathematical models to explore what kinds of changes can happen in populations, given certain initial conditions, mechanisms, and rates of change. For example, given a population of a certain size, given birth and death rates, and given rates of recombination during mating, the mutation rate, and the intensity of natural selection on one or more alleles, various models can be devised to show whether natural selection would be effective in producing evolutionary change in such populations. Later, empirical investigation would ask how closely the models developed by the mathematicians approximated what was actually observed in natural populations.

The founders of population genetics were principally three: Ronald A. Fisher (1890–1962), J. B. S. Haldane (1892–1964), and Sewall Wright (1889–1988). They did not always agree on their findings, but collectively they made great advances in understanding how natural selection and other processes could direct evolutionary change on a population level through actions on individuals. Fisher was relentlessly mathematical, Haldane had an excellent background in natural history, and Wright was more pluralistic about how evolution worked than most of his colleagues. Their models together, however, showed that both mutation and natural selection could work together to cause populational change.

One great advance was to translate Darwin's theory of natural selection into mathematical terms. It was both difficult and unrealistic to assign a theoretical "fitness advantage" to one allele or another, assuming that there is a given variation that will confer survival success on an individual. So, the theoretical population geneticists bypassed the measurement of "adaptation" and essentially redefined what Darwin called fitness. By "fitness," Darwin meant that an individual could outrun or outperform others of its species. So fitness, for Darwin, was explicitly an adaptive concept. Those organisms that were more fit were more likely to leave more offspring with their characteristics in the next generation. The population geneticists short-circuited this concept in order to make it quantifiable. They defined fitness as reproductive success. This was easier to quantify than some measure of adaptiveness. The problem, as C. H. Waddington (1905–1975) pointed out in the 1960s, was the assumption that the individuals that produced the most offspring were necessarily best fit for their environments, which is incorrect. Given this assumption, however, the models work.

Historically, the modern synthesis was a fusion of paleontology, population biology, and theoretical genetics. It affirmed that natural selection was the principal directional force in

populations. It backed up this assertion with laboratory studies that showed the effects of natural selection on fruit flies and other organisms. The pioneering work in this field was done by Theodosius Dobzhansky (1900–1975), who had emigrated from Russia, where he had begun his work with the great Russian geneticist Sergei Chetverikov (1880–1959). Dobzhansky gravitated to the United States and became the primary synthesizer of the theoretical population models with actual experimental data. The modern synthesis launched both theoretical and experimental studies of populational change. Its main precept was that natural selection on small mutations, changing gradually, was perfectly sufficient to explain nearly all evolutionary change. As for paleontology, it was recognized that the fossil record was too episodically deposited to give any reliable information about populational change. Therefore, it could be assumed that the patterns seen in the fossil record were consistent with what was seen at the populational level—as Darwin had inferred.

Since the 1930s, the modern synthesis has increasingly stressed that natural selection is the principal direction-giving force in evolution (effectively the only important one, for many or most of its supporters). As frequently summarized by its principal architects, such as Ernst Mayr, the basic precepts of the modern synthesis are these: (1) Evolution in populations is caused by numerous small mutations that are directed by natural selection; (2) populations diverge gradually from each other in response to environmental differences; and (3) species form in this way, and all macroevolutionary patterns seen in the fossil record are extrapolations of these same processes on a larger scale.

Is there a post-modern synthesis?

Since the mid-1970s, there have been challenges to what historians have called the orthodoxy of the modern synthesis. The challenges have not overturned the idea that natural selection is the main direction-giving force in evolution, but the importance of other forces is now more strongly acknowledged. For example, in the 1960s Jack Lester King and Thomas H. Jukes found that there was far more genetic variability in wild populations than had been imagined. Most of the variability was in the triplet codons of DNA: changes in the "letters" that represent the four bases symbolized by A, C, G, and T. This variability had no obvious adaptive value, because the third letters of the codon triplets could often be changed without any effect on which amino acid was coded for. This phenomenon was called "neutral evolution," and it countered the view that all evolutionary changes and variations were adaptive.

Another apparently nonrandom but not clearly adaptive phenomenon was the fact that in some taxa there was a stronger tendency for mutations toward coding C and T than toward A and G. Another was the discovery that in fact most DNA in cells is highly repetitive but is inactive. What could all that DNA be used for, why was it repeated, and was it ever used? The answers are still emerging.

A further part of the modern synthesis orthodoxy that was challenged had to do with modes of speciation and with

The Painted Hills in the John Day Fossil Beds are rich with the evidence of ancient habitats and the dynamic processes that shaped them. © Lukich, 2010/Shutterstock.com.

species concepts. The modern synthesis focused on the ability or inability of populations to breed in the wild as a critical concept in the definition of a species. Even when it was proposed, it was obvious to biologists that most organisms did not fit this "biological species concept" (BSC). First, most organisms are asexual. Second, some breed opportunistically, even though they tend to breed true to their kind. Plants in particular do not follow the BSC. Third, the concept could not be applied to extinct organisms, which make up 99.99 percent of the history of life.

The primary mode of speciation specified by the modern synthesis was allopatric, which means that populations diverge geographically until they are isolated. They then differentiate genetically and adapt locally, until they can no longer breed with their former relatives. This process is perfectly reasonable, and there is copious evidence that it has worked for many species. There were also findings, however, that populations could separate without so much geographic separation, that sometimes ecological differentiation was more important than geographic differentiation, and that what happened genetically was far more important than what was going on geographically.

There were challenges from the paleontological side as well. The emerging science of paleobiology quantified a great many macroevolutionary patterns and changes that could never have been predicted from ideas about population biology, genetics, or speciation. They were not in conflict with these fields; they just showed that evolution is a hierarchical process. The patterns seen at the macroevolutionary level, for example, cannot be readily predicted from the data of population biology. These studies also revealed a number of patterns that suggested explanations at the populational level that were previously unsuspected. For example, careful reevaluations of good sequences of fossils showed that for most of their durations species do not change significantly in one direction or another; instead, significant morphological change seems to happen in a very short time. This concept is called punctuated equilibria, and it suggests that "gradual" Darwinian change is not always slow and imperceptible, but largely steplike.

These discoveries and continuing issues should convey the impression that evolution is a very active and vibrant science. Scientists do not know all they need to know about it, and there is much healthy disagreement about how some processes work, what mechanisms explain certain patterns, and when some kinds of processes are more important than others. More rapport is needed among the different subfields of evolution, but recent progress is impressive. New discoveries in fields as diverse as genetics, biogeography, quantitative paleobiology, systematics, and developmental biology make evolution more exciting than it has been since the 1930s. Far from being a "theory in crisis," as it is sometimes depicted by antievolutionists, the field is active and stimulating, and its history continues to be written.

Resources

Books

Bowler, Peter J. 2003. *Evolution: The History of an Idea*. 3rd edition. Berkeley: University of California Press.

Browne, Janet. 2006. *Darwin's "Origin of Species": A Biography*. New York: Atlantic Monthly Press.

Darwin, Charles. 1859. *On the Origin of Species by Means of Natural Selection; or, The Preservation of Favoured Races in the Struggle for Life*. London: John Murray.

Darwin, Charles. 1871. *The Descent of Man, and Selection in Relation to Sex*. London: John Murray.

Desmond, Adrian. 1982. *Archetypes and Ancestors: Palaeontology in Victorian London, 1850–1875*. Chicago: University of Chicago Press.

Desmond, Adrian. 1989. *The Politics of Evolution: Morphology, Medicine, and Reform in Radical London*. Chicago: University of Chicago Press.

Desmond, Adrian, and James Moore. 1992. *Darwin: The Life of a Tormented Evolutionist*. New York: Warner Books.

Dewey, John. 1910. *The Influence of Darwin on Philosophy, and Other Essays in Contemporary Thought*. New York: Henry Holt.

Futuyma, Douglas J. 2005. *Evolution*. Sunderland, MA: Sinauer Press.

Gould, Stephen Jay. 2002. *The Structure of Evolutionary Theory*. Cambridge, MA: Harvard University Press, Belknap Press.

Malthus, Thomas Robert. 1998. *Essay on Population*. London: J. Johnson.

Mayr, Ernst, and William J. Provine. 1980. *The Evolutionary Synthesis*. Cambridge, MA: Harvard University Press.

Rudwick, Martin J. S. 1997. *Georges Cuvier, Fossil Bones, and Geological Catastrophes*. Chicago: University of Chicago Press.

Schopf, Thomas J. M., ed. 1972. *Models in Paleobiology*. San Francisco: Freeman, Cooper.

Thomson, Keith. 1995. *HMS Beagle: The Story of Darwin's Ship*. New York: Norton.

Thomson, Keith. 2005. *Before Darwin: Reconciling God and Nature*. New Haven, CT: Yale University Press.

Winchester, Simon. 2001. *The Map That Changed the World: William Smith and the Birth of Modern Geology*. New York: HarperCollins.

Young, David. 2007. *The Discovery of Evolution*. 2nd edition. Cambridge, UK: Cambridge University Press.

Periodicals

Padian, Kevin. 1999. "Charles Darwin's Views of Classification in Theory and Practice." *Systematic Biology* 48(2): 352–364.

Taquet, Philippe, and Kevin Padian. 2004. "The Earliest Known Restoration of a Pterosaur and the Philosophical Origins of Cuvier's *Ossemens Fossiles*." *Comptes Rendus Palevol* 3(2): 157–175.

Kevin Padian

· · · · ·

Charles Darwin: A life of discovery

Early years

Charles Robert Darwin was born on February 12, 1809, in Shrewsbury, England. He grew up at The Mount, the family home that overlooked the River Severn. His father, Robert Waring Darwin (1766–1848), was a well-respected and successful physician. Darwin's mother, Susannah (1765–1817), died when he was eight years old, and he was brought up by his older sisters who took charge of the household.

From 1818 to 1825 he attended Shrewsbury School run by the Reverend Samuel Butler. In his autobiography Darwin wrote, "Nothing could have been worse for the development of my mind than Dr. Butler's school, as it was strictly classical, nothing else being taught except a little ancient geography and history." (Darwin 1887). Darwin was more interested in the outdoors. At an early age, he developed a passion for collecting—shells, minerals, insects—and a love of fishing and hunting. But he was not a good student. At one point his father told him, "You care for nothing but shooting, dogs, and rat-catching, and you will be a disgrace to yourself and all your family." (Darwin 1887).

In 1825, hoping he would make something of himself, his father sent him off to Edinburgh University to study medicine. Darwin, however, was not cut out to be a doctor. He attended two operations, but he could not stay to see either finished (pre-anesthesia operations were grisly affairs).

In desperation, his father sent him to Cambridge University in 1827 to prepare him for the clergy, and it was there that he met John Stevens Henslow, Professor of Botany, who would become his mentor. They talked so often that Darwin became known as "the man who walks with Henslow."

Darwin later wrote, "No pursuit at Cambridge was followed with nearly so much eagerness or gave me so much pleasure as collecting beetles. It was the mere passion for collecting; for I did not dissect them, and rarely compared their external characters with published descriptions, but got them named anyhow. I will give a proof of my zeal: one day, on tearing off some old bark, I saw two rare beetles, and seized one in each hand; then I saw a third and new kind, which I could not bear to lose, so that I popped the one which I held in my right hand into my mouth. Alas! it ejected some

intensely acrid fluid, which burnt my tongue so that I was forced to spit the beetle out, which was lost, as was the third one." (Darwin 1887).

But beetles were just the beginning. Soon after attending Cambridge he set out on a voyage that opened his eyes to the incredible diversity of life. As he himself said later, "The voyage of the *Beagle* has been by far the most important event in my life and has determined my whole career." (Darwin 1887).

The voyage

History is full of famous sea voyages—from Christopher Columbus's 1492 journey to the new world to the first

Charles Darwin and his sister, Catherine (c. 1816). © Hulton Archive/ Getty Images.

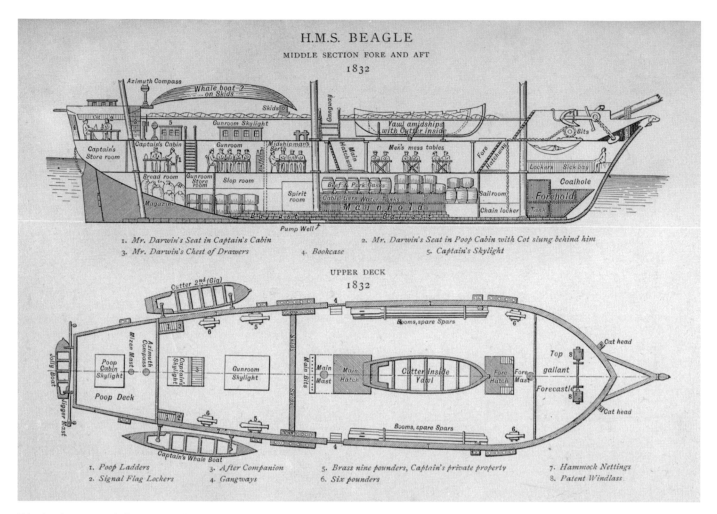

This drawing, not entirely accurate, first appeared in the 1890 edition of Voyage of the Beagle. It is based on a drawing by Darwin's shipmate, P. G. King. © Mary Evans Picture Library/Alamy.

circumnavigation of the Earth by Ferdinand Magellan's ship *Vittoria*. In modern atlases, colorful dotted lines crisscross the world's oceans, marking the routes of Vasco da Gama, Francis Drake, Captain Cook, and Jacques Cartier.

Their adventures fill our history books and fire our imaginations with the thunder of cannons, the horrors of scurvy, and the discovery of new lands. Yet within this rich tapestry of triumph and disaster, only a few voyages truly altered the course of history. The *Beagle* expedition is one.

Circling the world from 1831 to 1836, the *Beagle* discovered no new continents, fought no decisive sea battles, nor returned laden with gold doubloons, bolts of silk, or exotic spices. But onboard was Darwin. As the expedition's de facto naturalist, he explored unknown reefs and volcanoes, described new birds and reptiles, and unearthed mysterious fossils and shells. He hacked his way through the rain forests of Brazil and clambered to the top of the Andes Mountains. He experienced a devastating earthquake that shook the west coast of Chile and explored the tranquil coral islands of the Indian Ocean. From the Antarctic to the tropics, Darwin studied the world's geology, plants, and animals and, as a result, forged the

most far-reaching theory in the history of science: evolution by natural selection.

Riding the tide

In 1831 the British Admiralty commissioned *HMS Beagle*, under the command of Captain Robert FitzRoy (1805–1865), to conduct surveys of the South American coast. The voyage presented a rare opportunity for a naturalist to accompany the expedition, and Henslow recommended Darwin.

Getting Darwin onboard, however, was fraught with difficulties. First, Darwin's father objected—he thought it a waste of time. At one point, the position was offered to someone else. And as if there were not enough problems, FitzRoy did not like Darwin's nose. As Darwin wrote in his autobiography, "He [FitzRoy] was an ardent disciple of [Swiss physiognomist Johann Kaspar] Lavater, and was convinced that he could judge of a man's character by the outline of his features; and he doubted whether any one with my nose could possess sufficient energy and determination for the voyage. But I think he was afterwards well satisfied that my nose had spoken falsely." (Darwin 1887).

An illustration of the HMS *Beagle* in the Galápagos Islands. Trustees of John Chancellor.

The first week of September 1831 was tumultuous. Letters flew back and forth; interviews were scheduled and canceled; plans made and abandoned. Darwin overcame one obstacle only to face another, but his destiny prevailed. On September 5, with the details of the voyage finally settled, he wrote to his sister, "There is indeed a tide in the affairs of men, & I have experienced it." (The Darwin Correspondence Project, or DCP).

FitzRoy warned Darwin that space was tight, but nothing prepared him for what he found at Devonport on Tuesday, September 13: a ten-gun brig rebuilt as a three-masted bark (a third mast, the mizzenmast, had been added before the ship's first voyage), about 27.4 meters (90 feet) long with a beam of only 7.4 meters (24 feet). The *Beagle* carried more than seventy men, and in order to sleep at night, Darwin had to remove a drawer to make room for his feet. Lack of space, however, was the least of his problems. Although the voyage was originally scheduled to leave in October, there were many delays while the ship was refitted to FitzRoy's exacting specifications. Twice the crew departed only to be driven back by gales. Finally, on December 27, they set out for good on their five-year adventure. And the first thing Darwin learned was the agony of seasickness.

He was ill almost the entire voyage, scarcely able to get out of his hammock whenever the ship was at sea. He passed the

tortuous hours reading the books he had brought along—Alexander von Humboldt's *Personal Narrative*, John Milton's *Paradise Lost*, and a copy of the New Testament in Greek. Finding it impossible to even stand up without becoming seasick, Darwin wondered if he had made a serious mistake.

The *Beagle*'s first stop was Tenerife, in the Canary Islands, but upon arrival the crew faced a quarantine of twelve days because England was in the middle of a cholera epidemic. FitzRoy did not hesitate. "Up jib!" he ordered, and to Darwin's horror they sailed off immediately. Not only did Darwin want to visit the island, he desperately wanted to stand on dry land.

Fortunately, the break he needed was not far off. On January 16, 1832, the *Beagle* reached São Tiago in the Cape Verde Islands, about 480 kilometers (300 miles) off the African coast. Expecting it to be uninteresting, Darwin found it electrifying. He saw for the first time the lush tropical flora he had read about in Humboldt's *Narrative*, an account of Humboldt's visit to South America at the turn of the nineteenth century.

It was everything Darwin had dreamed of, a tangle of "Tamarinds, bananas & palms," a riot of bright colors and strange flowers in striking contrast to the island's black volcanic rocks. To his father he wrote, "It is utterly useless to say anything about the Scenery—it would be as profitable to

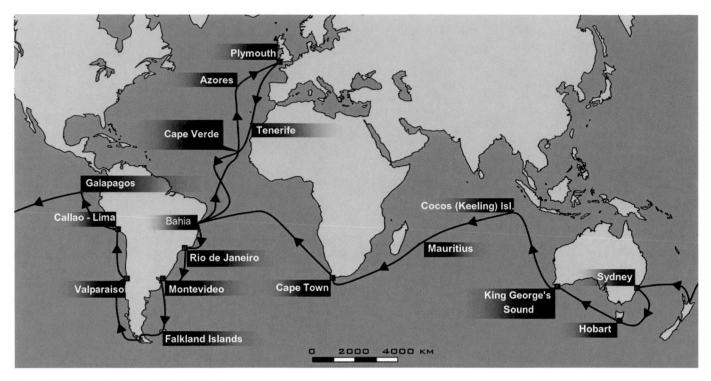

Route of HMS *Beagle* 1831—1836. Public domain.

explain to a blind man colours, as to a person, who has not been out of Europe, the total dissimilarity of a Tropical view." (DCP).

Law of the jungle

On February 16, they crossed the equator and on February 28 sailed into All Saints Bay at Bahia (Salvador), where Darwin took his first steps in South America. For the next year, the *Beagle* made its way down the coast, conducting surveys, taking soundings, and drawing charts, while Darwin collected insects, seashells, and rocks. He did not put the pieces together until he returned to England, but it was there—in the heart of South America—that Darwin made his first important discoveries.

Lost in the brilliance of Brazil's rain forest, surrounded by parrots, hummingbirds, and orchids, Darwin saw not only the incredible luxuriance and diversity of the Amazon but also the harsh reality of life within it. He watched a predatory wasp hunt down, kill, and drag off a spider—a fight to the death between two tiny monsters, a stark example of nature's first law: kill or be killed. Everywhere he looked was a ruthless struggle for survival: vampire bats attacking horses in the dead of night; an unstoppable column of army ants triggering panic throughout the forest. It was Darwin's first real glimpse of the never-ending battle between the hunters and the hunted.

In September at Bahía Blanca, south of Rio de Janeiro, Darwin excavated several huge skeletons, the remains of giant prehistoric beasts. One was a giant sloth similar to the present-day sloth—but much bigger. There were also bones of a giant llama and a giant armadillo. Darwin marveled at

their close resemblance to modern species. He later wrote in his *Journal of Researches*, "This wonderful relationship in the same continent between the dead and the living will, I have no doubt, hereafter throw more light on the appearance of organic beings on our earth and their disappearance from it, than any other class of facts." (Darwin 1845).

In October the ship returned to Montevideo, where Darwin received the second volume of Charles Lyell's *Principles of Geology* (1830–1833). Darwin had read volume one, but volume two conveyed a more profound message. Lyell argued that Earth was much older than most people imagined and that the same geological processes observed in modern times had been at work for millions of years. He said the world had come about through "causes now in operation," not catastrophic events such as the biblical Flood. In his second volume, Lyell also argued that species became extinct because they no longer fit their environments as the world changed.

Lyell was not an "evolutionist," but his observations must have made Darwin think: If there was an explanation for why species disappeared, there must be one for how they came about in the first place.

The ever-changing Earth

The *Beagle*'s crew spent the second year of the voyage (1833) surveying the east coast of South America, while Darwin explored the interior on horseback. Eventually, the ship made its way back to Tierra del Fuego, where Darwin encountered another key piece of information.

Earlier he had seen large rheas, ostrich-like birds, in the Pampas near Bahía Blanca and had heard of a smaller (and rarer) rhea to the south (which is now known unofficially as Darwin's rhea). Darwin was baffled by the presence of two similar kinds of birds in the same territory. While at Saint Gregory's Bay in the Straits of Magellan, he met the giant Patagonians and questioned them about the tiny rhea. He learned it lived south of the Río Negro, while the larger one lived only north of the river. Thus, Darwin acquired a small but important fact: Species appeared most similar to those in nearby, but geographically separated, areas (rheas are flightless birds).

After exploring the Santa Cruz River in April, the *Beagle* rounded Cape Horn for the last time in May. Fighting through the dangerous channels, lost in a world of "rugged snowy crags, blue glaciers…[and] rainbows," the *Beagle* made its way to the island of Chiloé in June 1834.

After two weeks the ship headed north to Valparaíso, Chile. From there Darwin struck out for the foothills of the Andes and reached Santiago on August 27. In September he fell seriously ill and barely got back to Valparaíso before collapsing for a month, unable to get out of bed. Upon recovery, the first news he heard was bad: FitzRoy had suffered a nervous breakdown and had given up command of the *Beagle*. Ever the perfectionist, the captain had pushed himself too far and had snapped. Pringle Stokes, the *Beagle*'s captain before FitzRoy, shot himself at Port Famine in 1828 during the *Beagle*'s first voyage, and FitzRoy appeared to be next. His officers, however, eventually persuaded him to retake command and complete the journey.

On November 21 the ship returned to Chiloé. On an excursion across the island, Darwin observed three volcanoes billowing smoke, and on January 19, 1835, he saw Mount Osorno erupt. At midnight the sentry reported a fire on the horizon. At three in the morning, Darwin and the rest of the crew stood on deck to watch the explosion of rock, fire, and lava—so bright it lit up the sky.

The *Beagle* then sailed north to Valdivia. On February 20, Darwin once again experienced nature's terrifying power. While exploring inland the ground shook as an earthquake struck the west coast. At Concepción, about 320 kilometers (200 miles) to the north, the cathedral was left in ruins and a 6.2-meter (20-foot) tidal wave hit the city, carrying a schooner into the center of town. Fires blazed everywhere. Amid the wreckage, however, Darwin made another important discovery: The beds of dead mussels were now above the high-tide mark. The ground had risen several feet—proof that Lyell was right. Indeed, over millions of years, the continents rise and fall, creating and destroying mountains and reshaping the world in small, imperceptible steps.

As winter approached, the ship again made its way north to Valparaíso and Darwin set out for the Andes with guides and mules. Making his way back to Santiago, he pushed on through the mountains to Mendoza, Argentina, shivering through night frosts at 3,950 meters (13,000 feet) and fighting against the thin air, freezing winds, and icy clouds.

He spent one night in a small village just south of the city and he remembered it well. He wrote in his *Journal of Researches*: "At night I experienced an attack (for it deserves no less a name) of the Benchuca [vinchuca], a species of Reduvius, the great black bug of the Pampas. It is most disgusting to feel soft wingless insects, about an inch long, crawling over one's body. Before sucking they are quite thin, but afterwards become round and bloated with blood, and in this state they are easily crushed." (Darwin 1839).

It is now known the vinchuca bug can transmit Chagas' disease, a debilitating (potentially fatal) disease that causes symptoms similar to many of those Darwin reported after he returned to England. His lifelong health problems may have started in Argentina.

Turning northwest he crossed back through Uspallata Pass and stumbled across fossilized trees, a petrified forest at the top of the world. The trees once must have stood on the coast, when the ocean had come up to the foot of the mountains. Buried with silt when the continent sank, and then thrust to the top when the continent rose up again, the trees were tilted at impossible angles, jutting out from the rock that had crumpled like paper. Darwin was slowly working out the puzzle. He fired off a letter to Henslow about his "absurd and incredible" discoveries. After Valparaíso, the *Beagle* visited Iquique, Peru, then set off for a destination now famously linked with Darwin's name—the Galápagos Islands.

Galápagos Islands

Although the *Beagle* stayed only five weeks in the Galápagos, it turned out to be an important stop for Darwin. The Galápagos Islands were a desolate, volcanic archipelago, ruled by giant tortoises and lizards. Located 965 kilometers (600 miles) off the coast of South America, the landscape was bleak and prehistoric, covered with black sand and lava, the islands cut in half by the equator. After visiting the islands in 1841, the American author Herman Melville wrote, "the chief sound of life is a hiss." For Darwin, however, the Galápagos were a microcosm of the larger world, and they held important secrets.

Exploring James Island (San Salvador), Darwin found a large salt lake and stumbled over the skull of a captain murdered by his crew years before. He rode on the backs of the giant tortoises and played with huge, iguana-like lizards, up to 0.9 meters (3 feet) in length, entertaining himself by throwing cactus branches into their midst, triggering little tugs-of-war between the miniature dragons, which grabbed the ends in their sharp teeth.

The great variety of birds—hawks, mockingbirds, and herons—were clearly related to birds of South America but with significant differences. This puzzled Darwin and later proved critical to the development of his theory. And then there were the finches. Unfortunately, Darwin did not record which island he collected them from, it seemed sufficient to him at the time to simply record that they were from the Galápagos Islands. It was not until 1838 that the British ornithologist John Gould sorted them out and

Birds Pl. 37.

Geospiza strenua.

Galápagos finches from the *Zoology of the Beagle*, Birds Private collection of Angus Carroll.

identified thirteen different species of finch in the *Beagle* collections.

In his journal Darwin observed, "It is very remarkable that a nearly perfect gradation of structure in this one group can be traced in the form of the beak, from one exceeding in dimensions that of the largest gross-beak, to another differing but little from that of a warbler." (Darwin 1839). But he would not realize the significance of this fact until much later. Darwin's finches, as they are now often called, would become one of the most famous examples of natural selection, but at the time Darwin did not grasp their full importance. In fact, the tortoises provided a better clue: Nicholas Lawson, the vice-governor of the islands, told Darwin he could "at once tell from which island any one was brought."

From the Galápagos the *Beagle* crossed the Pacific, visiting Tahiti on the way to New Zealand, Australia, and Tasmania. In Australia, Darwin came across eucalyptus trees, the kangaroo rat, and the bizarre duck-billed platypus. The plants and animals he saw were much different from anything he had seen before. He wrote, "An unbeliever in every thing beyond his own reason might exclaim, 'Two distinct Creators have been at work.'" (Darwin 1839). Ultimately, Darwin would not need to invoke two creators to explain the natural world—indeed, not even one.

Homeward bound

On March 14, 1836, the *Beagle* departed from King George's Sound, Australia, and headed north to the Keeling (Cocos) Islands. Here Darwin found giant clams, brightly colored corals, and emerald lagoons. At Keeling, Darwin tested his new theory of coral reefs. He had theorized they were formed when mountains sank back into the sea, and the coral reefs that originally surrounded the islands were left as rings around a lagoon—and he was right. Often thought of as an evolutionary theorist, Darwin was, in fact, an accomplished geologist, botanist, and zoologist.

The ship did not stay long in this tropical paradise before setting out across its third great body of water, the Indian Ocean. The *Beagle* arrived on Mauritius, east of Madagascar, on April 29. Captain J.A. Lloyd, the surveyor general, happened to have an elephant on the island and Darwin rode it back to the ship when they left. From Mauritius they sailed to the Cape of Good Hope and from there to the Ascension Islands in the middle of the Atlantic Ocean. After almost five years the *Beagle* was almost home.

When it set out on July 23, however, it did not head north to England, but west-southwest. Unbelievably, FitzRoy returned to Bahia in Brazil to double-check his measurements. Fortunately, it would be the last detour. After five ocean crossings (the Atlantic three times), it was finally time to go home. The *Beagle* landed at Falmouth, England, on October 2, 1836, almost five years after it set out. Darwin had literally sailed around the world. He had spent one and a half years at sea and three and a quarter years on land. On his return, he wrote: "As far as I can judge of myself I worked to the utmost during the voyage from the mere pleasure of investigation, and from my strong desire to add a few facts to the great mass of facts in natural science." (Darwin 1887). As for this last objective, it can be said that he certainly succeeded.

The making of a theory

Darwin set out in 1831, an aspiring naturalist headed for the clergy, but he stepped off the *Beagle* in 1836 a different man. Before the voyage he had read all of the major works of the English theologian and philosopher William Paley (1743–1805), including *A View of the Evidences of Christianity* (which was on the exams at Cambridge) and *Natural Theology*. In his autobiography (written in 1876, though not published until 1887) Darwin wrote, "The logic of this book [*Evidences*] and, as I may add, of his 'Natural Theology,' gave me as much delight as did Euclid. The careful study of these works, without attempting to learn any part by rote, was the only part of the academical course which, as I then felt and as I still believe, was of the least use to me in the education of my mind. I did not at that time trouble myself about Paley's premises; and taking these on trust, I was charmed and convinced by the long line of argumentation." (Darwin 1887). In *Natural Theology*, Paley invoked his famous watchmaker analogy: Any reasonable person, on finding a watch and seeing its complex and intricate design, would assume it had been made by a watchmaker. In short, design implies a designer.

But at least as early as June 1836—while still on the voyage—Darwin began to have doubts about the fixity of species. In his notes about how the birds and tortoises of the Galápagos varied island to island, he speculated, "If there is the slightest foundation for these remarks, the Zoology of Archipelagos will be well worth examining; for such facts would undermine the stability of species."

By the time he returned, the question in Darwin's mind was not *do* species evolve, but *how*. Evolution itself was not a new idea. In his 1809 book, Philosophie zoologique, the French naturalist Jean-Baptiste Lamarck had proposed that species evolved through the inheritance of acquired characteristics. Even Darwin's own grandfather had written on the subject: In Zoonomia (1794–1796), Erasmus Darwin had proposed that species adapt to their environment driven by "lust, hunger and danger," an idea at least superficially similar to the theory of natural selection. But these earlier ideas were either flawed or incomplete—few were convinced.

Malthus and population

In September 1838 Darwin read *An Essay on the Principle of Population* (1798) by the English economist Thomas Malthus (1766–1834), and all the pieces fell into place. Malthus argued that human population growth, unless somehow checked, would necessarily outstrip food production because an unchecked population would grow exponentially while food production could only increase arithmetically. If, for example, two parents had four children, each of whom had four children, in only two generations there would be sixteen grandchildren, and soon sixty-four, etc.. Darwin himself used elephants as an example, estimating that from only two original animals there would be over 15 million elephants in only 500 years (Darwin 1859). Given the food supply cannot increase at such a rate, there follows an inevitable competition for food.

Malthus was referring to human populations, of course—his objectives were sociopolitical, not scientific. But Darwin saw how the same principle could apply to the natural world. Far more offspring were born than could possibly survive, because there simply was not enough food to go around. Individuals with a slight advantage would do better. Over a long period of time, even the smallest advantage would prove decisive. (Thanks to then-new geological theories such as those expounded by Lyell in his *Principles of Geology*, Darwin had millions of years with which to work.)

From the Brazilian rain forest to the Galápagos Islands, Darwin had witnessed the considerable variation between individuals of the same species. True, he did not know what

caused such variations, but he did not need to—he theorized at a higher level. He contended that the struggle for existence acted on the smallest differences, however those differences came about. Forced to adapt to ever-changing environments, species evolved. In his autobiography Darwin wrote, "Here, then, I had at last got a theory by which to work." (Darwin 1887).

Marriage and family

In 1839 Darwin married Emma Wedgwood, his first cousin (a practice quite common in Victorian England among the propertied middle class). Although he had known Wedgwood all his life (she was one year older than he was, and the two families were close), it all happened rather quickly.

He visited Maer Hall, the Wedgwood family home, on his return from the voyage in late 1836 and made quite an impression on her. No longer the aimless young man who cared only for "shooting, dogs, and rat-catching," he was now to be taken more seriously—a young man with fantastic stories about far-off places who spoke of new scientific discoveries. Wedgwood wrote to her sister Fanny, "We enjoyed Charles's visit uncommonly.... we plied him with questions without any mercy. Harry and Frank made the most of him and enjoyed him thoroughly. Caroline [Darwin's sister] looks so happy and proud of him it is delightful to see her." (Healey 2001).

Wedgwood did not see him much over the next few months because he had to rush around organizing collections, manuscripts, and so on, but they met again at his brother's house in London in early 1837 and it confirmed her earlier impression. Darwin himself had begun to think about marriage. To decide, he made two columns headed "Marry" and "Not Marry" and wrote down pluses and minuses in each. If he married, he noted he would have "less money for books" and be "forced to visit relatives." On the other hand, he would have a "constant companion (& friend in old age)" and the "charms of music and female chit-chat," though he worried about the "*loss of time*." In the end, he concluded he best get married.

Wedgwood was the obvious choice, but when she saw him in London in 1838, he struck her as uninterested. Wedgwood wrote to her Aunt Jessie, "The week I spent in London on my return from Paris, I felt sure he did not care about me." (Healey 2001). This might have been nerves—Wedgwood was not only pretty but quite accomplished: She spoke French, Italian, and German; played the piano brilliantly (she had taken lessons from the Polish composer and pianist Frédéric Chopin); and was widely read. Furthermore, the Wedgwood family was connected to many famous people—one of Emma Wedgwood's aunts was a friend of the English nurse and hospital reformer Florence Nightingale, while an uncle was friends with the English poets Lord Byron, Samuel Coleridge, and William Wordsworth. To top it all off, Wedgwood had already turned down several suitors. Darwin may have thought his chances slim. Nevertheless, he summoned up his courage and proposed to her at Maer in November 1838. To his surprise, she said yes.

A page out of Darwin's notebook. Reproduced by permission of Cambridge University Library.

Emma and Charles Darwin's wedding portraits, 1840. © English Heritage Photo Library.

Over the next seventeen years they had ten children. Three died young: Mary Eleanor at three weeks (1842); Charles Waring at age one and a half (1856–1858); and Anne Elizabeth at age ten (1841–1851). Darwin was devastated when "Annie" died (he was very close to her), but the idea (often put forward) that her death influenced his work and/or religious views does not fit the facts. Darwin completed the first sketch of his theory in 1842. Two years later, he expanded it to about 200 pages. Clearly, Darwin had formulated his theory of natural selection long before Annie died. Nor did her death "drive him away from God." Although he believed in Christianity when he was young, by the late 1830s that was no longer true. In his autobiography he writes, "Disbelief crept over me [1836 to 1839] at a very slow rate, but was at last complete." (Darwin 1887). We know his views had changed before he got married because Wedgwood was concerned and said so in a number of letters she wrote him while they were still engaged. In November 1838 she wrote, "My reason tells me that honest & conscientious doubts cannot be a sin, but I feel it would be a painful void between us. I thank you from my heart for your openness with me & I should dread the feeling that you were concealing your opinions from the fear of giving me pain." (DCP). Darwin had leveled with her, and although she had reservations (and hoped he might yet come to a different conclusion), she could find

no fault in his goal—the search for truth. It was their mutual respect—he for her religious beliefs, she for his scientific worldview—that kept them together until the end.

In late 1838 Darwin took a flat on Gower Street in London, and he and Emma moved in together after their wedding on January 29, 1839. In July 1842 he moved the family (which then included William and Annie) to Down House, just outside the small village of Downe, Kent, about 24 kilometers (15 miles) southeast of London, where Darwin would spend the rest of his life.

A life of poor health

Darwin rarely left Down House because of ill-health. As noted earlier, he may have contracted Chagas' disease in South America. The idea that his illness was psychological, connected to his work, does not hold up because his health problems began on the voyage, long before he started theorizing about evolution. He did not record an encounter with the vinchuca bug on his first inland trip from Valparaíso to Santiago in South America (it was on his second excursion he recorded the "attack" outside Mendoza), but he was certainly exposed to the bug both times (the range of the vinchuca extends throughout the entire region), and at the end of the first excursion he barely made it back to Valparaíso before collapsing for a month. Although not all his symptoms match those of Chagas' disease, many do, including chronic fatigue, nausea, and abdominal pain.

His problems may have been aggravated by genetics. Both his grandfather (Erasmus) and father (Robert) were very large men. His father reportedly weighed over 350 pounds (he made his coachman—also a large man—walk through the houses he visited ahead of him to make sure the floors would hold). Both had health problems that Darwin may have inherited. The "cures" of the day may have hurt more than helped. At different times, Darwin was prescribed, given, or tried arsenic, opium, quinine, morphine, and even "batteries" (the height of quackery, which Darwin knew full well—though he tried it anyway in desperation—whereby one "galvanized" one's insides with electricity).

Whatever the cause, the effect was debilitating: He could work only a few hours a day. It is remarkable what he managed to accomplish under such circumstances: he wrote seventeen books, made major contributions to numerous others, and wrote dozens of important papers in geology, botany, and zoology. Often overlooked is the fact that had

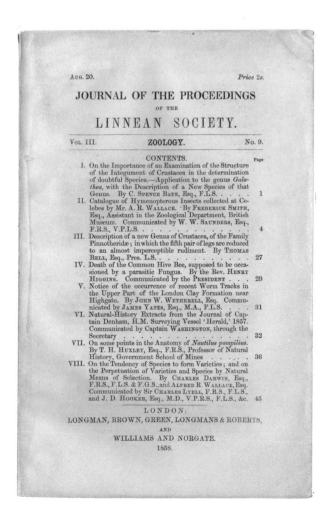

Darwin-Wallace Paper in the *Journal of the Linnean Society*, 1858. Private collection of Angus Carroll.

Darwin not published *On the Origin of Species by Means of Natural Selection*, he would still have been one of the leading scientists of his day, highly respected in several fields.

Delaying or delayed?

One of the great myths surrounding the *Origin* is that Darwin delayed publishing the book for twenty years because he feared the inevitable controversy. The facts, however, do not support this popular misconception. True, Darwin had the critical insight in 1838 after reading Malthus, and true, he wrote out a sketch of the theory in 1842. Nevertheless, Darwin had no intention of publishing anything until he had the facts to back it up. And before he could amass the facts he needed, he had to finish the projects he had underway.

Between 1838 and 1858, Darwin published his *Journal of Researches* from the voyage (1839), three volumes on the geology of the voyage (1842, 1844, and 1846), and four volumes on barnacles (two in 1851, two in 1854). He also edited the five-volume *Zoology of the Voyage of H.M.S.* Beagle (1838–1843). R. B. Freeman's bibliography of Darwin's works lists sixteen major scientific papers between 1838 and 1858, and according to the Darwin Correspondence Project he wrote 1,624 letters over the same period (that still survive). He also got married, moved twice, and had ten children. He was not delaying, he was *delayed*.

Research in the first decade of the twenty-first century has shown that the idea Darwin held back because he was afraid of what people might think is a relatively modern invention. To the contrary, Darwin was determined to publish regardless of what people thought. (He had discussed his ideas with many people—most of whom disagreed with him—and always made it clear that he intended to publish his theory despite their objections.) From the perspective of the early twenty-first century, it might seem incredible that anyone would sit on a groundbreaking theory for twenty years, but in mid-nineteenth-century England the situation was much different. Darwin was in no hurry. Nor, until Alfred Russel Wallace's letter arrived in 1858 (see below), did he think anyone else was on the same track. Not under any economic pressure (the Darwin family was quite wealthy), he was in no rush to publish. More to the point, he knew the theory would not be accepted unless he could amass substantial evidence to support it, and he was determined to do just that.

By 1846, Darwin had wrapped up his geological work and had only the invertebrates left from the voyage. He decided to undertake the barnacles himself. But what he thought would be a yearlong project stretched into eight years because the whole group had to be described, not just the specimens he had found and brought back himself. Near the end he would lament, "I hate a Barnacle as no man ever did before," (DCP) but he stuck it out and ultimately published two monographs so comprehensive they remain the definitive work on the subject. He finished the barnacles in 1854. Finally, he could give his full attention to his theory.

Darwin was not hesitant; he was busy. He did not "delay" twenty years; he was working. Only when he wrapped up the work from the voyage did he return to the "species question."

Not long after he did, a letter arrived that turned his world upside down.

Out of the blue

Darwin was still a long way away from publishing his "big species book," when a package arrived on June 18, 1858, from Alfred Russel Wallace (1823–1913), a naturalist working in the Malay Archipelago (modern-day Malaysia, Indonesia, and the Philippines). Wallace spent five years collecting on the Amazon before ending up in the Moluccas, or Spice Islands, and he, too, had read Malthus. As a result of his own observations he had reached the same conclusions as Darwin on the origin of species.

In early 1858, half-crazed in the grip of malaria, Wallace wrote a twenty-odd-page paper titled "On the Tendency of Varieties to Depart Indefinitely from the Original Type," and posted it to Darwin for his opinion. When Darwin read it he was stunned. Although there were important differences, Darwin wrote to Lyell, "If Wallace had my MS sketch written in 1842 he could not have written a better abstract." (DCP).

Darwin did not know what to do, but Lyell and the English botanist Joseph Dalton Hooker (1817–1911) took charge and arranged for Wallace's paper to be presented along with an extract of Darwin's 1844 essay and part of a letter Darwin had written to the American botanist Asa Gray (1810–1888) in 1857. The joint paper was read at a meeting of the Linnean Society on July 1, 1858, but it went largely unnoticed. It was not until the theory came out in book form the following year that it made headlines.

Galvanized by Wallace's paper, Darwin worked furiously on the *Origin*, finishing it in only one year.

The book that shook the world

On the Origin of Species by Means of Natural Selection was published on November 24, 1859. The first review, written by John R. Leifchild, appeared in the *Athenaeum* on November 19, 1859. It was negative but not scathing. Next came Thomas Henry Huxley's review in the *Times* of London, which was positive. There followed reviews in numerous periodicals and newspapers, some positive (Hooker in the *Gardeners' Chronicle*), some negative (Bishop Samuel Wilberforce in the *Quarterly Review*).

The first American review was written by Gray, and it appeared in the March 1860 issue of the *American Journal of Science and Arts*. It was critical, but positive. Darwin liked the review and wrote to Gray, "Your Review seems to me admirable; by far the best which I have read. I thank you from my heart both for myself, but far more for subject-sake." (DCP).

Like Huxley in the United Kingdom, Gray became Darwin's main supporter in the United States, and just as Huxley sparred with Darwin's opponents in England, Gray squared off against Louis Agassiz at home (both Gray and Agassiz were professors at Harvard University). Gray was not uncritical of Darwin's theory; his main goal was to get it a fair hearing.

Natural selection

Darwin did not, of course, discover "evolution." The idea that species "evolved" was not new. The problem was that no one—until Darwin—had offered a convincing explanation of *how* they evolved. Darwin called his theory "natural selection."

Considering its incredible explanatory power, the theory of natural selection is remarkably simple. Limited resources (not enough food for all the offspring produced) leads to competition. Some individuals will do better than others because they happen to have certain characteristics that give them an edge—speed, strength, and so on. Because those individuals are more likely to survive, they are more likely to reproduce and pass on their characteristics to their offspring. Thus the "population" of a species evolves as more and more individuals are born (and survive) who have inherited the characteristics that provide advantage.

It was a relatively simple idea, but it represented a challenge to both accepted wisdom and faith. In a famous episode, Wilberforce ridiculed the theory at the 1860 meeting of the British Association for the Advancement of Science, but it was not just religious leaders who found fault with Darwin's theory. Many leading scientists criticized it too, including Agassiz and Richard Owen, who had been a friend of Darwin's until the publication of the *Origin* (he had edited *Fossil Mammals*, Part I of the *Zoology of the Beagle*). Both were vehement critics, Agassiz ending his review of the *Origin* with, "I shall therefore consider the transmutation theory as a scientific mistake, untrue in its facts, unscientific in its methods, and mischievous in its tendency." Agassiz equated "species" to "thoughts of God." In his mind there was no need—or place—for the concept of evolution, let alone a theory to explain it.

Beyond challenging specific religious beliefs, many scientists and nonscientists alike could not bring themselves to accept Darwin's theory because it lacked purpose or direction. Not only did it eliminate humanity's special position in the natural order of things, but it made humanity's very existence a function of chance. In late 1859 Darwin wrote to Lyell, "I have heard by round about channel that [John] Herschel says my Book 'is the law of higgledy-piggledy.'" (DCP). The "random" nature of natural selection was simply too much for many people.

Darwin did not ignore his critics, but he did not put too much stock in them either. In 1859 he wrote to John Lubbock, "I should be grateful for any criticism. I care not for Reviews, but for the opinion of men like you & Hooker & Huxley & Lyell &c." (DCP).

The simplicity of natural selection is striking. When Huxley first read the *Origin* he later recalled thinking, "How extremely stupid not to have thought of that."

Darwin spent years gathering information to support it, but it is based on only three simple principles: the inevitability of competition, variability between individuals, and the effects of differential success. Add in an ever-changing world and from no more than that comes the history of life on Earth.

Later life

The *Origin* went through six editions during Darwin's life, and he made many small changes, but in detail only—none to the basic theory itself. One change he later regretted. In the closing paragraph of the first edition Darwin wrote, "There is grandeur in this view of life, with its several powers, having been originally breathed [by the Creator] into a few forms or into one; and that, whilst this planet has gone cycling on according to the fixed law of gravity, from so simple a beginning endless forms most beautiful and most wonderful have been, and are being, evolved."

Darwin added "by the Creator" in the third edition (1861). In 1863 he wrote to Hooker, "I have long regretted that I truckled to public opinion & used Pentateuchal term of creation, by which I really meant 'appeared' by some wholly unknown process. It is mere rubbish thinking, at present, of origin of life; one might as well think of origin of matter." (DCP).

But Darwin did more than make small edits and corrections to the *Origin* in his later years. Yet to come were several major works. In 1862 he published *On the Various Contrivances by which British and Foreign Orchids Are Fertilised by Insects*, known simply as *Orchids*. Darwin wrote to John Murray, his publisher, "I think this little volume will do good to the Origin, as it will show that I have worked hard at details." (DCP).

In *Orchids* Darwin applied the principles of natural selection to make a startling prediction. First, he described a remarkable flower from Madagascar called *Angraecum sesquipedale*: "A whip-like nectary of astonishing length hangs down beneath the labellum. In several flowers sent to me by Mr. Bateman I found the nectaries eleven and half inches long, with only the lower inch and a half filled with very sweet nectar." Then he added, "in Madagascar there must be moths with probosces capable of extension to a length of between ten and eleven inches!" (Darwin, 1862). No such moth was known, and some scientists ridiculed Darwin for suggesting it existed. But in the end he was proven right. The moth was

Darwin's hawk moth *Xanthopan morgani praedicta* has an extraordinary long tongue to pollinate flowers with long tubes, such as orchids. © Nick Garbutt/NPL/Minden Pictures.

found and described in 1903—forty-one years after Darwin's prediction. It had a wingspan of 13 to 15 centimeters (5 to 6 inches) and a proboscis 25 centimeters (10 inches) long. This new subspecies was named *Xanthopan morgani praedicta*—the "predicted" moth.

In 1871 Darwin published *The Descent of Man*. He had deliberately avoided human evolution in the *Origin*, but that had not worked—the whole controversy centered around the obvious implication for humans, and by the time the *Descent* was published, the controversy was largely over. Nevertheless, it was an important work because it also dealt with sexual selection, an important aspect of the larger evolutionary picture.

Darwin considered sexual selection a separate mechanism for explaining evolutionary change, though it is now regarded as but one aspect of natural selection. He explained sexual differences such as male antlers and the peacock's tail as the result of differential success in males either competing against other males or being chosen by females and therefore leaving more offspring.

Darwin explained why sexual selection occupied so much of the work: "During many years it has seemed to me highly probable that sexual selection has played an important part in differentiating the races of man. . . . When I came to apply this view to man, I found it indispensable to treat the whole subject in full detail. Consequently the second part of the present work, treating of sexual selection, has extended to an inordinate length, compared with the first part; but this could not be avoided." (DCP).

In 1872 came *The Expression of the Emotions in Man and Animals*, another important book in which Darwin showed that humans and animals expressed similar emotions in similar ways (pointing to common descent), in contrast to the Scottish anatomist Charles Bell (1774–1842), who claimed, in his 1824 work, *Essays on the Anatomy and Philosophy of Expression*, that humans had special facial muscles to express uniquely human emotions. Darwin wrote five more books between 1875 and 1881, working almost up to his death.

Darwin on religion

Much has been written about the reception of Darwin's theory in his lifetime (and after), less about what Darwin thought himself. Generally, Darwin tended to think of science and religion as two separate and distinct areas of inquiry. In 1866 he wrote to Mary Boole (the wife of the mathematician John Boole and a mathematician herself): "I am grieved that my views should incidentally have caused trouble to your mind but I thank you for your judgment & honour you for it, that theology & science should each run its own course & that in the present case I am not responsible if their meeting point should still be far off." (DCP).

Darwin viewed science and religion as separate and distinct—but not unconnected. His scientific worldview took precedence. He wrote to N. A. Mengden in 1879, "Science has nothing to do with Christ, except in so far as the habit of scientific research makes a man cautious in admitting evidence." This was the crux of Darwin's skepticism—by the end of the voyage, having seen so much evidence firsthand, he could no longer accept anything on "faith."

In his autobiography he wrote, "The old argument of design in nature, as given by Paley, which formerly seemed to me as conclusive, fails, now that the law of natural selection has been discovered. We can no longer argue that, for instance, the beautiful hinge of a bivalve shell must have been made by an intelligent being, like the hinge of a door by man. There seems to be no more design in the variability of organic beings and in the action of natural selection, than in the course which the wind blows. Everything in nature is the result of fixed laws." (Darwin 1887).

Not believing himself, however, did not lead him to speak out against religion—partly because he did not think doing so would make any difference, and partly because he did not want to offend his wife. Darwin wrote to Edward Aveling in 1880, "I am a strong advocate for free thought on all subjects, yet it appears to me (whether rightly or wrongly) that direct arguments against Christianity & theism produce hardly any effect on the public; & freedom of thought is best promoted by the gradual illumination of men's minds which follows from the advance of science. It has, therefore, been always my object to avoid writing on religion, & I have confined myself to science. I may, however, have been unduly biased by the pain which it would give some members of my family, if I aided in any way direct attacks on religion." (DCP).

Perhaps the deepest insight into Darwin's views on religion comes from a letter he wrote to John Fordyce in 1879, "In my most extreme fluctuations, I have never been an atheist in the sense of denying the existence of a God." He added, "I think that generally (and more and more as I grow older), but not always, that an Agnostic would be a more correct description of my state of mind." (DCP).

Death and funeral

Darwin died on April 19, 1882, at Down House, at age seventy-three. The funeral took place on April 26; the pallbearers included Huxley, Hooker, and Wallace. Darwin was buried at Westminster Abbey, next to John Herschel.

A man of great honesty and modesty, Darwin remains one of the giants in the history of science. *On the Origin of Species* is one of the most important books ever published—changing not only our understanding of the world around us, but of our place within it. But Darwin's legacy goes beyond the theory of natural selection. His story embodies the spirit of science itself: A keen observer with a love for natural history, he set out on a voyage of discovery with only an open mind and returned with the answer to one of life's greatest mysteries.

Resources

Works by (except as noted, published London: John Murray)

(ed.) *The Zoology of the Voyage of H.M.S.* Beagle, London: Henry Colburn, 1838–1843.

Journal of Researches, first published as Vol. 3 of *The Narrative of the Voyages of H.M. Ships* Adventure *and* Beagle, London: Henry Colburn, 1839; 2nd edition, 1845; definitive issue, 1860.

The Structure and Distribution of Coral Reefs, London: Smith, Elder, 1842.

Geological Observations on the Volcanic Islands Visited during the Voyage of H.M.S. Beagle, London: Smith, Elder, 1844.

Geological Observations on South America, London: Smith, Elder, 1846.

A Monograph of the Sub-class Cirripedia (Living Barnacles), 2 vols., London: Ray Society, 1851 and 1854.

A Monograph on the Fossil Lepadidae, or, Pedunculated Cirripedes of Great Britain (Fossil Barnacles), 2 vols., London: Palaeontographical Society, 1851 and 1854.

On the Origin of Species by Means of Natural Selection; or, The Preservation of Favoured Races in the Struggle for Life, 1859; 2nd edition, 1860; 3rd edition, 1861; 4th edition, 1866; 5th edition, 1869; 6th edition, 1st issue, 1872; 6th edition, 2nd (and definitive) issue, 1876.

On the Various Contrivances by which British and Foreign Orchids Are Fertilised by Insects, 1862.

The Movements and Habits of Climbing Plants, 1865. Published in book form, 1875.

The Variation of Animals and Plants under Domestication, 1868.

The Descent of Man, 1871; 2nd edition, 1874. The printing of 1877 is the definitive text.

The Expression of the Emotions in Man and Animals, 1872.

Insectivorous Plants, 1875.

The Effects of Cross and Self-Fertilisation in the Vegetable Kingdom, 1876.

The Different Forms of Flowers on Plants of the Same Species, 1877.

The Power of Movement in Plants, 1880.

The Formation of Vegetable Mould, through the Action of Worms, 1881.

Works about

"AboutDarwin.com." Available from http://www.aboutdarwin.com

Browne, Janet. 1995. *Charles Darwin: A Biography*, Vol. 1: *Voyaging*. New York: Knopf.

Browne, Janet. 2002. *Charles Darwin: A Biography*, Vol. 2: *Power of Place*. New York: Knopf.

Burkhardt, Frederic, ed. 2008. *The* Beagle *Letters*, by Charles Darwin. Cambridge, UK: Cambridge University Press.

"The Complete Works of Charles Darwin Online." Available from http://darwin-online.org.uk All works by Darwin referred to herein can be found in their entirety at this site.

"Darwin." American Museum of Natural History. Available from http://www.amnh.org/exhibitions/darwin

Darwin, Francis, ed. 1887. *The Life and Letters of Charles Darwin*. 3 vols. London: John Murray. Includes Darwin's autobiography.

"Darwin Correspondence Project." Cambridge University Library. Abbreviation: DCP. Available from http://www.darwinproject.ac.uk All letters quoted herein can be found in full at this site or in Life and Letters, 1887.

Dennett, Daniel C. 1995. *Darwin's Dangerous Idea: Evolution and the Meanings of Life*. New York: Simon and Shuster.

Desmond, Adrian, and James Moore. 1991. *Darwin: The Life of a Tormented Evolutionist*. New York: Warner Books.

di Gregorio, Mario A. 1984. *T. H. Huxley's Place in Natural Science*. New Haven, CT: Yale University Press.

"Evolution." University of California Museum of Paleontology. Available from http://www.ucmp.berkeley.edu/history/evolution.html.

Healey, Edna. 2001. *Emma Darwin: The Inspirational Wife of a Genius*. London: Headline Books.

Keynes, Richard. 2002. *Fossils, Finches, and Fuegians: Charles Darwin's Adventures and Discoveries on the* Beagle, *1832–1836*. London: HarperCollins.

Raby, Peter. 2001. *Alfred Russel Wallace: A Life*. London: Chatto and Windus.

Angus Carroll
John van Wyhe

Mechanisms of evolutionary change

How has the amazing diversity of organisms arisen? Darwin referred to this question as "that mystery of mysteries." Although scientists prior to Darwin and Wallace had proposed "mechanisms" for how evolutionary change could work to cause diversity, none had proposed a fully materialistic mechanism that was sufficient for explaining all patterns of evolutionary change. A suitable explanation for the heritable basis of variation had to await Mendel and later geneticists, but Darwin-Wallace selection has held up as a major mechanism for evolutionary change and the only one sufficient to explain adaptive diversity. This entry provides an introductory overview of how selection and genetics together provide mechanisms to explain evolution even more completely than either Darwin or Mendel themselves could have imagined.

The *Origin* and natural selection

The two main theses of Charles Darwin's most influential work, *On the Origin of Species by Means of Natural Selection* (1859), are: (1) species diversity has arisen through "descent with modification," and (2) the main mechanism for this pattern is "natural selection." Although it has referred to

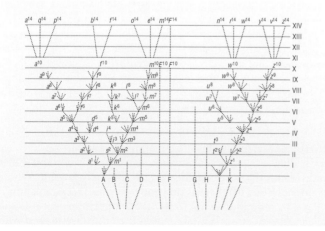

Figure 1. The phylogenies from Darwin's *Origin of Species*. The time axis progresses upward, with what could be interpreted as geological strata indicated by roman numerals in order of deposition. Degree of divergence is represented by the horizontal axis. Ancient ancestors are represented by uppercase letters; subsequent descendants are represented by lowercase letters. Reproduced by permission of Gale, a part of Cengage Learning.

different kinds of evolutionary ideas in the past, the modern meaning of the word "evolution" is synonymous with Darwin's "descent with modification," which encompasses two patterns: *anagenesis*, or change along a lineage, and *cladogenesis*, or branching into two or more lineages (i.e., *speciation*). The only figure in the *Origin* depicts both patterns as genealogies or *phylogenies* (see Figure 1). Darwin proposed that natural selection was responsible for both patterns.

Specifically, selection for a new advantageous variation within a population would most likely lead to the extinction of the previous, less competitive variation(s). Repeated cycles of such replacements would thus result in anagenetic change; the mode of selection responsible for this pattern is called *directional selection*. Additionally, Darwin predicted that selection would favor new variations that allow coexistence with the parental type, if, for example, such a variation allows a new resource to be used such that competition is reduced within the population. In such a case, selection is expected to drive divergence, because the more different two variants are, the more likely they are to avoid competition, thus allowing both variants to proliferate. The mode of selection responsible for this pattern of cladogenesis is called *disruptive* or *diversifying selection*. Both directional and disruptive modes of selection are often called *positive* or *Darwinian selection*. The absence of change, *stasis*, can be explained by *stabilizing selection*, in which no new variations are more advantageous than the parental type. But what is selection and how does it operate?

Selection is actually just a result of genetic processes and can be expected to operate in any system that displays the following characteristics: (1) the ability of an entity to reproduce (multiply), (2) variation between entities with respect to *Darwinian fitness* (the ability to survive and reproduce), and (3) heritability of this variation. How do these properties result in selection?

Reproduction

Darwin extrapolated to all organisms the recognition of the English economist Thomas Robert Malthus (1766–1834) that human populations have the capacity to reproduce at an exponential rate. Under conditions in which resources or other factors are not limiting, organisms should reproduce at their maximal "intrinsic" rate, r_m. Under such *density-independent growth*, the population size (N, number of individuals) at any generation time t can be predicted, given a starting population size, N_0, and the rate of population growth (r, which can be estimated as the per-capita birthrate minus the per-capita death rate) (Maynard Smith 1989): (equation 1) $N_t = N_0 e^{rt}$.

Most natural populations, however, are likely to suffer limiting amounts of resources most of the time. Under such *density-dependent growth*, Darwin predicted that there would be a "struggle for existence" stemming from competition among individuals. The *logistic equation* (equation 2) can be used to model population growth under conditions that limit population size to a certain *carrying capacity*, K (the maximum number of individuals that can be supported in a population) (Verhulst 1838):

$$\frac{dN}{dt} = r_m N \left(1 - \frac{N}{K} \right)$$

In this descriptive model, the rate of increase in population size (dN/dt) is highest when the population size is small; as N approaches K, the rate of increase approaches zero. In real populations, the population size may fluctuate around K. The logistic model can nevertheless be useful, for example, by showing how one variant may replace another (Maynard Smith 1989). In an asexually reproducing population made up of two variants, the one with the higher K (or r) will replace the other. Equations 3a and 3b (below) model these two variants using a version of the logistic equation in which both variants are competing for the same resources and the growth of each is thus affected by the other. As seen in equations 3a and 3b:

$$\frac{dN_1}{dt} = r_1 N_1 \left(1 - \frac{N_1 + N_2}{K_1} \right) \quad \frac{dN_2}{dt} = r_2 N_2 \left(1 - \frac{N_1 + N_2}{K_2} \right)$$

If $K_1 > K_2$, then N_1 will increase until $N_1 + N_2 = K_1$, at which point $N_1 + N_2 > K_2$, and dN_2/dt must be negative. In fact, N_2 will eventually decrease to zero (Maynard Smith 1989), and directional selection results. In contrast, if the variants do not compete for the same resources, equation 2 applies for each variant, and coexistence results. Thus, the ability of a heritable variation to spread throughout a population is conferred by the potential of a population to multiply at an exponential rate. Which variation is spread and replaces the others in the population (i.e., becomes *fixed*) depends on its fitness relative to other variants.

Variation with respect to fitness (*differential reproductive success*)

If all variations confer identical fitness (survivorship and fecundity), then which variant spreads throughout the population is a matter of chance alone, not selection. Yet, any variation that confers even the slightest competitive advantage in fitness by definition leaves more than the normal number of progeny (which themselves can carry the advantageous variation if it is heritable). The advantage need not be the sudden appearance of a "hopeful monster" with an optimal adaptation in all its glory (Goldschmidt 1940); as long as a variation is fitter than the previous one, the new variation will replace the previous one. Thus, positive Darwinian selection can be viewed as a directional, stepwise process (Dawkins 1986; Dennett 1995). Also, any new variation that confers lower fitness than the previous one will be unable to compete ("negative selection"), no replacement will occur, and stasis results until a more advantageous variation appears.

Heritability

If a new variation is not heritable, the advantageous trait is not passed on to progeny and thus will not spread through the population to replace previous variations.

The mistaken notion that selection is a random process that "therefore" cannot be responsible for complex adaptations and novelties probably stems from the common confusion between mutation and selection (as well as to obfuscatory literature and arguments promulgated by creationists, including those promoting "intelligent design"). *Genetic mutation* is the ultimate source of new variations and is a stochastic process; but heritability is quite the opposite of random, because it is highly predictable and ensures that a particular variation (and not some other random variation) is passed on to progeny. (That progeny look more like their own parents than like someone else's parents is obviously not random and arises from the genetic inheritance of traits.) Selection is also nonrandom because only those variations conferring higher reproductive success relative to other variations will subsequently spread through a population. Mutation alone cannot mold adaptations and is not considered to be a major force in shaping evolutionary change.

Genetic mechanisms of evolution

Because evolution involves heritable change, a major focus of research has been to understand the processes that are responsible for genetic change in populations. A gene is a segment of DNA at a particular locus in the genome or chromosome that encodes a protein or RNA product. Alternative forms of a gene are called *alleles*. Although a diploid organism usually has only two copies for each gene, and thus at most two different alleles, several alleles can coexist in the same population. At its most basic level, evolution (*microevolution*, or evolution within a species) can be described as changes in the relative frequencies of these alleles in a population. Although mutation is responsible for generating new alleles, mutation affects only one gene copy at a time and is thus not responsible for major shifts in allele frequencies in populations. Instead, the major factors include *genetic drift*, *gene flow* (*migration*), and selection. To understand how these factors play a role in evolution, it is first necessary to determine what dynamics are expected in the absence of these factors. For this purpose, the *null model* of a *Mendelian population* is used.

A Mendelian population is an idealistic model that makes five major assumptions: (1) mating is random (i.e., every individual has an equal chance of mating with any other individual); (2) the population size is infinitely large; (3) there is no mutation; (4) there is no gene flow (i.e., the population is closed to migration to or from other populations); and (5) all alleles have identical fitness effects (i.e., they are *neutral* with respect to selection; thus, selection does not operate). Because of random mating, the collection of all gene copies (one per gamete) can thus be treated as a "pool" from which two gene copies are randomly associated into zygotes. In this case, the probability (predicted frequency) of getting a particular genotype depends on the frequencies of the alleles in the population.

If the actual frequency of the A allele at gene A is p and that of the a allele is q, then the predicted frequencies of the genotypes can be calculated using the *Hardy-Weinberg theorem* (Hardy 1908; Weinberg 1908): (Equation 4) $(p + q)$ $(p + q) = p^2 + 2pq + q^2$ where $(p+q)$ is the probability of having either A or a at each of the two gene copies, p^2 is the predicted frequency of the AA genotype, $2pq$ is the predicted frequency of the Aa genotype, and q^2 is the predicted frequency of the aa genotype. Populations in which these predictions hold are said to be in Hardy-Weinberg equilibrium (HWE). A second prediction from this model is that allele frequencies do not change (i.e., there is no evolution in such an idealized population). The usefulness of this null model is in analyzing what happens in natural populations in which the assumptions of the Mendelian population are violated.

Nonrandom mating

Mating is often not random in natural populations, such that *assortative mating* can be common (i.e., matings more frequently involve similar genotypes than different genotypes). In such cases, the frequency of heterozygotes will be lower (and the frequency of homozygotes will be correspondingly higher) than expected by the Hardy-Weinberg equilibrium. Such populations are said to be *inbred*. The inbreeding coefficient, F, measures this difference between actual heterozygosity (H_F) in an inbred population and the heterozygosity expected if the population were in HWE ($H_0 = 2pq$) (Wright 1951). As seen in equation 5:

$$F = \frac{H_0 - H_F}{H_0}$$

Nonrandom mating by itself does not change allele frequencies and is thus not an evolutionary "force" per se. Inbreeding, however, shuttles alleles into homozygous genotypic states, exposing the effects of deleterious recessive alleles (normally hidden by heterozygosity). The resulting reduction in average fitness is called *inbreeding depression*. The frequencies of these alleles can then be driven down by selection. Another problem caused by inbreeding is the reduction of recombinant variation (haplotype diversity), which is produced only in genotypes that are heterozygous at multiple loci. That is, *linkage disequilibrium* will be maintained (particular alleles at different loci will be kept together). Such a reduction of recombinant variation should limit the effectiveness of selection, because selection depends on variation in a population.

Small population size

Just by virtue of being finite in size, populations will experience random shifts in allele frequencies, called *genetic drift*. Such drift results from sampling effects in the mating pool. In smaller populations, it is more likely for inbreeding to occur by chance alone. For any diploid individual in a randomly mating population of size N, the probability of getting the identical two gene copies by chance alone (i.e., of being inbred) is $F = 1/(2N)$. On average, heterozygosity is expected to decrease each generation by $1/(2N)$ as a result of this random inbreeding (Wright 1931). Thus, smaller populations will experience stronger and more rapid fluctuations in allele

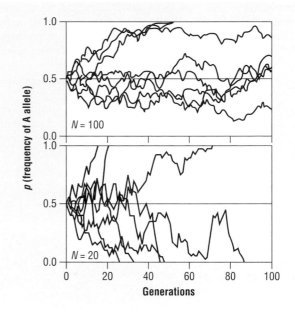

Figure 2. Different rates of genetic drift in larger (top panel, N = 100 individuals) versus smaller populations (bottom panel, N = 20 individuals). Eight separate populations were simulated for each population size for 100 generations, with a starting A allele frequency (p) of 0.5, no selection (all genotypes were equally fit, and both alleles thus selectively neutral), and no gene flow. (Source: Felsenstein, Joe. 2008. Available from http://evolution.gs.washington.edu/popgen/popg.html.) Whereas it was fixed in only two of the large populations over this time interval. Reproduced by permission of Gale, a part of Cengage Learning.

frequencies than larger populations. Because of genetic drift, an allele might become fixed or become extinct by chance alone; this result occurs more quickly in smaller populations (see Figure 2). In conclusion, evolution will occur, even in the absence of selection, by random chance alone, just by virtue of limitations on population sizes. Genetic drift, however, cannot build adaptations, which can result only from selection.

When small groups are founders of new populations, genetic drift can be a major factor in evolutionary divergence between populations, a result that has been confirmed in laboratory experiments (Buri 1956). Because of sampling effects in the founder group, the new population is unlikely to have the same genetic makeup (allele frequencies) as that of the parent population, a phenomenon known as the *founder effect*. Thus, genetic drift is also a mechanism that can potentially produce both cladogenetic as well as anagenetic patterns of evolutionary change.

Mutation

Mutations form new alleles, but only one gene copy at a time. In most populations (those that are not extremely small), mutation is thus a very weak "force" for evolution—that is, for changing allele frequencies. Because there are $2N$ copies of a gene in a diploid population, the initial frequency of a new allele is $1/(2N)$, which is also the probability that it will eventually be fixed if it is neutral with regard to fitness (Kimura 1983). The average time it takes for a new neutral mutation to become fixed by genetic drift depends on the size of the population that it must sweep through ($4N_e$; Kimura 1983). But the average time between fixation events (i.e., the average time it takes for one neutral allele to substitute for the previous allele in a population by drift alone) is $1/\mu$ generations, where μ is the mutation rate per gene copy (Kimura 1983). Because this rate of substitution is independent of population size, the accumulation of neutral substitutions over time is expected to be "clocklike," given constant generation times and mutation rates. This fairly even rate of molecular evolutionary change is called the *molecular clock* and has been used to provide rough estimates of divergence dates. Of course, if a new mutation is advantageous or disadvantageous for fitness, selection will be the primary determinant of the dynamics of its rise or fall in the population. Many mutations, however, do not seem to have large effects on fitness, such that drift may be sufficient to explain many evolutionary changes, at least at the molecular level (Ohta 1996). Of course, selection will cause new mutations to be fixed or to go extinct more rapidly than drift if they have a significant effect on fitness.

Gene flow

The migration of individuals or their gametes or spores results in the sharing of alleles between populations. There are many ways in which gene flow can occur, and the directions of flow may not be equal between two populations. For example, a small island population near a mainland may receive many more alleles than it returns. This can have important consequences. Even if a particular allele is favored by selection on the island, it may never rise to fixation if another allele continuously flows in from the mainland population. Migration also balances the loss of diversity (heterozygosity) stemming from genetic drift. Because gene flow tends to homogenize allele frequencies across populations, it tends to limit the divergence between them. This homogenization is why barriers to migration (isolation) are important for speciation, as recognized by Darwin.

Differential fitnesses of genotypes

As mentioned, selection is a determinative, stepwise process that leads to adaptations. At the population genetic level, it is also a potent "force" for changes in allele frequencies. Exactly how rapidly and to what extent an allele frequency rises and becomes fixed in a population by selection depends on its genetic interaction with other alleles in a genotype. For dominant alleles that confer a fitness advantage, the frequency is expected to rise very rapidly when the allele is rare, because every genotype that contains the allele has the fitness advantage. But when the allele becomes frequent in the population, the less-fit allele will occur primarily in heterozygotes, in which its recessive, deleterious effect is hidden. At this point, the less-fit allele will be very difficult to remove by selection, because the formation of recessive homozygotes will be rare. Similarly, recessive advantageous alleles that are rare will occur mostly in heterozygotes. Once recessive homozygotes form, however, the alleles can be selected. Because the less-fit allele in this case is dominant, it will be selected against, regardless of the genotype in which it is found.

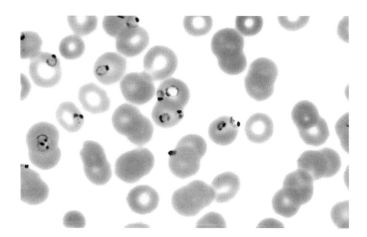

There are at least four types of human malaria caused by different parasites. This is the most dangerous, *Falciparum malaria*; it kills millions each year. Dr. Cecil H. Fox/Photo Researchers, Inc.

Balancing selection

Besides acting in a directional mode, which leads to reduced heterozygosity, selection can also maintain diversity, or *polymorphism*, in a variety of ways, including: (1) overdominance (heterozygote advantage), (2) frequency-dependent selection, and (3) selection under heterogeneous environmental conditions. Such maintenance of polymorphism is called *balancing selection*.

Overdominance

Overdominance refers to cases in which the heterozygote (Aa) has a higher fitness than either homozygote (aa or AA). Both alleles are thus maintained in the population. For example, in regions with high infection rates of the malaria parasite, the sickle-cell allele of hemoglobin provides resistance to disease. The allele in homozygous condition, however, produces anemia and other problems, resulting in lower fitness. The wild-type allele (standard or predominant) in the homozygous condition does not provide resistance to malaria. Thus, despite the high burden of sickle-cell disease, the sickle-cell allele is maintained in such populations by overdominance.

The opposite of overdominance is *underdominance*, in which the homozygotes have higher fitness than the heterozygote. Such a situation results in an unstable equilibrium of allele frequencies, and one allele or the other will rise to fixation. Which allele "wins," however, can be a matter of chance. By driving different alleles to fixation in different populations, underdominance may help drive diversification between populations.

Frequency-dependent selection

Frequency-dependent selection occurs when the fitness of a genotype depends on its frequency. Diversity of alleles is maintained when fitness depends inversely on frequency, such that fitness is higher when an allele is rarer than when it is more common. For example, for a butterfly species that is palatable to a potential predator and mimics the color pattern of another butterfly species that is unpalatable to the same predator, the predator will learn to associate the color pattern with palatability or unpalatability, depending on the relative frequencies of mimic and model. The lower the frequency of the mimic, the higher

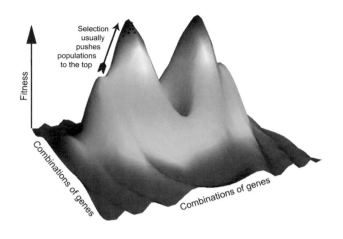

Each adaptive landscape is defined by a "height" (e.g. fitness) that exists for each point in a "plane" (e.g. combinations of genes). Computing or measuring this "height" and navigating the "plane" is extraordinary difficult for most biological systems. If fitness correlates such as survival or fecundity increase with "height," then selection will usually push a population towards the next locally reachable optimum. Courtesy of Laurence Loewe.

will be its fitness, because the predator will associate its color pattern more often with the unpalatable model and avoid it (Bates 1862). If the mimic evolves additional sets of alleles that mimic alternative models, the frequency of the mimicry patterns can be kept low even if the numbers of individuals in a population is large. Thus, inverse frequency-dependent selection can maintain high allelic polymorphism.

Heterogeneous conditions

Heterogeneous conditions in space and time may also result in the maintenance of polymorphism (Futuyma 1998). If there are different patches in a region where a species lives, and if these different habitats select for different genotypes, then allele diversity can be maintained (unless one type of patch is substantially more abundant). Different times of the year may select for different cryptic (camouflage) color patterns in a moth species (e.g., green for spring, brown for autumn). If these different colors are determined by different alleles at the same genes, the allele frequencies are likely to fluctuate with the seasons, but the cycling between seasons can maintain a balance of alleles. Research suggests, however, that heterogeneous temporal conditions may be insufficient to maintain polymorphism unless, for example, there is heterozygote advantage, or if there is a low rate of gene flow between subpopulations (*demes*) (Futuyma 1998).

Interactions between different mechanisms

Different modes of selection, genetic drift, migration, and mutation may all play roles simultaneously in changing allele frequencies or in maintaining variation in any particular population. Sewall Wright (1932) proposed an influential metaphorical model, called the *adaptive landscape*, to integrate such mechanisms. Although the model suffers from oversimplification, it is nevertheless a useful platform for formulating hypotheses and questions (Pigliucci 2008). In the 1932 version of the model (see

Figure 3), the peaks on the topographic surface represented fitness optima associated with particular allele combinations (genotypes). In other versions of the model, the peaks represent average population-wide fitnesses associated with particular allele frequencies. In "real life," the topography would also depend on how the genotype determines the phenotype and how the phenotype interacts with the environmental conditions, aspects that are ignored in the model but that could potentially be simulated. Similar genotypes or allele frequencies (which result in the fitness phenotypes represented by points on the surface) are "mapped" close together in the model such that evolutionary changes between genotypes or changes in allele frequencies can be traced as a "path" across the landscape.

Selection will always cause the path to "climb" toward a peak, because any variation of that genotype producing a lower fitness will not be competitive. Yet, which peak is reached on a multipeak landscape may be a matter of historical contingency (what was the starting genotype/allele frequency) and chance (which relatively fitter mutation occurred first). For example, if the starting point was in the central "valley," selection could drive the evolutionary changes toward any of the surrounding peaks, such as the peak in the middle. Once on the slope of this peak, stepwise selection by itself would not allow a different peak to be reached if such a "peak shift" would require the path to dip down into a valley, even if this other peak (e.g., the peak in the upper-right corner) were the optimal (highest fitness) adaptation. There are many examples of different species arriving at different adaptive solutions for the same problem. The African and Sumatran rhinoceros species have two horns, whereas the Indian and Javan species have single horns. Simply having horns is what was probably the important feature for selection, not the specific number of horns. One can think of these two different solutions as different peaks on an adaptive landscape.

Selection, however, is not the only factor that influences evolution. Gene flow is a strong force in shifting allele frequencies,

Figure 3. High-fitness "peaks" are represented by plus symbols and low-fitness "valleys" by minus symbols. Dotted lines represent isofitness contour lines. Source: Wright (1932). Reproduced by permission of Gale, a part of Cengage Learning.

Sumatran rhinoceros (*Diceros sumatrensis*) have two horns, whereas the Indian and Javan species have single horns. Terry Whittaker/Photo Researchers, Inc.

possibly in biased directions depending on the type of migration pattern. Genetic drift can also be a major factor in shifting allele frequencies, but this will be stochastic and its strength will depend on the population size. Thus, these selection-independent factors can do two things: (1) they can cause shifts to lower-fitness points, and (2) they can help to balance the loss of variation that results from directional selection. In the first case, it is possible that the shift would be to the slope of a different peak, in which case selection will drive evolution up a new peak (a *peak shift*). Thus, the action of all these mechanisms enables organisms to explore more fitness peaks and arrive closer to fitness optima than might be expected by selection alone.

Engineering, logistics, computational chemistry, drug design, and other fields also use the concept of a fitness landscape in *evolutionary optimization* or *genetic algorithms* to solve complex problems efficiently. It is often not possible to test every possible hypothesis or combination of parameters to determine which is the best solution, and just varying the parameters randomly may never achieve a near optimal solution. Instead, a "fitness function" is defined as a measure of how good a solution to the problem is (e.g., to minimize the distance needed for a delivery route). Although one starts with a good guess or a random set of parameters, the parameters of each successive solution can be "mutated" and only the next better solution kept. In this way, a local optimum solution can be reached

in which no changes provide better results. "Drift" can be introduced by allowing leaps between fitness peaks—again using selection to find the next optimum.

Selfishness and altruism

Selection has resulted in both cooperating and conflicting systems. There are many interesting phenomena involving conflict or cooperation, but there is space here to mention only a few.

In the phenomenon called *segregation distortion*, or *meiotic drive*, particular genetic elements are transmitted preferentially through meiosis or the products of meiosis (gametes). During normal meiosis, each allele has an equal chance of segregating into any gamete; normal genes can thus be thought of as "cooperating." In the case of *B chromosomes*, however, a supernumerary (extra) chromosome segregates preferentially into the oocyte. Alleles of genes that are carried on this B chromosome are thus more highly represented in the gametes than are any other alleles. This "selfish" genetic element can therefore spread throughout a population, even if it confers a fitness disadvantage on the individuals possessing it, as in the case of the B chromosome in mealy bugs (Nur and Brett 1987). In a real sense, such selfish genes behave like parasites. Like parasites, however, the host can also evolve means of evading or limiting their effects; different strains of mealy bugs, for example, have evolved mechanisms to reduce transmission of the B chromosome (Nur and Brett 1987).

Another well-understood example is the *t-haplotype* of the common house mouse, *Mus musculus* (Ardlie 1998). In this case, there are several "segregation distorter" genes (the *t*-haplotype) that can impair the function of gametes that carry a sensitive allele (+) at a "responder locus." Although heterozygous (*t*/+) males produce 50 percent sperm that have the *t*-haplotype and 50 percent that have the + allele, the sperm with the + allele are functionally impaired, presumably by the products of the *t* genes as the sperm are made. This preferential survival of the *t*-bearing sperm results in the spreading of the "selfish" *t*-haplotype throughout the population, even though there are associated lethal effects of the *t* genes on embryos that are homozygous *t/t*. Again, the meiotic drive is not as strong as predicted because other genes in the mice have also evolved to suppress this meiotic drive.

Surprisingly large fractions of the genomes of many animals, including nearly 50 percent of the human genome, are composed of transposable elements (TEs), mobile pieces of DNA that can "hop" around the genome, sometimes inserting into or near normal genes, whereupon they sometimes affect the regulation of gene expression (Slotkin and Martienssen 2007). Like other kinds of mutations, most TE-associated changes are likely to be deleterious. Yet, TE insertions may also be an important source of genetic variation. For example, they can act as sites of recombination events that cause gene duplications, an important source of new genetic material that can be modified into novel function (Fitch et al. 1991). As with other selfish elements, interesting mechanisms have evolved to suppress or regulate the activity of transposable elements. Such suppression can include *epigenetic* mechanisms (chemical modification of DNA) and/ or mechanisms that use RNA inhibition (a mechanism that uses double-stranded RNA to repress expression from RNA transcripts). These mechanisms and possibly the TEs themselves then appear to have been co-opted by organisms to regulate genes and chromosomal processes.

Another version of the gene as a "selfish" evolutionary entity (Dawkins 2006) is exhibited as apparently altruistic behavior between individuals in a population. In such cases, the effect of an allele may be to enhance the fitness of individuals in the population other than the one who actually expresses the allele. For example, the worker castes of social ants are sterile, with therefore zero fitness. Nevertheless, the genes that generate the worker caste continue to be inherited because fertile relatives who are carriers (queens) benefit strongly from the workers. Such relatedness between the "altruistic" individual and the "benefactor" is the basis of *kin selection* (Hamilton 1964). For example, assume an allele reduces the fitness of the "altruistic" individual by a particular cost C (e.g., C fewer offspring are produced), but its expression in this individual increases the fitness by a certain benefit, B, to others who are related to the altruistic individual by a certain degree, r. Then the allele frequency will increase in the population as long as $rB > C$.

Selection is not restricted to genes and individuals, but is *scale independent* (Leroi 2000). That is, selection is a property of any entity that can reproduce in a genealogical, treelike manner, displays variation that affects the ability to survive and reproduce, and inherits such variation. Thus, species themselves are potentially subject to selection, because they exhibit these conditions. Selection that applies to levels that are higher than the individual is known as *group selection* or *clade selection* (sometimes *species selection*). If a particular clade (a group of species that share a common ancestor) has a faster rate of speciation or a slower rate of extinction than a related clade because of some distinguishing feature of the clade, then selection will result in the maintenance or spreading of that feature in descendant species. As an example, in the group of rhabditid nematodes related to *Caenorhabditis elegans* (a major model genetic system used in biomedical research), self-fertilizing hermaphroditism has evolved multiple times, but such lineages appear to go extinct more rapidly than outcrossing male-female species (Kiontke and Fitch 2005). Of the species that have been tested, the self-fertilizing species have approximately 20 percent less genetic diversity than the outcrossing species (Graustein et al. 2002), which is consistent with the idea that self-fertilizing is associated with a higher rate of extinction than outcrossing (i.e., if variation allows populations to adapt to varying conditions and thus protects a species against extinction). Thus, although self-fertilizing may provide a short-term gain for an individual's fitness (a single self-fertilizing hermaphrodite can establish an entire colony on an ephemeral patch of food without having to mate), clade selection may result in the long-term survival of obligately outcrossing species.

Darwin (1871) recognized that some features of species, such as the ostentatious plumage of male pheasants and peacocks, do not appear to provide a survival advantage, either to the individuals that bear them or to their offspring. Instead, Darwin suggested that these features have evolved because they posed a reproductive advantage in obtaining mates. He called this *sexual selection*. Male–male competition occurred (resulting, for example, in selection for horns or antlers that would be advantageous in male–male combat), and/or female preference (resulting, for example, in strikingly colored male plumage) occurred. The latter would suggest the requirement for female preference to preexist the male structure selected as a result of this preference. Indeed, in a wonderful set of experiments, Alexandra L. Basolo (1990) found that females of fish species in which males did not have swordtails (a colored extension of the caudal fin) already have a preference for males with swordtails.

Mechanisms of speciation

Although this entry has dealt primarily with mechanisms of evolutionary change, at least a little must be said about the mechanisms of speciation, because of course evolutionary mechanisms are involved in speciation.

Darwin (1859) suggested that even experts had difficulties agreeing on what constituted the species of a particular genus and what were the number of varieties in a particular species. He explained that this difficulty stemmed from the human tendency to categorize things into discrete units, even if they were products of a continuous process, such as speciation.

Indeed, he called varieties "incipient species." Although there are many ways that biologists have of categorizing species, a common theme is that significant gene flow no longer occurs between the populations that are called species. The questions to address about the process of speciation then include how such barriers to gene flow arise, how populations become reproductively isolated, whether selection or drift or other evolutionary mechanisms have played the predominant role, and what are the genetic differences that arise and prevent hybridization between species.

Darwin and most modern authors favor the idea that the main mode of speciation is *allopatric* (Futuyma 1998). In allopatric speciation, a single species becomes divided into two populations by a geographical barrier that prevents significant gene flow between populations. Such barriers can include such processes as glaciation, the formation of rivers, and barriers resulting from tectonic plate movements (e.g., the formation of the Rocky Mountains or the Isthmus of Panama). During isolation, the two populations diverge from one another by evolutionary processes described above (selection and/or drift). As a consequence of genetic changes, hybrids between the populations have low fitness (e.g., are sterile or have low survivorship).

One famous model, the Dobzhansky-Muller model (see Figure 4), explains how such reproductive incompatibility can evolve between populations without causing reduced fitness within each population (Futuyma 1998). Assume that there are two genetic loci, gene A and gene B, and that the original population is homozygous for the a and b alleles. After isolation, population 1 evolves the A allele, which could become fixed by selection (as in the original model) or by drift. Population 2 evolves the B allele, which also becomes fixed in that population. If the A allele is incompatible with the B allele and this interaction negatively affects fitness (i.e., the loci are *epistatic* for fitness), then hybrids between populations 1 and 2 will have lower fitness than individuals within each population. Several examples of such loci have been discovered (e.g., Brideau et al. 2006), suggesting that the model could explain much of speciation. A similar type of model can be proposed for the evolution of chromosomal (karyotype) incompatibilities. Note that, according to the Dobzhansky-Muller model, although selection may have been involved in fixing the alleles, haplotypes, or chromosomes that cause incompatibility in hybrids, this variation did not arise in order to cause speciation or to maintain species boundaries. Thus, speciation per se may generally be a consequence of genetic divergence rather than its cause.

Sympatric speciation can also occur, although its prevalence is disputed (Futuyma 1998). In sympatric speciation, a biological barrier to gene flow evolves *within* a population, such that assortative mating occurs to isolate the phenotypes that are favored by disruptive selection. This can work if these

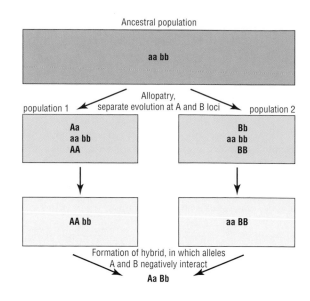

Figure 4. The Dobzhansky-Muller model explains how reproductive incompatibility can evolve between populations without causing reduced fitness within each population. (Modified from Futuyma [1998].) Reproduced by permission of Gale, a part of Cengage Learning.

phenotypes involve preference for particular resources that are segregated and if mating occurs in or near such resources. Sympatric speciation is also known to occur by *polyploidization*, in which the polyploid hybrid of two different species or of a single species is reproductively isolated from the diploid parental species (Futuyma 1998).

In conclusion, natural selection, genetic drift, and thus evolution are the result of genetic mechanisms. Selection is the only mechanism that can produce adaptations and is an extremely efficient "algorithm" for finding optimal solutions, particularly when combined with stochastic mechanisms, such as genetic drift and gene flow. Selection is also scale independent and can explain the evolution of "altruistic" as well as "selfish" characters. These mechanisms are sufficient for explaining both anagenetic change and cladogenetic diversification (speciation). Much further research is needed, however, to identify how genotypic change "maps" to phenotypic change, what are the relative roles of selection and drift in diversification, and what kinds of genetic changes are actually involved. With new and emerging technologies that allow entire genomes to be sequenced and resequenced for different populations, and with systems biology providing insights into genotype–phenotype mapping, the future of research into evolutionary mechanisms is very promising and exciting.

Resources

Books

Darwin, Charles. 1859. *On the Origin of Species by Means of Natural Selection; or, The Preservation of Favoured Races in the Struggle for Life*. London: John Murray.

Darwin, Charles. 1871. *The Descent of Man, and Selection in Relation to Sex*. 2 vols. London: John Murray.

Dawkins, Richard. 1986. *The Blind Watchmaker: Why the Evidence of Evolution Reveals a Universe without Design*. New York: Norton.

Dawkins, Richard. 2006. *The Selfish Gene*. 30th anniversary edition. Oxford: Oxford University Press.

Dennett, Daniel C. 1995. *Darwin's Dangerous Idea: Evolution and the Meanings of Life*. New York: Simon & Schuster.

Futuyma, Douglas J. 1998. *Evolutionary Biology*. 3rd edition. Sunderland, MA: Sinauer.

Goldschmidt, Richard. 1940. *The Material Basis of Evolution*. New Haven, CT: Yale University Press.

Kimura, Motoo. 1983. *The Neutral Theory of Molecular Evolution*. Cambridge, UK: Cambridge University Press.

Maynard Smith, John. 1989. *Evolutionary Genetics*. 2nd edition. Oxford: Oxford University Press.

Wright, Sewall. 1932. "The Roles of Mutation, Inbreeding, Crossbreeding, and Selection in Evolution." In *Proceedings of the Sixth International Congress of Genetics*, Vol. 1, ed. Donald F. Jones. Brooklyn, NY: Brooklyn Botanic Garden.

Periodicals

Ardlie, Kristin G. 1998. "Putting the Brake on Drive: Meiotic Drive of *t* Haplotypes in Natural Populations of Mice." *Trends in Genetics* 14(5): 189–193.

Basolo, Alexandra L. 1990. "Female Preference Predates the Evolution of the Sword in Swordtail Fish." *Science* 250(4982): 808–810.

Bates, Henry Walter. 1862. "Contributions to an Insect Fauna of the Amazon Valley: Lepidoptera: Heliconidae." *Transactions of the Linnean Society of London* 23(3): 495–566.

Brideau, Nicholas J., Heather A. Flores, Jun Wang, et al. 2006. "Two Dobzhansky-Muller Genes Interact to Cause Hybrid Lethality in *Drosophila*." *Science* 314(5803): 1292–1295.

Buri, Peter. 1956. "Gene Frequency in Small Populations of Mutant *Drosophila*." Evolution 10(4): 367–402.

Fitch, David H. A., Wendy J. Bailey, Danilo A. Tagle, et al. 1991. "Duplication of the γ-Globin Gene Mediated by L1 Long Interspersed Repetitive Elements in an Early Ancestor of Simian Primates." *Proceedings of the National Academy of Sciences of the United States of America* 88(16): 7396–7400.

Graustein, Andrew, John M. Gaspar, James R. Walters, and Michael F. Palopoli. 2002. "Levels of DNA Polymorphism Vary with Mating System in the Nematode Genus *Caenorhabditis*." *Genetics* 161(1): 99–107.

Hamilton, William D. 1964. "The Genetical Evolution of Social Behaviour," pts. I and II. *Journal of Theoretical Biology* 7(1): 1–16, 17–52.

Hardy, G. H. 1908. "Mendelian Proportions in a Mixed Population." *Science* 28(706): 49–50.

Leroi, Armand M. 2000. "The Scale Independence of Evolution." *Evolution and Development* 2(2): 67–77.

Nur, Uzi, and Betty Lou H. Brett. 1987. "Control of Meiotic Drive of B Chromosomes in the Mealybug, *Pseudococcus affinis (obscurus)*." *Genetics* 115(3): 499–510.

Ohta, Tomoko. 1996. "The Current Significance and Standing of Neutral and Nearly Neutral Theories." *BioEssays* 18(8): 673–677.

Pigliucci, Massimo. 2008. "Sewall Wright's Adaptive Landscapes: 1932 vs. 1988." *Biology and Philosophy* 23(5): 591–603.

Slotkin, R. Keith, and Robert Martienssen. 2007. "Transposable Elements and the Epigenetic Regulation of the Genome." *Nature Reviews Genetics* 8(4): 272–285.

Verhulst, P. F. 1838. "Notice sur la loi que la population poursuit dans son accroissement" [Note on the law by which population size increases]. *Correspondance mathématique et physique* 10: 113–121.

Weinberg, Wilhelm. 1908. "Über den Nachweis der Vererbung beim Menschen" [On the demonstration of heredity in man]. *Jahreshefte des Vereins für vaterländische Naturkunde in Württemberg* 64: 368–382.

Wright, Sewall. 1931. "Evolution in Mendelian Populations." *Genetics* 16(2): 97–159.

Wright, Sewall. 1951. "The Genetical Structure of Populations." *Annals of Eugenics* 15: 323–354.

Other

Kiontke, Karin, and David H. A. Fitch. 2005. "The Phylogenetic Relationships of *Caenorhabditis* and Other Rhabditis." *WormBook: The Online Review of* C. elegans *Biology*. Available from http://www.wormbook.org/chapters/www.phylogrhabditids/phylorhab.html.

David H.A. Fitch

Origins of the universe

A remarkable fact about the universe we find ourselves in is that it is capable of sustaining a planet such as Earth and the complex chemistry of life. It was only as recently as the 1920s that we began to get a glimpse of the vastness of the universe of galaxies. The discovery of the microwave background radiation and the realization that the universe began in a hot "big bang" date back only to the mid-1960s. And it is only since the beginning of the new millennium that cosmology has become a precision science, with a strong consensus emerging about what kind of universe we inhabit.

First steps on the distance ladder

Aristotle (384–322 B.C.E.) was the first to estimate the size of the Earth, using the angle of the shadow of a pole at noon at a location 100 miles south of the equator. Eratosthenes and Posidonius later used a similar method. These latter estimates are within about 10 percent of the modern value. In the second century B.C.E. Hipparchus used an eclipse method to estimate the distance from Earth to the Moon and deduced a value of 59 Earth radii, compared to the modern value of 60.3. Aristarchus of Samos (fl. c. 270 B.C.E.) tried to estimate the distance from Earth to the Sun using an eclipse method, but was off by a factor of 20. The Greeks also gave us Euclidean geometry (Euclid [fl. c. 300 B.C.E.]); the idea of absolute, uniform time (Aristotle); and the idea of an infinite physical frame (the atomists, Epicurus [341–270 B.C.E.]). Interestingly, and contrary to the picture held by medieval thinkers, Aristotle believed that the stars were at a range of distances.

The Aristotelian geocentric universe reached its epitome in the detailed system of Ptolemy (second century C.E.), described in *The Almagest*. The title reflects the crucial role of Arab scientists in preserving and extending the achievements of the Greeks. Certainly Nicolaus Copernicus (1473–1543) was aware of Arab work in his development of a heliocentric model of the solar system. A discovery of Copernicus that is less well known is that he gave, for the first time, the correct relative distances of the Sun and planets. His values were within 5 percent of the modern values. The absolute scale of the solar system was not determined accurately till the nineteenth century. The Copernican system also implied a huge increase in the minimum distance to the stars.

The 1610 discovery of the four largest moons of Jupiter by Galileo Galilei (1564–1642) lent weight to Copernicus's picture of the planets orbiting the Sun. Galileo's discovery of mountains on the moon showed this was another world like Earth. And his resolution of the Milky Way into stars was the first step into the universe of galaxies. Galileo's work on kinematics demonstrated the limitations of Aristotle's physics and paved the way for the Newtonian system. In his interesting dialogue with Richard Bentley, Isaac Newton (1642–1727) discussed the concept of an infinite universe of stars.

The first step on the distance ladder outside the solar system was taken by Bessel (1784–1846) in 1838 when he measured the parallax of the nearby star 61 Cygni, its change in apparent direction on the sky due to the orbit of Earth around the Sun. This was the final proof of the Copernican system. James Bradley (1693–1762) had discovered aberration, the elliptical motion of all stars on the sky due to Earth's motion, a century earlier.

In 1781 the French comet-finder Charles Messier (1730–1817) made a list of 110 objects that looked fuzzy or extended through a small telescope. Using much larger telescopes William Herschel (1738–1822) showed that many of these are in fact clusters of stars. The advent of spectroscopy in the nineteenth century showed that others were hot clouds of gas heated by young stars. The nature of one remaining class of nebulae, the spiral nebulae, remained uncertain. Were they simply gas clouds in our Milky Way system, or could they be distant island universes, as suggested by Christopher Wren, Thomas Wright, and Immanuel Kant?

Cepheids, M31, and the universe of galaxies

The next crucial step on the distance ladder, still of prime importance today, was the discovery by Henrietta Leavitt in 1912, working at the Harvard Observatory, that the periods of variation of Cepheid variable stars in the Small Magellanic Cloud are related to their luminosity, the period–luminosity relation. In 1926 Edwin Hubble used Leavitt's discovery to estimate the distance of Messier 31, the Andromeda Galaxy. It clearly lay far outside the Milky Way, thus resolving the long-standing controversy about the spiral nebulae and opening up the universe of galaxies.

Hubble classified galaxies as spiral, elliptical, or irregular. Modern spectroscopic studies showed that many galaxies, especially the spirals, are rotating systems, but the orbital speeds of the stars were found to be too large for the amount of visible matter they contain. Some other form of dark matter must be attracting the stars gravitationally. The velocities of galaxies seen in clusters of galaxies also imply dark matter. The final line of evidence for dark matter is provided by an effect known as gravitational lensing. When we see a distant background galaxy behind a foreground galaxy or cluster of galaxies, the gravitational bending of light—predicted by Albert Einstein's general theory of relativity—distorts the image of the galaxy, acting like a lens. This lensing effect allows us to map the dark matter in clusters of galaxies.

Hubble's law and the expansion of the universe

In 1929 Hubble announced, based on the distances of eighteen galaxies, that the more distant a galaxy, the faster it is moving away from us: velocity/distance = constant (Hubble's law).

The constant of proportionality is known as the Hubble constant, written H_0, and is measured in km/s/Mpc. This is often called the redshift-distant law, because the recession velocities of galaxies are measured by the shifting of their spectral lines toward the red end of the visible spectrum. This is just what would be expected in an expanding universe. Alexander Friedmann had shown in 1922 that expanding universe models are what would be expected according to Einstein's general theory of relativity, if the universe is (1) homogeneous (everyone sees the same picture) and (2) isotropic (the universe looks the same in every direction).

This unlikely assumption, the cosmological principle, had been introduced by Einstein in 1917 when he derived a static

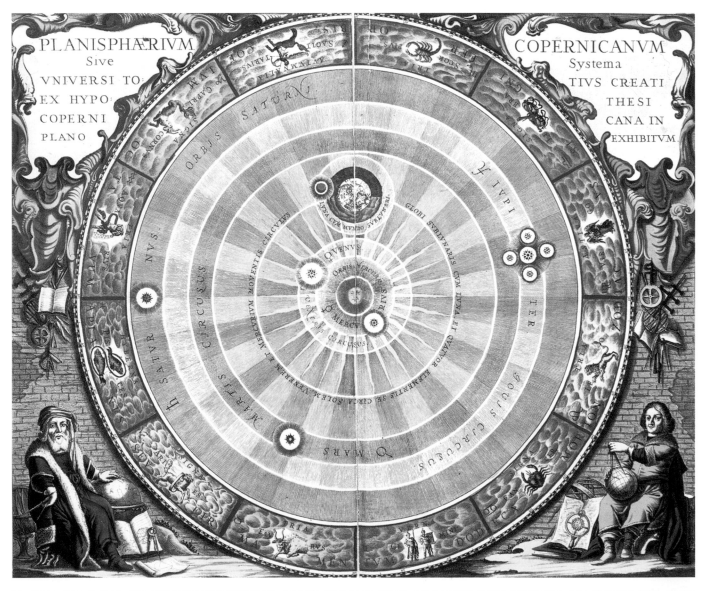

The Copernican System, 'Planisphaerium Copernicanum', c.1543, devised by Nicolaus Copernicus (1473–1543) from 'The Celestial Atlas, or the Harmony of the Universe' (Atlas coelestis seu harmonia macrocosmica) Amsterdam, c.1660. © British Library Board. All Rights Reserved/The Bridgeman Art Library.

Galileo Galilei discovered the four largest moons of Jupiter in 1610.
© Science Photo Library/Alamy.

formation of the solar system. The elements of Earth thus reflect the whole history of star formation, evolution, and death in our galaxy.

Controversy over the Hubble constant, H_0

Hubble's 1929 estimate of H_0 was 500 km/s/Mpc, where 1 megaparsec (Mpc) = 3.26 light years. Now H_0 has the dimensions of $time^{-1}$ and so 1/H_0 is the expansion age of the universe—the age the universe would have if no forces were acting on it. Hubble's value for H_0 implied an age of the universe of 2 billion years, and it was soon realized this was shorter than the age of Earth, derived from radioactive isotopes. From 1929 to 2001 the value of the Hubble constant was a matter of fierce controversy. Walter Baade pointed out in 1952 that there were two different types of Cepheid variable star, so Hubble's calibration had been incorrect. This reduced H_0 to 200 km/s/Mpc. In 1958 Allan Sandage recognized that objects that Hubble had thought were the brightest stars in some of his galaxies were in fact clouds of hot gas and arrived at the first recognizably modern value of 75 km/s/Mpc. During the 1970s there was an acute disagreement between Sandage and Gustav Tammann, on the one hand, favoring H_0 = 50 km/s/Mpc, and Gérard de Vaucouleurs, on the other, favoring 100 km/s/Mpc.

Hubble space telescope key program

Following the launch of the Hubble Space Telescope (HST) in 1990, and the subsequent repair mission, substantial amounts of HST time were dedicated to measuring Cepheids in galaxies out to distances of 20 Mpc, to try to measure the Hubble constant accurately and to give the different distance methods a secure and consistent calibration. The HST Key Program soon split into two teams, one led by Wendy Freedman, Jeremy Mould, and Robert Kennicutt, and the other by Sandage and Tammann. In 2001 Freedman and colleagues announced their final result: H_0 = 72 +- 8 km/s/Mpc

This as we shall see agreed extremely well with the first results from the WMAP cosmic microwave background mission (72 +- 5 km/s/Mpc). It gave an age of the universe for an Einstein–de Sitter model of 9.1 Gyr (1 Gyr = 10^9 years = 1 billion years), which meant that a positive cosmological constant would be required for constancy with the age of the oldest stars.

Direct evidence for a positive cosmological constant came in 1995 from studies of Type Ia supernovae in distant galaxies, using both ground-based telescopes and HST. Type Ia supernovae arise in binary star systems when the more massive star starts to dump gas on a companion white dwarf, the relic of a star like the sun when it reaches the end of its life, and this white dwarf then reaches a critical mass and explodes. These supernovae are found to form a homogenous population, and their luminosity at maximum light can be used as a distance estimator. The way this distance varies with redshift points to the need for a cosmological constant to get the required geometry for the universe. The modern interpretation of Einstein's cosmological constant is that it represents the energy density

model of the universe in which gravity is balanced by a new force, the cosmological repulsion. Einstein's inspired guess that the universe must be very simple (homogeneous and isotropic) has since been confirmed to very high degree of accuracy.

Formation of the elements and planetary systems

The growing understanding of nuclear synthesis processes in the Sun and stars from the 1930s onward allowed the development of detailed models of how stars evolve. A star like the Sun is fusing hydrogen to form the next simplest element, helium. Later in its life it will become a red giant star and fuse helium to carbon, nitrogen, and oxygen. More massive stars continue this sequence on to the formation of elements such as silicon, magnesium, and sulfur, through to iron. In a classic paper of 1957, E. Margaret Burbidge and colleagues showed that almost all elements could be made either in normal stars or during supernova explosions at the end of the life of massive stars. Through winds and explosions stars spread the elements they manufacture through the interstellar gas between the stars. New stars and planetary systems then form by condensation out of dense clouds of interstellar gas. A crucial first step in planetary formation is believed to be the formation of planetesimals, aggregates of dust and ice, within the disc-shaped cloud surrounding the forming star. In the outer reaches of the solar system, the Oort cloud of comets may be a relic of this early stage in the

M31, the Andromeda Galaxy, in a "deep" (long) exposure, showing "triangular appendages" at either end, possibly clouds of stars. This galaxy is the closest spiral galaxy to our own, and is the largest in the Local Group, and was crucial in the discovery that other galaxies were separate from ours. © Tony Hallas/Science Faction/Corbis.

of the vacuum, loosely called "dark energy," but a problem is that the observed value is 120 orders of magnitude (powers of ten) smaller than the expected theoretical value from particle physics.

The hot big bang

In 1965 Arno Penzias and Robert Wilson announced the discovery of the cosmic microwave background radiation. This radiation is a relic of the "fireball" phase of the hot big bang universe, when the dominant form of energy was radiation. The universe appears to have formed in a single explosive event, the initial "singularity," with essentially infinite density and temperature. As the universe expands the temperature and density drop. Initially the universe is very simple, consisting of photons (particles of light), light particles (electrons and neutrinos), and quarks, which are the building blocks of heavier particles such as protons and neutrons. When the temperature drops to 10^{12} K, the quarks are confined to making protons and neutrons. When the temperature reaches 10^{10} K, about 1 second after the big bang, nuclear reactions begin and neutrons and protons fuse together to make deuterium, helium, and lithium. 150,000 years after the big bang, when the temperature has dropped to 3000 K, electrons and protons combine together to make hydrogen, and the universe becomes transparent to radiation for the first time. This is the moment, called the epoch of recombination, we are looking at when we view the cosmic microwave background.

Formation of galaxies

During the fireball phase, the ordinary matter that the Earth and humans are made of—protons, neutrons, and electrons—is extremely smoothly distributed, to better than one part in 100,000. For there to be planets, stars, and galaxies today there do have to be some small fluctuations in density present, and these are believed to be randomly distributed through the universe. To make galaxies, much stronger fluctuations must have already developed in the dark matter, which ceased to be controlled by radiation at a much earlier epoch. After recombination the ordinary matter can respond to the gravitational attraction of the dark matter lumps and falls toward them, making protogalaxies consisting of ordinary matter concentrations, in which stars start to form, embedded in halos of dark matter. These protogalaxies then merge

Just over a thousand years ago, the stellar explosion known as supernova SN 1006 was observed. A composite image of the SN 1006 supernova remnant, which is located about 7000 light years from Earth. © Stocktrek Images, Inc./Alamy.

together in a hierarchical assembly of the galaxies we see today, these mergers being accompanied by vigorous formation of new stars. The gravitational assembly process also eventually collects galaxies together into groups and clusters.

Cosmic microwave background fluctuations

The remarkable isotropy of the cosmic microwave background was the first real evidence that Einstein's cosmological principle is a good approximation for the universe. The first deviation from perfect isotropy was found in the 1970s, with the discovery of the cosmic microwave background "dipole" anisotropy, that the background is slightly hotter in one direction of the sky and cooler in the opposite direction. This is due to our galaxy's motion through the cosmic frame at a velocity of about 600 km/s as a result of the combined gravitational pull of galaxies and clusters within about 300 million light years of us. In 1992 the team working with COBE, the Cosmic Background Explorer satellite, found fluctuations in the cosmic microwave background, at a level of about one part in 100,000. These were the crucial clue to how galaxies and clusters of galaxies formed in the universe. A later mission, WMAP, made precision measurements of the cosmic microwave background fluctuations from which they were able to determine several cosmological parameters (density of ordinary and dark matter, the cosmological constant, age of the universe, and Hubble constant) rather precisely. So today a rather tight consensus exists on the kind

The Hubble Space Telescope. Fuse/Jupiterimages/Getty Images.

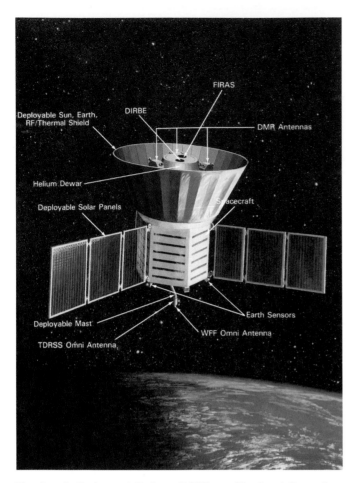

The Cosmic Background Explorer (COBE) satellite found fluctuations in the cosmic microwave background, at a level of about one part in 100,000. LAMBDA/NASA.

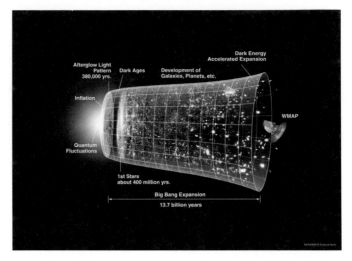

This illustration shows how astronomers believe the universe developed from the "Big Bang" 13.7 billion years ago to today. They know very little about the Dark Ages from 380,000 to about 800 million years after the Big Bang, but are trying to find out. © NASA/WMAP Science Team/MCT.

of universe we inhabit, with just 4 percent being in the form of ordinary matter, 21 percent in the form of dark matter, and 75 percent in the form of dark energy.

What was before the big bang?

It remains a puzzle why the universe is so isotropic and, at least in its early stages, so uniform in density. When we look at the cosmic microwave background in opposite directions on the sky, the regions we are looking at seem to have had no causal contact with each other, each lies far beyond the other's horizon, and yet they appear identical. A second problem is that the universe today seems to be quite close to being spatially flat. This requires the universe to have been incredibly close to flatness in its early stages. The horizon and flatness problems are partially solved by the idea that the universe went through a period of exponential expansion, "inflation," at or close to the big bang itself, driven by a phase with a very strong cosmological constant (dark energy). This period of inflation can drive the universe closer to isotropy and uniformity. Some cosmologists speculate that prior to the big bang the universe exists as a shadowy, chaotic quantum world. Some small region undergoes a fluctuation and finds itself with a very high vacuum energy, which triggers the

exponential expansion of that region and the emergence of the universe we see today. In this picture there would be other expanding universes beyond our horizon, the multiverse. This allows scope for the anthropic principle—that our universe has to be as it is so that life could have arisen in it. Most of the parallel universes would be devoid of life.

Because the physics that we really know about extends back to the era of quark confinement, but no further, these ideas are essentially pure speculation.

Life in the universe

We think we understand how the universe has evolved, how galaxies were assembled, and how stars and planetary systems formed. About one-third of stars are single stars, and it is likely that many of these have planetary systems. We know that there are many cases of stars with a Jupiter-sized planet close to the star, and we can confidently expect to find stars with planetary systems like our own, with small rocky planets, in the decades ahead. In principle there should be hundreds of thousands of stars like the Sun, with planets such as Earth, in our galaxy. We do not know how life formed on Earth, and so we do not know exactly what astronomical conditions are required. Is Jupiter important to shield Earth from asteroids and comets? Is the Moon crucial for maintaining a stable axis of rotation? It is plausible, however, that there are many potential earths out there, and there is no reason that some of them may not have formed several billion years before Earth did.

If as some argue, life-forms inevitably arise if conditions are right, then there may be many planets in our galaxy with at least simple forms of life on them. Evolution to intelligent creatures capable of developing a technological civilization is, however, less inevitable. And if it has happened at hundreds of thousands of locations in our galaxy, and often with several billions of years start on us, where are They? If humanity survives on Earth for another billion years, it is hard to believe we will not have begun the exploration and colonization of our galaxy. Perhaps we should face the possibility that we are alone.

The origin and evolution of the early universe appears to have been extremely simple. The emergence of structure such as galaxies, stars, and planets seems to be the result of random processes. The physics we have discovered on Earth is enough to understand the whole observed universe and the origin of galaxies, stars, the elements, and Earth itself. The origin and very early stages of the universe require physics beyond the range we can at present explore on Earth.

Resources

Books

Rowan-Robinson, Michael. 1985. *The Cosmological Distance Ladder: Distance and Time in the Universe*. New York: W. H. Freeman.

Rowan-Robinson, Michael. 1999. *Nine Numbers of the Cosmos*. Oxford: Oxford University Press.

Silk, Joseph. 1980. *The Big Bang: The Creation and Evolution of the Universe*. San Francisco: W. H. Freeman.

Weinberg, Steven. 1993. *The First Three Minutes: A Modern View of the Origin of the Universe*. Rev. edition. New York: Basic Books.

Michael Rowan-Robinson

Origins of life on Earth

Biology is sometimes viewed as the field within chemistry that seeks to understand chemical systems able to support Darwinian evolution (Joyce 1994). From this perspective, questions related to life's origins are fundamentally chemical in nature. They ask how a collection of inanimate organic molecules that might have been present on early Earth might have spontaneously assembled to give a system able to support Darwinian evolution.

Such a system must have had three properties. First, the system must have been able to direct the synthesis of more of itself. This ability to reproduce is well known in the chemistry of inanimate molecules; crystal growth is perhaps the most familiar example.

But simple replication is not sufficient to support Darwinian evolution. The original Darwinian system must also have had the ability to generate replicates *imperfectly*; mutation must have been possible. Further, those imperfections must themselves have been replicable.

Any chemical system that has these properties should be able to spontaneously improve itself via Darwinian mechanisms. Unfortunately, scientists have little idea of what chemical structures might support these properties generally, let alone what structures might have actually supported these properties on early Earth. No examples are known of Darwinian systems other than the ones found on Earth after four billion years of biological evolution. Furthermore, these are hardly "primitive." Nor do we have any kind of artificial life working in the laboratory, something that might better define models for the first Darwinian molecular systems.

Thus, ongoing efforts to understand life's origins seek to define molecular structures that support Darwinian evolution that are more primitive than those present on Earth today. This includes efforts to generate in the laboratory artificial chemical systems that evolve as they exploit these structures. This entry illustrates some of these efforts after providing the basic rules of chemical theory needed to allow biologists to understand the chemical questions associated with the origin of life.

The entry begins by introducing some tools for analyzing *reactivity* in organic molecules. Here, "reactivity" concerns ways in which molecules are transformed into other molecules, especially in water, and especially at habitable temperatures. These tools are then applied to ask simple questions about how amino acids and nucleotides, the building blocks for proteins

and nucleic acids, might be created abiologically. Finally, these tools are applied to address the historical question about origins: What chemistry might *actually have occurred* four billion years ago that led to the emergence of the life on Earth that we know?

Carbon and biochemistry

Carbon atoms form four bonds. A single bond joining two carbon atoms is strong, on the order of 400 kilojoules (about 100 kilocalories) per mole. This means that a typical pair of carbon atoms joined by a single bond will remain joined for many thousands of years at temperatures where water is a liquid. Single bonds between carbon and hydrogen, carbon and oxygen, and carbon and nitrogen are similarly strong.

As a consequence of their strength, bonds between two carbon atoms break only when an energetically favorable path exists that allows them to do so. At temperatures where water

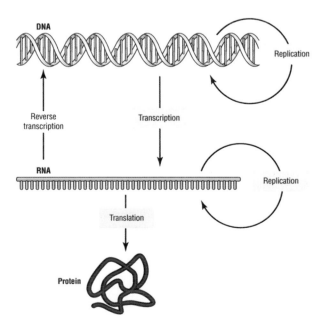

The central dogma of molecular biology. This diagram helps us visualize the relationships among DNA, RNA, and proteins in the cell. Reproduced by permission of Gale, a part of Cengage Learning.

is a liquid, such a path generally exists only when the carbon can form another bond to another atom soon after the original bond breaks, or (possibly) at the same time as when the original bond breaks. Understanding the possibilities of new bond formation at the same time as old bond breaking is a key to understanding reactivity.

Reactivity

Reactivity in organic molecules is based on "structure theory." Structure theory holds that the behavior of all organic molecules, including those in living matter, can be understood in terms of the structures of collections of atoms held together by bonds.

Practitioners of structure theory develop an understanding of the relationship between chemical structure and reactivity by studying thousands of reactions that organic molecules undergo. From this, chemists acquire an intuition about reactivity. From that intuition and the surrounding skills, chemists evaluate the plausibility that a reaction, perhaps one that they have never seen before, might occur. They are also able to propose new chemical reactions. This section offers an introduction to the basic skills needed to assess the plausibility that various chemical pathways might have led to the components of early life.

Pairs of electrons form bonds between atoms

Covalent bonds between two atoms are formed by a pair of electrons. Thus, in molecular hydrogen (dihydrogen, written as H–H or H_2), a line joining the two hydrogen atoms represents a pair of electrons, or a single bond. In water (H–O–H), the lines between the hydrogen atoms and the oxygen atom each represent a pair of electrons that form a single bond holding the hydrogens to the oxygen. In formaldehyde ($H_2C=O$), the double line between the carbon and the oxygen represents two pairs of electrons, four electrons in all.

Chemists often abbreviate the structure of organic molecules, neglecting to represent all of the atoms and electrons in a molecule by letters, lines, or dots. They do so primarily to save time.

Therefore, the first skill required in analyzing reactivity in organic chemistry requires the completion of the structure of the organic molecule. This ensures that the locations of all of the valence electrons are identified. A completed structure is known as a *Lewis structure* and must include the electrons present in many molecules *not* involved in bonding. For example, in the Lewis structure of water, oxygen carries two pairs of unshared electrons. Each valence electron not involved in a bond is represented by a dot. Likewise, the oxygen in formaldehyde carries two pairs of unshared electrons.

A nucleophilic center brings a pair of electrons to form a new bond

Because a pair of electrons is needed to form a chemical bond, any unshared pair of electrons is available in principle to form a new bond. Atoms that contain pairs of electrons that can form a new bond are called *nucleophilic centers*. To form a bond, the electron pair on the nucleophile must find an atom

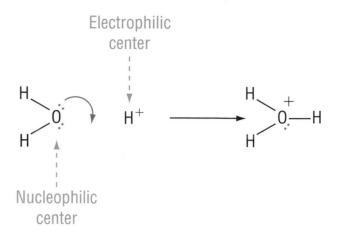

Figure 1. Reaction of the nucleophilic center on the oxygen of water with an electrophilic center, H^+;. The movement of a pair of electrons in the reaction is illustrated using a curved arrow. The result is $H_3;O^+$, the hydronium ion. Reproduced by permission of Gale, a part of Cengage Learning.

that lacks a bond, an atom that is called an *electrophilic center*. This center has a free valence unoccupied by either other atoms or electrons. A simple electrophile, for example, is a proton (H^+). H^+ is not bonded to anything and can intrinsically form one bond. H^+ is therefore looking for a single partner with which to bond. But because H^+ itself has no electrons that it can use to form a bond, it must wait for a nucleophilic center as that partner.

Curved arrows describe the movement of pairs of electrons in reactions that form and break bonds between atoms

Organic chemists use curved arrows to describe reactions between nucleophilic and electrophilic centers to produce a new bond. Correctly drawn, the curved arrow begins with a pair of electrons on the nucleophile, the pair that will form the new bond in the product. It ends at a position (on the structures of the reactants) where the electron pair *will be* after the bond is formed. Figure 1 shows the reaction of the unshared pair of electrons on the oxygen of water (the nucleophilic center) with H^+ (the electrophilic center) to give H_3O^+ (hydronium ion).

While nucleophilic centers are often easy to spot (they bear their electrons prominently), electrophilic centers are frequently less so. This is especially true when the electrophilic center is a carbon atom. For example, the carbon of formaldehyde (H_2CO) has all of its four valences filled. That carbon does not seem to have a valence available to form a new bond with anything.

If, however, one of the two bonds between carbon and oxygen breaks, with the electron pair moving from a position between the carbon and the oxygen to a new position on the oxygen, then the carbon center has a valence free. It then welcomes a nucleophilic attack from the oxygen from any nearby H_2O molecule.

This process is shown using curved arrows in Figure 2. Here, a bond between carbon and oxygen is broken at the same time

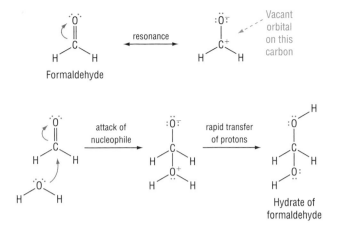

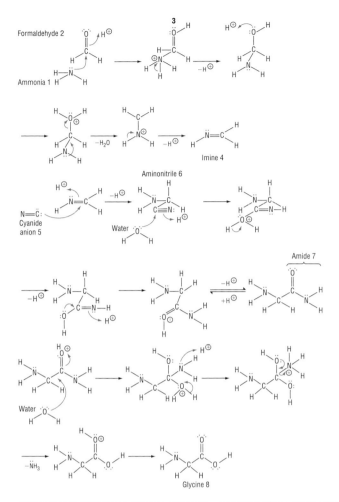

Figure 2. Reaction of the nucleophilic center on the oxygen of water with an electrophilic center, the carbon atom of formaldehyde. The movement of a pair of electrons in the reaction is illustrated using a curved arrow. Reproduced by permission of Gale, a part of Cengage Learning.

as the carbon forms a new bond to an incoming oxygen. As the energy in the second C–O bond is lost through breakage, the energy of a new C–O bond is gained. The resulting product is known as the hydrate of formaldehyde (H_4CO_2).

Curved arrows describe the synthesis of biological molecules

The curved arrow tool can be used to describe most reactions of organic molecules under standard conditions. This includes transformations that might have converted formaldehyde, present in a prebiotic world, into molecules that are characteristic of contemporary terran life.

Curved Arrow Mechanisms for the Formation of Amino Acids

Amino acids are the building blocks of proteins, and the curved arrow formalism can be used to describe the synthesis of a simple amino acid, glycine (NH_2CH_2COOH), from formaldehyde, ammonia, and water. Formaldehyde, ammonia, and water are all compounds that are observed by astronomers in gas clouds in the process of forming planetary systems today, which strongly suggests that they were available on early Earth.

As illustrated in Figure 3, ammonia has three hydrogen atoms and one nitrogen atom. A Lewis structure shows that the nitrogen atom in ammonia also carries an unshared pair of electrons. The nitrogen atom is therefore a nucleophilic center. Ammonia should therefore react with formaldehyde (2) for the same reason that water does.

In this reaction, the unshared pair of electrons on ammonia forms a new bond between its nitrogen and the carbon of formaldehyde, just as the pair of electrons forming the second carbon–oxygen bond leaves to form a new bond to H⁺. This generates an "amino alcohol" (3), after various H⁺ units are gained or lost.

After the transfer of the hydrogens, the nitrogen of the amino alcohol again has an unshared pair of electrons. This is

Figure 3. The Strecker synthesis of glycine. Reaction of the nucleophilic centers on the nitrogen of ammonia, the carbon of the cyanide anion and the oxygen of water with electrophilic centers on formaldehyde and key intermediates. The movements of pairs of electrons in the reactions are illustrated using curved arrows. Reproduced by permission of Gale, a part of Cengage Learning.

able to form a second bond between the carbon atom and the nitrogen atom of the amino alcohol. The resulting compound is known as an imine (4), which contains a unit (C=N) having a carbon atom bonded twice to a nitrogen atom.

The carbon of the C=N unit is also an electrophilic center. This sets the stage for the next reaction, where the carbon of the cyanide anion (5) attacks the imine carbon to form an aminonitrile (6). The nitrile has a C≡N unit, where the nitrogen is bonded three times to the carbon. This carbon is again an electrophilic center. If a pair of electrons forming one of the bonds between carbon and nitrogen goes away, then the carbon has a free valence. It is therefore available to form a bond with a nucleophilic oxygen atom from water.

The product, again after H⁺ atoms are transferred, has another C=O group in a unit known as an *amide* (7). The carbon of the amide is again an electrophilic center. It therefore can now be attacked by the nucleophilic oxygen of another water molecule. This leads to the hydrolysis of the

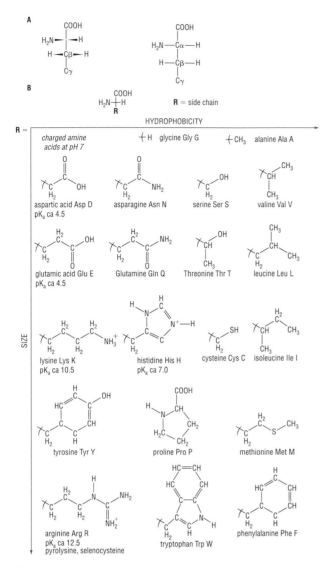

Figure 4. A) The general structure of an α-amino acid. The α refers to the position on the carbon chain that carries the amino group H_2N; other possibilities are the β and γ carbons. The four atoms attached to each carbon are indicated by the wedged lines. On the right is a Fischer projection, which places bonds that project above horizontally and those below vertically. B) The side chains of the 20 predominant amino acids for Earth life. In general, carbon atoms are not explicitly designated by a C; unlabelled vertices in a structure are assumed to be carbon carrying four bonds, where bonds not explicitly shown are assumed to be carbon atoms. Recent discoveries have increased the number of encoded amino acids to 22, with the inclusion of the "minor" amino acids selenocysteine and pyrrolysine (Rother et al., 2000; Berry et al., 2001; Hao et al., 2002; Srinivasan, James &; Krzycki, 2002). Reproduced by permission of Gale, a part of Cengage Learning.

amide and the formation of the amino acid glycine, together with an ammonia molecule.

The net process is the reaction of one molecule of formaldehyde, one molecule of hydrogen cyanide, and two molecules of water to create one molecule of glycine (8). Ammonia is used in the first step and is released in the last step. Therefore, ammonia is a catalyst for the reaction, being consumed and formed in equal amounts in the reaction cycle.

This sequence of reactions is known as the *Strecker synthesis of amino acids*, named after the German chemist (Adolph Strecker) who developed it in the 1850s. The Strecker synthesis is driven by the rules of nucleophilicity and electrophilicity and proceeds spontaneously and in reasonable yield. Further, the Strecker synthesis is quite general. It can be used to prepare any amino acid for which the corresponding aldehyde is available. The molecular structures of the amino acids found in proteins from terran life are shown in Figure 4. For example, if one starts with acetaldehyde (CH_3CHO) rather than formaldehyde, the amino acid alanine ($CH_3CH(NH_2)COOH$) is formed (see Figure 5).

The types of organic reactivity shown in the Strecker process and the presumed availability of many compounds involved in the process make a compelling case that simple amino acids were available on early Earth. This is confirmed by the presence of simple amino acids in meteorites (see below). More complex amino acids are not as readily accessible, however.

Curved arrow mechanisms for forming nucleobases for RNA and DNA

Curved arrow mechanisms can be used to generate nonbiological routes for the synthesis of many molecules in biology. For example, the Oró–Orgel synthesis of adenine ($C_5H_5N_5$) exploits the reactivity of HCN to make this key component of DNA and RNA. The cyanide anion again reacts as a nucleophile, this time with HCN, whose carbon atom serves as an electrophilic center.

Curved arrow mechanisms for forming sugars for RNA

The Swiss chemist Albert Eschenmoser took into consideration the special reactivity of hydrogen cyanide to generate the sugar ribose, also a part of RNA. An intriguing sequence of reactions from an interesting starting material (derived from HCN) is the Eschenmoser synthesis. This process occurs occurs in the laboratory (Müller et al. 1990). It is not known to occur in nature, and the starting material has not been detected naturally in the cosmos.

Extraterrestrial sources of organic building blocks

This raises the next question of whether it is possible to create an inventory of all materials that might have been available in the cosmos or on Earth to support the origin of life. The cosmos offers several places to look directly. One is the meteorites that fall to Earth continuously. Another is the interstellar dust in clouds surrounding stars that are forming. The organic molecules in these clouds can be detected by seeing what kinds of light they absorb and emit.

Organic molecules from meteorites

Many meteorites that fall to Earth are believed to originate in or near the asteroid belt. Others may be remnants of comets that spent most of the time since the Sun's formation out past the orbit of Neptune. Particularly interesting are carbonaceous chondrites, which have evidently suffered exposure to liquid water when within their original parent body.

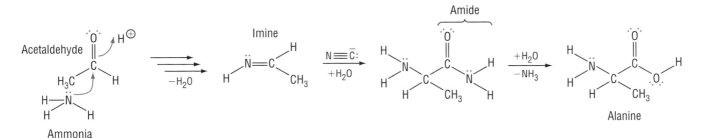

Figure 5. Strecker synthesis of alanine, starting from acetaldehyde. Reproduced by permission of Gale, a part of Cengage Learning.

What organic molecules do meteorites contain? Some of the most reliable studies to address this question exploited a meteorite that fell in 1969 near Murchison, Australia. This meteorite was quickly recovered and therefore suffered little contamination from biological molecules coming from Earth.

The Murchison meteorite contains a large amount of organic material, summarized in Tables 1 and 2. The amino acids found in the meteorite have been classified into two general groups. Within these groups, some seventy specific amino acids have been reported. Many of these are not known to occur in contemporary Earth life, diminishing the likelihood that they arose in the meteorite as a result of contamination from a terran source.

Several of these amino acids contain *stereogenic centers*. These are atoms in which carbon atoms are bonded to four different substituents. Compounds having stereogenic centers can give rise to *enantiomers*. A pair of enantiomers has a mirror-image structure; that is, the enantiomers related to each other like the left hand is related to the right hand.

Enantiomers are isomers, meaning the left-handed and right-handed forms are different. In natural processes, left- and right-handed molecules are often formed in equal amounts. To form the regular structures when these monomers are assembled together, however, it is convenient to have the same handedness in all molecules. Thus, all amino acids in terran proteins are left-handed; all sugars in terran nucleic acids are right-handed. Enantiomeric enrichment, the preference for one enantiomer over the other (a property often also called chirality, or handedness), is therefore believed to be a universal feature of chemistry derived from living systems.

In this light, researchers were surprised to find that several of the amino acids found in the Murchison meteorite are in fact enantiomerically enriched. In several cases, a small excess of one of two enantiomers was observed (Pizzarello and Cronin 2000). Especially notable were enantiomeric enrichments of 7 to 9 percent in three specific amino acids not known in forms of life currently living on Earth. This makes contamination from terrestrial sources unlikely.

If one assumes that the amino acids in the Murchison meteorite were *not* generated by living processes, this suggests that enantiomeric enrichment can be achieved by processes that are independent of life. Several of these are known in the laboratory, but not in nature. This weighs against the usual thinking that enantiomeric enrichment is a unique signature of life. The data available, however, are also consistent with a biotic origin. This would be the case if the Murchison meteorite originated in a body that supported life based on α-methyl amino acids that preferred one enantiomer to the other. While few would argue so, this result is consistent with life having emerged in the asteroid belt.

It is not known how well Murchison organics reflect what was available on early Earth before life emerged. For example, only a few amino acids (glycine, alanine, α-aminoisobutyric acid, α-amino-n-butyric acid, γ-aminobutyric acid) are found in the meteorite that fell in 2000 onto frozen Tagish Lake in Canada.

Some of the chemical fragments of DNA and RNA likewise can be found in meteorites. For example, some meteorites have been reported to contain small amounts of adenine, one of the four nucleobases found in RNA and DNA. The current view is that the Murchison meteorite contains adenine, guanine, and their hydrolysis products hypothanine and xanthine, as well as uracil. The reported concentration of adenine, however, is low, approximately 1.3 parts per million. Murchison and other meteorites may also contain ribitol and ribonic acid, the reduced and oxidized forms (respectively) of ribose (Cooper et al. 2001).

Total carbon	2.12% (Jarosevich, 1971), 1.96% (Fuchs, Olsen & Jensen, 1973)
Carbon as interstellar grains	
Diamond	400 ppm (Lewis *et al.*, 1987)
Silicon carbide	7 ppm (Tang *et al.*, 1989)
Graphite	<2 ppm (Amari *et al.*, 1990)
Carbonate minerals	2–10% of total carbon (Grady *et al.*, 1988)
Macromolecular carbon	70–80% of total carbon

Table 1. Carbon in the Murchison meteorite. Reproduced by permission of Gale, a part of Cengage Learning.

Amino acids	60 ppm	Purines and pyrimidines	1.3 ppm
Aliphatic hydrocarbons	>35 ppm	Basic N-heterocycles	7 ppm
Aromatic hydrocarbons	15–28 ppm	Amines	8 ppm
Carboxylic acids	>300 ppm	Amines	55–70 ppm
Dicarboxylic acids	>30 ppm	Alcohols	11 ppm
Hydroxycarboxylic acids	15 ppm	Aldehydes and ketones	27 ppm

Table 2. Organic compounds in the Murchison meteorite. (Cronin & Pizarello, 1988). Reproduced by permission of Gale, a part of Cengage Learning.

Phosphorus is abundant on Earth, both as an element (the eleventh most abundant atom in Earth's crust) and as phosphate. Meteorites hold a variety of phosphate-containing minerals and some phosphide minerals (Moore 1971). Unfortunately, a clear source of phosphate as a precursor component for RNA and DNA is not yet documented for the early Earth.

While these building blocks might have been available on early Earth, meteorites provide no evidence for their assembly. No examples are known where a nucleobase and a sugar have been found in a meteorite joined together to form a nucleoside. Nor have any proteins or nucleic acid strands been found. If proteins or nucleic acids were needed to generate the first Darwinian molecular system, then the puzzle remains as to how they might have emerged spontaneously.

Organics from the interstellar medium

The interstellar medium is an alternative place to look for organic molecules that may have been delivered to the primitive Earth. Organic molecules are assembled from the elements in interstellar space, without the need for a planet. The inventory of organic molecules found in interstellar space now numbers 120. These molecules include formaldehyde, cyanide, acetaldehyde, water, and ammonia, all of which are building blocks for the creation of the amino acids mentioned above.

Laboratory simulations

In a number of cases, laboratory environments designed to model conditions on early Earth have generated amino acids and/or components of nucleosides from species known or suspected to be present in the cosmos. For nucleic acids, work by Joan Oró and Leslie E. Orgel showed that the key nucleobase adenine can be prepared from hydrogen cyanide, as discussed above (Oró 1960; Sanchez, Ferris, and Orgel 1967).

Stanley L. Miller, working in Harold C. Urey's laboratory at the University of Chicago in 1953, made the first conscious attempt to model organic synthesis under prebiotic conditions. Amino acids were found to be generated after electrical discharges from electrodes were passed through an atmosphere of hydrogen, methane, and ammonia over water. These amino acids presumably arose from compounds generated in the discharge that later self-assembled in the water (Miller 1955). The electrical discharge was necessary simply to generate the formaldehyde and cyanide needed as starting materials for the synthesis. Once these precursors were formed, the synthesis of amino acids occurred, presumably via the Strecker synthesis, without the need for any energy at all. Here, the final step, the conversion of the CN group to the COOH group, may have occurred in the workup up of the procedure.

The simulated atmosphere chosen by Miller for his laboratory experiments was considered at the time to approximate the atmosphere of early Earth. This view is no longer accepted. It is now believed that the amount of methane on early Earth was much smaller than that used in the Miller experiments, and that Earth's early carbon inventory was present more as carbon dioxide (Kasting 1993).

Accordingly, much effort has been devoted to seeking nonbiological syntheses of biomolecules under conditions presumed to better fit those of early Earth. Some success has been achieved, reflecting the general reactivity of organic molecules. In general, when simple carbon-containing compounds are treated with energy sources, they form more complex organic molecules. These energy sources include electrical discharge, ultraviolet light, impact, and ionizing radiation. One example of this is "tholin," a red-brown collection of organic molecules made by irradiating simple molecules (where carbon is present largely as carbon dioxide). Tholins have been proposed as an important part of the atmosphere of Saturn's moon Titan.

Analogous work has been extended to simulate interstellar environments. For example, Louis J. Allamandola and his group at NASA's Ames Research Center have synthesized the amino acids glycine, alanine, and serine in their laboratory model of icy interstellar grains irradiated by ultraviolet light as an energy source (Bernstein et al. 2002). This suggests that at least some amino acids may have arrived on early Earth as a product of interstellar photochemistry, rather than through formation in liquid water on an early solar system body.

Problems in origins and their partial solution

These examples illustrate how concepts of nucleophilicity and electrophilicity unify processes that might create amino acids, nucleobases, and sugars from prebiotic mixtures to become components of the first form of life on Earth. The Miller experiments are cited in many texts as providing laboratory support for these ideas.

These and other examples over the years have generated a view that the spontaneous emergence of Darwinian chemistry from inanimate matter in nonbiological environments is easy, particularly in environments that may have been present on the early Earth. In this view, life is a natural consequence of organic chemistry. Organic molecules, if provided with the right kind of energy in the right amounts under the right conditions, will self-organize to create chemical systems capable of Darwinian evolution.

This view, as well as a hint that it might have problems, is reflected by a fictional exchange between a computer called Mnemosyne and the journalist John Hockenberry, recorded for a National Public Radio program:

Mnemosyne: Well, chemists think that if you could recreate the conditions of the earth about four and a half billion years ago . . . you'd see it [life] happen spontaneously. You'd just see DNA . . . just pop out of the mix.

John Hockenberry: Can they do that in a laboratory?

Mnemosyne: Well, no. Actually, they've tried, but so far they can't seem to pull it off. In fact chemists have a little joke about that, you know: they say that life is impossible. Experience shows that it can't happen. That we're just imagining it. Hahahaha. . . .

Hockenberry: Right. Those . . . those chemists.

(SoundVision Productions 1998)

As Mnemosyne noted, the chemists' objection to the notion that life is a natural consequence of organic reactivity is simple, and it comes from experience that many individuals have themselves had, in an organic chemistry laboratory perhaps, or more commonly in the kitchen. When one bakes a cake too long, it chars. With each additional minute at 300°F, the material resembles less and less something that one might call "living." When one applies random energy to mixtures of organic molecules, one gets tar. Robert Shapiro (1986) has provided a detailed discussion of how the intrinsic propensity of organic matter to form tar creates challenges for any model for the formation of life based on macromolecules.

From this perspective, existing prebiotic chemistry experiments are not particularly compelling. In a complex chemical mixture generated under prebiotic conditions, one may indeed be able to find trace amounts of amino acids and perhaps nucleobases. Some of these might indeed catalyze reactions that might have some utility. But other compounds in these types of mixtures may well inhibit this catalysis or may catalyze undesired reactions. For example, Gerald F. Joyce and Orgel (1999) pointed out that the clay-catalyzed condensation of nucleotides that results in oligonucleotides must have only one enantiomer; if both are present, the desired reactions are inhibited.

Even crystallization, a well-known method of obtaining order through self-assembly, is not particularly powerful as a way of separating mixtures of organic chemicals into their constituents. Normally, an organic compound must be relatively pure before crystallization occurs. Salts crystallize better, which may explain why crystals are more common in the mineral world than the organic world. But even organic salts can have problems crystallizing from an impure mixture.

These facts generate the central paradox in prebiotic chemistry. Spontaneous self-organization of organic matter is not known to be an intrinsic property of most organic matter, at least as scientists understand it in the laboratory setting. Obtaining the chemical order that is (presumed to be) necessary for life from complex mixtures of organic compounds seems as unlikely to a chemist as a ball rolling uphill seems to a child. It opposes the apparent natural tendency of organic molecules to become tar.

Nucleophilic and electrophilic reactions can destroy as well as create

Tar formation is well exemplified by processes that might have, under prebiotic conditions, generated ribose—the *R* in RNA, and a molecule often cited as being a possible early genetic molecule for the first life on Earth. In such a scenario, ribose might be made by a process that exploits the natural nucleophilicity of the enediolate of glycolaldehyde, also known in interstellar dust clouds (Hollis, Lovas, and Jewell 2000; Hollis et al. 2001), and the natural electrophilicity of formaldehyde. Glycolaldehyde first enolizes to give an *enediolate*. This species reacts as a nucleophile with formaldehyde (acting as an electrophile) to give glyceraldehyde. Reaction of glyceraldehyde with a second equivalent of the enediolate generates a pentose sugar, ribose, arabinose, xylose, or lyxose (depending on stereochemistry). A curved arrow mechanism describes this process as well.

Unfortunately, under conditions where these reactions occur, ribose reacts further, becoming a brown, complex mixture of organic species. Further reaction arises because ribose itself has both electrophilic and nucleophilic sites, at the aldehyde carbon and the carbon directly bonded to the aldehyde, respectively. Molecules having both reactivities are, as expected, prone to polymerization as the nucleophilic sites and electrophilic sites react with each other, with more formaldehyde, with water, and so on.

These reactivities undoubtedly cause the rapid destruction of ribose formed under formose conditions. Based on this reactivity, Miller and colleagues concluded in 1995 that "ribose and other sugars were not components of the first genetic material" (Larralde, Robertson, and Miller 1995, p. 8160).

For these reasons, some have suggested that life may have begun not with RNA but with an alternative organic compound as a genetic material, one based on molecules that are less fragile (Schöning et al. 2000; Nielsen 1999). Underlying this concept is the notion of a "genetic takeover," where delicate RNA and/or DNA molecules later supplanted a hardier genetic molecule that founded life (Cairns-Smith 1982).

Thus, the reactivity of nucleophilic and electrophilic centers can convert molecules that were plausibly present on early Earth into biologically interesting products. But these products themselves often have nucleophilic and electrophilic centers. Therefore, they can react further to create uninteresting products. This is the central paradox associated with the origin of life even given plausible mechanisms to create its components.

Water as a substance antithetical to life

Water creates its own set of problems for prebiotic synthesis. Water is an essential ingredient in the formation of amino acids from aldehydes and cyanide. Yet many biological molecules, once they are formed in water, are unstable in the water in which they formed. For example, adenine spontaneously hydrolyzes in water to form inosine. Other nucleobases also hydrolyze in water. For example, cytosine hydrolyzes to form uracil, while guanine hydrolyzes to form xanthine.

This is true for the next step as well. Polypeptide chains do not spontaneously form in water through the condensation of amino acids. Rather, polypeptide chains spontaneously hydrolyze in water to yield their constituent amino acids.

The same is true for RNA and DNA. Oligonucleotide chains spontaneously hydrolyze in water to generate individual nucleotides. Thus, even if the building blocks for proteins and nucleic acids are obtained, water and environments rich in water mean that their assembly is thermodynamically uphill.

In this regard, it is remarkable that water is viewed as essential for life. Scientists believe that some liquid is needed for life, as a solvent within which to hold chemical reactions. Certainly in modern terran life, water is a very useful solvent. It is also clear, however, that water is inimical to the stability of many biopolymers. Every minute, for example, perhaps ten cytidines in the genome of each human cell suffer spontaneous hydrolysis to yield uridine. These must constantly be repaired; if they are not, the information stored in a DNA molecule is irretrievably lost.

Minerals as a possible solution to the instability of ribose

To find solutions to individual prebiotic problems, one needs to consider the full range of chemical reactivity. For example, to make ribose, a prebiotic way is needed to remove the nucleophilicity of glyceraldehyde and to remove both the nucleophilicity and electrophilicity of ribose. If this could be done, then the formose reaction might generate ribose as a stable end product under prebiotic conditions.

Borate is known to form anionic complexes with 1, 2-dihydroxy units in organic molecules. The borate complex carries a negative charge. The anionic nature of the complex should prevent glyceraldehyde from losing a proton to create a nucleophilic enolate, but not prevent glyceraldehyde from reacting as an electrophile with the enediolate of glycolaldehyde to generate pentoses.

Further, the 1,2-dihydroxy unit of the cyclic form of ribose should form a stable complex with borate. This will stabilize the cyclic form of ribose at the expense of the aldehyde form. This should render ribose largely unreactive as either a nucleophile or an electrophile, as the cyclic form lacks a C=O carbonyl group, which is the center of this molecule's electrophilicity when ribose is in the ring-opened aldehyde form.

Experiments confirm this reasoning. In the presence of a base at temperatures ranging from 25^+ to 85^+, a solution of ribose rapidly turns brown over periods of one hour to five minutes (respectively). Analysis of the brown mixture showed a loss of the sharp signals characteristic of ribose and the emergence of a complex pattern of other products, a tar. When the same incubation was done in the presence of borate, however, the solution did not turn brown even at 50^+ and over the course of days.

Boron is not a particularly abundant element on Earth. Borate is, however, concentrated in igneous rocks in tourmalines, well known as gemstones. As tourmalines weather, borate is delivered into runoff water, where they are concentrated in pools where the water evaporates. As a consequence, colemanite and other borate-containing minerals are found in deserts and other dry environments, often under alkaline conditions. These are close to the conditions that generate pentoses in laboratory experiments.

This synthesis of pentose in the presence of the mineral colemanite is therefore plausibly prebiotic. Indeed, in the presence of borate, and given that ribose is the first compound in the formose product progression that offers a non-aldehydic cyclic form with unhindered *cis*-diols, the formation of ribose appears to be the natural consequence of the intrinsic chemical reactivity of compounds available from the interstellar medium under alkaline, calciferous conditions. As these conditions are not excluded from the early Earth, it is also not possible to exclude the availability of pentoses at the time when life originated.

This example of how minerals can productively control organic reactivity serves as a reminder that prebiotic chemistry is occurring on a planet, in the context of a larger geology. Minerals must be considered as we constrain models for the origin of life. For example, Orgel and his colleagues compiled a substantial literature demonstrating the preparation of oligonucleotides by template-directed polymerization, where clay acts as a catalyst (Ferris et al. 1996). Jack W. Szostak and his colleagues have shown how clays might have helped isolate catalytically interesting biological macromolecules (Hanczyc, Fujikawa, and Szostak 2003).

Some workers have taken the notion of minerals and their involvement in early life one step further. They have suggested that minerals themselves may have provided the genetic material for early forms of life (Cairns-Smith 1982).

Final thoughts

From these considerations, one can construct a "best-case scenario" for a prebiotic chemistry that might lead to life or, perhaps, conditions favorable for life. The starting points are the inventories of organic compounds in the cosmos, in the forming solar system, and perhaps on early Earth. The starting material consisted of:

1. Interstellar medium: molecular hydrogen, cyanide, water, formaldehyde, ammonia, and other organics;

2. meteoritic components, including amino acids and aromatic hydrocarbons;

3. chemicals from planetary chemistry, exemplified perhaps by the chemistry on Titan;

4. comets, which appear to be rich in organic materials;

5. organic molecules created on Earth as the consequence of shock, heat, lightning, or other sources of energy.

In the best-case scenario, a combination of mineral fractionation and evaporation allowed these organic compounds to be stabilized, separated, and concentrated. This generated usable concentrations of these compounds. This scenario places the assembly of the components of life in a desertlike environment, where evaporite minerals such as colemanite and nonaqueous solvents such as formamide accumulated. Under these conditions, the condensation of monomeric units would have been possible. Then, water would have been introduced to allow the biomolecules to fold and act.

This is a best-case scenario. As is discussed by Shapiro (1988), these isolated steps have never been shown to occur together without human intervention. Such a scenario does not solve other problems in the area of origins, such as the origin of chirality.

The literature contains many theoretical papers that outline possible syntheses of life's organic molecules from possible precursors based on a general knowledge of organic chemical reactivity. A second literature covers experimental work in a contemporary laboratory that is presumed to model prebiotic conditions. A third literature criticizes the first two because they inadequately consider prebiotic conditions, or because the molecules and processes that they generate would not have produced the intended result (life).

Scientists are pursuing a number of opportunities to gather additional information. NASA's Stardust mission, for example, collected a sample of material from a comet and returned it to Earth for analysis in 2006. Comets are believed to contain substantial reservoirs of organic compounds, and they are black in color because of this. Analysis of organics in a comet may generate an "Aha!" experience, where compounds discovered clarify issues of origins.

Among the activities of the Cassini–Huygens mission, which reached Saturn in 2004, was an investigation of Titan, a moon of Saturn containing a large reservoir of organic matter, evidently undergoing chemical transformations on a large scale. Experiments from this mission may generate insight into the fate of organic molecules undergoing transformation on a planetary scale.

In addition to these space probes, information from Earth may be directly relevant to the theory that life is a natural consequence of organic reactivity. If structures in 3.5-billion-year-old rocks from Australia and carbon-containing material in 3.8-billion-year-old rocks from Greenland are indeed biogenic, such findings would indicate that life emerged on Earth "soon" (that is, within a few hundred million years) after the surface of Earth cooled (Schopf 1993). A rapid emergence of life is consistent with an easy emergence of life, which in turn suggests that one might see life happen spontaneously if one could only reproduce the conditions of early Earth. If, however, these structures and materials are not biogenic, then this argument is weakened (Brasier et al. 2002). Further studies are needed to address the ongoing dispute regarding the biogenicity of these structures.

Another outcome that may emerge from exploratory missions is a better understanding of the nature of early oceans on Earth. It is not yet clear whether the concentrations of salt in these oceans was high or low (Knauth 1998). The activity of water is greatly different at the extremes of salinity, and this has an impact on the constraints imposed on prebiotic chemistry by the nature of water.

Further progress is needed in understanding the potential for mineral catalysis in the transformation of organic molecules. Much more work remains to construct an inventory of the minerals present on the early Earth, and very little is known about what kinds of reactions minerals catalyze. It is plausible that a particular mineral might catalyze a useful transformation on the inventory of organics that were present. Greater knowledge about the history of the planet will lead to an improved understanding of this inventory (and its evolution over time). Current models for the accretion of continents speak of an evolving mineralogy, and it may be that life emerged only when the minerals arose that could catalyze key prebiotic reactions.

Perhaps the greatest insight into how life might have originated on Earth, however, might be gleaned from laboratory work attempting to generate artificial chemical systems capable of Darwinian evolution. Much progress has been made in this direction (Yang et al. 2010). Even when it is separated from the cosmochemistry and geology of the early Earth, these efforts will better define the basic science behind chemical systems that might be capable of supporting Darwinian evolution.

Resources

Books

Cairns-Smith, A.G. 1982. *Genetic Takeover and the Mineral Origins of Life*. Cambridge, UK: Cambridge University Press.

Cronin, John R., Sandra Pizzarello, and Dale P. Cruikshank. 1988. "Organic Matter in Carbonaceous Chondrites, Planetary Satellites, Asteroids, and Comets." In *Meteorites and the Early Solar System*, ed. John F. Kerridge and Mildred Shapley Matthews. Tucson: University of Arizona Press.

de Duve, Christian. 1998. "Clues from Present-Day Biology: The Thioester World." In *The Molecular Origins of Life: Assembling Pieces of the Puzzle*, ed. Andre' Brack. Cambridge, UK: Cambridge University Press.

Joyce, Gerald F. 1994. Foreword to *Origins of Life: The Central Concepts*, ed. David W. Deamer and Gail R. Fleischaker. Boston: Jones and Bartlett Publishers.

Joyce, Gerald F., and Leslie E. Orgel. 1999. "Prospects for Understanding the Origin of the RNA World." In *The RNA World: The Nature of Modern RNA Suggests a Prebiotic RNA*, 2nd edition, ed. Raymond F. Gesteland, Thomas R. Cech, and John F. Atkins. Cold Spring Harbor, NY: Cold Spring Harbor Laboratory Press.

Moore, Carleton B. 1971. "Phosphorus." In *Handbook of Elemental Abundances in Meteorites*, ed. Brian Mason. New York: Gordon and Breach.

Shapiro, Robert 1986. *Origins: A Skeptic's Guide to the Creation of Life on Earth*. New York: Summit Books.

Periodicals

Amari, Sachiko, Edward Anders, Alois Virag, and Ernst Zinner. 1990. "Interstellar Graphite in Meteorites." *Nature* 345(6272): 238–240.

Bernstein, Max P., Jason P. Dworkin, Scott A. Sandford, et al. 2002. "Racemic Amino Acids from the Ultraviolet Photolysis of Interstellar Ice Analogues." *Nature* 416(6879): 401–403.

Brasier, Martin D., Owen R. Green, Andrew P. Jephcoat, et al. 2002. "Questioning the Evidence for Earth's Oldest Fossils." *Nature* 416(6876): 76–81.

Cooper, George, Novelle Kimmich, Warren Belisle, et al. 2001. "Carbonaceous Meteorites as a Source of Sugar-Related Organic Compounds for the Early Earth." *Nature* 414(6866): 879–883.

Cronin, John R., and Carleton B. Moore. 1971. "Amino Acid Analyses of the Murchison, Murray, and Allende Carbonaceous Chondrites." *Science* 172(3990): 1327–1329.

Cronin, John R., and Sandra Pizzarello. 1986. "Amino Acids of the Murchison Meteorite. III. Seven Carbon Acyclic Primary Alpha-Amino Alkanoic Acids." *Geochimica et Cosmochimica Acta* 50(11): 2419–2427.

Ferris, James P., Aubrey R. Hill Jr., Rihe Liu, and Leslie E. Orgel. 1996. "Synthesis of Long Prebiotic Oligomers on Mineral Surfaces." *Nature* 381(6577): 59–61.

Fuchs, Louis H., Edward Olsen, and Kenneth J. Jensen. 1973. "Mineralogy, Mineral-Chemistry, and Composition of the Murchison (C2) Meteorite." *Smithsonian Contributions to the Earth Sciences* 10: 1–39.

Grady, Monica M., I.P. Wright, P.K. Swart, and C.T. Pillinger. 1988. "The Carbon and Oxygen Isotopic Composition of Meteoritic Carbonates." *Geochimica et Cosmochimica Acta* 52(12): 2855–2866.

Hanczyc, Martin M., Shelly M. Fujikawa, and Jack W. Szostak. 2003. "Experimental Models of Primitive Cellular Compartments: Encapsulation, Growth, and Division." *Science* 302(5645): 618–622.

Hollis J.M., F.J. Lovas, and P.R. Jewell. 2000. "Interstellar Glycolaldehyde: The First Sugar." *Astrophysical Journal* 540: L107–L110.

Hollis J.M., S.N. Vogel, L.E. Snyder, et al. 2001. "The Spatial Scale of Glycolaldehyde in the Galactic Center." *Astrophysical Journal* 554(1): L81–L85.

Jarosewich, Eugene. 1971. "Chemical Analysis of the Murchison Meteorite." *Meteoritics* 6(1): 49–52.

Kasting, J.F. 1993. "Earth's Early Atmosphere." *Science* 259(5097): 920–926.

Khare, N.B., Carl Sagan, Eric L. Bandurski, and Batholomew Nagy. 1978. "Ultraviolet-Photoproduced Organic Solids Synthesized under Simulated Jovian Conditions: Molecular Analysis." *Science* 199(4334): 1199–1201.

Knauth, L. Paul. 1998. "Salinity History of the Earth's Early Ocean." *Nature* 395(6702): 554–555.

Kvenvolden, Keith; James Lawless; Katherine Pering; et al. 1970. "Evidence for Extraterrestrial Amino-Acids and Hydrocarbons in the Murchison Meteorite." *Nature* 228(5275): 923–926.

Larralde, Rosa, Michael P. Robertson, and Stanley L. Miller. 1995. "Rates of Decomposition of Ribose and Other Sugars: Implications for Chemical Evolution." *Proceedings of the National Academy of Sciences of the United States of America* 92(18): 8158–8160.

Lewis, Roy S., Tang Ming, John F. Wacker, et al. 1987. "Interstellar Diamonds in Meteorites." *Nature* 326(6109): 160–162.

Miller, Stanley L. 1955. "Production of Some Organic Compounds under Possible Primitive Earth Conditions." *Journal of the American Chemical Society* 77(9): 2351–2361.

Müller, Daniel, Stefan Pitsch, Atsushi Kittaka, et al. 1990. "Chemie von a-Aminonitrilen" [Chemistry of alpha-amino-nitriles]. *Helvetica Chimica Acta* 73(5): 1410–1468.

Nielsen, Peter E. 1999. "Peptide Nucleic Acid: A Molecule with Two Identities." *Accounts of Chemical Research* 32(7): 624–630.

Oró, Joan. 1960. "Synthesis of Adenine from Ammonium Cyanide." *Biochemical and Biophysical Research Communications* 2(6): 407–412.

Pietrogrande, M.C., P. Coll, R. Sternberg, et al. 2001. "Analysis of Complex Mixtures Recovered from Space Missions: Statistical Approach to the Study of Titan Atmosphere Analogues (Tholins)." *Journal of Chromatography* A 939(1–2): 69–77.

Pizzarello, Sandra, and John R. Cronin. 2000. "Non-racemic Amino Acids in the Murray and Murchison Meteorites." *Geochimica et Cosmochimica Acta* 64(2): 329–338.

Pizzarello, Sandra, Yongsong Huang, Luann Becker, et al. 2001. "The Organic Content of the Tagish Lake Meteorite." *Science* 293(5538): 2236–2239.

Sagan, Carl, and N.B. Khare. 1979. "Tholins: Organic Chemistry of Interstellar Grains and Gas." *Nature* 277(5692): 102–107.

Sanchez, Robert A., James P. Ferris, and Leslie E. Orgel. 1967. "Studies in Prebiotic Synthesis. II. Synthesis of Purine Precursors and Amino Acids from Aqueous Hydrogen Cyanide." *Journal of Molecular Biology* 30(2): 223–253.

Schöning, K.-U., P. Scholz, S. Guntha, et al. 2000. "Chemical Etiology of Nucleic Acid Structure: The α-Threofuranosyl-(3'→2') Oligonucleotide System." *Science* 290(5495): 1347–1351.

Schopf, J. William. 1993. "Microfossils of the Early Archean Apex Chert: New Evidence of the Antiquity of Life." *Science* 260(5108): 640–646.

Shapiro, Robert. 1988. "Prebiotic Ribose Synthesis: A Critical Analysis." *Origins of Life and Evolution of Biospheres* 18(1–2): 71–85.

Tang, Ming, Edward Anders, Peter Hoppe, and Ernst Zinner. 1989. "Meteoritic Silicon Carbide and Its Stellar Sources: Implications for Galactic Chemical Evolution." *Nature* 339 (6223): 351–354.

Yang, Zunyi, Fei Chen, Stephen G. Chamberlin, and Steven A. Benner. 2010. "Expanded Genetic Alphabets in the Polymerase Chain Reaction." *Angewandte Chemie* 122(1): 181–184.

Zhai, Mingzhe, and Denis M. Shaw. 1994. "Boron Cosmochemistry," Pt. I: "Boron in Meteorites." *Meteoritics* 29(5): 607–615.

Other

Pizzarello, Sandra. 2001. "Soluble Organics in the Tagish Lake Meteorite: A Preliminary Assessment." Paper presented at the annual Lunar and Planetary Science Conference, Houston, Texas. Available from http://www.lpi.usra.edu/meetings/lpsc2001/pdf/1886.pdf.

SoundVision Productions. 1998. "DNA and Evolution: Where Did We Come From? Where Did We Go?" *The DNA Files*. National Public Radio. Transcript available from http://www.dnafiles.org/node/551.

Steven A. Benner

• • • • •

Speciation: The origins of diversity

Introduction

All creatures, past and present, either have gone or will go extinct. Yet, as each species vanished over the past 3.8-billion-year history of life on Earth, new ones inevitably appeared to replace them or to exploit newly emerging resources. From only a few very simple organisms, a great number of complex, multicellular forms evolved over this immense period. The origin of new species, which the nineteenth-century English naturalist Charles Darwin once referred to as "the mystery of mysteries," is the natural process of speciation responsible for generating this remarkable diversity of living creatures with whom humans share the planet. Although taxonomists presently recognize some 1.5 million living species, the actual number is possibly closer to 10 million. Recognizing the biological status of this multitude requires a clear understanding of what constitutes a species, which is no easy task given that evolutionary biologists have yet to agree on a universally acceptable definition.

Darwin (1859) himself viewed speciation as a continuous process whereby varieties, or races as scientists now call them, are transformed over time by natural selection into distinct species during the course of adapting to an ever-changing environment. Because the way new species arise in nature differs from group to group, Darwin emphasized that it is unnecessary, indeed likely impossible, to define a species. He was well aware that the variability and diversity of the natural selective forces acting on diverging populations make it impossible to determine whether sister races have crossed that elusive tipping point when both have irrevocably established lineages sufficiently different to ensure independent futures.

In some respects, the scientists of the early twenty-first century are no closer to resolving this "species problem" than those of Darwin's time. Although various species concepts, including ones based on morphology, behavior, genetic differences, and ecology, have been favored by one group or another, in recent decades many have adopted Ernst Mayr's *biological species concept*, which states that species are "groups of actually or potentially interbreeding natural populations... reproductively isolated from other such groups" (Mayr 1942, p. 120). Thus, two groups are distinct species if they are incapable of exchanging genes or forming viable hybrids. The biological species concept, however, applies only to *sympatric* populations, that is, those that share the same locations and habitats. This concept is obviously difficult to apply to

geographically isolated or *allopatric* populations. Unless brought together under natural conditions, such populations cannot pass the test of sympatry, and experimental crosses made under artificial conditions are often ambiguous. Obviously the biological species concept cannot be applied to either fossils or asexually reproducing forms.

In practice, biologists often encounter coexisting natural situations in which sympatric *sister species*, which share an immediate common ancestor, may interbreed to varying degrees yet still maintain their morphological and genetic distinctiveness indefinitely. Several Galápagos finches, for example, interbreed frequently, yet they are still recognized by ornithologists as distinct species (Grant and Grant 2008). To recognize species in such cases or in situations in which populations are allopatric requires the subjective application of the reproductive isolation rule. Not surprisingly, the same populations may be arbitrarily designated races by one investigator, and species by another, thus reinforcing the subjectivity of the classification process.

Biologists, therefore, either explicitly or implicitly employ a more flexible definition that recognizes any group or connected set of related groups a species if they have evolved a recognizable lineage independent of other sister groups (de Queiroz 2005). For example, if two sympatric sister taxa interbreed and exchange genetic information, yet maintain their phenotypic and genetic identity, they are "good species." Furthermore, they represent important units of biodiversity. They must not be ignored simply because they fail to meet the arbitrary criteria of complete reproductive isolation prescribed by the biological species concept.

Natural selection and speciation

Darwin's central theme emphasized the central role of natural selection in the process of species formation as populations adapt to different environmental conditions. Dolph Schluter (2009) and others have emphasized that species formation by natural selection usually occurs in one of two ways: ecological speciation or mutation-order speciation.

In ecological speciation, selection acts on ecologically relevant traits that favor different alleles in different environments, causing populations to diverge as they adapt to different ecological conditions. Alleles advantageous in one

The Origin of Species

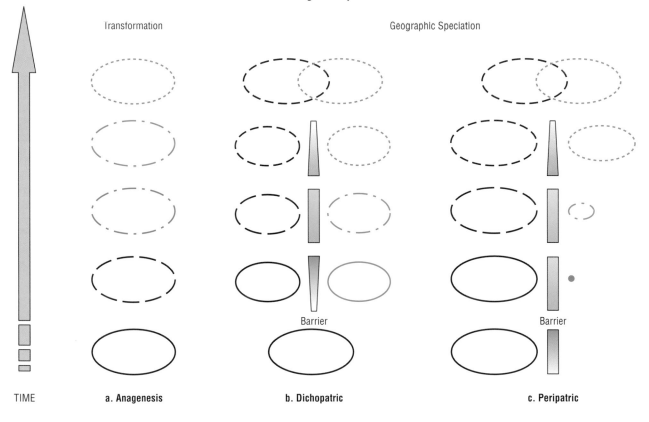

Figure 1. Anagenesis results in the transformation of a species over time into another species with no increase in number. Dichopatric and peripatric speciation, alternative forms of geographic species formation, result in an increase in the number of species. Reproduced by permission of Gale, a part of Cengage Learning.

environment are disadvantageous in another, which results in fixation of contrasting alleles and the emergence of distinct lineages. Genomes of these separate lineages are sufficiently reproductively independent of one another to ensure each continued autonomous evolution even in the face of continued gene flow during the early phase of the speciation process. Ecological speciation predicts that independent lineages evolve between populations adapting to different environments but not between populations adapting to similar environments (Schluter 2009). For example, a cichlid fish may establish a population that feeds on a new resource in response to divergent selection pressures experienced in the new habitat; a new lineage may evolve independent of the one still adapted to the original habitat.

Mutation-order speciation, in contrast, involves species formation (the development of separate lineages) by the *chance* occurrence and fixation of different alleles between isolated populations experiencing and adapting to separate but *similar* environments (Schluter 2009). Which gene and the order in which it is mutated will differ by chance in each population. Under these conditions, divergence in mate recognition and reduction in actual or potential gene flow between diverging populations evolve as each accumulates and fixes *different* mutations that are advantageous to each in *both* environments.

Independent lineages in a cichlid fish under these conditions would evolve when isolated populations inhabiting similar habitats accumulate different advantageous mutations over time sufficient to have evolved separate mate recognition systems. When such populations reestablish contact, they would compete directly for similar resources. The outcome would be either extinction of one species, fusion, or elimination of competition by subdivision of habitat preference. The relative importance of these two general categories of species formation is unknown (Schluter 2009).

Species can arise in other ways that do not involve either ecological or mutation-order natural selection, such as by hybridization and by polyploidy of the genome. These will be discussed later. Except in the case of hybridization and polyploidy, speciation is a continuous process of genetic divergence.

How new species evolve

Speciation in nature has three outcomes. *Anagenesis* results in no change in species number, *cladogenesis* brings about an increase in the number of species, and the *fusion* of two species decreases species number.

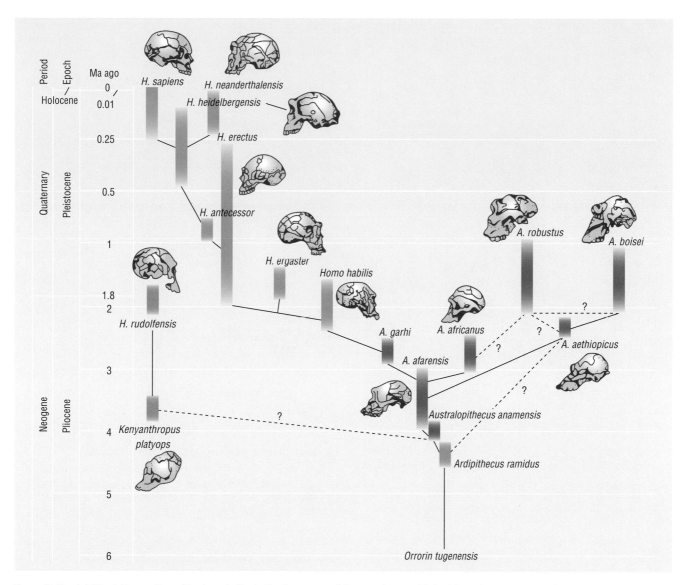

Figure 2. Hominid Evolutionary Tree. The bars indicate the time span of the species established from fossil evidence. Solid lines connect likely sister species. Dotted lines indicate relationship is provisional. From: LearningSpace. The Open University, UK. Reproduced by permission of Gale, a part of Cengage Learning.

Anagenesis (see Figure 1), sometimes called "phyletic gradualism" by paleontologists, happens when a species slowly transforms over time into another species without increasing the number of species. Paleontologists and paleoanthropologists, for example, sometimes encounter a well-preserved fossil series that spans millions of years. A single ancestral species at the beginning of such a series may appear to slowly transform morphologically into distinct descendant species one or more times before the series ends. Even with well-preserved material, however, it is often difficult to determine whether the observed morphological changes actually involved transformation of a single species into another or whether observed changes include several episodes of ecological speciation and lineage splitting followed by the eventual extinction of one species.

An example of anagenesis cited by some paleoanthropologists is the apparent transformation of archaic *Homo erectus*, which first appeared about 2 million years ago, into modern *H. sapiens*. In the process, cranium size doubled (Bräuer 2008). Nevertheless, others recognize several forms morphologically intermediate between these two taxa as species (see Figure 2). Thus, do *H. antecessor*, *H. heidelbergensis*, and *H. neanderthalensis* actually represent only stages in a single transformational lineage from *H. erectus* to *H. sapiens* or, as a 2008 study from Ian Tattersall and Jeffrey H. Schwartz suggests, are they the products of independently evolved lineages that coexisted at various times with each other and with *H. erectus* before *H. erectus* went extinct? In her 2009 report on sequencing the *H. neanderthalensis* genome, for instance, Elizabeth Pennisi indicated that *H. neanderthalensis* and *H. sapiens* diverged some 500,000 years ago. These two thus represent independent, non-interbreeding sister species that coexisted, sometimes in close proximity, until about 30,000 years ago. How important anagenesis actually has been in evolution of hominids and

Allopatric Speciation

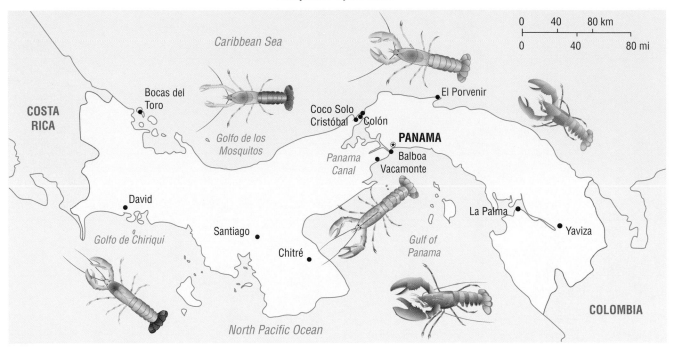

Figure 3. Dichopatric speciation in Alpheus snapping shrimp across the Isthmus of Panama. Reproduced by permission of Gale, a part of Cengage Learning.

other animals and plants is unclear and controversial as it is usually impossible to unequivocally reconstruct events from the distant past.

Cladogenesis (see Figure 1), or "true speciation," refers to the splitting of a lineage that results in multiplication of species. This process is responsible for most of the world's biodiversity. Paleontologists sometimes link cladogenesis to the theory of "punctuated equilibrium," which asserts that a long period of morphological stasis follows the sudden appearance of a new, morphologically distinct fossil species. Most morphological change, they believe, takes place during or shortly after speciation. Yet, morphological change may or may not accompany cladogenesis. Comparison of genomes between morphologically similar species in several groups of organisms reveals that little morphological change may accompany several rounds of speciation (Bickford, Lohman, Sodhi, et al. 2006). Such speciation events would be invisible in the fossil record. Whether punctuated equilibrium is a real phenomenon is still debated and its use challenged.

The fusion of two taxa previously recognized as species into a single interbreeding population occasionally takes place in nature. When mate recognition is contingent primarily on habitat preference, an ecological disturbance may result in the breakdown of reproductive isolation and local or widespread interspecific hybridization. For example, an ecologically differentiated pair of Darwin's Galápagos finches derived from a common ancestor, *Geospiza fortis* and *G. scandens*, began hybridizing and converging on Daphne Major island in response to a shift in ecological conditions (Grant and Grant 2008). Various other examples of species fusion are detailed in Mayr (1963).

Genetic drift versus natural selection

For much of the twentieth century, most evolutionary biologists agreed that when a geographic barrier subdivides a population, the gene pools of the separated groups slowly drift apart by the chance fixation of different mutations. In their view, cladogenesis was complete when the genomes of these isolated sister populations became sufficiently incompatible such that they could no longer successfully interbreed if contact was reestablished. Many now believe divergent natural selection rather than random genetic drift is the major cause of population divergence. Whether or not biogeography plays a necessary role in the evolution of species formation depends on biological factors intrinsic to the organism.

Biogeography and species formation

While natural selection is the main engine that powers the speciation process, biogeography plays a major role in the way diverging populations cope with gene flow and natural selection during speciation. Populations adapting to different ecological conditions must somehow limit the homogenizing effects of gene exchange to allow the emergence of independent lineages. How important spatial isolation is in this process depends on the organism's inherent biological characteristics. In some animals and plants, complete geographic isolation is required before divergence can proceed and *allopatric speciation* is completed, as shown diagrammatically in Figure 1. In others, two sister populations inhabiting adjacent but different habitats may be in contact and overlap along a common border yet diverge and eventually undergo *parapatric speciation* (see Figure 3). In extreme cases, *sympatric speciation* may occur in the absence of geographic isolation (Figure 3) when sister populations adapt to different habitats or resources in response to divergent natural selection.

Awareness of the geographical context influencing speciation provides an important framework for understanding speciation. Differences in inherent biological traits between different groups of organisms, however, may also favor one mode of speciation over another. Allopatric and sympatric modes of speciation discussed here actually represent extremes in a continuum of geographical isolation that may seldom be absolute in nature. Modeling and measuring quantities, such as gene flow and selection, rather than consigning diverging populations to geographically based allopatric or sympatric categories, can provide a valuable perspective for evaluating the contribution of these biological traits to the speciation process.

Geographic speciation

Species may evolve in two ways under conditions of *geographic isolation*. The most widely accepted and least controversial is *dichopatric speciation* (see Figure 1), which comes about when a geographic barrier separates a relatively large and widespread species into isolated populations. Examples of such physical barriers include mountain ranges, large bodies of water, and uninhabitable or uncrossable terrain. Once isolated from one another, sister populations are free to track and adapt to local ecological changes in their respective environments. Because mutations and the order in which they appear will inevitably differ in the two populations, over time each population will accumulate an increasing number of different adaptive genetic changes. How fast they diverge depends mostly on how much their ecological conditions differ or change and the intensity of natural selection. Even if the isolated regions remain essentially the same, each population will eventually accumulate adaptive genetic differences, though at a much slower rate, and eventually acquire sufficient differences in their genomes to ensure that each has established a distinct and independent lineage.

Populations separated for known lengths of time by *vicariant events*, those that create geographic barriers separating sister groups, provide convincing examples of dichopatric speciation in nature. One of the best examples is that of seven pairs of sister species of *Alpheus* snapping shrimp (see Figure 3) that diverged after becoming isolated on either side of the Isthmus of Panama (Knowlton et al. 1993). Hybridization tests under laboratory conditions found that only 1 percent of matings between trans-Isthmus sister species yielded fertile clutches. Although the connection between the Caribbean Sea and Pacific Ocean closed completely about 3 million years ago, geological and molecular-based dating indicates that some sister pairs of shrimp that inhabit lower depths diverged up to 10 million years before the final closing. Other clear examples of dichopatric speciation may be found in Jerry A. Coyne and H. Allen Orr's 2004 book, *Speciation*.

A number of other examples of dichopatric speciation rely on varying degrees of circumstantial evidence where dates of isolation are uncertain or the degree of reproductive isolation is

unknown. Some credible examples involve speciation after colonization of distant islands, although few have actually been subjected to rigorous investigation. The real test of species status comes when and if such isolated populations reestablish contact either naturally or by manipulation. If crosses are sterile, or produce sterile or inviable offspring, the populations are obviously distinct species. If populations maintain independent lineages once they become sympatric, even in the face of some gene flow, they are considered species. In a few cases in which the populations remain isolated, however, crosses under laboratory conditions may be attempted. To be reliable, such crosses must be made under conditions that accurately simulate the natural environment suitable for normal mate recognition. This is because with many organisms such a test may circumvent important mate recognition (physical or behavioral) cues that would inhibit interspecific mating.

More controversial is *peripatric speciation* (see Figure 1), sometimes referred to as "founder event" speciation and first proposed by Ernst Mayr (1963). It occurs when one or only a few individuals colonize a new region or distant new habitat. The most likely occurrences involve colonization of a distant island, lake, or habitat beyond the periphery of a wide-ranging species. Colonizers under these conditions may encounter radically different environmental conditions to which they must rapidly adapt. As their initial population size is very small, possibly only a single fertilized female, Mayr assumed that a newly established colony would be subject to a combination of strong genetic drift and divergent selection, resulting in rapid genetic reorganization of the genome—a "genetic revolution" and the emergence of a new species. Experimental and theoretical studies, however, have shown that such founder-effect genetic revolutions are actually unlikely to be important in colonization unless populations remain very small for several generations and meet other unlikely restrictions. Yet, several good examples exist of peripatric speciation initiated by founder events involving a few colonizing individuals (Templeton 2008). Whether the founder event or the exploitation of a new resource free of competitors and predators is responsible for an increase in the rate of speciation and adaptive radiation in these cases is yet unclear.

A widely recognized example involving the colonization of an island by what is assumed to be one fertilized female or a few individuals is the Hawaiian picture-winged *Drosophila*. They provide a classic example of peripatric speciation and adaptive radiation (see Figure 4). At least 400 and possibly as many as 800 recognized species of *Drosophila* live on the Hawaiian Islands, all of which have descended from a single common ancestor. The members of the large "picture-winged group" are particularly well studied by Hampton L. Carson and his colleagues (see Coyne and Orr [2007] and Templeton [2008] for overview). These researchers used the banding pattern observed in their giant polytene chromosomes to reconstruct the order, estimate the number of founder events, and confirm species status of 97 of the currently recognized 101 species. Only two species occur on more than one island, while the rest live as single-island endemics as a result of colonization events and speciation within each island. Migration has been predominantly from the oldest island in the northeast to the youngest island in the southwest. Of the ninety-seven pictured winged species distributed throughout the islands, forty-five interisland founder events are inferred. The remaining fifty-two species diverged within islands, with many of these speciation events involving shifts in habitats and use of new host-plant resources (Kambysellis et al. 1995).

Yet, evidence of the increased homozygosity (having two identical alleles of the same gene) expected if these species had experienced a founder *effect* (i.e., a genetic bottleneck) is not apparent. Apparently either each interisland colonization event involved several individuals or the colonization of a new habitat or resource was followed by a rapid expansion in population size with little or no loss of genetic variation. In the case of ecological speciation events within an island, gene flow probably continued for some time after a new lineage was established, or newly isolated populations were relatively large.

Making species the nonallopatric way

More controversial is the proposal by Darwin and others later on that animal and plant species can split into two independent lineages without spending extended periods geographically isolated from one another. Most evolutionary biologists in the latter half of the twentieth century argued that gene flow between lineages would continuously disrupt the evolution of divergent adaptive gene complexes as they form. Research on natural populations, theoretical studies, and laboratory experiments over the past twenty years has shown that under certain conditions divergent natural selection can lead to a lineage splitting into two genetically independent distinct sister species. This can occur in sexually reproducing organisms in four major ways: parapatric speciation, sympatric speciation, homoploid hybrid speciation, and polyploid hybrid speciation. The first two of these, parapatric and sympatric speciation, share some biological attributes but differ primarily in their pattern of geographic isolation. The second two involve hybridization between related species or races (Mallet 2007).

Parapatric speciation

Two sister populations are parapatric when they occupy adjoining habitats and their distributions overlap along a narrow ecotone, or transitional zone between two habitats. Parapatric speciation (see Figure 3) may originate in its simplest form when a species encounters a new habitat at the margin of its normal range. Certain individuals at the periphery, however, may be sufficiently genetically preadapted to colonize the new ecotone that they eventually spread into and well beyond the margin of the new habitat. Evidence from population genetics shows that over time the level of adaptive divergence of the predominantly isolated population beyond the ecotone reaches a point at which reinforcement, a process by which divergent natural selection enhances reproductive isolation, begins to operate within the overlap zone. As populations continue to adapt to different habitats outside the contact zone and thereby diverge, selection intensifies against hybrids formed within the contact zone. Such hybrids display reduced fitness in either parental habitat and are at a competitive disadvantage for mates and resources.

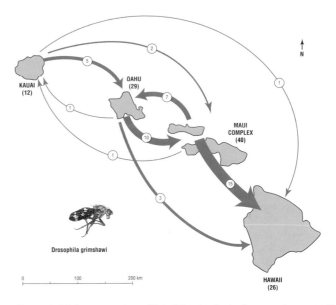

Drosophila grimshawi

Figure 4. Minimum number of interisland colonization events accounting for the distribution of 97 Hawaiian picture-winged *Drosophila*. Arrows indicate direction of colonization. Thickness of arrows is proportional to the number of colonization events. Numbers in parentheses include all species inhabiting an island (Carson 1983). Reproduced by permission of Gale, a part of Cengage Learning.

As a result, continued divergent natural selection will favor increased assortative mating among individuals sharing similar ecologies and mate preferences, and a gradual reduction and eventual elimination of gene flow between the diverging independent lineages will occur. Care must be taken to determine whether the hybrid zones between parapatric populations represent *primary hybrid zones*, which result from populations that remain and diverge while in constant contact, or *secondary hybrid zones*, which develop after two populations reestablish contact following periods of complete isolation. This is often difficult or impossible to establish, particularly in animals and insects because events of the past cannot usually be accurately determined.

Good examples of parapatric species evolution involve plants that have diverged across recently established ecotones. Speciation has occurred rapidly in some cases in which an endemic specialized edaphic plant adapted to growing on a toxic soil has its parent sister species growing in an adjacent area on normal soil. For example, the edaphic monkey flower *Mimulus cupriphilis* grows only on the toxic mine tailings of an old abandoned copper mine in California (Macnair and Gardner 1998). Most other surrounding plant species, including its sister species *M. guttatus*, cannot grow on copper-laced tailing soils. The two species differ in a number of ways. *M. cupriphilis*, an obligate annual, flowers earlier and produces more and smaller flowers that differ in shape and color from those of *M. guttatus*. On dry copper mine areas, *M. cupriphilis* is fitter than *M. guttatus* as it flowers more efficiently and produces more seed under xeric conditions. The species are also reproductively isolated. The early flowering *M. cupriphilis* is self-fertilizing and thus not visited by pollinators, which are rare at that time of year. *Mimulus guttatus*, in contrast, blooms later and is an outcrosser that is fertilized by

Common monkey flowers (*Mimulus guttatus*) grow only on the toxic mine tailings of an old abandoned copper mine in California. Maurice Nimmo/Photo Researchers, Inc.

bumblebees, which promotes mating between unrelated individuals.

At the genetic level, the species differ in genes controlling flowering time, flower size, corolla spot number, and general size. Two major genes are homozygous for alleles that govern petal shape and stem color, and, with the exception of flowering time, all the genetic systems are recessive. In the original outbreeding parent species, recessive alleles would be spread by natural selection. In inbreeders such as *M. cupriphilis*, heterozygotes are infrequent and recessive genes can evolve at effectively the same rate as dominant genes and spread once inbreeding is established.

Although events leading to the invasion and adaptation to mine tailings are unknown, the following steps were probably required. A primary adaptation, probably involving a dominant mutation, shifted flowering time earlier in response to the more xeric habitat. Because pollinators would be rare, cross-pollination is reduced and selection would favor alleles for self-fertilization. This was apparently accomplished by reducing flower size, which brought the stigma and anthers into close proximity. Corolla size and associated structures were reduced, which freed up resources for seed production.

These changes resulted in the establishment of a new species that has evolved in a very short period of time. No evidence of other major genetic changes suggests that gene flow occurred during species formation. In this case, new selection pressures on normal *M. guttatus* achieved speciation as it colonized an unusual habitat.

Genome mapping of floral traits responsible for reproductive isolation studied in other *Mimulus* species supports this view (Schemske and Bradshaw 1999). Other examples of parapatric speciation may be found in Coyne and Orr's 2004 book, *Speciation*.

Sympatric speciation

Populations are sympatric when they occur within the same region, and gene flow between them is obstructed as a result of the biology of the organisms rather than because of physical barriers that prevent their meeting (Bush and Butlin 2004). When new species evolve in the absence of geographic isolation the process is called sympatric speciation (see Figure 3) Although sympatric speciation is theoretically possible, it is the most controversial form of speciation and some evolutionists believe that it is extremely rare. Coyne (2007) contends

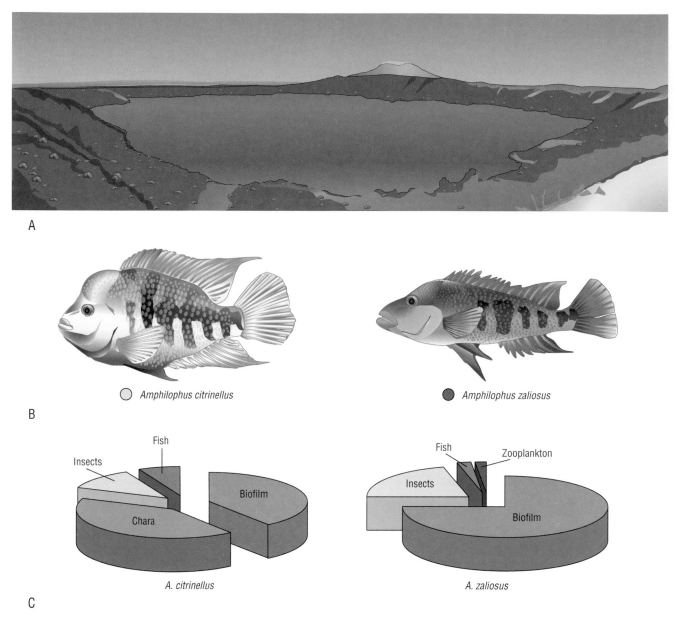

A

B

Amphilophus citrinellus

Amphilophus zaliosus

C

Figure 5. A: Lake Apoyo, Nicaragua. B: The Midas cichlid (*Amphilophus citrinellus*) and the Arrow cichlid (*A. zaliosus*) are morphologically distinct. C: Stomach content analyses reveal clear-cut differences between the two Lake Apoyo species. Both species forage on biofilm, but whereas winged insects were found in all individuals of *A. zaliosus*, insects were recovered in only about half of the *A. citrinellus* stomachs, which feed on the algae *Chara* and other plant material that is found close to the lake's shore. The feeding niche of *A. citrinellus* is wider than that of *A. zaliosus*. From Barluenga, et al. (2006). Reproduced by permission of Gale, a part of Cengage Learning.

that four conditions must be met before a compelling case can be made that a pair of species has diverged only under sympatric conditions. First, the species must be sympatric; second, they must show substantial reproductive isolation; third, the sympatric taxa must be sister species; and fourth, existence of an allopatric phase must be *very unlikely*. These criteria are so stringent, few examples meet all four conditions.

Many recognized examples of sympatric speciation meet the first three of the four conditions, but lack sufficient evidence that sister species have never been geographically isolated sometime in the past. In fact, the geographical history of most examples of speciation—including birds, fish, and many

parasitic and phytophagous (plant-feeding) insects and other invertebrates—is unlikely to ever be known. With the exception of instances of vicariant speciation, the same is true for many purported cases of allopatric speciation in which parapatric or sympatric speciation cannot be excluded because the past history of their origin and distribution cannot be known or has not been studied. Nevertheless, many evolutionary biologists believe that sympatric speciation is common in certain groups of organisms whose biological properties predispose them to speciate sympatrically but rare in others.

Two convincing examples of sympatric speciation that meet all four criteria involve cichlid fish in a small Nicaraguan crater

Lord Howe Island and its palms (*Howea forsteriana*) in Australia, New South Wales. Tom Till/Photographer's Choice/Getty Images.

lake and palm trees on isolated Lord Howe Island in the South Pacific Ocean. Marta Barluenga and colleagues (2006) studied two members of the Midas cichlid species complex (*Amphilophus* sp.) living in Lake Apoyo, a small, 23,000-year-old, volcanic crater lake in Nicaragua (see Figure 5). One species, the widespread benthic (bottom-dwelling) Midas cichlid, *A. citrinellus*, was possibly blown into the lake during a hurricane after the volcano went extinct and filled with rainwater. As a bottom dweller, this fish uses its powerful pharyngeal jaws to feed on algae along the lake margin. A second endemic species, the elongated arrow cichlid, *A. zaliosus*, is limnetic, dwelling in open water where it feeds on insects and other shallow-water organisms. Detailed phylogenetic, population genetic (mitochondrial and nuclear DNA), morphometric, and ecological analyses showed that Lake Apoyo had clearly been colonized only once by the Midas cichlid. In less than 10,000 years, possibly as recently as 2,000 years, the arrow cichlid diverged from the Midas cichlid and in the process became more elongated and slender as it became more and more efficient at feeding on resources available in the open water of the lake.

The two *Amphilophus* species in Lake Apoyo are now clearly two morphologically distinct sister species that have speciated sympatrically as a result of divergent habitat preference, resource partitioning, and assortative mating. Ecological divergent selection has played an important role in this speciation process. This example is similar to sympatric speciation reported in other tilapine cichlid species flocks found in crater lakes in Cameroon in Africa (Schliewen, Tautz, and Pääbo 1994) and in stickleback fish in postglacial lakes of British Columbia in Canada (Schluter 2009).

Two endemic wind-pollinated sister species of palms occur sympatrically on Lord Howe Island (Savolainen et al. 2006). This small, subtropical island of less than 12 square kilometers (4.6 square miles) is located about 580 kilometers (360 miles) from the east coast of Australia. It was formed by volcanic activity around 6.4 to 6.9 million years ago. The Kentia palm *Howea forsteriana*) grows on sandy sedimentary soils and reaches more than 15 meters (50 feet), whereas the curly palm (*H. belmoreana*), which prefers soils formed from volcanic rock,

Apple maggot fly (*Rhagoletis pomonella*) perched on a leaf. © Bill Beatty/ Visuals Unlimited, Inc.

partners has been random, a characteristic that distinguishes sympatric speciation from other modes of speciation.

The organisms possibly most widely cited as candidates for sympatric speciation are parasites that feed and often mate on plants and animals. These include many insects as well as mites and nematodes and others that inhabit diverse terrestrial, soil, and aquatic habitats. They not only represent a large percentage of living multicellular organisms but also have biological attributes that predispose them to sympatric ecological speciation (Drès and Mallet 2002). Many meet and mate only on their hosts and are known to rapidly form new genetically distinct populations on a new host when it becomes available. Their biology is often tightly synchronous with that of their host. Colonization of a new host is clearly more likely to occur when old and new hosts are within the dispersal range of the parasite. This provides the parasite an extended opportunity for individuals genetically preadapted to establish a population on the new host.

One such example extensively studied since the 1970s is the origin of a new host race or possibly species in the tephritid fruit fly genus *Rhagoletis*. In early 1860s larvae of a fly morphologically indistinguishable from the apple maggot fly, *R. pomonella* (see Figure 8), whose native host are fruits of several hawthorn species (*Crataegus* spp.), were discovered infesting apples (*Malus pumila*) in an orchard along the Hudson River in New York. The Dutch had introduced apples when they colonized the island of Manhattan in the early seventeenth century and planted many orchards in close proximity to *Crataegus*. Once established, the apple-infesting race spread rapidly and within thirty to forty years infested apples throughout eastern North America. No opportunity for the apple-infesting race to remain isolated from hawthorn populations existed for any length of time.

Although the two races continue to hybridize at a low level, they maintain significant genetic differences in host preference, fruit-odor perception, habitat-specific mating preference, host-associated fitness trade-offs for developmental rates and larval survival on their respective host plants, and time of emergence of adult from the pupal case. Races emerge from diapause (dormancy) in summer on average three weeks apart in response to different pre-winter periods in the soil as pupae.

A key factor that facilitates sympatric speciation in plant and animal parasites is that many use their host as a rendezvous for mating (Bush and Butlin 2004). In *Rhagoletis*, host choice is determined primarily by host odor, which is under genetic control, and thus males and females of the same *Rhagoletis* host race seek out one another and mate on their respective host. Host shifts and host-race formation in other insects are observed most frequently following introduction of a new host plant free of competing insect specialists. The stages of host-race formation and speciation are presented below.

Stages of host-race formation

1. A new host relatively free of competitors becomes available for colonization because of changes in the host or insect range, or because of changes in host ecology or physiology.

grows to only 6 meters (20 feet). Using two different molecular dating methods, Vincent Savolainen and colleagues estimated that *Howea* and *Laccospadix*, *Howea*'s Australian monotypic sister genus, diverged about 4.57 to 5.53 million years ago, while the two endemic *Howea* species split about 1 to 1.92 million years ago, dates that are younger than the estimated age of the island.

A study of flowering times of the two sister species revealed that peak flowering time for *H. forsteriana* is six weeks before *H. belmoreana* and that flowering time differences appear to be influenced by substrate-induced physiological changes. Soon after erosion exposed calcareous soils about 2 million years ago, the flowering times of early colonizers were presumably shifted in response to induced physiological changes. The resulting reduced gene flow between those early colonizers and *H. belmoreana*, already long ensconced on volcanic soils, allowed divergent ecological selection to initiate sympatric divergence and eventually speciation. Given that both sister species are wind pollinated, that Lord Howe Island is extremely isolated and very small, and that the two soil types are interdigitated like fingers of two clasped hands, the two sister species are unlikely to have ever been geographically isolated. Mating with respect to place of birth of mating

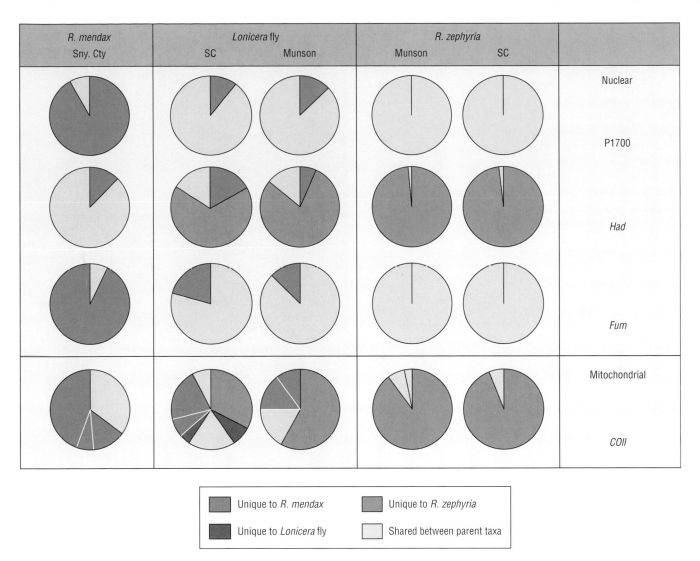

R. mendax	Lonicera fly		R. zephyria			
Sny. Cty	SC	Munson	Munson	SC		

Figure 6. Allele frequencies at hybrid diagnostic nuclear and mitochondrial loci in Lonicera fly and parental populations from central Pennsylvania. (From Schwarz, et al., 2005) Reproduced by permission of Gale, a part of Cengage Learning.

2. Mutation and/or recombination generate individuals able to utilize the new host from within the parental species (if appropriate genotypes are not already available).

3. A number of males and females that exhibit a genetically based preference for and/or ability to survive on the new host colonize the new host over successive generations and establish a population. Low efficiency of host utilization may be compensated by low competition.

4. Assortative mating occurs among the earliest colonists of the new host (mating occurs on the preferred host). This reduces gene flow between the original and new host-associated populations and allows adaptive gene combinations to be maintained in each population.

5. During the course of adaptation to the new host, a genetically distinct host race evolves as host-associated differences in fitness and host fidelity increase over time between the host-associated populations.

6. For loci not involved in host-associated adaptations, genetic similarity between the original species and the new host race is maintained by continued low levels of gene flow, particularly during the early stages of host-race formation.

7. Speciation is completed by reinforcement or because continued divergence incidentally reduces gene exchange to negligible levels(Bush and Butlin 2004).

Do the *Rhagoletis* populations on apple and hawthorn represent races or species? There seems to be no simple way to draw a clear distinction between the two categories. Host races are generally viewed as populations that specialize on alternative hosts that differ genetically from one another in host preference and host fidelity but still exchange genes (Berlocher and Feder 2002). Some low level of hybridization may still occur between the host races even though they maintain distinct genetic differences. In these phytophagous parasites, sympatric speciation is a progressive process that occurs over a protracted period depending on the intensity of divergent natural selection and the degree of host preference and assortative mating. Until these populations become completely reproductively isolated, it may be impossible to establish when they pass some threshold of isolation beyond which they will never reunite into one species again.

Hybridization

In the absence of geographic isolation, new species can arise rapidly in two other ways as a result of hybridization between closely related species. In *recombinational* or *homoploid hybrid speciation*, two related species give rise to a third species by hybridization without a change in chromosome number. *Polyploid hybrid speciation* (also called allopolyploid speciation), by contrast, results in a species that combines a complete set of chromosomes from each parental species. The level of genetic differentiation and divergence in chromosomal rearrangements between the hybridizing parental species influences the pattern of divergence of the hybrid species. Hybridization between genetically distinctive species may provide recombinant offspring that can shift to and exploit a new habitat or resource not used by either parental species.

Homoploid hybrid speciation   Homoploid hybrid speciation was at one time considered not to occur or be extremely rare in animals. Examples in certain insects suggest, however, that it may be far more common than previously thought. Dietmar Schwarz and colleagues' 2005 study discovered that offspring resulting from interspecific mating of two tephritid host-specific fruit fly species, the snowberry maggot, *Rhagoletis zephyria*, and the blueberry maggot, *R. mendax*, were a new species that infests berries of a nonnative, brushy honeysuckle, *Lonicera* spp. These flies are almost indistinguishable from the apple maggot fly. The newly infested host plants represent a mixture of *L. morrowii* and several described hybrids and recombinational forms introduced from Asia over the past 250 years. Although all three species of *Rhagoletis* are interfertile (capable of interbreeding), they remain genetically distinct in complete sympatry (see Figure 6). No naturally occurring hybrid offspring have been found.

Because *Rhagoletis* species mate on their respective host fruits, the hybrid *Lonicera* fly species has escaped major competition from its putative parental species, which use very different hosts to locate mates. In laboratory experiments, Schwarz and colleagues (2005) demonstrated that although parental species discriminate against each other's native host fruit, both accept honeysuckle fruit, suggesting that snowberry and blueberry flies probably met, hybridized, and oviposited on the introduced nonnative honeysuckle fruit. The invasive *Lonicera* spp., introduced by horticulturists as ornamentals, served to break down the ecological reproductive isolation barrier between the native parent species. Although this form of homoploid speciation involving colonization of an introduced plant by a resident parasitic insect or mite is not often studied, such animals represent a large percentage of all organisms. Recent genetic research indicates that homoploid speciation is more common in such organisms than previously supposed (Mallet 2007).

Blue flag iris (*Iris hexagona*) in the Audubon Corkscrew Swamp Wildlife Sanctuary, Florida. © Fritz Polking/PhotoLibrary.

Another form of recombinational hybrid speciation that is mostly limited to plants occurs when, in the progeny of a chromosomally sterile or semisterile species hybrid, a new, structurally homozygous recombination type is formed that is fertile within its own descendent line but is isolated from other lines and from the parental species by a chromosomal sterility barrier. Computer modeling (McCarthy, Asmussen, and Anderson 1995) indicates that speciation in such plants requires certain biological prerequisites. It is most likely to occur when (1) the parapatric hybrid interface is long; (2) the organisms involved are predominantly selfing (undergo self-pollination); (3) the hybrids are relatively fertile; and (4) the number of differences in chromosomal structure between parental species is small. Under these conditions, speciation is preceded by a period of long-term stasis followed by an abrupt transition to a new, reproductively isolated species.

Irises in the bayous of Louisiana provide a classic example of recombinational hybrid speciation (Cornman et al. 2004). Hybrid swarms produced by natural crosses between *Iris fulva* and *I. hexagona* (see Figure 10) are frequently observed in Louisiana and resemble a third species, *I. nelsonii*. Populations of *I. fulva* from the Mississippi River often interdigitate with *I. hexagona*, a species adapted more to intercoastal, freshwater

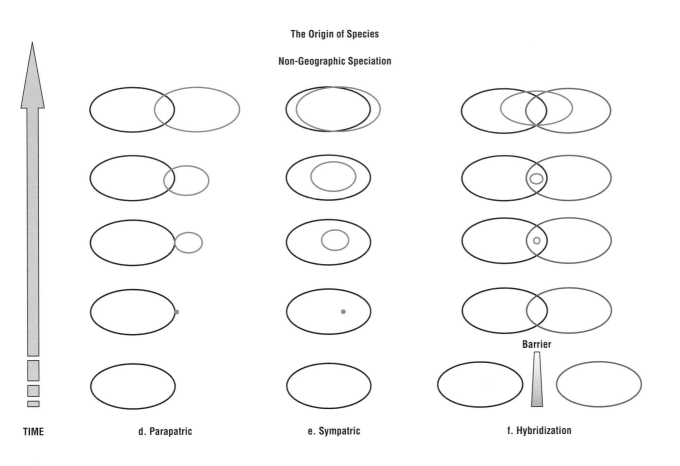

Figure 7. Parapatric and sympatric speciation are modes of speciation that require no period of complete geographic isolation during the speciation process. Speciation by hybridization can occur in a number of ways discussed in the text. Reproduced by permission of Gale, a part of Cengage Learning.

swamps and marshy environments. *I. nelsonii* is adapted to heavily shaded areas with deep water, a habitat intermediate between that of the two other species of iris.

All three *Iris* species share the same diploid chromosome number but differ somewhat in chromosomal rearrangements that have contributed to speciation. Allozyme and polymerase chain reaction analysis of chloroplast and nuclear DNA reveals substantial gene flow between *I. fulva* and *I. hexagona* in zones of secondary contact where they hybridize, and alleles of both parent species can be found in the derivative species. These studies support the conclusion that *I. nelsonii* is a recombinational hybrid species derived from *I. fulva* and *I. hexagona* and isolated from both by a weak chromosomal sterility barrier. Michael L. Arnold, James L. Hamrick, and Bob D. Bennett (1990) point out that the rhizomatous habit and clonal reproduction in *Iris* probably facilitated the establishment of *I. nelsonii*.

One of the most detailed studies of recombinational hybrid speciation is that carried out by Loren H. Rieseberg, Chrystal Van Fossen, and Andrée M. Desrochers (1995) on wild sunflowers (*Helianthus* spp.). *Helianthus annus* and *H. petiolaris* are widespread variable species distinguished from one another by several morphological and chromosomal features. Based on

chloroplast DNA and nuclear ribosomal DNA variation, they occur in divergent clades, each having different ecological requirements. *H. annus* is restricted to heavy, clay soils, whereas *H. petiolaris* prefers predominantly dry, sandy soils.

The two are sympatric and grow in close proximity throughout the western United States. Their F$_1$ hybrids are semisterile, with pollen viabilities of less than 10 percent and seed set at less than 1 percent, and with F$_2$ pollen viability highly variable, ranging from 13 percent to 97 percent. A third species, *H. anomalus*, resulted from natural hybridization between *H. annus* and *H. petiolaris*. *H. anomalus* is a rare species, endemic to xeric habitats in northern Arizona and southern Utah well within range of parental species, but it is locally well segregated from the parental species by soil preference. Not only is *H. anomalus* morphologically distinct from both parental species, it also combines DNA and genetic markers of *H. annus* and *H. petiolaris* as predicted for diploid hybrid species and the chloroplast DNA haplotypes of *H. annus* and *H. petiolaris* rather than a unique one.

By comparing the genomic location and linear order of homologous DNA markers, chromosomal structural relationships among the three species revealed a great deal of structural reorganization had rapidly occurred during and

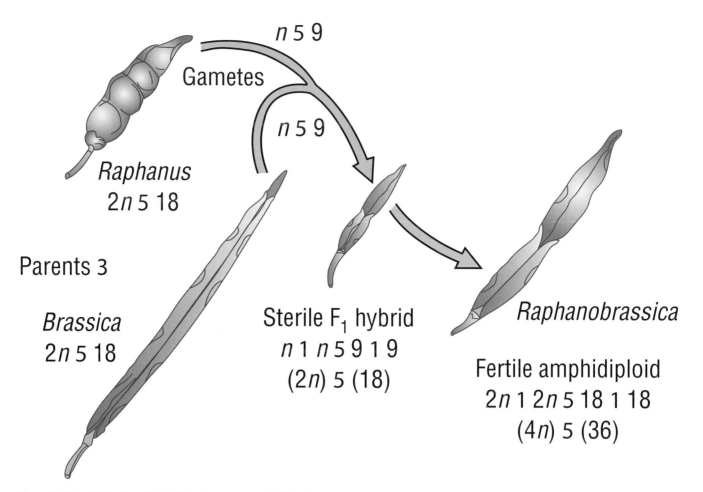

Figure 8. Tetraploidization of hybrid *Raphanobrassica*. © Fritz Polking/PhotoLibrary.

following the formation of the new species. Although six linkage groups (chromosomes) remained unchanged in all three species, gene order in the remaining eleven linkage groups was not conserved. The parental species differ from *H. anomalus* by at least ten separate structural rearrangements— three inversions and at least seven interchromosomal translocations. The genome of *H. anomalus* therefore underwent extensive rearrangement relative to both parents. This included the merger of pre-existing structural differences between the parents along with major chromosomal rearrangements apparently induced by recombination. These chromosomal structural differences enhance reproductive isolation and facilitate speciation as F_1 hybrids mating with parental species are partially sterile because in such crosses meiosis will be abnormal. Also, the hybrid species occupies a different habitat further reducing competition and level of gene flow from parental species. For further details and a discussion of similar patterns of habitat divergence found in two other *Helianthus* hybrid species, see the 2007 article by Rieseberg and John H. Willis.

Polyploid hybrid speciation   Polyploid hybrid speciation, in which the entire genome is duplicated, occurs frequently in plants but rarely in animals (Rieseberg 1997). Although less important in animals, it could have played an important role in

some groups such as salmonid fish, some frogs, and salamanders (White 1978).

Polyploidy in plants is often associated with rapid speciation promoted by an ecological shift, the development of self-fertilization or asexual reproduction, and changes in other biological traits that enhance reproductive isolation from parental species. Polyploid speciation may take place in several ways, which are treated in detail by Verne Grant (1981). Only the most frequently encountered are discussed here.

Allopolyploid (amphiploid) speciation   Interspecific hybridization can sometimes unite two or more complete chromosome sets to form a new species. When two established but related species hybridize they may produce a sterile F_1 hybrid because its chromosomes lack sufficient homology to pair during meiosis, resulting in the production of sterile pollen. Because plants may reproduce asexually for long periods of time, chromosomes may undergo somatic doubling in a flower or after the rare union between two unreduced gametes. This somatic doubling restores fertility and establishes a new, sexually reproducing species at once isolated from both parental species. For this reason, allopolyploid speciation was not known to occur in mammals until it was discovered in the red vizcacha rat, *Tympanoctomys barrerae*, in Argentina (Gallardo, González, and Cebrián 2006). Examples in plants are well known and provide details on how such allopolyploid speciation events take place.

A new species of perennial salt marsh grass, *Spartina anglica* (Raybould et al. 1991), originated along the southern coast of Great Britain toward the end of the nineteenth century. After one of its parental species, the North American *S. alterniflora* (2n = 62), was introduced by shipping, it hybridized with the native *S. maritima* (2n = 60) growing along the British coast. The hybridization produced sterile offspring, *S. townsendii* (2n = 62), whose chromosomes were too dissimilar to pair during meiosis. *S. townsendii* survived and reproduced asexually for many years. Eventually the chromosome number doubled in offspring and a new, sexually reproducing species, *S. anglica* (4n = 120, 122, 124), appeared and rapidly spread widely again by shipping throughout Europe and the rest of the world. It is now considered a noxious weed that rapidly outcompetes and replaces other *Spartina* species.

Another example, representing the first intentionally human-made species, involves the development of an intergeneric allopolyploid hybrid resulting from a cross between radish (*Raphanus*) and cabbage (*Brassica*). In the 1920s the Russian agronomist and geneticist Georgii D. Karpechenko dedicated considerable time and effort to produce a plant that combined the edible part of the radish, namely the root, with the head of the cabbage in one plant (Karpechenko 1927). He first produced an artificial hybrid crossing the radish with the cabbage. Both had the same number of chromosomes (2n = 18). Although hybrids had a complete set of chromosomes from each parental species, the plants were sterile. Over time, a few of Karpechenko's hybrids spontaneously doubled their chromosome number (4n = 36) restoring fertility and producing a genetically stable hybrid "rabbage," often referred to as *Raphanobrassica*. Unfortunately for food lovers, the resulting plant combined only the nonedible parts—the head of the radish and the root of the cabbage. For his efforts and his early seminal work on allopolyploids, Karpechenko was arrested, convicted of anti-Soviet activity, and executed in 1941.

Several other forms of polyploid speciation have been recorded in both animals and plants that are rare or less frequently encountered (Bullini 1994; White 1978). One that should be mentioned is *reticulate* speciation encountered in some insects such as stick insects (*Pasmatodea*), in which individuals of a *unisexual* hybrid taxa hybridize with their *bisexual* relatives giving rise to new, unisexual species. Usually these species have a higher ploidy level than the parental species. Often they are triploid hybrid parthenogens reproducing asexually whose individuals have highly heterozygous genomes. They may display heterosis or hybrid vigor and thus possess a demographic advantage over one or both parental species and will sometimes replace them.

Summary

Clearly, many different modes of speciation have contributed to the remarkable diversity of animal and plant forms that have occupied Earth since the origin of life some 3.8 billion years ago. With the introduction of molecular tools to probe the very nature of the genomes of species coupled with ever-expanding knowledge within population ecology, scientists' understanding of how species evolve has greatly improved. Nevertheless, much is yet to be learned about the speciation process, and many surprises are likely ahead.

Resources

Books

Bush, Guy L., and Roger K. Butlin. 2004. "Sympatric Speciation in Insects." In *Adaptive Speciation*, ed. Ulf Dieckmann, Michael Doebeli, Johan A. J. Metz, and Diethard Tautz. Cambridge, UK: Cambridge University Press.

Coyne, Jerry A., and H. Allen Orr. 2004. *Speciation*. Sunderland, MA: Sinauer.

Darwin, Charles. 1964 (1859). *On the Origin of Species*. Cambridge, MA: Harvard University Press.

Grant, Peter R., and B. Rosemary Grant. 2008. *How and Why Species Multiply: The Radiation of Darwin's Finches*. Princeton, NJ: Princeton University Press.

Grant, Verne. 1981. *Plant Speciation*. 2nd edition. New York: Columbia University Press.

Macnair, Mark R., and Mike Gardner. 1998. "The Evolution of Edaphic Endemics." In *Endless Forms: Species and Speciation*, ed. Daniel J. Howard and Stewart H. Berlocher. New York: Oxford University Press.

Mayr, Ernst. 1942. *Systematics and the Origin of Species*. New York: Columbia University Press.

Mayr, Ernst. 1963. *Animal Species and Evolution*. Cambridge, MA: Harvard University Press, Belknap Press.

White, M. J. D. 1978. *Modes of Speciation*. San Francisco: W. H. Freeman.

Periodicals

Arnold, Michael L, James L. Hamrick, and Bob D. Bennett. 1990. "Allozyme Variation in Louisiana Irises: A Test for Introgression and Hybrid Speciation." *Heredity* 65(3): 297–306.

Barluenga, Marta, Kai N. Stölting, Walter Salzburger, et al. 2006. "Sympatric Speciation in Nicaraguan Crater Lake Cichlid Fish." *Nature* 439(7077): 719–723.

Berlocher, Stewart H., and Jeffrey L. Feder. 2002. "Sympatric Speciation in Phytophagous Insects: Moving beyond Controversy?" *Annual Review of Entomology* 47: 773–815.

Bickford, David, David J. Lohman, Navjot S. Sodhi, et al. 2006. "Cryptic Species as a Window on Diversity and Conservation." *Trends in Ecology and Evolution* 22(3): 148–155.

Bräuer, Günter. 2008. "The Origin of Modern Anatomy: By Speciation or Intraspecific Evolution?" *Evolutionary Anthropology: Issues, News, and Reviews* 17(1): 22–37.

Bullini, Luciano. 1994. "Origin and Evolution of Animal Hybrid Species." *Trends in Ecology and Evolution* 9(11): 422–426.

Carson, Hampton L. 1983. "Chromosomal Sequences and Interisland Colonizations in Hawaiian *Drosophila*." *Genetics* 103(3): 465–482.

Cornman, R. Scott, John M. Burke, Renate A. Wesselingh, and Michael L. Arnold. 2004. "Contrasting Genetic Structure of Adults and Progeny in Louisiana Iris Hybrid Population." *Evolution* 58(12): 2669–2681.

Coyne, Jerry A. 2007. "Sympatric Speciation." *Current Biology* 17(18): R787–R788.

de Queiroz, Kevin. 2005. "Different Species Problems and Their Resolution." *BioEssays* 27(12): 1263–1269.

Drès, Michele, and James Mallet. 2002. "Host Races in Plant-Feeding Insects and Their Importance in Sympatric Speciation." *Philosophical Transactions of the Royal Society* B 357(1420): 471–492.

Gallardo, Milton H., C. A. González, and I. Cebrián. 2006. "Molecular Cytogenetics and Allotetraploidy in the Red Vizcacha Rat, *Tympanoctomys barrerae* (Rodentia, Octodontidae)." *Genomics* 88(2): 214–221.

Kambysellis, Michael P., Kin-Fan Ho, Elysse M. Craddock, et al. 1995. "Pattern of Ecological Shifts in the Diversification of Hawaiian *Drosophila* Inferred from a Molecular Phylogeny." *Current Biology* 5(10): 1129–1139.

Karpechenko, Giorgii D. 1927. "Polyploid Hybrids of *Raphanus sativus* x *Brassica oleracea* L." *Bulletin of Applied Botany* 17: 305–408.

Knowlton, Nancy, Lee A. Weigt, Luis Aníbal Solórzano, et al. 1993. "Divergence in Proteins, Mitochondrial DNA, and Reproductive Compatibility across the Isthmus of Panama." *Science* 260(5114): 1629–1632.

Mallet, James. 2007. "Hybrid Speciation." *Nature* 446: 279–283.

McCarthy, Eugene M.; Marjorie A. Asmussen; and Wyatt W. Anderson. 1995. "A Theoretical Assessment of Recombinational Speciation." *Heredity* 74(5): 502–509.

Pennisi, Elizabeth. 2009. "Neandertal Genomics: Tales of a Prehistoric Human Genome." *Science* 323(5916): 866–871.

Raybould, Alan F., Alan J. Gray, M. J. Lawrence, and D. F. Marshall. 1991. "The Evolution of *Spartina anglica* G. E. Hubbard (Gramineae): Genetic Variation and Status of the Parental Species in Britain." *Biological Journal of the Linnean Society* 44(4): 369–380.

Rieseberg, Loren H. 1997. "Hybrid Origin of Plant Species." *Annual Review of Ecology and Systematics* 28: 359–389.

Rieseberg, Loren H., Chrystal Van Fossen, and Andrée M. Desrochers. 1995. "Hybrid Speciation Accompanied by Genomic Reorganization in Wild Sunflowers." *Nature* 375(6529): 313–316.

Rieseberg, Loren H., and John H. Willis. 2007. "Plant Speciation." *Science* 317(5840): 910–914.

Savolainen, Vincent, Marie-Charlotte Anstett, Christian Lexer, et al. 2006. "Sympatric Speciation in Palms on an Oceanic Island." *Nature* 441(7090): 210–213.

Schemske, Douglas W., and H. D. Bradshaw Jr. 1999. "Pollinator Preference and the Evolution of Floral Traits in Monkeyflowers (*Mimulus*)." *Proceedings of the National Academy of Sciences of the United States of America* 96(21): 11910–11915.

Schliewen, Ulrich K., Diethard Tautz, and Svante Pääbo. 1994. "Sympatric Speciation Suggested by Monophyly of Crater Lake Cichlids." *Nature* 368(6472): 629–632.

Schluter, Dolph. 2009. "Evidence for Ecological Speciation and Its Alternative." *Science* 323(5915): 737–741.

Schwarz, Dietmar, Benjamin M. Matta, Nicole L. Shakir-Botteri, and Bruce A. McPheron. 2005. "Host Shift to an Invasive Plant Triggers Rapid Animal Hybrid Speciation." *Nature* 436(7050): 546–549.

Tattersall, Ian, and Jeffrey H. Schwartz. 2008. "The Morphological Distinctiveness of *Homo sapiens* and Its Recognition in the Fossil Record: Clarifying the Problem." *Evolutionary Anthropology: Issues, News, and Reviews* 17(1): 49–54.

Templeton, Alan R. 2008. "The Reality and Importance of Founder Speciation in Evolution." *BioEssays* 30(5): 470–479.

Other

American Museum of Natural History. "Online Educator's Guide." Available from http://www.amnh.org/education/resources/rfl/web/hhoguide/tree.html.

Guy L. Bush

· · · · ·

Sexual reproduction

Sexual reproduction is the process by which two individuals create offspring with a genotype that is a mix of parental genotypes, where each parent contributes exactly 50 percent of the offspring's genes. In this process, two haploid gametes, generated by the two parents, are combined to create a diploid zygote. Later in the life cycle, and before the next sexual reproduction, the diploid genome would be divided into two haploid gametes during meiosis. In this process the two alleles at each locus are separated into different gametes (segregation), and alleles at different loci on the two homologous chromosomes are mixed as chromosomes are broken and rejoined (recombination). The life cycle of a sexual organism is thus an alteration of haploid and diploid stages. The common definition of an organism as "haploid" or "diploid" usually depends on the relative length and importance of the relevant stage.

The two haploid gametes can be identical in size (isogamy) or, as is the case in most organisms, different in size (anisogamy). When the two gamete types are of different size and individuals produce only one gamete type, then the term female refers to the sex with the larger and less mobile gametes. While offspring get exactly half their chromosomes from each parent, this is not the case for cytoplasm. An offspring gets most of its cytoplasm, including mitochondria, from its mother. As a result, various traits can be maternally inherited—either through the genetic material in maternally inherited cell organelles or from nongenetic material. In hermaphroditic organisms, the same individual can produce both male and female gametes, whereas in most organisms, an individual is either male—producing only male gametes; or female—producing only female gametes.

Sexual reproduction allows for sexual selection: differential success of individuals of the same sex with respect to reproduction. In most cases one sex, the sex that invests less in the offspring (hereafter referred to as "males" for convenience, but note that the situation reverses when males invest more in the offspring and females compete, as is the case, for example, with the seahorse), is subject to stronger sexual selection. Sexual selection can be direct—in the form of male-to-male competition; or indirect—through female choice.

Sexual reproduction is practiced by the majority of plants and animals. Some organisms (including yeast, daphnia, aphids, and most plants) can reproduce either sexually or asexually (vegetatively), whereas others (including most higher animals and some higher plants) reproduce only through sexual reproduction. Some eukaryotic species (e.g., the gecko *Lepidodactylus lugubris*) are parthenogenic and reproduce asexually all the time. Interestingly, the vast majority of these asexual species are relatively young (but note the bdelloid rotifers, a type of microscopic invertebrate, as a very rare exception), suggesting that sexual reproduction has some long-term advantages leading to early extinction of asexual species.

Bacteria do not reproduce sexually in the strict sense but practice other forms of genetic mixing, even between species, including transformation (actively taking in external genetic material and recombining with it) and conjugation (transfer of genetic material through a direct cell-to-cell contact, initiated by a conjugative virus or plasmid). These differ from classic sexual reproduction in the asymmetry between the "parents."

Sex is closely related to the central biological concept of species. Two populations of sexually reproducing organisms are considered of the same species when they can interbreed and have fertile offspring. Speciation is the process by which a reproductive barrier appears between two populations that had previously belonged to the same species.

The evolution of sex

Sex is a very costly process. First and foremost, it incurs a twofold "cost of males" (Maynard Smith 1978): If males do not contribute any resources other than genes to the offspring, then—everything else being equal—a sexual population would produce half the number of offspring of an asexual population of the same size in any given generation and is likely to go extinct rapidly when competing with an asexual population. Additional costs for a sexually reproducing individual include finding a mate (directly or indirectly, e.g., through the attraction of pollinators), courtship, increased risks of pathogen transmission and predation, and loss of time and energy in the mating process. Furthermore, sex opens the door for conflicts between the sexes and between different parts of the genome. If reproduction is asexual, different alleles in the genome are "stuck together" forever, and their "evolutionary interests" are therefore very similar. Once different alleles might move into a different genetic background in the next generation, conflicts can arise, including genetic parasites, segregation distorters, and so on.

Mourning geckos (*Lepidodactylus lugubris*) are parthenogenic and reproduce asexually all the time. Gilbert S. Grant/Photo Researchers, Inc.

Thus, to understand the prevalence of sex in the world, one needs to understand the advantages it confers. These advantages need to compensate for all the costs mentioned above, for a very wide range of organisms and environments. The evolution of sex remains one of the major open questions in the theory of evolution, despite extensive research since the 1930s (for reviews see Barton and Charlesworth 1998; Hadany and Comeron 2008; Otto and Lenormand 2002; Rice 2002).

Most explanations for the evolution of sex are based on the potential of sexual reproduction to increase genetic variability. Sexual reproduction involves segregation—separation and recombination of the two chromosome copies, leading to the mixing of genotypes; segregation acts within a locus, while recombination acts between loci. Either way, sex tends to break down existing associations between alleles, bringing allelic distributions closer to random assortment.

Evaluating the advantage of sexual reproduction thus involves two questions: (1) When is sex likely to increase genetic variation? and (2) When would an increase in genetic variation be favored? Consider a simple example in a single locus, where AA is the best genotype, aa the worst, and the heterozygote has intermediate fitness. A population consisting of 50 percent AA and 50 percent aa has the maximal variation in fitness, meaning that the effectiveness of selection is also maximal. If such a

population reproduces sexually, the amount of variation (in fitness) would *decrease*, because the frequencies of the genotypes AA, Aa, aa would now change to 0.25, 0.5, 0.25, respectively. More generally, sexual reproduction brings the population closer to linkage equilibrium (with respect to recombination) and Hardy–Weinberg equilibrium (with respect to segregation)—a situation in which the frequency of a combination of alleles is the product of the frequencies of each of the alleles. This might either increase or decrease the amount of variation in the population, depending on the starting point. Discussed in the next sections are factors that can generate an excess of genotypes with intermediate fitness and thus can promote the evolution of sexual reproduction—epistasis, deleterious mutations, genetic drift, and environmental changes.

Selection and associations between alleles

The effect of multiple mutations on an individual's fitness cannot be predicted based solely on the effect of each mutation alone. If the combined effect of multiple deleterious mutations is larger than the product of their individual effects (a condition known as "synergistic epistasis"), then genotypes with intermediate fitness would be overrepresented. In such a case, selection would favor sex and recombination in the long term, because it would result in increased variation and therefore increased average fitness in the population (Kondrashov 1988).

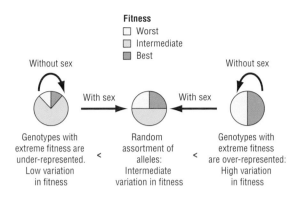

Fitness
☐ Worst
◻ Intermediate
■ Best

Figure 1. The effect of sex on variation depends on the starting point of the population with respect to variation. When the population has an excess of genotypes with extreme fitness (right) compared with what one would expect by random assortment of alleles (center), sex would result in decreased variation. In contrast, when the population has an excess of genotypes with intermediate fitness (left)—a situation known as negative linkage disequilibrium—sex would result in increased variation. Only in the latter case does sex have an advantage in the long term, and this advantage needs to be large enough for sex to evolve. Reproduced by permission of Gale, a part of Cengage Learning.

Long- and short-term advantages of sex

A mean fitness argument such as the one above is sufficient when comparing two populations that cannot mix. As such, it reveals possible long-term advantages of sex. To discuss short-term effects of sex on population dynamics, more delicate treatment is required involving analysis of "modifier alleles"—alleles that affect the frequency of sex in an organism in which they appear. The short-term effect of sex depends on the interactions between alleles at different loci in the genome and is often negative. This result can be intuitively understood given that selection favors certain allelic combinations, turning them more common than random. Sex breaks down these overrepresented combinations that were favored by selection, resulting in decreased fitness for the offspring. A striking example of this result occurs when selection is the only force acting on the population, and there are no mutations, random genetic drift, or environmental changes. The only modifiers of recombination that can be favored under these conditions are ones that *reduce* the rate of recombination (Feldman, Christiansen, and Brooks 1980), indicating that recombination is always harmful in the short term under these assumptions.

In general, for sex to evolve it should either have an advantage both in the short and long term or have long-term advantages that outweigh the short-term disadvantage. Why this is true for such a wide range of organisms and environments is still an open question.

The effect of small population size

Consider an asexual population exposed to mutations, each carrying a small harmful effect. In a small population, a day would come when no individual remains free of deleterious mutations. From that point on, the best genotype would be one

with at least one deleterious mutation. This process, known as "Muller's ratchet" (Muller 1964), would repeat itself, resulting in a slow deterioration of the small asexual population. Sex can resolve this problem, because it regenerates good genotypes that had been lost because of random drift.

Now consider a population subject to advantageous mutations (Maynard Smith 1978). If a new advantageous mutation appears in a single individual, then without sex this new allele would remain 100 percent associated with the genetic background it first appeared in. A second beneficial mutation would be 100 percent associated with its background, and an individual containing two beneficial mutations would appear only if the second mutation occurred in an individual already containing the first one. Sex can help in such a case by combining multiple beneficial mutations that occurred in different genotypes.

The combination of common deleterious mutations and rare beneficial mutations can result in an even greater advantage for sex and recombination: In an asexual population, a beneficial mutation can arise on various backgrounds. Usually, it would arise on a background that includes too many deleterious mutations, resulting in a genotype that has almost no chance of spreading in the population—a phenomenon known as "background trapping" (Peck 1994). From time to time, it would appear on a background that contains relatively few deleterious mutations, resulting in a genotype that does have a chance to take over. But if that happens, both the beneficial mutation and the weakly deleterious background would increase to fixation, resulting in increased mutational load in the asexual population (Hadany and Feldman 2005).

Generally, in a small population, random genetic drift creates associations between alleles—some of them with increased variation and some with decreased variation. Selection acts to break associations with increased variation—leaving associations with decreased variation to linger much longer—creating the exact situation in which sex can be advantageous (Barton and Otto 2005).

Changing environments

Another factor that might play a role in the evolution of sex is environmental changes (Otto and Michalakis 1998)—either in time or in space. A change in the environment can have two types of effects on the evolution of recombination: First, environmental changes favor new alleles and create room for beneficial mutations, a situation that can be advantageous as mentioned above. Second, an environmental change can favor new genetic associations ("changing epistasis"). In such a case, recombination would break down existing associations that have become unfavorable in the new environment. Every such change results in a temporary advantage for recombination, but it turns out that the environment needs to fluctuate quite rapidly to favor sex. Scenarios that might result in common fluctuating epistasis usually involve biotic interactions: predator and prey, competition between species, and especially the coevolution of hosts and parasites.

The so-called red queen hypothesis suggests that the coevolution of interacting species is a general scenario that might favor the evolution of sex. Assuming that the matching between host and parasite is genetically determined and that parasites are much shorter lived than their hosts, parasites would tend to evolve to best infect the most common host genotype at any given time, thus turning this genotype into the least favorable in the next host generation. Such frequency-dependent selection has been demonstrated to favor sex in various models, but its generality is still largely controversial. Interestingly, red queen models seem to work best in simulations using small populations (Hamilton, Axelrod, and Tanese 1990; Howard and Lively 1994), where coevolution of hosts and parasites can result in a combination of factors that might favor the evolution of sex in addition to fluctuating epistasis: strong selection, beneficial mutations, and a significant increase in genetic drift as different genotypes go through "bottlenecks" of dramatically reduced population size. In some cases, the "synergistic" interaction of different factors (e.g., drift, selection, and mutation) results in a further increased advantage for sex and recombination, leading as well to the need for pluralistic models.

Sex and phenotypic plasticity

Numerous organisms can reproduce either sexually or asexually. This includes animals from yeast to daphnia, certain worms, and most plants. In classical theories of the evolution of sex, it was usually assumed that the tendency to reproduce sexually in such organisms is constant. But in fact this tendency varies significantly according to the individual's condition: Various stresses induce a shift from asexual to sexual reproduction (Bell 1982), including starvation, DNA damage, and parasites. This variation within the population can help explain the evolution and maintenance of sex.

Recent works have considered the effect of a modifier that increases sex when the individual's condition is poor. Such a fitness-associated modifier can evolve even under conditions that make classical modifiers highly harmful and even when the cost of sex is high (Hadany and Otto 2007). This can be intuitively understood by considering these modifiers as offering an "abandon-ship" mechanism: They break away from unfit genetic backgrounds by recombination

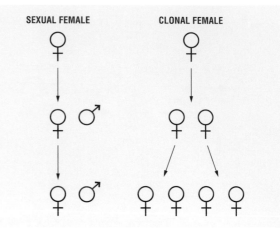

The cost of males. Reproduced by permission of Gale, a part of Cengage Learning.

or segregation and link to better genetic backgrounds. An association thus appears between the modifier allele for condition-dependent sex and beneficial alleles at other loci. As a result, the modifier has a short-term advantage even if it confers no long-term advantage for the population as a whole. This allows the evolution of some level of sexual reproduction or recombination over a much wider parameter range than in classical theory. In fact, condition-dependent sex often results in an advantage in terms of average population fitness as well, so it combines short- and long-term advantages.

The evolution of obligatory sex

A particularly puzzling aspect in the evolution of sexual reproduction is the abundance of species, especially among higher eukaryotes, in which sex is the only mode of reproduction (e.g., all birds and mammals). In theory, most advantages of sex should still accrue when only a small proportion of the offspring are produced sexually, while most of the cost of sex would be avoided in such a case (Green and Noakes 1995; Hurst and Peck 1996). To account for the abundance of obligate sexuals, one thus needs to demonstrate the advantage of sexual reproduction when competing directly with a population of facultative sexuals, which can switch between sexual and asexual modes of reproduction.

Sexual selection and the evolution of sex

Sexual selection is another factor that can turn sex into a favorable mode of reproduction, as it can make natural selection more effective: Successful males can produce many more offspring than successful asexuals, whereas males carrying deleterious alleles are less likely to reproduce than asexuals. These effects can outweigh the twofold cost of males under common scenarios of deleterious mutations and environmental change (Agrawal 2001). In particular, sexual selection might help explain the evolution of obligate sex when competing with facultative sex (Hadany and Beker 2007). For sexual selection to be advantageous, overall fitness

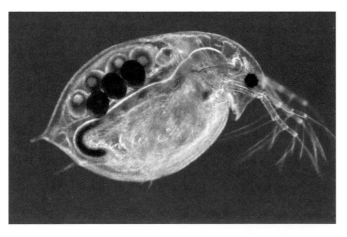

Light microscopy of a female *Daphnia pulex*, the most common species of water flea, with eggs. M. I. Walker/Photo Researchers, Inc.

should be positively correlated with mating success. Such associations have been documented in various organisms, but do not apply genome-wide, and sometimes genes that are advantageous for females conflict with ones that benefit males. All in all, why sexual reproduction is so abundant in nature remains an intriguing question.

Resources

Books

Bell, Graham. 1982. *The Masterpiece of Nature: The Evolution and Genetics of Sexuality*. Berkeley: University of California Press.

Maynard Smith, John. 1978. *The Evolution of Sex*. Cambridge, UK: Cambridge University Press.

Periodicals

Agrawal, Aneil F. 2001. "Sexual Selection and the Maintenance of Sexual Reproduction." *Nature* 411(6838): 692–695.

Barton, N. H., and B. Charlesworth. 1998. "Why Sex and Recombination?" *Science* 281(5385): 1986–1990.

Barton, N. H., and Sarah P. Otto. 2005. "Evolution of Recombination Due to Random Drift." *Genetics* 169(4): 2353–2370.

Feldman, Marcus W., Freddy B. Christiansen, and Lisa D. Brooks. 1980. "Evolution of Recombination in a Constant Environment." *Proceedings of the National Academy of Sciences of the United States of America* 77(8): 4838–4841.

Green, Richard F., and David L. G. Noakes. 1995. "Is a Little Bit of Sex as Good as a Lot?" *Journal of Theoretical Biology* 174(1): 87–96.

Hadany, Lilach, and Josep M. Comeron. 2008. "Why Are Sex and Recombination So Common?" *Annals of the New York Academy of Sciences* 1133: 26–43.

Hadany, Lilach, and Marcus W. Feldman. 2005. "Evolutionary Traction: The Cost of Adaptation and the Evolution of Sex." *Journal of Evolutionary Biology* 18(2): 309–314.

Hadany, Lilach, and Sarah P. Otto. 2007. "The Evolution of Condition-Dependent Sex in the Face of High Costs." *Genetics* 176(3): 1713–1727.

Hamilton, William D., Robert Axelrod, and Reiko Tanese. 1990. "Sexual Reproduction as an Adaptation to Resist Parasites. (A Review)." *Proceedings of the National Academy of Sciences of the United States of America* 87(9): 3566–3573.

Howard, R. Stephen, and Curtis M. Lively. 1994. "Parasitism, Mutation Accumulation, and the Maintenance of Sex." *Nature* 367(6463): 554–557.

Hurst, Laurence D., and Joel R. Peck. 1996. "Recent Advances in Understanding of the Evolution and Maintenance of Sex." *Trends in Ecology and Evolution* 11(2): 46–52.

Kondrashov, Alexey S. 1988. "Deleterious Mutations and the Evolution of Sexual Reproduction." *Nature* 336(6198): 435–440.

Muller, Hermann Joseph. 1964. "The Relation of Recombination to Mutational Advance." *Mutation Research* 106: 2–9.

Otto, Sarah P., and Thomas Lenormand. 2002. "Resolving the Paradox of Sex and Recombination." *Nature Reviews Genetics* 3(4): 252–261.

Otto, Sarah P., and Yannis Michalakis. 1998. "The Evolution of Recombination in Changing Environments." *Trends in Ecology and Evolution* 13(4): 145–151.

Peck, Joel R. 1994. "A Ruby in the Rubbish: Beneficial Mutations, Deleterious Mutations, and the Evolution of Sex." *Genetics* 137(2): 597–606.

Rice, William R. 2002. "Experimental Tests of the Adaptive Significance of Sexual Recombination." *Nature Reviews Genetics* 3(4): 241–251.

Other

Hadany, Lilach, and Tuvik Beker. 2007. "Sexual Selection and the Evolution of Obligatory Sex." *BMC Evolutionary Biology* 7: 245. Available from http://www.biomedcentral.com/1471-2148/7/245.

Lilach Hadany

Species nomenclature, classification and problems of application

Imagine Carolus Linnaeus's dilemma in the eighteenth century as he examined the specimens that had been brought back to Sweden during the age of discovery. Specimens would have been laid out in neat rows on tables and grouped on the basis of overall morphological similarity. While some specimens were so clearly unique that they could be confidently classified as a "species," there were other groups of specimens that possessed sufficient overlap in characteristics that a judgment call would have been required. And as more specimens were added to the collections, from more locations, the problem would have been compounded as the number of "gray areas" multiplied accordingly. The continuum of variation in nature, from highly unique to practically indistinguishable, did not escape Linnaeus's eyes. Species, a category initially thought by Linnaeus to be fixed and immutable, turned out to be highly variable and sometimes difficult to define.

Linnaeus's contribution to the history of biological science was a binomial system of taxonomic nomenclature that is universally applied. The taxonomies of organisms themselves, however, have remained in a state of constant flux. Although much of this taxonomic instability can be attributed to new information obtained from improved methods—such as advances in genetics, morphometry, behavior, bioacoustics, physiology, and ecology; increased sample sizes, and a deeper understanding of evolutionary processes—the evolving definitions of species have contributed as well. Species concepts have proliferated (twenty-seven by one count), yet all have been hindered in some way by vague definitions, notable exceptions, or issues of practicality. Although most of these are derivatives of the biological species concept (Mayr 1970) or the phylogenetic species concept (see Agapow et al. 2004 for a review), none has proven to be universally applicable and some authors have advocated abandoning the term *species* altogether (Hendry et al. 2000). Like Linnaeus, scientists continue to contend with the dilemma of trying to draw lines through continua of natural variation and are sometimes faced with conflicting information. No matter what term is applied to the units, the dilemma remains.

Why taxonomy matters

Although taxonomy was once the obscure domain of curators and naturalists, times have changed. Legitimate concerns over the loss of biodiversity have led to the passage of laws and treaties that regulate the impact of human activities on species and subspecies, bringing new legal, financial, and social relevance to the classification of organisms. Although the first species protection laws that relied on taxonomies primarily regulated hunting and harvests, later laws focusing on endangered species protection have had much broader implications and cover a greater diversity of taxa. These include the U.S. Endangered Species Act (ESA) of 1973, the Convention on International Trade in Endangered Species of Wild Fauna and Flora (CITES), Australia's Environment Protection and Biodiversity Conservation Act of 1999, Canada's Species at Risk Act of 2002, and South Africa's National Environmental Management: Biodiversity Act of 2004. The scope of protection from each of these laws and treaties also extends to subspecies and a sub-subspecies category of "distinct populations." (The ESA uses multiple sub-subspecies categories: distinct vertebrate population segment [DPS], evolutionarily significant unit [ESU], management unit [MU], and recovery unit [RU].)

The descriptions of many of the species and subspecies listed as threatened or endangered predate these laws and were based on information that would not be acceptable in the early twenty-first century: small sample sizes, unknown inheritance of traits, and little or no quantitative basis or statistical analysis. Many are largely based on opinion. Essentially, a species or subspecies was what a taxonomist said it was. While modern taxonomists typically rely on much more sophisticated tools, including DNA sequence analysis and larger sample sizes, a substantial body of the work suffers from the same epistemological shortcomings as in the past: a reliance on post hoc interpretations of information and gross discrepancies in how taxonomic assignments are made (Avise and Johns 1999; Zink 2004).

The opinions of taxonomists, therefore, can carry the force of law, influence the expenditures of hundreds of millions of dollars for the protection and recovery of taxa listed as threatened or endangered, and affect the future direction of conservation efforts worldwide. Yet, the troubling aspect of this situation is that species and subspecies remain "in the eye of the beholder."

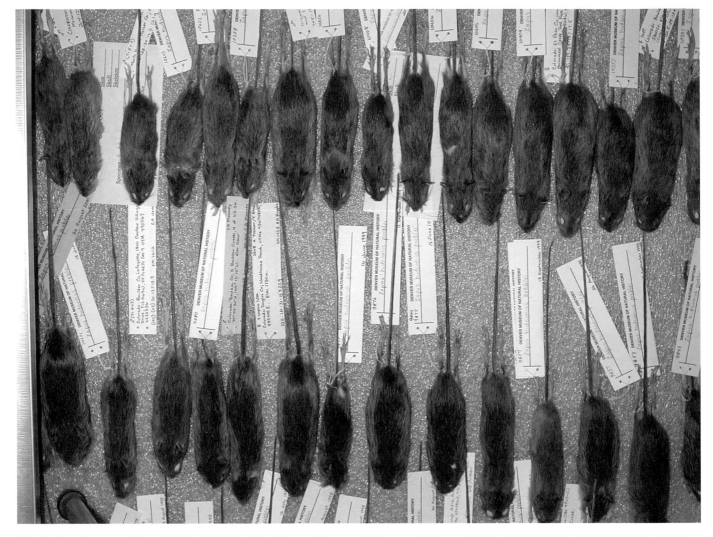

Meadow jumping mouse (*Zapus hudsonius*) skins at the Denver Museum of Nature and Science. EDGE rank #2,114. Courtesy of Rob Roy Ramey.

Taxonomic inflation

As those who followed in the footsteps of Linnaeus described more species, there was an increasing tendency to split existing species into several additional "new" species. This trend may have reached its high point with C. Hart Merriam's 1918 description of eighty-two species of North American brown bears and one new genus.

The category of "subspecies" was introduced as a way to describe geographic variation often found within species, and as a result many of those newly described species were demoted to the rank of subspecies. The addition of this subspecific level of classification resulted in the trinomial genus-species-subspecies nomenclature still in use in the early twenty-first century.

In the early to mid-twentieth century, subspecies designations became a fertile ground for splitting into smaller and smaller units. Many new subspecies were described, and frequently these designations were based on minor superficial traits having no known genetic basis or were based on small sample sizes. Subspecies were split further into additional

subspecies. During this time, the number of recognized species declined as some were reclassified as subspecies. The high tide of the subspecies trend was exemplified by E. Raymond Hall and Keith R. Kelson's 1959 recognition of 214 subspecies of southern pocket gopher (*Thomomys umbrinus*) in the southwestern United States and northern Mexico. Practically every valley held a different subspecies. These same authors reported that the number of North American mammal species decreased from 1,399 in the year 1923 to 1,003 in the year 1957, whereas the number of subspecies more than doubled, from 1,155 to 2,676, during the same period.

In more recent decades, this trend has reversed, and an increasing number of "new" species are being described from former subspecies because of changing definitions. As noted by Shai Meiri and Georgina M. Mace, "the total number of mammal species [worldwide] has risen from 4,659 in 1993 to 5,418 in 2005" (2007, p.1385). Yet, this trend is not the result of a renaissance of discovery in the biological hotspots of the world; it is instead primarily attributable to many subspecies and populations being elevated to the level of species, a process known as "taxonomic inflation."

Hawaiian monk seal (*Monachus schauinslandi*), sleeping on Kalalau Beach, Kauai, USA. EDGE rank #144. Courtesy of Laura MacAlister Brown, Wildlife Science International, Inc.

From a practical standpoint, taxonomic inflation both increases the perception of widespread endangerment while devaluing the basic currency of conservation (species). The more that species are subdivided into subspecies and sub-subspecies categories (in the United States, as noted above, these are DPSs, ESUs, MUs, or RUs), the smaller their ranges will be and the smaller the number of individuals found in the supposed endangered "population"—and thus, the more endangered each appears to be (see Agapow et al. 2004). The more "species" that are perceived to be in trouble, however, the less money is available for helping each of them. The United States allocates the majority of its endangered species budget to nondistinct but presumably threatened or endangered populations of common species (listed as DPSs and ESUs). It is clear, however, that this conservation approach comes at the expense of many "full" species that are far more endangered. With many full species endangered worldwide, and limited resources to save them, other nations may not find the ESA model to be a desirable or sustainable approach to conservation. Fully one-fifth of ESA-listed "species" are really subspecies or DPSs (Ramey et al. 2005), and these tend to be the most controversial and costly to protect (e.g., salmon ESUs, wolf DPSs, and presumed subspecies of otherwise widespread meadow jumping mice, beach mice, and California gnatcatchers).

Why has taxonomic inflation proliferated to the point that in 2007 *The Economist* published an editorial opinion commenting on the implications of this upward trend (*The Economist* 2007)? First, the application of the phylogenetic species concept (PSC) can lead to the splitting of taxa into smaller "diagnosable" units, if each shares at least one unique heritable trait (shared derived character). Thus a single shared nucleotide substitution may be sufficient evidence of monophyly and species status. Widespread use of genetic data in conservation genetics, especially relatively easy to obtain mitochondrial DNA sequence data, has led to broader application of the PSC (Agapow et al. 2004). Subspecies are not typically a part of this scheme because PSC purists consider any diagnosable unit a full species.

A second contributing factor can be a premature assumption of reproductive isolation under the biological species concept (BSC). Admittedly, testing for reproductive isolation between presumed species, especially in the wild, has been a challenge for the practical application of the BSC (e.g., when the presumed species exist on different continents, making breeding studies impractical). Such constraints necessitate an indirect approach to inferring reproductive isolation, typically involving some measure of chromosomal or genetic differences that would preclude successful interbreeding; adaptive differences in morphology, behavior, or physiology that would reduce fitness of hybrids; and/or traits involved with mate recognition such that successful mating is unlikely. While most evolutionary biologists concur that some level of interbreeding is possible among species, it is generally limited. A problem emerges, however, when low thresholds for accepting reproductive isolation are used without rigorously testing (and rejecting) the alternative hypotheses (e.g., that interbreeding does occur or is likely to occur if populations overlapped).

For example, in some cases allopatry (geographic separation) among vertebrate populations is due to natural causes of recent origin or is the result of human-caused isolation (e.g., urban or agricultural development since the early twentieth century). Hence, the evolution of genetic differences that could lead to reproductive isolation is unlikely. Similarly, if statistical significance is uncritically equated with biological significance, then even slight but statistically significant differences in allele frequencies (or minor behavioral and/or morphological variation) are considered evidence of reproductive isolation, thereby elevating many subspecies to the level of species and many populations to the level of subspecies. Such loose interpretations of the BSC are becoming increasingly common (Isaac, Mallet, and Mace 2004; Agapow et al. 2004; Meiri and Mace 2007).

A third cause of taxonomic inflation is the desire to elevate some populations or subspecies to the level of species out of concern for their conservation status, the logic being that erring on the side of caution requires erring on the side of species status, an "ends justify the means" rationale (Chaitra, Vasudevan, and Shanker 2004; Isaac, Mallet, and Mace 2004; Meiri and Mace 2007). As noted above, the listing of taxa or populations brings regulatory actions and funding associated with one of the world's most powerful environmental laws, the ESA.

Fourth and finally, taxonomic inflation has been further exacerbated by a convention of referring to species, subspecies, and even populations as if they were "full" species. This practice has become commonplace recently as a result of a linguistic loophole in the ESA. The loophole occurs because the ESA defines "species" in an unusual way. The statute and the regulations that flow from it allow for the federal listing of species, subspecies, and DPSs as threatened or endangered; and because the statute refers to all three categories as *species*, it has become increasingly common for conservation biologists, environmental advocates, the news media, and lawmakers to refer to all of these as *species*. At best, this is a confusing practice. At worst, it fosters an erroneous perception that all such

Platypus (*Ornithorhynchus anatinus*) underwater. EDGE rank #2. Tom McHugh/Photo Researchers, Inc.

ESA-listed entities are all *full* species. This perceptual problem is compounded when common names are used rather than Linnean nomenclature—one type of misunderstanding that Linnean nomenclature was intended to avoid.

Species and the law

The plethora of species concepts and definitions, as well as a general lack of operational definitions with thresholds to test species, has meant that species descriptions are often based on post hoc interpretations of data rather than on a hypothetico-deductive approach using thresholds set in advance of data collection, that is, the scientific method (Popper 1962; Platt 1964). If species concepts and definitions can be selected post hoc to fit any set of observations, then just about any group of organisms could potentially qualify (or not qualify) as a species depending on the investigator's whim (Hull 1965; Ramey et al. 2006; Wilson and Brown 1953; Hennig 1966). Such an approach renders *species* a meaningless term, and reliance on information derived from such an approach puts conservation and regulatory decisions on shaky epistemological grounds.

Disputable taxonomic classifications are anathema to governments that prefer some level of scientific certainty prior to committing to difficult regulatory decisions. A lack of consensus on operational definitions, the absence of consistently applied quantitative thresholds, data sets that were never made public, and the continued use of antiquated taxonomic descriptions all contribute to the problem (Cronin 1997, 2006, 2007; Fischman and Meretsky 2001; Quarles 2001). As a result, governmental agencies such as the U.S.

Fish and Wildlife Service (USFWS) and National Marine Fisheries Service often rely on the opinions of taxonomists, peer-review committees, interest groups, professional societies, and internal agency reviewers to determine if a taxon is valid. The problem with such reviews is that they are not held to any threshold of uniqueness or standards of evidence that could be consistently used to accept or reject a species, subspecies, or DPS (Ramey 2007). Because of that, even peer reviews can become de facto opinion surveys or be based on less than a complete record, leaving species very much "in the eye of the beholder." Using opinion rather than falsification as a basis for resolving taxonomic disputes has the implication of effectively putting these decisions outside the realm of science (the fundamental distinction between science and nonscience being the criterion of potential falsifiability [Popper 1962]).

While such an approach may lead to disputable taxonomic decisions with far-reaching regulatory implications, the discretion of the USFWS is generally shielded by the so-called Chevron deference—a legal precedent that requires the courts to defer to agency interpretations of ambiguous statutory provisions, including what constitutes a species. (The relevant cases are *Chevron U.S.A., Inc. v. Natural Resources Defense Council, Inc.*, from 1984, and *United States v. Mead Corp.*, from 2001.) Nevertheless, while the USFWS may insist that deciding what constitutes a species, subspecies, or DPS is a "complex scientific issue" and therefore afforded the Chevron deference, such a viewpoint ignores the ambiguity of defining these taxonomic categories and has led the USFWS to apply inconsistent criteria to the recognition of subspecies and DPSs. In the words of one endangered species legal expert testifying before the U.S. Senate on the ambiguous language used to define DPSs: "With this amount of discretion [by the USFWS], the eyesight of the beholder can be quite poor, and yet suffice" (Quarles 2001).

With a long-running disagreement over "Just what is a species?" (Isaac, Mace, and Mallet 2005; Hey 2006; Mallet 2007a, 2007b), it does not seem likely that taxonomists will be the ones to ultimately decide the issue for the rest of society. Nor is it likely that governmental agencies, executive branches, or legislation branches will decide, for similar reasons. Therefore, it will ultimately be the courts that decide what constitutes a "species," simply because they are the ultimate arbiter in matters of law and public policy.

In anticipation of such an eventuality, it behooves scientists to advance well-argued alternatives from which the courts might make informed choices, even though the scientists themselves may passionately disagree with the notion of an operational species definition or other approach that can be applied across a wide range of taxa. If scientists step up to the challenge, it may be possible for the courts to arrive at biologically meaningful and legally defensible definitions of species that everyone can live with.

Potential solutions

An answer to the problem might be to set a minimum threshold for species and subspecies or, for taxonomic groups

African savannah elephant (*Loxodonta africana*) foraging in the Northern Namib Desert, Namibia. EDGE rank #77. Courtesy of Laura MacAlister Brown, Wildlife Science International, Inc.

to be ranked based on measures of phylogenetic divergence. These two approaches are not mutually exclusive.

The minimum thresholds approach

If there is disagreement on what a species *is*, then maybe it is possible to set explicit criteria for rejecting those entities that are *not*. And if criteria were set in advance of data collection, this hypothesis-testing approach would transform a subjective system of classification into an objective one. While it may not be possible to develop thresholds that can be applied across all species or subspecies, it would be possible to arrive at standards within major groups (e.g., mammals). The primary caveat is analogous to type I and II statistical error: If the bar is set too high, then some unique organisms may fail to be recognized; yet, if the bar is set too low, then any population could potentially qualify as a species, contributing to taxonomic inflation. The criteria themselves are subject to debate, but the need for explicit criteria is obvious (Sites and Marshall 2003; Ramey et al. 2005; Crandall 2006). While appropriate thresholds can be

debated and revised, the first step in establishing standards is to state them explicitly.

In 2005 Rob Roy Ramey II and colleagues addressed the issue of quantitative criteria for subspecies and DPS by employing an explicit minimum thresholds approach with multiple data sets, complete spatial sampling, and criteria set in advance of data collection (Ramey et al. 2005, 2006, 2007). Each of these used threshold levels for various tests that have some conventional history below the level of species (e.g., Worley et al.'s 2004 use of $q > 0.90$ as a standard in assignment tests, Wehausen and Ramey's 2000 use of $> 90\%$ correct assignment using posterior probabilities of $P > 0.95$ in linear discriminant analysis for morphometric data). The goal was to establish reasonable threshold levels for these sorts of tests where they have often been absent, adapting quantitative criteria from the conceptual approaches of John C. Avise and R. Martin Ball Jr. (1990), Ball and Avise (1992), and Crandall and colleagues (2000). Congruence among multiple data sets suggests a higher

Short-nosed echidna (*Tachyglossus aculeatus*) in Australia. EDGE rank # 359. Tom McHugh/Photo Researchers, Inc.

degree of confidence in genetic uniqueness and thus a higher conservation priority.

The phylogenetic ranking approach

One way of avoiding the species debate is to prioritize the rank of groups regardless of what they are called (e.g., species, subspecies, DPS) based on their degree of evolutionary distinctiveness. The Zoological Society of London (ZSL) has developed a quantitative, phylogenetic approach to ranking species and their conservation priorities based on measures of evolutionary distinctiveness and level of endangerment (Isaac et al. 2007; ZSL). This method, dubbed "evolutionarily distinct and globally endangered" (EDGE), gives higher rankings to species that are monotypic genera (a single species in the genus) and lower rankings to species that are one of many found in a genus. For example, in ranking 4,173 mammal species, the top three ranking evolutionarily distinct species were found to be the duck-billed platypus (*Ornithorhynchus anatinus*), aardvark (*Orycteropus afer*), and short-beaked echidna (*Tachyglossus aculeatus*). Clearly, all of these represent evolutionarily distinct lineages, like those that Linnaeus would have had confidence in. In contrast, speciose groups that are widespread, such as the murid rodents (mice), tend to rank low on the list.

This approach is similar in principle to the temporal ranking scheme proposed by Avise and Glenn C. Johns (1999). The

attractiveness of this approach, which is based on phylogenetic measures of uniqueness and hierarchical ranking, is that it solves some of the inherent problems associated with drawing lines through a continuum by ranking groups *across* the entire range of genetic variation in major taxonomic groups (e.g., class). It also provides an objective means for setting conservation priorities and therefore some degree of regulatory consistency under biodiversity laws and treaties. If this approach were expanded to include species and presumably distinct populations, setting conservation priorities would acquire a level of scientific rigor and repeatability that is presently lacking. Such a ranking system is robust and can readily adapt to the introduction of new information (Isaac, Mallet, and Mace 2004).

Should we look before we leap?

Would conservation efforts be more effective if species, subspecies, and populations were tested against consistent threshold(s) of genetic uniqueness prior to being listed as endangered? Or would a ranking of taxa and populations based on phylogenetic uniqueness similar to the EDGE program provide a more workable approach? Either approach could provide a consistent basis for allocation of more conservation resources to evolutionarily distinct taxa while winnowing out erroneously designated ones. Clearly, there are now more precise conceptual and analytical tools that can be used to objectively distinguish or rank species, subspecies, and distinct population segments than there were in the early twentieth century (e.g., Crandall et al. 2000; Sites and Marshall 2003; Baker and Bradley 2006). In addition, the application of a consistent method could remove much of the subjectivity that has dominated taxonomic decisions over the last several hundred years. Until such as a paradigm shift takes place, species will likely remain "in the eye of the beholder."

Resources

Books

Avise, John C., and R. Martin Ball Jr. 1990. "Principles of Genealogical Concordance in Species Concepts and Biological Taxonomy." In *Oxford Surveys in Evolutionary Biology*, Vol. 7: *1990*, ed. Douglas Futuyma and Janis Antonovics. Oxford: Oxford University Press.

Darwin, Charles. 1859. *On the Origin of Species by Means of Natural Selection; or, The Preservation of Favoured Races in the Struggle for Life*. London: John Murray.

Popper, Karl R. 1962. "Science: Conjectures and Refutations." In *Conjectures and Refutations: The Growth of Scientific Knowledge*. New York: Basic Books.

Periodicals

Agapow, Paul-Michael, Olaf R. P. Bininda-Emonds, Keith A. Crandall, et al. 2004. "The Impact of Species Concept on Biodiversity Studies." *Quarterly Review of Biology* 79(2): 161–179.

Avise, John C., and Glenn C. Johns. 1999. "Proposal for a Standardized Temporal Scheme of Biological Classification for Extant Species." *Proceedings of the National Academy of Sciences of the United States of America* 96(13): 7358–7363.

Baker, Robert J., and Robert D. Bradley. 2006. "Speciation in Mammals and the Genetic Species Concept." *Journal of Mammalogy* 87(4): 643–662.

Ball, R. Martin, Jr., and John C. Avise. 1992. "Mitochondrial DNA Phylogeographic Differentiation among Avian Populations and the Evolutionary Significance of Subspecies." *Auk* 109(3): 626–636.

Chaitra, M. S., Karthikeyan Vasudevan, and Kartik Shanker. 2004. "The Biodiversity Bandwagon: The Splitters Have It." *Current Science* 86(7): 897–899.

Chevron U.S.A., Inc. v. Natural Resources Defense Council, Inc., 467 U.S. 837 (1984).

Crandall, Keith A. 2006. "Advocacy Dressed Up as Scientific Critique." *Animal Conservation* 9(3): 250–251.

Crandall, Keith A., Olaf R. P. Bininda-Emonds, Georgina M. Mace, and Robert K. Wayne. 2000. "Considering Evolutionary Processes in Conservation Biology." *Trends in Ecology and Evolution* 15(7): 290–295.

Cronin, Matthew A. 1993. "Mitochondrial DNA in Wildlife Taxonomy and Conservation Biology: Cautionary Notes." *Wildlife Society Bulletin* 21(3): 339–348.

Cronin, Matthew A. 1997. "Systematics, Taxonomy, and the Endangered Species Act: The Example of the California Gnatcatcher." *Wildlife Society Bulletin* 25(3): 661–666.

Cronin, Matthew A. 2006. "A Proposal to Eliminate Redundant Terminology for Intra-species Groups." *Wildlife Society Bulletin* 34(1): 237–241.

Cronin, Matthew A. 2007. "The Preble's Meadow Jumping Mouse: Subjective Subspecies, Advocacy, and Management." *Animal Conservation* 10(2): 159–161.

Fischman, Robert L., and Vicky J. Meretsky. 2001. "Endangered Species Information: Access and Control." *Washburn Law Journal* 41(1): 90–113.

Hall, E. Raymond, and Keith R. Kelson. 1959. *The Mammals of North America*, Vol. 2. New York: Ronald Press.

Hendry, Andrew P., Steven M. Vamosi, Stephen J. Latham, et al. 2000. "Questioning Species Realities." *Conservation Genetics* 1(1): 67–76.

Hennig, Willi. 1966. *Phylogenetic Systematics*, trans. D. Dwight Davis and Rainer Zangerl. Urbana: University of Illinois Press.

Hey, Jody. 2006. "On the Failure of Modern Species Concepts." *Trends in Ecology and Evolution* 21(8): 447–450.

Hull, David L. 1965. "The Effect of Essentialism on Taxonomy: Two Thousand Years of Stasis." *British Journal for the Philosophy of Science* 15(60): 314–326.

Isaac, Nick J. B., Georgina M. Mace, and James Mallet. 2005. "Response to Agapow and Sluys: The Reality of Taxonomic Change." *Trends in Ecology and Evolution* 20(6): 280–281.

Mallet, James. 2007a. "Species, Concepts of." In Vol. 5 of *Encyclopedia of Biodiversity*, ed. Simon Asher Levin. 2nd edition. San Diego, CA: Academic Press.

Mallet, James. 2007b. "Subspecies, Semispecies, Superspecies." In Vol. 5 of *Encyclopedia of Biodiversity*, ed. Simon Asher Levin. 2nd edition. San Diego, CA: Academic Press.

Mayr, Ernst. 1970. *Populations, Species, and Evolution*. Cambridge, MA: Harvard University Press, Belknap Press.

Meiri, Shai, and Georgina M. Mace. 2007. "New Taxonomy and the Origin of Species." *PLoS Biology* 5(7): 1385–1386.

Merriam, C. Hart. 1918. *Review of the Grizzly and Big Brown Bears of North America (Genus* Ursus*) with the Description of a New Genus*, Vetularctos. North American Fauna 41. Washington, DC: U.S. Government Printing Office.

Platt, John R. 1964. "Strong Inference: Certain Systematic Methods of Scientific Thinking May Produce Much More Rapid Progress than Others." *Science* 146(3642): 347–353.

Ramey, Rob Roy, II, Hsiu-Ping Liu, Clinton W. Epps, et al. 2005. "Genetic Relatedness of the Preble's Meadow Jumping Mouse (*Zapus hudsonius preblei*) to Nearby Subspecies of *Z. hudsonius* as Inferred from Variation in Cranial Morphology, Mitochondrial DNA, and Microsatellite DNA: Implications for Taxonomy and Conservation." *Animal Conservation* 8(3): 329–346.

Ramey, Rob Roy, II, John D. Wehausen, Hsiu-Ping Liu, et al. 2006. "Response to Vignieri et al. (2006): Should Hypothesis Testing or Selective Post Hoc Interpretation of Results Guide the Allocation of Conservation Effort?" *Animal Conservation* 9(3): 244–247.

Ramey, Rob Roy, II, John D. Wehausen, Hsiu-Ping Liu, et al. 2007. "How King et al. (2006) Define an 'Evolutionary Distinction' of a Mouse Subspecies: A Response." *Molecular Ecology* 16(17): 3518–3521.

Sites, Jack W., Jr., and Jonathon C. Marshall. 2003. "Delimiting Species: A Renaissance Issue in Systematic Biology." *Trends in Ecology and Evolution* 18(9): 462–470.

The Economist. 2007. "Hail Linnaeus. Conservationists—and polar bears—should heed the lessons of economics." *The Economist* 383(8529): 13.

United States v. Mead Corp., 121 S. Ct. 2164 (2001).

Wehausen, John D., and Rob Roy Ramey II. 2000. "Cranial Morphometric and Evolutionary Relationships in the Northern Range of *Ovis canadensis*." *Journal of Mammalogy* 81(1): 145–161.

Wilson, E. O., and W. L. Brown Jr. 1953. "The Subspecies Concept and Its Taxonomic Applications." *Systematic Zoology* 2(3): 97–111.

Worley, K., C. Strobeck, S. Arthur, et al. 2004. "Population Genetic Structure of North American Thinhorn Sheep (*Ovis dalli*)." *Molecular Ecology* 13(9): 2545–2556.

Zink, Robert M. 2004. "The Role of Subspecies in Obscuring Avian Biological Diversity and Misleading Conservation Policy." *Proceedings of the Royal Society* B 271(1539): 561–564.

Other

Isaac, Nick J. B., Samuel T. Turvey, Ben Collen, et al. 2007. "Mammals on the EDGE: Conservation Priorities Based on Threat and Phylogeny." *PLoS ONE* 2(3): e296. Available from http://www.plosone.org/article/fetchArticle.action?articleURI=info%3Adoi%2F10.1371%2Fjournal.pone.0000296.

Ramey, Rob Roy, II. 2007. Written testimony before the Committee on Natural Resources, U.S. House of Representatives, for the hearing *Crisis of Confidence: The Political Influence of the Bush Administration on Agency Science and Decision-Making*. 110th Cong., 1st sess., July 31, 2007. Available from http://lamborn.house.gov/UploadedFiles/07%2007%20Ramey_Testimony.pdf.

Quarles, Steven P. 2001. "Issues Concerning the Listing of Distinct Population Segments of Vertebrates under the Endangered Species Act of 1973." Testimony before the Subcommittee on Fish, Wildlife, and Water, Environment and Public Works Committee, U.S. Senate, May 9, 2001. Available from http://epw.senate.gov/107th/qua_0509.htm.

Wilkins, John S. 2007. "Species." ScienceBlogs. Available from http://scienceblogs.com/evolvingthoughts/2007/01/species.php.

Zoological Society of London (ZSL). "EDGE of Existence: Evolutionarily Distinct and Globally Endangered." Available from http://www.edgeofexistence.org.

Rob Roy Ramey, II

The fossil record: A window to the past

Paleontology is the study of ancient life through the remains or traces (*fossils*) preserved in the Earth's crust either in rocks or in unconsolidated sediments. Paleontology is closely allied with geology, sharing interests in rock sequence correlation and reconstruction of ancient environments; and with biology, sharing interests in biological structures and the evolution of life on Earth. No fossil can be "complete" in comparison to the original organism, so the *fossil record* is the sum total of what scientists reconstruct on the basis of the variably preserved remains.

Greek scholars such as Xanthos of Sardis recognized as early as 500 B.C.E. that fossils were the remains of ancient animals, and Xenophanes of Colophon viewed shells in rock as evidence that the sea had once covered the land. Fossils of marine organisms likely had contributed even earlier to ancient beliefs that a great flood had once covered the Earth. In the Middle Ages, European beliefs turned to the idea that fossils had grown in the rocks through some force of change that may or may not have been supernatural. Leonardo da Vinci (1452–1519) echoed the classical view, interpreting fossils as remains of ancient animals and concluding that the sea had once covered the land. In the early twenty-first century, scientists recognize that fossils are of many different ages and that such raising of marine fossils above sea level, even to rocks on mountainsides, has involved not only changes in sea level but also tectonic processes such as folding, faulting, and localized uplift.

Geological roots

The fundamental principle underlying the science of geology is the *principle of uniformity*, formulated by the Scottish geologist James Hutton (1726–1797). This principle holds that the processes acting today are sufficient to explain past Earth history, given sufficient time. Several lines of information indicate that Earth is about 4.6 billion years old. An opposing perspective, catastrophism, is based in a literal interpretation of the Bible, holding that Earth is only about 6,000 years old. If the latter were true, then Earth would have had to have been formed by processes unimaginably different from those seen operating today.

Even before radiometric dating had provided absolute numbers for the age of the Earth, *stratigraphers* had been able to arrange sedimentary rock strata into a chronological sequence using principles established in the mid-seventeenth

century by the Danish geologist Nicolaus Steno (1638–1686) from his studies in Italy. Steno recognized that sediments tended to be deposited in horizontal layers, or strata, with the oldest layers at the bottom and the youngest layers at the top of a given sequence. Fossils could occur in such strata, and Steno's English contemporary, Robert Hooke (1635–1703), suggested that fossils might be used to establish the ages of rocks, and that the Earth must have moved to bring fossil shellfish to mountain tops. William Smith (1769–1839), in England, initiated geological mapping in the late eighteenth century and recognized that specific strata had distinctive *assemblages* of co-occurring fossils. The strata could be arranged in relative chronological sequences, and the fossil assemblages provided evidence to allow correlation of rock strata over great distances.

The giant ground sloth fossil (*Eremotherium laurillardi*) is the most spectacular of the four North American ground sloths. This huge animal weighed as much as a mammoth, could rear up as high as a giraffe, and had claws the size of a man's forearm. Courtesy of John H. Lienhard

It was soon recognized that the fossil assemblages could be used to establish a global *geological timescale*, a sequence of subdivisions of Earth time—eons, eras, periods, and epochs—from larger to smaller subdivisions. Time boundaries were set up on the basis of observed changes in the fossil record. With fossils as the key basis for global stratigraphic correlation, paleontology became an indispensable aspect of exploring expeditions, from the spread of European empires to the exploration of the U.S. and Canadian west. The second U.S. president, Thomas Jefferson (1743–1826), took a personal interest in paleontology and participated in the discovery of the giant ground sloth that bears his name: *Megalonyx jeffersonii*. He instructed the explorers Meriwether Lewis and William Clark to collect natural history specimens, including fossils, and keep a careful watch for creatures that might or might not be extinct, such as the mammoth.

The advent of radiometric dating in 1950 provided absolute ages for established geological events. Radiometric dating worked best with minerals in igneous rocks, while fossils allowed correlation of sedimentary rocks. The realization that microfossils were both abundant and widespread meant that thousands of specimens could be obtained from a small sample of rock and that their larger counterparts, macrofossils, were no longer as important in developing stratigraphic correlations. Macrofossils remain important in the early twenty-first century, however, as direct evidence of evolutionary changes over time. Studies of modern DNA clearly support the common ancestry of existing organisms, but investigations of fossils go beyond this to reveal many lineages that did not survive to the present day, and hence their relationship to other living and/or extinct life-forms cannot be evaluated on the basis of genetic material.

How fossils are formed

Fossilization is not an event but an ongoing process, closely linked to the *rock cycle*. The rock cycle was established largely through the studies of Hutton in the late eighteenth century. With molten magma as a starting point, cooling and solidification produce igneous rocks. Crystalline igneous rocks are fine-grained if magma cooled rapidly on the Earth's surface as lava, or coarse-grained if the magma cooled slowly in the subsurface. Volcanic rocks also include pyroclastics, formed by consolidation and cooling of explosively ejected droplets and particles. Igneous rocks, once exposed at the surface, break down through weathering into mineral grains and dissolved ions. These materials can be transported (as sediment) by air, water, or ice and deposited far from their point of origin. Sedimentary deposits are compacted and cemented into sedimentary rocks. Being of Earth-surface origin, these rocks can contain fossils. In the subsurface, both igneous and sedimentary rocks can be altered by heat and pressure (short of melting) into metamorphic rocks. Melting returns metamorphic rocks to magma, thus completing the cycle.

Modification of sediments into sedimentary rocks involves processes of alteration that are grouped together as *diagenesis*. Fossilization is just a special example of diagenesis: The same processes that affect the surrounding sediments also alter the buried remains of organisms. A fossil may undergo a complex series of changes over a long period of time.

Dr. Robert T. Bakker, Curator of Paleontology at the Houston Museum of Natural Science, prepares to remove the humerus bone of a Dimetrodon found on the Craddock Ranch in Seymour, Texas. © Rex C. Curry/Dallas Morning News/MCT.

Fossils are grouped as *body fossils*, direct remains of the organism or its parts; or *trace fossils*, indirect indications of an organism's former presence, such as footprints or filled burrows. This distinction is usually clear, but there can be ambiguity. For example, a fish skeleton or a tree leaf could completely decompose leaving only an impression where it once rested. How is this different from a footprint? One argument is that living animals made the footprint and burrow, whereas the others were impressions formed after death, during fossilization. Not all body fossils represent the death of organisms, either: Living deer shed their antlers and crabs molt by shedding their skins, and these can in turn become body fossils.

Fossil *preservation types* depend upon the diagenetic environment. Some are not mutually exclusive, whereas others are distinctive. *Direct preservation* involves the presence of little-altered remains, such as the frozen meat and hair of a woolly mammoth or the chitinous bodies of insects trapped in amber. *Desiccation* can occur in dry environments such as deserts or caves, in which case natural "mummification" can result. Mummified duck-billed dinosaur remains were discovered as

Baltic amber with mosquito fossilized, 10 million years old. © Katrina Brown, 2009/Shutterstock.com.

early as 1908 in Wyoming by Charles H. Sternberg, and others have since been found, allowing detailed study of the scale patterns of their skins.

Carbonization involves the destructive distillation of organic tissues, leaving behind a direct residue as a carbon film, usually dark gray or black. The carbon may be remobilized through bacterial action in iron carbonate (sidcrite), resulting in an orange "stain." Carbonized remains often include leaves and wood but can also involve organic tissues of an animal, preserved as a "stain" around a skeleton. *Recrystallization* also involves the original materials of the organism. In this case, a biomineral compound is altered in the form of its mineral lattice with minor (or no) chemical change. The best example involves the minerals aragonite and calcite, both with the chemical formula $CaCO_3$. Many animals build calcareous skeletons, as do coralline algae, but the crystal lattice of an aragonite skeleton will convert under moderate burial pressure to calcite. Such a change destroys the fine microstructure that initially existed in the skeleton. Aragonite is the iridescent "mother-of-pearl" in mollusks but in recrystallizing to calcite becomes a white and often powdery material. During the process of change, however, the fossil will have an unstable crystal lattice of aragonite and calcite, providing spectacular plays of color.

Permineralization, the classic "petrification" (or petrifaction) process, is frequent for porous materials such as bone, in which microscopic pores or branching tubules allow groundwater to penetrate deeply. Permineralization ("per-" means "through") begins as pore spaces are filled with minerals such as calcite, hematite, or silica, precipitated from percolating groundwater. As and after this happens, the surrounding bone mineral is being replaced on an atom-by-atom basis by other precipitated minerals. Wood can also be permineralized, with the cells being filled with precipitated mineral, followed by replacement of cell walls. Through this process finely detailed microstructures can be preserved.

Replacement involves the loss, through dissolution, of the original material from which an organism was made, and the substitution of another mineral of inorganic origin. It can

occur atom-by-atom in the permineralization process or through the more extensive dissolution, leaving a cavity or mold in the rock as a "negative" of the original specimen. Both external and internal molds (*steinkerns*) can be created if the original fossil is hollow (as with a snail or clam shell). The mold is filled by sediments or by growth of precipitated mineral crystals, producing a cast or replica ("positive") of the original. Mold-and-cast replacement generally destroys internal microstructural details.

Lagerstätten are fossil occurrences with unusual and sometimes spectacular preservation involving evidence of soft parts as well as hard parts, as a result of rare combinations of geological and fossilization processes. Such sites provide assemblages of soft-bodied organisms not otherwise known from the fossil record; therefore their richness by far offsets their rarity. Such sites and faunas as the Burgess Shale (Cambrian, western Canada), the Chengjiang fauna (Cambrian, China), the Mazon Creek fauna (Pennsylvanian, Illinois), the Solnhofen Limestone (Jurassic, Germany), the Green River formation (Eocene, Wyoming), and the Messel Pit (Eocene, Germany) are world famous because of the insights they provide about past biodiversity and detailed anatomy.

Discovering and preparing fossils

A popular view is that scientists "stumble across" fossils, but discoveries usually involve diligent searches. Fossil invertebrates are far more abundant than vertebrates, for which a search can be very painstaking. A region is chosen based on the presence of rocks of appropriate type (sedimentary; marine or nonmarine) and age. Rock exposures are searched for fossils revealed by erosion, and rocks showing such signs are, in some cases, further broken open to reveal hidden fossils. Many large vertebrate fossils are recovered from "badlands" where sedimentary rocks rich in expanding clays derived from volcanic ash are rapidly

Fossilized shells of the brooch clam (*Myophorella elisae*) set in sandstone. They flourished in the deep seas of the Mesozoic era. © DK Limited/ Corbis.

Paleontologist carefully excavates fossil bones of a ceratopian dinosaur, dated about 70 million years ago, during an excavation in south central New Mexico, east of the Rio Grande. Ceratopians were large, heavy-bodied horned dinosaurs, which include the well-known triceratops, that lived during the Late Cretaceous period. © Ray Nelson/MedNet/Corbis.

eroded from repeated wetting and drying Examples include the Red Deer River badlands at Dinosaur Provincial Park in Alberta and the Judith River badlands in Montana. Despite popular images, dinosaurs did not live in badlands: Their bones are revealed because of rapid modern erosion.

Vertebrate paleontologists will often walk along foot slopes, searching for bone fragments carried downslope by rainstorms. When one is found, a search is made upslope along erosional rills until remnants of the eroding bone are found in place. It may, in turn, be part of a still-buried skeleton. In many cases, and particularly with larger specimens, fossils are not removed from their matrix in the field, because this could result in damage. Small fossils are easily carried in matrix, but a dinosaur skeleton must be isolated in one or more blocks of bedrock that will be encased in protective fabric and plaster jackets. The large blocks are then removed with the aid of machinery and shipped to the laboratory. There, careful preparation through the delicate removal of enclosing rock may take months of work, using small chisels and vibrating tools, as well as airbrushes with fine grit. Fossils in limestone may be prepared through the use of acid to dissolve the surrounding matrix, and specialized solvents may be used to help disperse other types of matrix.

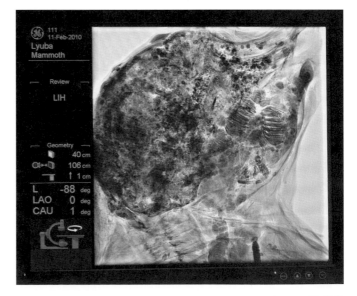

This image shows a three-dimensional x-ray of a head from a 42,000-year-old wooly mammoth baby at GE Healthcare Institute in Waukesha, Wisconsin. Although her soft tissue had dried somewhat, the mammoth was surprisingly well-preserved. © Mark Hoffman/Milwaukee Journal Sentinel/MCT.

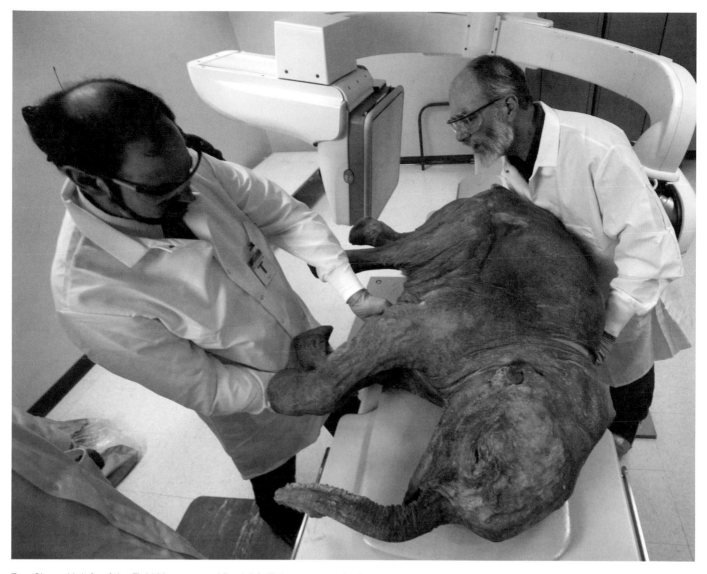

Tom Skwerski, left, of the Field Museum, and Daniel C. Fisher, curator of paleontology at the University of Michigan Museum of Paleontology, position the remains of a 42,000-year-old wooly mammoth baby. Mark Hoffman/Milwaukee Journal Sentinel/MCT.

How old is this fossil?

The age of a fossil can be determined in both relative and absolute terms. Relative dating involves placing a fossiliferous rock unit as part of a stratigraphic sequence, as was the basis for the sequence of geological periods. Absolute dating involves placement in terms of a calibrated system of time, measured in years. Absolute dating of ancient rocks awaited Marie and Pierre Curie's discovery and documentation of radioactivity in 1898. Many elements have unstable isotopes that "decay" to stable isotopes (often of other elements) through the emission or absorption of nuclear particles, which produces measurable radioactivity at a known rate. A ratio of the amount of the daughter isotope to the amount of the remaining parent isotope can be converted into an absolute date. Radiometric techniques are most readily applied to igneous rocks, dating the time of crystallization of mineral grains from molten magma, and encompass the full 4.6 billion years of Earth history. Radiocarbon dating is a

radiometric method applied to biological materials, but the short half-life of carbon-14 limits its use to materials less than about 45,000 years old. A good fossil record of complex organisms extends throughout the Phanerozoic eon, the last 544 million years of Earth history, and it is now clear that less complex life-forms came into existence as early as 3.5 billion years ago.

Some people reject such findings and argue for the literal Biblical interpretation of a young Earth. If this is the case, truth must itself be relative, because the biblical story is only one of many different creation stories that vary in detail from culture to culture. Geological relativism in this sense would require a different interpretation of Earth history and a different age of the Earth for each human cultural group, according to their varied beliefs. Science is an attempt to provide testable explanations that are not culture-bound and differs from religion in that the test of an explanation's validity cannot be on the basis of faith. In science new observations can challenge and refine the written record of knowledge; in religion, by contrast, observations that conflict with prevailing thought must be challenged as a test of faith. These are different systems of thought that share the pervasive human desire to explain the natural world; but they deal with different questions. Science tends to deal with "how" questions (Earth and life processes), whereas religion leans to culturally relative "why" questions (ultimate purpose and meaning of life).

Interpreting fossils

Ivan A. Efremov introduced taphonomy in 1940 as the "science of embedding": the study of what has happened to an organism from the point of death through its burial (fossilization) in the geological environment to its ultimate exposure and collection by scientists. A taphonomic perspective ensures that interpretation does not simply jump from fossil to reconstruction of form, behavior, or ancient environments. The circumstances of recovery, discovery, exposure, burial, postmortem modification, and death must be investigated sequentially (in reverse order from their history). This stepwise contextual study parallels the methods of forensics as applied in crime scene investigation. For example, it is unwise to reconstruct an ancient animal without first assessing whether a fossil was deformed (crushed or sheared) when or after it was buried.

Once changes due to erosion, diagenesis, scavenging, and other factors are ruled out, a fossil specimen can be compared with others of similar character. Fossil species are classified and named in similar manner to modern species, but the two are not strictly comparable. A modern species (*biospecies*), where possible, is established on the basis of a biological test: It includes all individuals capable of interbreeding and producing viable offspring under natural conditions. Such a test cannot be carried out for fossils, which are grouped into *morphospecies* ("form-species") based upon close similarity of form and structure.

Once fossils are described, samples are compared across space and time to reveal geographic variation and evolutionary relationships. When no direct biological evidence of linkage can be found, lineages are inferred on the basis of closest morphological (phenotypic) similarity. Each statement of evolutionary relationships is therefore, in a perspective of science, a testable *hypothesis*. Before Charles Darwin's time the French comparative anatomist Georges Cuvier (1769–1832) had recognized that fossil faunas had changed through geologic time and that extinctions had occurred. Darwin (1809–1882), in his 1859 work, *On the Origin of Species by Means of Natural Selection*, provided an evolutionary explanation for such sequences, arguing that all life had evolved from a distant common ancestor and that the fossil record documented the branching lineages leading up to modern organisms.

Paleontologists find the remains of a ten million year old rhinoceros at a dig site in Orchard, Nebraska. © Annie Griffiths Belt/Corbis.

Darwin's logic in *Origin of Species* was based on observations from living organisms. All modern species exhibit variability in form and physiology, including tolerances. Following Thomas Robert Malthus (1766–1834), Darwin saw that organisms produce more young than can possibly survive, given the carrying capacity of the environment. Therefore certain individuals die without reproducing, and Darwin suggested that this was due to natural selection favoring some over others. Natural selection could therefore cause a population to drift, over time, toward the favored type. Darwin also knew from studies of selective breeding in domestic animals that new and unusual characteristics could sometimes appear. He did not understand the mechanisms by which such mutations could arise, but their presence was further evidence that populations could change in character over time: That is, they could evolve. The fossil record, in turn, provided Darwin with an illustration of evolutionary change as expressed in a variety of lineages. Others over the ensuing 150 years would find many such lineages. Fossil horses provided a well-documented series showing the transition from a five-toed foot to an emphasis on a single toe, much more efficient in terms of running. Early snake fossils have now been found that show highly reduced limbs, confirming evolution from a four-limbed ancestor. Fossils now document much of the transition from lobe-finned fish to early amphibians.

Early studies of evolutionary sequences relied too much on generalized similarity and stratigraphic position as a basis for reconstruction of lineages (the *stratophenetic* technique). This approach fell short of scientific testability and was strongly subject to the experience of the analyst. Modern *cladistic* approaches look not for generalized similarity but for shared evolutionary novelties, meaning that a single characteristic may be enough to show relationships. Some similarities result from convergence (e.g., body form in fish, ichthyosaurs, and whales), so careful consideration must be given to this possibility. Cladistics is a properly scientific technique that documents all possible evolutionary alternatives for a given group, given the observable characteristics; these alternatives become formal evolutionary hypotheses, amenable to further testing. The principle of parsimony ("Occam's razor") is used to select the most plausible alternative based on the highest number of shared novelties and the smallest number of evolutionary convergences necessary to explain the relationships within a given group.

Where to view fossils

There are many museums with excellent fossil collections as well as protected sites that can be visited to see fossils in place in the rock. Many of the most famous museum collections were assembled during the period of expanding empires and western American exploration in the late nineteenth century. As interest in paleontology, particularly of the vertebrates, grows stronger and as human impact increases on the environment, more and more of the original sites are being protected and interpreted for their heritage values.

Resources

Books

Cutler, Alan. 2003. *The Seashell on the Mountaintop: A Story of Science, Sainthood, and the Humble Genius Who Discovered a New History of the Earth.* New York: Dutton.

Darwin, Charles. 1859. *On the Origin of Species by Means of Natural Selection; or, The Preservation of Favoured Races in the Struggle for Life.* London: John Murray.

Gould, Stephen Jay. 1989. *Wonderful Life: The Burgess Shale and the Nature of History.* New York: Norton.

Gould, Stephen Jay. 1998. *Leonardo's Mountain of Clams and the Diet of Worms: Essays on Natural History.* New York: Harmony Books.

Mayor, Adrienne. 2000. *The First Fossil Hunters: Paleontology in Greek and Roman Times.* Princeton, NJ: Princeton University Press.

Monroe, James S.; Reed Wicander; and Richard Hazlett. 2007. *Physical Geology: Exploring the Earth.* 6th edition. Belmont, CA: Thomson Brooks/Cole.

Olson, Everett C. 1971. *Vertebrate Paleozoology.* New York: Wiley-Interscience.

Prothero, Donald R. 2003. *Bringing Fossils to Life: An Introduction to Paleobiology.* 2nd edition. Boston: McGraw-Hill.

Winchester, Simon. 2001. *The Map That Changed the World: William Smith and the Birth of Modern Geology.* New York: HarperCollins.

Periodicals

Efremov, Ivan A. 1940. "Taphonomy: A New Branch of Paleontology." *Pan-American Geologist* 74: 81–93.

Hutton, James. 1788. "Theory of the Earth; or, An Investigation of the Laws Observable in the Composition, Dissolution, and Restoration of Land upon the Globe." *Transactions of the Royal Society of Edinburgh* 1(2): 209–304.

Osborn, Henry Fairfield. 1912. "Integument of the Iguanodont Dinosaur *Trachodon.*" *Memoirs of the American Museum of Natural History*, n.s., 1(2): 33–54.

Petrovich, Radomir. 2001. "Mechanisms of Fossilization of the Soft-Bodied and Lightly Armored Faunas of the Burgess Shale and of Some Other Classical Localities." *American Journal of Science* 301(8): 683–726.

Other

Spamer, Earle E., and Richard M. McCourt. 2006. "Lewis and Clark's Lost World: Paleontology and the Expedition." In *Discovering Lewis and Clark Web site*, ed. Joseph A. Mussulman. Available from http://www.lewis-clark.org/content/content-channel.asp?ChannelID=372.

Michael C. Wilson

What does the fossil record tell us about evolution?

The fossil record of life on Earth extends back 3.5 billion years to a time when the most complex organisms were cyanobacteria (Schopf 1999). Microfossils representing filamentous strings of cells occur in the Apex Chert in Australia, dating to 3.465 billion years ago, during the Archean eon. The fossil record reveals a generalized pattern of increasing diversity and the addition of increasingly complex organisms through time. This is consistent with evolutionary theory, as stated by Charles Darwin in 1859, that all life-forms share a common ancestor, that they diversified over time, and that complex organisms came from less complex ancestors. While Darwin saw the underlying process as gradual accretion of small increments, analogous to the role of compounding in population growth or interest-bearing investments, Darwin accepted that rates of evolution could vary over time.

The fossil record is nevertheless marked by major pulses of speciation and extinction, which might seem inconsistent with a gradualist model of descent with modification. Even before Darwin, some scholars cited these apparent times of major change as evidence against a uniformitarian view that the processes operating on Earth today are sufficient to explain the events of the past. The persistence of this rejection in some circles into the twenty-first century reflects a misunderstanding of the difference between uniformitarian and catastrophist viewpoints. The major events are now seen as involving modern processes but having long average recurrence intervals, consistent with a uniformitarian model. They are tied to impacts of celestial bodies, effects of continental drift, major climatic developments such as glaciations and global warming, and biological factors including the interspecific impact of evolutionary innovations such as organs of vision. Evolutionary rates may have also varied in response to fluctuations in selective pressures (natural selection) and rates of mutation. Clarification of these processes requires both neontological and paleontological studies (i.e., studies of modern organisms as well as ancient ones).

Though Darwin brought an evolutionary perspective to the public eye, the concept of continual change had been discussed by Roman philosophers, including Lucretius, who argued in *De rerum natura* that all organisms (including people) have undergone changes through time. This view was set aside as the Christian faith grew in influence in western Europe. Literal interpretation of the book of Genesis carried the view that the Creation produced the Earth precisely as we see it today, obviating the explanatory need for evolution. Such a need arose anew from increasingly detailed, direct observations of the natural world, including a dramatic increase in understanding of the fossil record from the 1700s onward. Although Leonardo da Vinci (1452–1519) had suggested that fossil shells found in rocks on mountaintops were evidence of changes in sea level, those changes could be accommodated by the story of the Noachian flood, a story that could well have been reinforced through observation by the ancients of in-place fossils.

Fossil succession: A key discovery

Recognition of fossil succession escalated the need for an evolutionary explanation. The French scholar Jean-Baptiste de Lamarck (1744–1829) had preceded Darwin in observing variation within species and in positing the interconnectedness of past and present life on Earth; he believed, however, in an ongoing process of creation, with existing "lower" life-forms having been more recently created and therefore having evolved less than "higher" life-forms such as humans. The fossil record was soon recognized to comprise a sequence of assemblages that showed continuities in group composition and body plans despite differences in detail from one taxonomic group to the next.

Other French scholars, such as Georges Cuvier (in the early 1800s) and Alcide d'Orbigny (in the 1840s), saw the fossil record as documenting a series of distinct creations and extinctions ("revolutions"), though simple logic showed that each one led to more complex forms than the last. Cuvier, whose view comprehended the entire sweep of geologic history, defined relatively few such cycles. Based on this worldview, Orbigny, making detailed studies of individual species' first and last appearances in western European strata, perceived the need for as many as twenty-seven creations and extinctions ("stages") within the Jurassic alone, so the total number for all of geologic time must have been much larger. This explanation of convenience became increasingly unacceptable to both geologists and theologians and was eventually ridiculed as the evidence accumulated (Berry 1987). Oddly, it seemingly required the *Creator* to have evolved, with each creation being different from the last. But it can be viewed more positively as part of a "calculus" of faunal succession, leading consistently toward an evolutionary model. To use an analogy, the area

The Berlin specimen of *Archaeopteryx* is one of the most famous fossils in the world. Seemingly half dinosaur and half bird, it has been called a fossil caught in the act of evolution. © Louie Psihoyos/Corbis.

within a circle is difficult to measure but can be estimated readily by summing the areas of complete squares nested within the circle: If one makes the squares infinitely small, the estimate verges on a true value. Similarly, as the number of creation-extinction stages increased, the explanation verged on an evolutionary model—not a reduction to absurdity but a reduction to evolution. The sequence of fossil assemblages was thus established not by scientists trying to "prove evolution" but by creationist geologists before Darwin's monograph appeared, and their efforts contributed directly to the growth of his body of evidence supporting evolutionary theory.

Neontology and paleontology

Questions regarding evolutionary mechanisms that involve genetics and physiology can be addressed best by neontologists, scientists who study now-living organisms. The oft-cited contrast in evolutionary models as to inheritance of acquired characteristics is not readily testable through the fossil record. Paleontologists do have access to genetic information from the relatively recent past and therefore can test certain scenarios involving evolutionary relationships, as with fossil bison (Shapiro et al. 2004). For fossils more than a few hundred thousand years old, however, such information remains elusive: What is available relates to observation of morphological changes and the assessment of relationships on the basis of shared morphological novelties.

Paleontologists lagged behind neontologists in employing and testing evolutionary theory, echoing Darwin's lament that the fossil record was too incomplete to provide a compelling test. The British paleontologist Richard Owen (1804–1892), who named the Dinosauria and studied the transitional reptile-bird *Archaeopteryx*, was a firm antievolutionist. Lamarckian perspectives as to inheritance of acquired characteristics persisted among paleontologists, including the American Edward Drinker Cope (1840–1897), into the early twentieth century. The American paleontologist Henry Fairfield Osborn (1857–1935) offered the sociopolitically influenced concept of "aristogenesis," a view that organisms share an innate force leading them "to strive for greater and higher achievements" (Prothero 2004, p. 68). This view was orthogenetic, the overall trajectory viewed as a series of steps directed consistently toward an evolutionary end point. Similar views have contaminated studies of human evolution, with fossil hominids discussed retrospectively in terms of their degrees of "hominization" as if that was the predetermined "goal" of human evolution.

The neo-Darwinian synthesis

Not until the 1940s and 1950s were compelling efforts made to integrate fossil evidence with developing views of population genetics and allopatric speciation. Ernst Mayr (1904–2005), George Gaylord Simpson (1902–1984), Julian Huxley (1887–1975), and others brought forth a synthesis that came to be called "neo-Darwinism," showing that each organism had a genetic code (genotype) expressed physically as its outward characteristics (phenotype). Natural selection acting upon small-scale phenotypic variations within populations created gradual incremental changes in gene frequency (microevolution) over many generations, which could lead to speciation. All changes were taken to be adaptive, as responses to natural selection for optimizing reproductive fitness in a given environment, and natural selection was pervasive (panselectionism). The fossil record typically provides examples only of phenotype, so testing of mechanisms and explanations was based on laboratory breeding and selection of modern short-generation organisms such as fruit flies. Even here, the link between microevolution and macroevolution (speciation events and the appearance of major adaptive features) remained unclear: Accumulation of gradual changes was not enough to account for the origin of new species, and subspecies, which were based upon regional variation, could not simply be viewed as incipient species.

While evolutionary theory itself is well established, new views regarding evolutionary mechanisms challenge the neo-Darwinian perspective. Panselectionism excluded the possibility that characteristics could be of neutral selective value or even maladaptive. Ironically, rejection of the Biblical concept of a "perfect" Creation gave way to "perfect" (in the sense of pervasively positive) adaptation. Yet, evolutionary transitions were far from perfect and were limited, if not hampered by, the range of possibilities provided by a given species' ancestors (phylogeny): ground sloths carried the legacy of arboreal life in their foot structures, and the human spine (with its attendant structural weaknesses and associated pain) is modified from a structure previously adapted for quadrupedal locomotion and support. Such a situation has been called a *kluge*, from Jackson Granholm's computer-related term for "an ill-assorted collection of poorly matching parts, forming a distressing whole" (Marcus 2008, p. 4).

If genetic changes can be neutral, then they might accumulate to the point that they are morphologically expressed and only then subject to natural selection. Understanding of this is clouded by the fact that genes can interact and some produce more than one protein (pleiotropy); hence, selection for one important character may carry along other neutral or even mildly negative characters. Some genes code for no proteins at all, and it is understood that the genome contains both structural genes, which code for morphological structures, and regulatory genes, which influence the expression of other genes. Thus some "disappearances" of morphological structures may involve the blocking action of regulatory genes, and through other changes to the regulatory gene the structure can reappear. Reappearances of previously suppressed characters include premolars in the diastema (anterior mandibular gap) of cattle and bison, teeth in chickens, flexible tails in humans, and lateral toes in horses. Similar observable structures in the paleontological record could be misinterpreted as documenting poorly visible parallel lineages that maintained primitive characters, rather than as reappearances in a single lineage. Their proper recognition can provide a measure of environmental stress where large samples of fossil populations are well controlled chronologically, as in the case of dental characters in Great Plains mass-kills of bison by human hunters over the past 10,000 years (Wilson 1988).

Challenging neo-Darwinism: New perspectives

The mechanisms that produced evolutionary novelty were a problem for Darwin and remained so for another century (Müller and Wagner 1991). The discovery of homeotic genes (those that control the transition of one body part into another) shows how major morphological changes (macromutations) can occur through alteration of the basic body plan or repositioning of structures. For example, the *Hox* gene complex controls basic segmentation of the body in both arthropods and vertebrates. Changes to this complex can produce new body plans, extra appendages, or extra segments in a single generation. This finding does not invalidate the Darwinian view but provides an alternative to gradualism.

The geological record shows that global change is ongoing with few, if any, long periods of environmental stasis. Consequently, living organisms are like players in a game with ever-changing rules. Simpson, in *Tempo and Mode in Evolution* (1944), argued that the fossil record did not provide special insights beyond those of the fruit-fly laboratory, but this understated its value as a test of evolutionary theory (Kidwell and Flessa 1995; Kidwell and Holland 2002). The immense time depth provided by the fossilized remains of long-dead or extinct creatures reveals long-term consistency in patterns of speciation and adaptation, supporting a single underlying set of laws. Paleontology also provides many examples of major speciation events, for which environmental or other correlates can be sought.

One aspect of the paleontological record underplayed by gradualists was the presence of times of morphological stasis for many fossil species, sometimes for millions of years, even in the face of well-documented environmental changes. Stephen Jay Gould and Niles Eldredge (1977) advanced the concept of *punctuated equilibrium* in species progression, arguing from fossil evidence that species underwent periods of relative stasis punctuated by times of rapid change (speciation events). Here, then, was a facet of evolutionary theory developed directly from the stratigraphic and fossil records. In fact, Darwin had suggested such a thing could occur (Penny 1983), but he focused more strongly on a gradualistic model. In the sense of punctuated equilibria, species could be said to have their own "life histories," which suggests the need for a body of theory addressing evolution above the population level: "species selection." Steven M. Stanley (1979) argued that in the case of two species in direct competition, differential survival is based upon general species properties and not natural selection at the individual or population level. The tendency toward speciation (as expressed in speciation rate) varies from one group to another, suggesting inherent species-level tendencies such as degree of variation or rate of gene flow.

Although many stratigraphic sequences of fossils do seem to show punctuated trajectories, times of nondeposition or erosion could appear to reflect apparent times of rapid change. Indeed, rapid change has been used as evidence for hidden stratigraphic breaks. This introduces the danger of circularity, requiring examples to be established not merely on the basis of sequences of fossils but also on detailed lithostratigraphic and geochronological studies, an arduous test. An additional matter is time perspective: North American bison over the past 12,000 years have shown gradual change to smaller body size and proportionally smaller horn cores, but from the perspective of geological time, this still seems sudden (Wilson 1980). Therefore, punctuated equilibria must be properly documented for each group on the basis of observed times of stasis and significant contrasts in *rates* of change.

Three and one-half billion years of life

Modern global warming is one aspect of a greater issue, that of ongoing global change. If the global record had been one of stability, it would have been easy to detect human-induced changes in the past two centuries. The past record has been one of nearly constant change, however, and it is more difficult to recognize "changes upon changes." Earth has repeatedly experienced greenhouse climates, when atmospheric carbon dioxide levels more efficiently trapped heat near the surface, as well as glaciations or even "icehouse" conditions when carbon dioxide levels fell. Organisms, in order to survive, have needed to deal with these climatic extremes. Adherents of James E. Lovelock's Gaia hypothesis (Lovelock 1979) have argued that environments on Earth are marvelously tuned to the preservation of life, with homeostasis providing just the right mix of factors. An evolutionary viewpoint takes a complementary tack: that organisms on Earth are well adapted to available environments through natural selection. The green pigment of plants (chlorophyll)

The deterioration of glaciers is an effect of global warming. This has already impacted species such as the polar bear. © Seth Resnick/ Science Faction/Corbis.

exploits the color region of highest light intensity in the filtered light spectrum that reaches Earth's surface. It is the process of adaptation that provides the fascinating linkage, but that also makes life itself vulnerable should conditions on Earth too rapidly change beyond the current tolerances of specific organisms. The fossil record shows that past events have produced several major extinctions involving the majority of complex life-forms that have existed on Earth.

Environments have changed dramatically over the 4.6 billion years of Earth's existence. For the past 3.5 billion years, life processes have played a significant role in the structuring of those physical environments. The early atmosphere was deficient in oxygen, because molecular oxygen outgassed from volcanoes was sequestered in iron oxides precipitated in Archean oceans. A surplus of oxygen resulted from the appearance of photosynthetic organisms, at first unicellular. Stromatolites are finely layered structures that grew upward in globular or columnar form from shallow Archean and Proterozoic seafloors, suggesting photosynthetic organisms capable of secreting carbonates upon their substrate. Cyanobacterial mats produce identical structures in restricted lagoonal environments today where evaporitic hypersaline conditions prevent other organisms from grazing and disrupting them. Elsewhere, modern cyanobacterial mats produce localized carbonate coatings on varied substrates, though grazers prevent these from developing into stromatolites. Despite their seeming simplicity, these organisms produced structures that would serve as a substrate for other organisms, and through photosynthesis they had a global effect on the prehistoric composition of the atmosphere.

Plate-tectonic movements involving continents and ocean floors began in the Archean and have exerted a strong influence on speciation. Fossils, geologic units, structural trends, and economic mineral deposits show that the modern continents were previously assembled into a vast supercontinent known as Pangaea, and that it was not the first such occurrence. Pangaea was ephemeral, having accumulated during the Paleozoic era through a series of collisions and approaches of moving plates. The periodic assembly of supercontinents (named the Wilson cycle after the Canadian geophysicist J. Tuzo Wilson) is rendered inevitable by the continual movement of plates on the surface of a sphere. Three such supercontinents are now documented: Pangaea (Permian, 245 million years ago [mya]), Rodinia (Neoproterozoic, 1 billion years ago), and Columbia or Nuna ("pre-Rodinia," Paleoproterozoic, 1.9 to 1.8 billion years ago) (Zhao, Sun, and Wilde 2002). Each was complemented by a single, more widespread ocean than any of today's, with correspondingly distinctive circulation patterns for oceanic waters. The recurrent separation and joining of continents provided opportunities for the separation and mixing of faunas and floras in the sea and on land, and therefore influenced evolution through isolation or competition.

Variations in rates of seafloor spreading and of subduction along ocean margins have caused changes in ocean basin volumes, in turn resulting in global patterns of shoreline transgressions and regressions: the basis for the modern study of sequence stratigraphy. Rising sea levels increased the areas of continental shelves (platforms) but correspondingly decreased the areas of exposed lands; falling sea level had the opposite effect. Thus times of evolutionary opportunity afforded by increased available territory were asynchronous between land and sea. Sea levels were highest in the Ordovician, a time of extensive epeiric seas (large but shallow seas that lie over a continent) and great marine faunal diversification. Some extinctions may be attributable to significant reductions in platform or land area and increased competition. Such extinctions would also be asynchronous between land and sea, so synchronous extinctions would require a different explanation.

By the Neoproterozoic era metazoans had appeared, comprising a varied assemblage called the Ediacara fauna (572 to 542 mya). Some resembled sea pens and were anchored to the seafloor, whereas others appear to have been mobile, segmented crawlers. Divergence times calculated on the basis of genetic differences between modern groups, calibrated to other dated divergence events in the fossil record, suggest considerable metazoan diversity prior to the "Cambrian explosion" that originally defined the base of the Cambrian and the start of the Phanerozoic eon. The relationships between Ediacaran and Cambrian groups are debated, but the Ediacara fauna may include early representatives of the Cnidaria, Annelida, and Arthropoda. In the minority view of Adolf Seilacher, most Ediacaran taxa were members of an extinct kingdom, Vendobionta, the body plans of which were based in repeated and interlinked pillowlike units (Fedonkin et al. 2008; Vickers-Rich and Komarower 2007). Studies of their morphological diversity suggest a rapid adaptive radiation (the "Avalon explosion") followed by a decline. The rapid diversification resembles the pattern of the so-called Cambrian explosion and may have been influenced by a preceding Neoproterozoic glaciation, oxygenation, and/or ecological interactions (Shen et al. 2008). The Ediacara fauna appeared after the "icehouse" climate associated with Neoproterozoic glaciation (750 to 580 mya) that might have caused extinctions and a genetic "bottleneck" constraining the diversity of taxonomic groups that survived into the warmer aftermath.

Oxygenation of the atmosphere appears itself to have been influenced by the initial appearance of a terrestrial biota, likely facilitated by the symbiosis of a photosynthetic organism and a fungus: an innovation sufficient to foster global change (Heckman et al. 2001; Fairchild and Kennedy 2007). Oxygenation by life processes may have reduced the greenhouse effect so much as to have brought on glaciation, but much remains to be learned about this sequence of events and their causation.

The Ediacaran "experiment" thus becomes an early complement to the Phanerozoic evolutionary faunas defined by John J. Sepkoski Jr. (1981). The Cambrian explosion initiated the first of these faunas and involved all of the major phyla that survive to this day, though their relative abundances were different then. Molecular divergence times and probable bilaterian fossils from Ediacaran sites suggest that the diversification occurred in the late Neoproterozoic, but there remains the relatively sudden appearance in the early Cambrian of diverse groups with hard skeletal parts, readily preserved as fossils. Multiple hypotheses for the appearance of hard parts need not be mutually exclusive. Hard parts can provide protection from predators (of which they are therefore evidence), attachment sites for muscles, mechanisms for attack or for excavation of the substrate, or protection from potentially damaging ultraviolet rays. For the appearance of such parts to have been simultaneous in many groups, an external cause must be sought that affected all, such as the appearance of predators with vision (Parker 2003), a change in the atmospheric filtering of ultraviolet light, or the opening up of significant shallow oceanic habitats by transgression. The last is documented for the Early Cambrian, and the presence of eyes is documented for the trilobites and in the poorly represented but likely predatory anomalocarids.

Sepkoski used factor analysis to define three temporally overlapping marine evolutionary faunas, each with its own time of maximum dominance and diversity. The "Cambrian fauna" peaked early in the Paleozoic era and included such archaic forms as archaeocyathids, trilobites, inarticulate brachiopods, archaic mollusks, and eocrinoids. Gould (1989) argued that the spectacularly preserved Burgess Shale fauna (Middle Cambrian) documented greater diversity in body plans than survived to later times, based for arthropod-like groups on the character of appendages and the placement of gills; more recent studies, however, tend to link the oddities to known groups (Briggs, Erwin, and Collier 1994). These groups went into decline after the Ordovician period and a Late Ordovician mass extinction event, as the "Paleozoic fauna" rose in importance. This second fauna, which dominated until another mass extinction at the end of the Permian, included rugose and tabulate corals and saw the proliferation of metazoan reefs. Other members of the fauna included articulate brachiopods, stenolaemate bryozoans, cephalopods, ostracods, crinoids, starfish, and graptolites. The succeeding "modern fauna" saw the dominance in the Mesozoic and Cenozoic eras of demosponges, gymnolaemate bryozoans, crustaceans, bivalves, gastropods, echinoids, and vertebrates. While coral reefs were again present, a different group, the scleractinians, constructed them. The Paleozoic fauna did include bivalves and gastropods in more sediment-rich nearshore environments, with the filter-feeding but sediment-sensitive brachiopods, crinoids, corals, and bryozoans dominating farther offshore. The Permian extinction more strongly affected the offshore groups, which were largely replaced in the Triassic as bivalves and gastropods expanded into the offshore environments as well.

The evolutionary faunas differed in that the degree of tiering increased over time: The Cambrian fauna was made up largely of organisms that lived on the seafloor, whereas the later faunas saw increasing numbers of attached organisms that extended well above the seafloor as well as organisms that burrowed into the substrate. Also increasing over time was the use of one organism as a substrate by others, as exemplified by reefs and by benthic (seafloor-dwelling) shelled organisms that were colonized by epifauna. Another factor driving evolutionary change and affecting diversity was escalation, as various groups of organisms took part in predator–prey "arms races." In response to the characteristics of the predator, the prey developed defensive mechanisms such as spines or thickened shells, or defensive behaviors. Different strategies might be developed by related species, resulting in their morphological divergence, as is illustrated by Miocene heart urchins of the genus *Lovenia*. Predatory cassid snails would cut disks out of the urchins to allow access to the soft parts. The urchins had large defensive spines to combat this, but over time developed fewer defensive spines and more digging spines, as selective pressure favored species that could burrow more deeply. Some urchins also moved to deeper waters, away from the snails. The urchins increased in size through time, but so did the predatory snails. The result was increased diversity of defensive types, reflected also by increasing diversity in the predators (McNamara 1991).

Fossils might suggest that animals (e.g., scorpions) preceded plants onto land, but the molecular record indicates the opposite (Pisani et al. 2004). Colonization of the land by invertebrate animals took place at least by the Ordovician, facilitated by a much earlier (Neoproterozoic) colonization sequence involving symbiosis of a photosynthetic organism and

Fossil worm (*Ottoia* sp.) from the Burgess Shale area. This priapulid worm lived in the Middle Cambrian era, over 500 million years ago. The Burgess Shale is important because many soft-bodied animals were trapped and became fossilized in mud in the deep-sea. Alan Sirulnikoff/ Photo Researchers, Inc.

The purple heart urchin (*Spatangus purpureus*) has developed a spiny covering to deter predators. © blickwinkel/Alamy.

a fungus (cryptogamic crusts). Body fossils are few for the earliest terrestrial animals, but trace fossils (burrows, possibly of millipedes) indicate their presence by the Late Ordovician (Retallack and Feakes 1987). Trilete spores and plant cuticle fragments from the Late Ordovician (if not earlier) document the presence of vascular plants. By the Early Devonian, communities of standing plants up to 50 centimeters tall were present, and by the Carboniferous there were forests of lycophytes with trees up to 35 meters tall.

Vertebrates were also taking to the land by the Late Devonian, the fossil record of which documents the transition from crossopterygian fish to early amphibians. The genus *Tiktaalik*, from the Canadian Arctic, is one of the finest examples of an evolutionary "intermediate" ever discovered. *Tiktaalik* was a fish with scales and gills but had a flattened, amphibian-like skull, a tetrapod pattern of fins, and a functional neck (Daeschler, Shubin, and Jenkins 2006). Amphibians remained tied to the water for egg laying, but the evolution of the amniote egg broke that link and allowed the reptiles to be truly terrestrial by the Early Carboniferous period. The divergence between synapsid reptiles (including the ancestors of mammals) and archosaur ancestors had

already occurred in the Permian, so much of the differentiation of major vertebrate groups had occurred before the end of the Paleozoic era. The transition from synapsids to mammals occurred during the Triassic and provides one of the more compelling examples of evolutionary change, particularly in the transformation of the lower jaw articulation and associated structures. The synapsid lower jaw had several bony elements and was hinged from the movable quadrate bone of the skull. Advanced synapsids show an expansion of the dentary (the tooth-bearing element of the lower jaw) and an extreme reduction of the postdentary elements, one of which was the articular. The articular and the quadrate, greatly reduced in size, eventually became the malleus and incus of the mammalian inner ear.

Despite their early differentiation, mammals did not dominate in the Mesozoic; that role was played by the dinosaurs, comprising two archosaur groups, the Saurischia and the Ornithischia. The Ornithischia were dominantly herbivorous and diversified into such groups as the ornithopods (including the duck-billed dinosaurs), the stegosaurs (plated dinosaurs), and the ceratopsians (horned dinosaurs).

University of Chicago scientist Neal Shubin shows a model of *Tiktaalik roseae*, the "missing link" fish whose fossils he discovered in 2004. © Chris Walker/Chicago Tribune/MCT.

Tiktaalik roseae is hailed as a missing link between fish and land animals. The 380-million-year-old fossil suggests that the animal had limb-like fins, which could have been used haul itself out of the water, scientists say. © Shawn Gould/National Geographic/KRT.

"harmony" in which changes to one will result in changes to the surrounding ecosystem and to other members of the community. This sort of zero-sum game also holds true for coevolution of competing or otherwise interacting species—for example, predators and prey. Leigh Van Valen (1973) referred to this as the "Red Queen hypothesis," after the experience of Alice in Lewis Carroll's *Through the Looking Glass* (1872), when the Red Queen ordered her to keep running to stay in the same place. Grasses have developed defenses against grazing animals, in the form of abrasive silica phytoliths ("plant stones") in their cell walls. These can rapidly wear down teeth, so over time grazing animals such as horses have been selected for higher and higher crowned grinding teeth, allowing them to live longer despite rapid tooth wear. Such animals do not die of "old age": Rather, as ranchers know, they die when their teeth wear out and can survive longer if fed less abrasive diets. Similarly, pursuit predators are more successful if they run faster, but this introduces selective pressure on the prey species also to do so: Selection for cursoriality (running ability) is a form of "arms race" between predator and prey. Predator guilds could experience evolutionary convergences through a relationship with a limited variety of prey species, as could guilds of prey species influenced by a narrow range of predators. Evolutionary novelties are the basis for modern cladistic studies of relationships, but the problem is how to distinguish them from convergences. If guilds of related prey species evolve in parallel in response to predator cursoriality, evolutionary novelties could well co-occur, building upon similar ancestral forms. Similar Red Queen considerations are involved in the relationships between host and parasitic species.

The Saurischia included the herbivorous sauropods, such as *Diplodocus* and *Camarasaurus*, and the largely carnivorous theropods, ranging from small chicken-sized creatures to the giant *Tyrannosaurus*. Recent finds show that several if not many theropods were feathered, and a close relationship with the birds is indicated. The birds arose during the Jurassic, completing the list of modern vertebrate classes. The dinosaurs offer many well-documented examples of evolutionary sequences, one of the best of which is the transition of the ceratopsians from bipedal forms such as *Psittacosaurus* to quadrupedal, heavy-skulled forms such as *Triceratops*. All ceratopsians share the expansion of the posterior margin of the skull into a shelflike frill, a lateral flaring of the jugal (cheek) bone, and the presence of a beaklike structure involving the predentary (typical of all ornithischians) and the rostral bone, an evolutionary novelty.

Coping with change and surviving extinction

Evolution must be viewed in light of ongoing global change: Constant evolutionary change is needed simply for an organism to "maintain its place" as conditions change around it. Organisms therefore do not exist in longstanding "balanced" relationships, though they do coexist in a form of

The *Triceratops* is the best-known of the ceratopsids, which were characterized by horns over the eyes and on the nose. © Louie Psihoyos/Corbis.

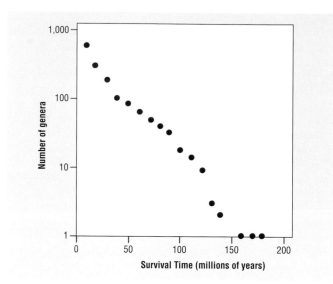

The Macroevolutionary Red Queen. Reproduced by permission of Gale, a part of Cengage Learning.

Some 90 percent or more of all species that have existed are now extinct, a figure likely in the low billions. More than anything else, perhaps, the fossil record provides direct evidence of past extinctions—of organisms that no longer exist. Many of these "extinctions" were phyletic; that is, they reflect evolutionary changes within persisting lineages. But others tell of entire lineages that have ended, and some authors have even called them "failed experiments" of evolution. By Cuvier's time it was realized that unusual fossils such as mastodons and mammoths could not simply be assumed to represent modern animal groups still to be found in some exotic corner of the world. Extinction was thought of, in the Darwinian perspective, as a result of evolutionary processes and the "struggle for survival." It seemed axiomatic that if one species outcompeted another, the latter would become extinct. Studies of island biogeography clearly show how this can happen when introduced species cause extinctions of endemics. There is evidence for an ongoing "background count" of nonphyletic extinctions, one species here and another there, each with its own last appearance datum. But are all of these disappearances really extinctions? *Lazarus taxa* are groups that seemingly disappear from the fossil record, only to reappear in later strata; this might reflect survival in restricted habitats or taphonomic factors, that is, how remains were buried and/or preserved. Corals were absent in the Early Triassic, and it appears that the Paleozoic coral groups became extinct, later to be replaced by another Anthozoan group. But Early Triassic oceans may have been too acidic for corals to secrete calcareous skeletons, allowing an alternative hypothesis that the organisms survived and evolved in soft-bodied form until conditions once again allowed precipitation of calcium carbonate (Fine and Tchernov 2007).

It is now realized that there were extinctions both from minor "background events" and major extrinsic events, the latter providing opportunities in lottery-like fashion for surviving groups. In the case of extinction through competition there is immediate replacement of one species by another; but with the extrinsic events, when Earth systems deviate from their typical limits, extinction can occur without immediate replacement. Extinction of both kinds has been as vital in constructing ecosystems as evolution has been (Stanley 1987). Mass extinctions, in the perspective of Earth history and global change, are direct evidence of evolution because they record times during which environmental parameters deviated significantly from long-term trends, exceeding the tolerances of many organisms. Of the "Big Five" mass extinctions in the Phanerozoic record, the Cretaceous-Tertiary (K-T) boundary event (65.1 mya) is best known to the general public because it involved the extinction of all dinosaurs except for their offshoot, the birds. But the greatest extinction of all was at the Permian-Triassic boundary (245 mya), at which time some 96 percent of marine species were wiped out, along with 75 percent of the families of land vertebrates. There is good evidence that the K-T event resulted from the impact of a large meteorite, and the same may have been true of the Permian-Triassic event. Such a bolide, exploding on impact, would have created a global atmospheric dust cloud, blocking sunlight for months, interfering with photosynthesis, and causing the breakdown of many food chains on land and in the oceans. Many vertebrate survivors of the K-T event were relatively small and could feed on decaying material or on the eaters of decaying material, and could burrow or remain for long periods underwater.

The fossil record of the ammonoid cephalopods provides a fascinating example of repeated adaptive radiation and extinction spanning the latter half of the Paleozoic and all of the Mesozoic era. Ammonoids arose from nautiloid ancestors and possessed similarly chambered shells but were distinctive in exhibiting a trend toward increasingly complex sutures, the contact between chamber walls and the inner surface of the shell wall (Kröger 2005). Suture lines changed over time from the simple curving nautiloid pattern to the zigzag angular pattern of the goniatites, then to the increasingly crenulated (frilled) patterns of the ceratites and ammonites. The frilling was an adaptation that reinforced the shell against external pressures without adding significant mass (De Blasio 2008). The ammonoids diversified to 30 families by the late Devonian, but the Frasnian-Famennian extinctions reduced these to only 2 families with 3 genera. In the Famennian they again radiated to 80 genera, but after another extinction there were only 2 at the close of the Devonian. Adaptive radiation produced 25 families with 180 genera by the late Mississippian, but extinctions left only 9 families surviving into the Pennsylvanian; again they radiated to 30 families. A slow decline began in the Middle Permian. With the Permian-Triassic boundary extinction only 2 families (each with 1 genus) survived; but in the Triassic these again diversified to 80 families with 500 genera. After another extinction event at the end of the Triassic, the group again expanded to 90 families in the Jurassic, and then saw a slow decline to 11 families in the Late Cretaceous (Prothero 2004). All were wiped out by the Cretaceous-Tertiary boundary event, though other cephalopod groups such as nautiloids and coleoids did survive. It is evident from this that the trajectory of the ammonoids had as much to do with the selective influence of mass extinctions as with

Ammonite (*Promicroceras planicosta*) 186 million years ago, from the Early Jurassic, England. © Ken Lucas/Visuals Unlimited/Corbis.

incremental evolutionary change. Mass extinctions, then, are natural selection "writ large."

Concluding remarks

In closing, the fossil record presents many examples of evolving lineages with evolutionary intermediates, or so-called missing links. *Tiktaalik* is becoming a "poster child" for this fact, and in the face of such evidence, an argument that we still do not have satisfactory evidence of intermediate forms is untenable. Evolution is to a degree testable by the fossil record, in that the "missing links" can often be

predicted before they are found, whether they be fossil snakes or whales with functional legs, birds with teeth, or dinosaurs with feathers. Prediction (in the form of hypotheses) and testing through observation or experiment are the hallmarks of science, and the great time depth of the fossil record allows scientists to see evolutionary processes playing out in a broad tableau populated by more extinct species than survivors. The very fact of extinction itself is strong evidence for natural selection and evolution. Gaia has not always been a congenial hostess and the story of life on Earth is organized around the theme of what is necessary in order to be a survivor.

Resources

Books

Berry, William B. N. 1987. *Growth of a Prehistoric Time Scale: Based on Organic Evolution*. Rev. edition. Palo Alto, CA: Blackwell Scientific.

Briggs, Derek E. G., Douglas H. Erwin, and Frederick J. Collier. 1994. *The Fossils of the Burgess Shale*. Washington, DC: Smithsonian Institution Press.

Darwin, Charles. 1859. *On the Origin of Species by Means of Natural Selection; or, The Preservation of Favoured Races in the Struggle for Life*. London: John Murray.

Gould, Stephen Jay. 1989. *Wonderful Life: The Burgess Shale and the Nature of History*. New York: Norton.

Fedonkin, Mikhail A., James G. Gehling, Kathleen Grey, et al. 2008. *The Rise of Animals: Evolution and Diversification of the Kingdom Animalia*. Baltimore: Johns Hopkins University Press.

Huxley, Julian. 1974. *Evolution: The Modern Synthesis*. 3rd edition. London: Allen and Unwin.

Lovelock, James E. 1979. *Gaia: A New Look at Life on Earth*. Oxford: Oxford University Press.

Lucas, Spencer G. 2007. *Dinosaurs: The Textbook*. 5th edition. Boston: McGraw-Hill.

Lucretius. 1916. *De rerum natura* [On the nature of things], trans. William Ellery Leonard. New York: Dutton.

Marcus, Gary. 2008. *Kluge: The Haphazard Construction of the Human Mind*. Boston: Houghton Mifflin.

Mayr, Ernst. 1942. *Systematics and the Origin of Species*. New York: Columbia University Press.

McNamara, Ken, and John Long. 2007. *The Evolution Revolution: Design without Intelligence*. 2nd edition. Carlton, Australia: Melbourne University Press.

Parker, Andrew. 2003. *In the Blink of an Eye*. Cambridge, MA: Perseus Publishing.

Prothero, Donald R. 2004. *Bringing Fossils to Life: An Introduction to Paleobiology*. 2nd edition. Boston: McGraw-Hill.

Schopf, J. William. 1999. *Cradle of Life: The Discovery of Earth's Earliest Fossils*. Princeton, NJ: Princeton University Press.

Sepkoski, J. John, Jr., and Arnold I. Miller. 1985. "Evolutionary Faunas and the Distribution of Paleozoic Benthic Communities in Space and Time." In *Phanerozoic Diversity Patterns: Profiles in Macroevolution*, ed. James W. Valentine. Princeton, NJ: Princeton University Press; San Francisco: American Association for the Advancement of Science, Pacific Division.

Simpson, George Gaylord. 1944. *Tempo and Mode in Evolution*. New York: Columbia University Press.

Stanley, Steven M. 1979. *Macroevolution: Pattern and Process*. New York: W. H. Freeman.

Stanley, Steven M. 1987. *Extinction*. New York: Scientific American Books.

Vickers-Rich, Patricia, and Patricia Komarower, eds. 2007. *The Rise and Fall of the Ediacaran Biota*. London: Geological Society.

Periodicals

Daeschler, Edward B., Neil H. Shubin, and Farish A. Jenkins Jr. 2006. "A Devonian Tetrapod-Like Fish and the Evolution of the Tetrapod Body Plan." *Nature* 440(7085): 757–763.

De Blasio, Fabio V. 2008. "The Role of Suture Complexity in Diminishing Strain and Stress in Ammonoid Phragmocones." *Lethaia* 41(1): 15–24.

Fairchild, Ian J., and Martin J. Kennedy. 2007. "Neoproterozoic Glaciation in the Earth System." *Journal of the Geological Society* 164(5): 895–921.

Fine, Maoz, and Dan Tchernov. 2007. "Scleractinian Coral Species Survive and Recover from Decalcification." *Science* 315(5820): 1811.

Gould, Stephen Jay, and Niles Eldredge. 1977. "Punctuated Equilibria: The Tempo and Mode of Evolution Reconsidered." *Paleobiology* 3(2): 115–151.

Granholm, Jackson. 1962. "How to Design a Kludge." *Datamation* 8(2): 30–31.

Heckman, Daniel S., David M. Geiser, Brooke R. Eidell, et al. 2001. "Molecular Evidence for the Early Colonization of Land by Fungi and Plants." *Science* 293(5532): 1129–1133.

Kidwell, Susan M., and Karl W. Flessa. 1995. "The Quality of the Fossil Record: Populations, Species, and Communities." *Annual Review of Ecology and Systematics* 26: 269–299.

Kidwell, Susan M., and Steven M. Holland. 2002. "The Quality of the Fossil Record: Implications for Evolutionary Analyses." *Annual Review of Ecology and Systematics* 33: 561–588.

Kröger, Björn. 2005. "Adaptive Evolution in Paleozoic Coiled Cephalopods." *Paleobiology* 31(2): 253–268.

McNamara, Ken. 1991. "Murder and Mayhem in the Miocene." *Natural History* 100(8): 40–47.

Müller, Gerd B., and Günter P. Wagner. 1991. "Novelty in Evolution: Restructuring the Concept." *Annual Review of Ecology and Systematics* 22: 229–256.

Penny, David. 1983. "Charles Darwin, Gradualism, and Punctuated Equilibria." *Systematic Zoology* 32(1): 72–74.

Pisani, Davide, Laura L. Poling, Maureen Lyons-Weiler, and S. Blair Hedges. 2004. "The Colonization of Land by Animals: Molecular Phylogeny and Divergence Times among Arthropods." *BMC Biology* 2(1).

Retallack, Gregory J., and Carolyn R. Feakes. 1987. "Trace Fossil Evidence for Late Ordovician Animals on Land." *Science* 235(4784): 61–63.

Sepkoski, J. John, Jr. 1981. "A Factor Analytic Description of the Phanerozoic Marine Fossil Record." *Paleobiology* 7(1): 36–53.

Shapiro, Beth, Alexei J. Drummond, Andrew Rambaut, et al. 2004. "Rise and Fall of the Beringian Steppe Bison." *Science* 306(5701): 1561–1565.

Shen, Bing, Lin Dong, Shuhai Xiao, and Michao Kowalewski. 2008. "The Avalon Explosion: Evolution of Ediacara Morphospace." *Science* 319(5859): 81–84.

Van Valen, Leigh. 1973. "A New Evolutionary Law." *Evolutionary Theory* 1: 1–10.

Wilson, Michael C. 1978. "Archaeological Bison Kill Populations and the Holocene Evolution of the Genus *Bison*." In *Bison Procurement and Utilization: A Symposium*, ed. Leslie B. Davis and Michael Wilson. Lincoln, Nebraska: Plains Anthropologist.

Wilson, Michael C. 1980. "Morphological Dating of Late Quaternary Bison on the Northern Plains." *Canadian Journal of Anthropology* 1(1): 81–85.

Wilson, Michael C. 1988. "Bison Dentitions from the Henry Smith Site, Montana: Evidence for Seasonality and Paleoenvironments at an Avonlea Bison Kill." In *Avonlea Yesterday and Today: Archaeology and Prehistory; A Plains Conference Symposium*, ed. Leslie B. Davis. Saskatoon, Canada: Saskatchewan Archaeological Society.

Zhao Guochun, Sun Min, and S. A. Wilde. 2002. "Reconstruction of a Pre-Rodinia Supercontinent: New Advances and Perspectives." *Chinese Science Bulletin* 47(19): 1585–1588.

Michael C. Wilson

Biogeography: The distribution of life

Animals occur across Earth—from the bottom of the oceans to the tops of mountain ranges, from tropical rain forests to dry savannahs and to the Arctic and Antarctic tundra. No species, however, can live everywhere. No species is distributed across such obviously different environments as, for example, the deep sea and a terrestrial desert. Even in the terrestrial realm, very few species occur on all continents. Why, for example, out of the 10,000 or so bird species are only a very few, such as the peregrine falcon (*Falco peregrinus*), true cosmopolitans (meaning they occur on almost all continents)? Why do most birds inhabit only a restricted area? The house wren (*Troglodytes aedon*), for instance, is found in North and South America, while the Yucatan wren (*Campylorhynchus yucatanicus*) inhabits only the northern coast of the Yucatan peninsula of Mexico. This pattern—a few species having large and many having small ranges—is actually very common across regions, scales, and taxa (Willis 1922; Gaston 2003). The question of what determines a species' distribution is fundamental to the scientific discipline of biogeography. Biogeography is the study of the distribution of life. It tries to explain why organisms occur where they do, and it aims to discover the reasons why some places harbor more species than others. In sum, the discipline of biogeography seeks to explain the variation of biological diversity over the surface of Earth (Lomolino, Riddle, and Brown 2006; Cox and Moore 2005).

So, what factors affect geographic distributions? To answer this key question, biogeography needs to integrate many other disciplines, such as geology and geomorphology, climatology and paleontology, evolutionary biology and ecology, and genetics and physiology. In general, no single factor determines a species' distribution. Instead, one has to consider the interplay of various abiotic and biotic factors to explain why a particular organism is currently found where it is. All these factors affect three processes, though, that are fundamental in biogeography—speciation, extinction, and dispersal. As comprehensively phrased in the formidable textbook of Mark V. Lomolino, Brett R. Riddle, and James H. Brown (2006), "these are the means by which biotas respond to spatial and temporal dynamics of the geographic template. Thus, all of the biogeographic patterns that we study derive from the effects of these processes" (p. 141).

The current variation of environmental conditions across Earth's surface is, obviously, a major factor affecting species' distributions. Because of its morphological and physiological characteristics, no fish species, for instance, will be able to survive without water in the long run. More specifically, polar bears (*Ursus maritimus*), for example, have evolved seemingly perfect adaptations to the climate of high Arctic environments, but they would not be able to withstand the climate of more temperate or even tropical regions. Beyond abiotic environmental conditions, the biotic environment strongly influences the distribution of species. The availability of photosynthetic plants is one of the most important factors for survival for many animals, particularly for the huge number of herbivores (plant eaters). Again, more specifically, the larvae of many butterfly species (Lepidoptera) are specialized in feeding on very few species of plants. Without those plants, the butterfly will not be able to persist.

Of course, the distribution of animals does not depend only on plants. Key factors in determining species' occurrences are the presence and/or absence of predator, prey, competitor, and mutualist species (Krebs 2001). An important example is the principle of competitive exclusion, which postulates that two species with identical requirements cannot coexist (Hardin 1960). In his studies of island bird communities of the Bismarck Archipelago east of New Guinea, Jared M. Diamond (1975) documented several mutually exclusive distributions for closely related species. The whistler species

The current variation of environmental conditions across Earth's surface is a major factor affecting species' distributions. Polar bears (*Ursus maritimus*), for example, have evolved seemingly perfect adaptations to the climate of high Arctic environments, but they would not be able to withstand the climate of more temperate or even tropical regions. © Steve Ringman/The Seattle Times/KRT.

Peregrine falcons (*Falco peregrinus*) occur on almost all continents. Jim Zipp/Photo Researchers, Inc.

Pachycephala melanura and *P. pectoralis*, which belong to the same genus, have very similar habitat and resource requirements, occur in close geographical proximity, but are never found on the same island. Biotic and abiotic factors in turn often interact in many ways, and their importance depends on spatial scale (a prominent feature to consider in biogeography): While on local levels, species interactions are an important factor influencing the presence or absence of a species, broad climatic conditions may be more influential on a continental scale (MacArthur 1972; Pearson and Dawson 2003). The study of contemporary spatial variation of all abiotic and biotic conditions and their effects on species distributions is commonly referred to as ecological biogeography (Lomolino, Riddle, and Brown 2006).

Historical biogeography

For a comprehensive explanation of why a certain animal species is found where it is, biogeographers have to consider another important factor—time—or the history of the abiotic and biotic environment. Neither Earth's surface nor its climate have remained constant over time, nor have species entities remained unchanged. Species evolve (speciate), disperse, and finally go extinct—the ultimate fate of every species. To explain species' distributions or the spatial

variation of biological diversity, biogeography needs to explore the evolutionary history of the organism under study, and within the context of the changes in the geographical template over time. This "temporal branch" of the field has been referred to as historical biogeography (e.g., Funk 2004). As with the spatial component, timescales must be considered in biogeography. This can be understood with three examples. First, for evolutionary processes as well as events of mass extinctions, the geological timescale (up to hundreds of millions of years) is of great significance. Second, the retreat of glaciers in the Northern Hemisphere after the last glacial maximum (about 20,000 years ago) had severe impacts on species' distributions in Europe and North America. Finally, current, seasonal climatic variations affect temporal changes in distributions of migratory birds over as little as one year.

An epochal scientific breakthrough of the first half of the twentieth century was the theory of continental drift. After several centuries of speculation about potential changes in the positions of Earth's landmasses, the German meteorologist Alfred Wegener (1880–1930) finally presented the theory of continental drift in his 1915 book *Die Entstehung der Kontinente und Ozeane* (The origin of continents and oceans). Wegener's theory was not widely accepted, however, until the 1960s. By that time, among other evidence, the former connectivity of the continents was documented by the fact that they indeed fit together if the contours of the continental shelves are used to delineate them (Bullard, Everett, and Smith 1965). Convection forces in Earth's mantle have been carrying landmasses over its surface (plate tectonics). This resulted in fragmentation of supercontinents, such as Pangaea, Gondwana, and Laurasia, or in collisions of major landmasses, as in the case of the Indian subcontinent and the Asian mainland approximately 50 to 55 million years ago. Movements, fragmentations, and collisions also caused other phenomena of major importance, such as the uplift of mountain ranges (the rise of the Himalaya and the Tibetan Plateau, for instance, was caused by the collision of India and Asia), the opening of pathways for sea currents (the Tethys Sea, for example, came into existence as a result of the breakup of Pangaea), and volcanic activity (for which the emergence of the Hawaiian or Canary islands are examples). Furthermore, the change in the latitudinal position of the landmasses resulted in climatic changes within the continents.

Of course, all these changes in geographical and environmental settings resulting from plate tectonics heavily impacted biotic distribution patterns. Groups of organisms that occurred together on the supercontinents of Gondwana or Laurasia, for example, were separated by splits of the land bodies into smaller fragments. Conversely, formerly separated biotas were connected when their continents merged. A classic example for this is the Great American Interchange: After having been isolated for about 130 million years, South America became connected to North America via the Central American land bridge around 3 million years ago. This caused a wave of biotic interchange, with many species from North America, such as rabbits, horses, deer, tapirs, cats, and peccaries, just to name some mammalian examples, invading the Neotropics, and also many, but fewer, species, including porcupines, armadillos, and opossums, crossing the isthmus from South America toward the Nearctic (Marshall et al. 1982).

Numerous studies have shown how the breakup of the formerly connected landmasses of Gondwana, which proceeded in several stages between approximately 180 and 50 million years ago, is still reflected in disjunct distribution patterns. According to one of these studies, the work of Lars Zakarias Brundin on the distributions of midges (family Chironomidae) in the Southern Hemisphere (1966), about 600 to 700 Chironomid species occur in the cool mountain streams of temperate South America, southern Africa, southeastern Australia, Tasmania, and New Zealand. Based on this pattern of distribution and also their phylogenetic relationships, Brundin concluded that "the chironomid midges give clear evidence that their transantarctic relationships developed during periods when the southern lands were directly connected with each other" (pp. 451–452).

Vicariance and dispersal

This work is also a good example of a concept of overall importance: vicariance. Vicariance biogeography aims to explain disjunct distributions by the appearance of a barrier that divides a once continuous range into separate units (Platnick and Nelson 1978). Many vicariant distribution patterns are well documented. One of them, also referring to

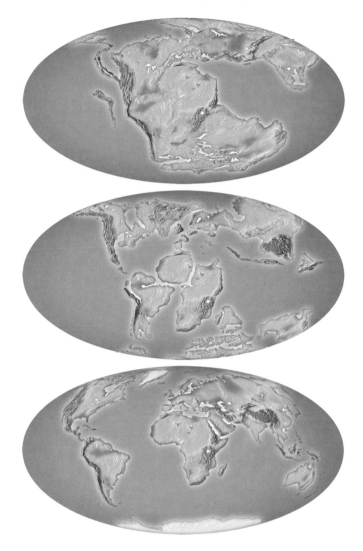

Continental drift (200 mya, 100 mya, modern era). Mikkel Juul Jensen/Bonnier Publications Photo Researchers, Inc.

the breakup of Gondwana, is about birds—tinamous and flightless ratites, the latter including ostriches, rheas, emus, cassowaries, and kiwis. All members of this group, the Palaeognathae, occur on the continents that made up the supercontinent of Gondwana, with only the tinamous' current distributions reaching into Central America. Most likely, their common ancestor(s) were widely distributed across Gondwana, and their ranges became disjunct when it fragmented. Then, in geographic isolation, populations evolved independently from each other into the six contemporary lineages (Cracraft 1974, 2001).

The Palaeognathae and Chironomidae examples demonstrate a link between the evolutionary process of speciation and geographic isolation. In fact, the latter has been proposed to be of major importance or even necessary to initiate speciation (Mayr 1942, 1963). Geographic isolation imposes barriers that inhibit the movement of individuals from one population to another, that is, the process of dispersal. Dispersal is defined as the movement of offspring away from their birthplaces to new sites. In the biogeographical context, dispersal also embraces the processes of immigration, emigration, and colonization. Dispersal barriers can be of various origin and nature. Water bodies separate islands from continents or fragment grasslands or forests into distinct habitat patches. Similarly, terrestrial barriers may separate populations of aquatic organisms. Mountain ranges impose major barriers to dispersal, as many lowland species are unable to cross them. No matter the kind of isolation, geographic and dispersal barriers result in a reduction of gene flow among populations. Thus, populations will evolve independently, and given a certain amount of time, populations will eventually become reproductively isolated—even if the populations merge again, individuals may not be able to mate successfully.

Two major modes can be distinguished in allopatric speciation (speciation due to independent evolution in geographic isolation; Platnick and Nelson 1978). In the first mode, vicariance, as already mentioned above, changes in the environment impose a barrier within the range of a species, isolating formerly connected and interbreeding populations. In the second mode, dispersal, a few individuals can disperse across a barrier, colonizing an uninhabited area and thus founding a new population. For a successful colonization to occur, however, individuals of a species must travel to the new area, during which travel they must survive unfavorable conditions, and finally they must establish a viable population after arrival. Because of these significant challenges, many accidental barrier crossings do not result in the colonization of new areas and, thus, neither do they result in the evolution of new species.

The dispersal mode of allopatric speciation is documented by many examples of occasional, random events (also known as jump dispersal or founder events), with subsequent speciation. Galápagos finches (genera *Geospiza*, *Camarhynchus*, *Certhidea*, and *Pinaroloxias*) and Hawaiian honeycreepers (family Drepanidae) are two examples. For both these groups, individuals of an ancestral mainland species colonized one or several islands, which was then followed by evolutionary changes (Grant and Grant 1997; Grant 1986, Pratt 2005). Another striking, more recent, and direct case of a rare founder event are Caribbean floating iguanas (Censky, Hodge, and Dudley 1998). In 1995 at least fifteen individuals of the green iguana (*Iguana iguana*) arrived on the West Indian island of Anguilla, floating on a mat of logs and uprooted trees, which most likely was propelled by a hurricane from some other iguana-inhabited island, estimated as much as 320 kilometers (200 miles) away from Anguilla.

In the course of biogeographic history, the two extremes of allopatric speciation—vicariance versus dispersal—were a matter of heavy and vigorous debates, mainly concerning which of these processes is the prevailing one in shaping the distribution of species and diversity. Both are important, however, and their predominance varies from case to case. The fundamental factors involved—geographical isolation, dispersal, and evolution—are actually the same in both modes, and the only remaining question is the timing of dispersal, that is, if it occurs before or after a barrier arises (Lomolino, Riddle, and Brown 2006).

Dispersal (or its absence because of geographic isolation) determines the distribution of species via its influences on speciation. Extinction is also strongly influenced by dispersal. Climatic changes, such as during the Pleistocene glaciation cycles or the current anthropogenically caused global climate change, may render the region where a species currently occurs unsuitable for its survival. If this species cannot adapt locally or "escape" (disperse) to an area where climatic conditions are suitable, it will go extinct. Another example of the interaction of extinction and dispersal, at a smaller temporal and spatial scale, are freshwater ponds. If a small pond dries up because of high rates of evaporation, animals inhabiting it die unless they can disperse to another pond or have evolved strategies to cope with the ephemeral character of their habitats.

Climatic changes that have influenced the distributions of species in space and time have also occurred outside the influence of the movement of tectonic plates. Narrowing the temporal focus down to the last 2 million years reveals that, despite a relative constancy of the spatial position of the continents, Earth's climate has fluctuated dramatically. This has had major impacts on biota's ranges still mirrored by current distributions of animals and plants, such as in Europe (e.g., Araújo et al. 2008; Hof, Brändle, and Brandl 2008). At least twenty major glacial periods (during which global mean temperatures were at least 4°C lower than the current, interglacial temperature) have been documented for the Pleistocene (Lomolino, Riddle, and Brown 2006; Pielou 1991). During these cycles, huge Arctic ice sheets reached as far southward as latitude 40° north, and even further along mountain ranges of Eurasia and North America. Glaciers forced many animals to shift their distributions into non-glaciated areas, known as refugia (de Lattin 1957). Species occurring in nonglaciated areas were also affected by climatic changes: During cooling periods, they had the "choice" to adapt to the new conditions, shift their distributions into warmer areas (refugia), retreat to small refugia with suitable microclimates, or go extinct. For Europe, for instance, the Iberian, Italian, and Balkan peninsulas of the Mediterranean have been identified as refugia where species were able to endure (Hewitt 1999, 2000). Refugia also occurred, however, in more northern areas (e.g., Stewart and Lister 2001; Deffontaine et al. 2005). This topic remains a viable area of research, but all authors agree that the current patterns of species distribution—in particular in the Northern Hemisphere—still bear the mark of the Pleistocene ice ages.

The vast glaciers and temperature fluctuations also led to profound sea level fluctuations. During the last glacial maximum, sea levels dropped as much as 100 meters (330 feet) below current levels. This provided opportunities for the dispersal of species between terrestrial regions that were separated when water levels rose during the Holocene deglaciation period. Alaska and the eastern part of Eurasia, for example, were connected by the Bering Land Bridge, which allowed a faunal exchange between Eurasia and North America (e.g., bears, wolves, moose, ravens, hawks, owls and crossbills), which now occurs on both continents (Lomolino, Riddle, and Brown 2006). The Bering Land Bridge also allowed humans (*Homo sapiens*) to colonize North and South America.

Another famous example of the biotic consequences of the periodic lowering of sea levels during the Pleistocene centers

Marine iguanas (*Amblyrhynchus cristatus*) found on the Galápagos Islands do not occur on any of Earth's major landmasses. Novastock/Photo Researchers, Inc.

on Southeast Asia. Here, many islands of the Malay Archipelago, including Borneo, Sumatra, Java, and Bali, located on the Sunda Shelf, were connected to the Asian mainland. New Guinea, however, was linked to Australia and Tasmania. About 150 years ago, the English naturalist and biogeographer Alfred Russel Wallace (1823–1913)—now often called the father of modern biogeography—recognized distinct differences in distribution patterns in this region. He became particularly interested in locating a line that divided the Australian and Asiatic faunas. In his monumental 1876 work, *The Geographical Distribution of Animals*, Wallace drew the line (later referred to as "Wallace's line") between Bali and Lombok and, further north, between Borneo and Celebes. Later it turned out that rather than a distinct demarcating line, there is a transition zone of distributional overlap between the two faunistic realms. Nevertheless, this zone has been aptly named "Wallacea," because Wallace's observations of an area where species from different zoogeographic realms come together were striking, and they actually coincided with the patterns of terrestrial connectivity during the Pleistocene.

Island biogeography

Islands and their fascinating floras and faunas have drawn the attention of biogeographers for centuries. The English

naturalist Charles Darwin (1809–1882) was intrigued by the creatures of the Galápagos Archipelago, which he visited in 1835, during the famous voyage of the HMS *Beagle*. There, he had important insights inspiring him to develop his revolutionary theory of evolution through natural selection (Darwin 1859). As mentioned above, at about the same time, Wallace studied the island faunas of the Malay Archipelago, from which he drew conclusions on biogeographic patterns and theory that remain valid today (Wallace 1876), and he also developed—independently of Darwin—a theory of natural selection. Indeed, islands harbor fascinating and peculiar species such as the kiwis of New Zealand (genus *Apteryx*) and the marine iguanas (*Amblyrhynchus cristatus*) of the Galápagos, that do not occur on any of Earth's major landmasses. This phenomenon—the occurrence of taxa in only a limited area—is called endemism. However, geographic scope always has to be taken into account when talking about endemic taxa, as they can be endemic to such different areas—from a small forest patch to an island archipelago to a continent. The marine iguana is endemic to the Galápagos, whereas the hummingbirds (family Trochilidae) are endemic to the Americas, and macropods (family Macropodidae) are endemic to Australia, New Guinea, Tasmania, and some smaller islands in close proximity. Islands have also been called "natural laboratories" because their simplified ecosystems, limited size, discrete character, and huge numbers across the globe enable biogeographers to study processes that supposedly underlie observed phenomena (Whittaker and Fernández-Palacios 2007).

Among the most influential works for biogeography in the twentieth century were the studies of Robert H. MacArthur and Edward O. Wilson. In their famous "equilibrium theory of island biogeography" (ETIB; MacArthur and Wilson 1963, 1967), they brought together many important concepts, such as geographic isolation, dispersal, and extinction, and integrated them within one theoretical framework. In one of its key underpinnings, the theory draws on the relationship between species richness and geographical area. This relationship had been recognized as early as the 1700s (e.g., Forster 1778). In 1921 the Swedish agricultural chemist Olof Arrhenius (1896–1977) clearly stated that "the number of species increases continuously as the area increases" (Arrhenius 1921, p. 99) and quantified this rule in a formal equation. One of the classic examples of this species–area relationship, which is actually considered to be one of the few laws in ecology and biogeography, is the reptile species of the West

Indian islands, which follow almost perfectly the quantitative equation of Arrhenius: The small island of Redonda is inhabited by the smallest number of species, while the comparatively larger Hispaniola and Cuba harbor the richest reptile faunas (Darlington 1957). Another pattern MacArthur and Wilson made use of was the tendency of species numbers to decrease when island isolation increases –a pattern also already recognized by the early biogeographers of the nineteenth century.

Species turnover comprised a third pattern that influenced the development of the ETIB. A study system for this phenomenon is the Krakatau Islands located in the Sunda Straits between Sumatra and Borneo in the Malay Archipelago. The original island of Krakatau was destroyed by a tremendous volcanic eruption in 1883, resulting in several remnant islands where all life was wiped out. This catastrophic event gave rise to a natural experiment for island biogeographers investigating processes of colonization and extinction. In fact, recolonization of the islands was rapid: After only fifty years, the development of a tropical rain forest, harboring many species of animals and plants, was reported from the islands of Rakata

Deforestation is a serious danger to threatened species and the global extinction rate is still accelerating. © David Hyde, 2010/Shutterstock.com.

and Sertung—two of the remaining Krakatau fragments. Studies of bird occurrences showed that species richness rapidly increased until 1920, but then remained relatively constant. The composition of species did not remain constant, however, because of species turnover: New species arrived from the mainlands of Borneo and Sumatra, some of which became residents, while some earlier colonizers became extinct (MacArthur and Wilson 1963, 1967).

MacArthur and Wilson's theory brought these phenomena—species–area relationships, species–isolation relationships, and species turnover—together with the fundamental processes of immigration and extinction into one unifying quantitative theory. Only some of its basic elements will be explained here. One of the theory's general assumptions is that rates of immigration and extinction depend on the number of species present on an island. The island's species number can vary from zero to a maximum value representing the species number in the "mainland pool." On an island without any species, the immigration rate (defined as the rate of arrival of individuals belonging to species that are not yet present) reaches its maximum, as each species is new to the island. Thus, with increasing species richness, the immigration rate decreases. The extinction rate (defined as the rate of loss of species present on the island) is zero with an island species number of zero, and it increases with increasing species numbers. When the immigration and extinction rates are equal, the theory holds that an equilibrium in the number of species will be reached, as will an equilibrial rate of turnover.

This simple model describes relationships among immigration, extinction, turnover, and richness, and it also explains why species richness increases with increasing island area and with decreasing isolation. Larger islands have larger populations than smaller ones. Because species with larger populations are less prone to extinction, extinction probability should be smaller on larger islands, resulting in higher levels of species richness. Isolation, on the contrary, has strong influences on immigration: Species will be less likely to colonize remote islands from a mainland source than close-in islands. Hence, immigration rates decrease with increasing isolation, resulting in higher species numbers on islands closer to mainland sources. Combining the effects of area and isolation on the numbers of species present on an island, species richness increases from the smallest, most remote islands toward large islands that are close to a mainland source.

As with every new theory, the equilibrium theory of island biogeography had many strengths, but also weaknesses. The model was tested extensively, and although many authors found corroborating results for the theory, others did not, and the debate on both empirical and conceptual evidence for the ETIB continues (see, e.g., Heaney 2000; Walter 2004; Whittaker and Fernández-Palacios 2007). Most importantly, many islands do not reach an equilibrium between immigration and extinction because of events of disturbance or natural catastrophes, such as the volcanic eruption on Krakatau. Furthermore, the theory is less successful in explaining processes operating on evolutionary and geological timescales. Recently, a general dynamic model of oceanic island biogeography has been proposed which includes the

fundamental biogeographical processes through time as well as in relation to island ontogeny—the changing characteristics of the island during its life cycle (Whittaker, Triantis, and Ladle 2008). Yet, despite many potential criticisms of MacArthur and Wilson's theory, no biogeographer would eventually doubt their enormous impact, which reached far beyond island biogeography. The young discipline of conservation biology, for instance, was influenced significantly by the ETIB's findings. The dependency between extinction and area is important for conservation planning because reserves should be large enough to maintain healthy populations of animals and plants, minimizing their chance of extinction.

Human impacts on species distribution

To explain species' distributions and the spatial variation of life, modern biogeography must take into account the distribution of one species in time and space in particular—that is, *Homo sapiens*. The spread of humans around the globe (we are one of the few true cosmopolitans) has greatly influenced the key biogeographical factors of extinction and dispersal. This began early. Some extinctions of large mammals after the end of the Pleistocene, for instance, have been assigned to human impacts. A prime example is the woolly mammoth (*Mammuthus primigenius*) in Eurasia. Although climatic changes probably had a negative impact on mammoth populations, few contemporary researchers doubt that humans played a major role in the extinction of these creatures about 3,700 years ago (Nogués-Bravo et al. 2008). Other extinctions are unambiguously attributable to humans alone. The most famous one may be the extinction of the dodo (*Raphus cucullatus*)—a large flightless bird that occurred in huge numbers on the islands of Mauritius and Réunion in the Indian Ocean. After Mauritius was discovered

The extinction of the dodo (*Raphus cucullatus*) can be attributed to humans alone. F Hart/The Bridgeman Art Library/Getty Images.

Since they were accidentally imported from Eastern Europe in the late 1980s, zebra mussels (*Dreissena polymorpha*) have spread throughout the Great Lakes, wrecking havoc on the ecosystem. © Jim West/Alamy.

the demands of traditional Asian medicine that lead to the poaching of rhinos, bears, and tigers. Recently, numerous studies have documented that anthropogenically caused global climate change has already led to shifts in and shrinkages of species distributions, and many species may go extinct within the next decades or centuries as a result (Thomas et al. 2004).

Beyond other factors, which cannot all be listed in detail here, invasive species—the human-assisted dispersal of species to places where they did not occur before—have also had severe consequences for species survival and distribution. As already mentioned, prominent examples of the negative impacts of invasive species are—again—islands. For example, no mammals except bats inhabited the islands of New Zealand until humans arrived there. Thus, for birds until very recently there was no need to escape from mammalian predators, and in the course of their evolutionary history, many bird species, including the kakapo (*Strigops habroptila*) and the Chatham rail (*Cabalus modestus*), lost their ability to fly. *Homo sapiens* brought along dogs, cats, pigs, and rats, all of which preyed on flightless birds or their eggs and young. The kakapo so far has not yet gone extinct, thanks to major efforts of local conservationists, but the number of individuals left is very low, ranging between 50 and 100. The Chatham rail went extinct around 1900, as did about fifty other species of New Zealand's birds during the period after human colonization.

Species invasions and their negative impacts, however, are not known only for remote islands. The Eurasian zebra mussel (*Dreissena polymorpha*), for example, was first noticed in the North American Great Lakes in 1988, after which it spread rapidly and caused significant damage not only to native mussel populations, but also to boats, piers, and other kinds of structures in the water. In the nineteenth century, the Colorado potato beetle (*Leptinotarsa decemlineata*)—a severe pest for potato crops—was accidentally introduced to Europe. It spread rapidly and reached the former Soviet Union around 1960. In Africa, the Nile perch (*Lates niloticus*), which was introduced to Lake Victoria in the 1950s, is one of the best-known examples of the negative effects of invasive species. Not only does it pose a major threat for the hundreds of species of endemic cichlids, of which several are thought to be extinct already, but its introduction has also had devastating consequences for the entire ecosystem (Baskin 1992).

by the Portuguese in 1507, the first European report of the species' existence dates back to 1601 and came from Dutch sailors. Due to the extensive hunting and introduction of nonnative dogs, cats, pigs, and monkeys, which ate the birds, their eggs, and their young, the dodo became extinct as early as 1662 (Quammen 1996).

During the age of exploration, numerous species on remote islands went extinct. This wave of extinction, which started in the fifteenth or sixteenth century, continues, however, and the global extinction rate is still accelerating. Many different human impacts are responsible for this. Among the most important ones are habitat destruction, degradation, and fragmentation stemming from land use, as in the deforestation of vast areas of tropical rain forests (Sala et al. 2000). Direct exploitation still imposes a serious threat for many species, be it for food resources, as in industrial fisheries and the African bushment trade, or from

These examples illustrate the consequences of human impacts for the processes that shape species distributions—namely extinction, dispersal, and speciation. Without being aware of these mechanisms, putting them into an evolutionary framework, and taking into account the variation of the environment over space and time, we will not understand the distribution of species, and neither will we be able to reduce the impacts threatening them with extinction. Therefore, a deep knowledge of biogeography is fundamental for all efforts to protect the world's biodiversity.

Resources

Books

Brundin, Lars Zakarias. 1966. *Transantarctic Relationships and Their Significance, as Evidenced by Chironomid Midges*. Stockholm, Sweden: Almqvist & Wiksell.

Cox, C. Barry, and Peter D. Moore. 2005. *Biogeography: An Ecological and Evolutionary Approach*. 7th edition. Malden, MA: Blackwell.

Darlington, Philip J., Jr. 1957. *Zoogeography: The Geographical Distribution of Animals*. New York: Wiley.

Darwin, Charles. 1859. *On the Origin of Species by Means of Natural Selection; or, The Preservation of Favoured Races in the Struggle for Life*. London: John Murray.

Diamond, Jared M. 1975. "Assembly of Species Communities." In *Ecology and Evolution of Communities*, ed. Martin L. Cody and Jared M. Diamond. Cambridge, MA: Harvard University Press, Belknap Press.

Forster, John Reinhold. 1778. *Observations Made during a Voyage Round the World, on Physical Geography, Natural History, and Ethic Philosophy*. London: G. Robinson.

Funk, Vicki A. 2004. "Revolutions in Historical Biogeography." In *Foundations of Biogeography: Classic Papers with Commentaries*, ed. Mark V. Lomolino, Dov F. Sax, and James H. Brown. Chicago: University of Chicago Press.

Gaston, Kevin J. 2003. *The Structure and Dynamics of Geographic Ranges*. Oxford: Oxford University Press.

Grant, Peter R. 1986. *Ecology and Evolution of Darwin's Finches*. Princeton, NJ: Princeton University Press.

Krebs, Charles J. 2009. *Ecology: The Experimental Analysis of Distribution and Abundance*. 6th edition. San Francisco: Benjamin Cummings.

Lomolino, Mark V., Brett R. Riddle, and James H. Brown. 2006. *Biogeography*. 3rd edition. Sunderland, MA: Sinauer.

MacArthur, Robert H. 1972. *Geographical Ecology: Patterns in the Distributions of Species*. New York: Harper & Row.

MacArthur, Robert H., and Edward O. Wilson. 1967. *The Theory of Island Biogeography*. Princeton, NJ: Princeton University Press.

Marshall, Larry G., S. David Webb, J. John Sepkoski Jr., and David M. Raup. 1982. "Mammalian Evolution and the Great American Interchange." *Science* 215(4538): 1351–1357.

Mayr, Ernst. 1942. *Systematics and the Origin of Species, from the Viewpoint of a Zoologist*. New York: Columbia University Press.

Mayr, Ernst. 1963. *Animal Species and Evolution*. Cambridge, MA: Harvard University Press, Belknap Press.

Pielou, E. C. 1991. *After the Ice Age: The Return of Life to Glaciated North America*. Chicago: University of Chicago Press.

Pratt, H. Douglas. 2005. *The Hawaiian Honeycreepers*. Oxford: Oxford University Press.

Quammen, David. 1996. *The Song of the Dodo: Island Biogeography in an Age of Extinctions*. New York: Scribner.

Wallace, Alfred Russel. 1876. *The Geographical Distribution of Animals*. London: Macmillan.

Wegener, Alfred. 1915. *Die Entstehung der Kontinente und Ozeane* [The origin of continents and oceans]. Brunswick, Germany: Vieweg.

Whittaker, Robert J., and José Maria Fernández-Palacios. 2007. *Island Biogeography: Ecology, Evolution, and Conservation*. 2nd edition. Oxford: Oxford University Press.

Willis, J. C. 1922. *Age and Area: A Study in Geographical Distribution and Origin of Species*. Cambridge, UK: Cambridge University Press.

Periodicals

Araújo, Miguel B., David Nogués-Bravo, José Alexandre Diniz-Filho, et al. 2008. "Quaternary Climate Changes Explain Diversity among Reptiles and Amphibians." *Ecography* 31(1): 8–15.

Arrhenius, Olof 1921. "Species and Area." *Journal of Ecology* 9 (1): 95–99.

Baskin, Yvonne. 1992. "Africa's Troubled Waters." *BioScience* 42 (7): 476–481.

Cracraft, Joel. 1974. "Phylogeny and Evolution of Ratite Birds." *Ibis* 116(4): 494–521.

Cracraft, Joel. 2001. "Avian Evolution, Gondwana Biogeography, and the Cretaceous–Tertiary Mass Extinction Event." *Proceedings of the Royal Society* B 268(1466): 459–469.

Bullard, Edward, J. E. Everett, and A. Gilbert Smith. 1965. "The Fit of the Continents around the Atlantic." *Philosophical Transactions of the Royal Society of London* A 258(1088): 41–51.

Censky, Ellen J., Karim Hodge, and Judy Dudley. 1998. "Over-Water Dispersal of Lizards Due to Hurricanes." *Nature* 395 (6702): 556.

Deffontaine, V., R. Libois, P. Kotlík, et al. 2005. "Beyond the Mediterranean Peninsulas: Evidence of Central European Glacial Refugia for a Temperate Forest Mammal Species, the Bank Vole (*Clethrionomys glareolus*)." *Molecular Ecology* 14(6): 1727–1739.

de Lattin, G. 1957. "Die Ausbreitungszentren der holarktischen Landtierwelt" [The dispersal centers of the Holarctic terrestrial fauna]. *Verhandlungen der Deutschen Zoologischen Gesellschaft Hamburg* 1956: 380–410.

Grant, Peter R., and B. Rosemary Grant. 1997. "Genetics and the Origin of Bird Species." *Proceedings of the National Academy of Sciences of the United States of America* 94(15): 7768–7775.

Hardin, Garrett. 1960. "The Competitive Exclusion Principle." *Science* 131(3409): 1292–1297.

Heaney, Lawrence R. 2000. "Dynamic Disequilibrium: A Long-Term, Large-Scale Perspective on the Equilibrium Model of Island Biogeography." *Global Ecology and Biogeography* 9(1): 59–74.

Hewitt, Godfrey M. 1999. "Post-Glacial Re-colonization of European Biota." *Biological Journal of the Linnean Society* 68(1–2): 87–112.

Hewitt, Godfrey M. 2000. "The Genetic Legacy of the Quaternary Ice Ages." *Nature* 405(6789): 907–913.

Hof, Christian, Martin Brändle, and Roland Brandl. 2008. "Latitudinal Variation of Diversity in European Freshwater Animals Is Not Concordant across Habitat Types." *Global Ecology and Biogeography* 17(4): 539–546.

MacArthur, Robert H., and Edward O. Wilson. 1963. "An Equilibrium Theory of Insular Zoogeography." *Evolution* 17 (4): 373–387.

Pearson, Richard G., and Terence P. Dawson. 2003. "Predicting the Impacts of Climate Change on the Distribution of Species: Are Bioclimate Envelope Models Useful?" *Global Ecology and Biogeography* 12(5): 361–371.

Platnick, Norman I., and Gareth J. Nelson. 1978. "A Method of Analysis for Historical Biogeography." *Systematic Zoology* 27 (1): 1–16.

Sala, Osvaldo E., F. Stuart Chapin III, Juan J. Armesto, et al. 2000. "Global Biodiversity Scenarios for the Year 2100." *Science* 287(5459): 1770–1774.

Stewart, John R., and Adrian M. Lister. 2001. "Cryptic Northern Refugia and the Origins of the Modern Biota." *Trends in Ecology and Evolution* 16(11): 608–613.

Thomas, Chris D., Alison Cameron, Rhys E. Green, et al. 2004. "Extinction Risk from Climate Change." *Nature* 427(6970): 145–148.

Walter, Hartmut S. 2004. "The Mismeasure of Islands: Implications for Biogeographical Theory and the Conservation of Nature." *Journal of Biogeography* 31(2): 177–197.

Whittaker, Robert J., Kostas A. Triantis, and Richard J. Ladle. 2008. "A General Dynamic Theory of Oceanic Island Biogeography." *Journal of Biogeography* 35(6): 977–994.

Zschokke, F. 1908. "Die Beziehungen der mitteleuropäischen Tierwelt zur Eiszeit" [The relationships of the Central European fauna to the ice age]. *Verhandlungen der Deutschen Zoologischen Gesellschaft* 1908: 21–77.

Other

Nogués-Bravo, David, Jesús Rodríguez, Joaquín Hortal, et al. 2008. "Climate Change, Humans, and the Extinction of the Woolly Mammoth." *PLoS Biology* 6(4): e79. Available from http://biology.plosjournals.org/perlserv/?request=get-document&doi=10.1371/journal.pbio.0060079.

Christian Hof

• • • • •

Genetics: The blueprint of life

Blueprinting, a photochemical process, was used from the middle of the nineteenth century through the twentieth century to make multiple copies of engineering drawings, architectural plans, and the like. The term *blueprint* has come to mean any detailed plan. To say that genetics is the blueprint of life is to mean that genetics explains the plan of life. In particular, scientists believe that genetic material, which is made of a polymer called *deoxyribonucleic acid* (DNA), contains the instructions for making, at the right time and place, the parts of whatever organism the genetic material comes from.

DNA was first isolated by Friedrich Miescher (1871), but he did not know what its function was. This did not become clear until Oswald T. Avery, Colin M. MacLeod, and Maclyn McCarty (1944) demonstrated that DNA from a pathogenic strain of bacteria could be absorbed by nonpathogenic bacteria, thereby rendering some of them (and their descendants) pathogenic themselves. This means that the DNA contained the instructions (*gene*) for pathogenicity. The process whereby DNA converts a heritable trait, or *genotype*, in an organism from one form to another is called *transformation*, and it is still used routinely in genetic engineering research and development to move traits from one organism to another.

Of course a single case of transformation does not prove that DNA is the blueprint for life. The assignment of this role to DNA is the result of a great deal of research in genetics and molecular biology.

How genes are inherited

Certain diseases run in families. Medical doctors were the first to report about this and to record the different patterns of inheritance of some of these diseases in the nineteenth century. For example, Huntington's disease, a severe neurological malady characterized by exaggerated spontaneous movements, is inherited directly from either an affected mother or father. Children of couples with one affected parent have a 50 percent chance of inheriting this disease and developing symptoms, usually in middle age. A child of an affected person who does not develop the disease by late middle age will almost never be the parent of an affected child. Hemophilia, which interferes with proper blood coagulation leading to excessive bleeding, on the other hand, is inherited almost exclusively by boys through mothers who have either an affected father or a mother descended from an affected

man. These patterns, which are now understood, were simply unexplained facts in the nineteenth century.

Humans have exploited heredity in a practical way for thousands of years: Ancient farmers selectively bred several strains of crop plants and domestic animals that had particular traits. Also, everyone recognizes resemblances—and differences—among family members. Apart from these practical understandings, however, there was little to explain the underlying mechanisms until Gregor Mendel (1866) provided the first great insight. His idea was to study hybridization of pea plants using quantitative methods, that is, by counting offspring and determining definitely their kinds and relative numbers, a novel approach at that time. Mendel insisted on having true breeding lines of plants differing in very distinct characteristics. In this way he could begin with a sure knowledge of the properties of the parents in his crosses, and recognize clearly the different types of offspring. In early experiments Mendel found that hybrids between two true-breeding lines (tall versus dwarf) were tall. When he allowed the hybrid plants to self-fertilize, which is the normal way for peas to produce the next generation, he found that among 1,063 progeny, 787 were tall but 277 were dwarf. This can be understood if one supposes the following: (1) each adult plant has two genes (Mendel called them *elementen*) controlling height, one (D), specifying the normal height, and the other (d), which in the absence of D specifies a dwarf growth pattern; and (2) only one of these two genes is contributed at random by each parent to form a new individual (see Figure 1).

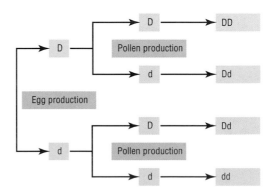

Figure 1. Branched line diagram showing basis for the 3:1 ratio in Mendel's work. Heavy lines indicate the tall plants. Reproduced by permission of Gale, a part of Cengage Learning.

The principle of inheritance discovered in these experiments is called Mendel's first law, or the law of segregation: Each form of the gene, D and d, is now called an *allele* of the gene controlling height in this plant. The alleles, being attached to separate chromosomes within the same cells, are carried along as chromosomes separate during formation of the egg and pollen cells, so that each egg or pollen cell comes to possess either one allele or the other but not both. Then when fertilization occurs, an organism that contains two alleles in each cell begins development. In this example the allelic composition would be either DD, Dd, or dd, depending on which kind of egg cell is fertilized by which kind of pollen. The allele (D) that controls the appearance of the hybrid (Dd) is called *dominant*, while the other (d) is called *recessive*. A recessive allele is in effect hidden within each cell of the hybrid organism but can emerge in the next generation if it is joined in an embryo by another recessive allele. Mendel observed essentially the same behavior in other true-breeding lines that varied with regard to six other characteristics of pea plants, such as seed shape or color, or placement of flowers.

Mendel also investigated another question by looking simultaneously at two different characteristics in the same plants. He was able to investigate true breeding lines that had either (1) round and yellow seeds (both because of dominant alleles) or (2) wrinkled and green seeds (both recessive). After making hybrids and allowing these to self-fertilize he observed that offspring were of four different kinds in a significant numerical ratio: 315 round and yellow, 101 wrinkled and yellow, 108 round and green, and 32 wrinkled and green. This ratio can be reduced essentially to 9:3:3:1. Assuming genes for shape and color are not coupled to each other, there is a 3/4 chance for each offspring to be round seeded; of these offspring, there is a 3/4 chance for each to be also yellow seeded. So overall 9/16 will be round and yellow seeded. Similar arguments account for the two groups of progeny that have only one dominant characteristic (3/16 each) and the fourth group that has two recessive characteristics (1/4 x 1/4 = 1/16). Mendel also followed up these results by predicting and verifying characteristics of descendants of offspring from this experiment. His conclusion is known as Mendel's second law: Stated in modern terms, alleles controlling different traits are inherited independently of one another.

Mendel had to be able to observe directly which of two character states was present in each plant, for each of several (actually seven) different characteristics. Also, he would not have been able to get these exact results if the genes were linked closely to each other. Scientists now believe that the seven genes Mendel studied are located on only four of the seven pairs of pea chromosomes. During egg and pollen formation, however, the chromosomes switch segments (*cross over*) often enough to obscure the linkage (Fairbanks and Rytting 2001). Mendel's 1866 paper probably has been studied as critically as any in the history of science. Even though his data fit theory somewhat better than expected, there is no evidence that Mendel did anything but honestly interpret his results using the best methods available at the time (Fairbanks and Rytting 2001).

Mendel's 1866 paper remained rather obscure for some thirty-five years. For example, Charles Darwin (1809–1882)

Flower color and placement were other traits studied in pea plants by Gregor Mendel. © Wally Eberhart/Visuals Unlimited/Corbis.

apparently never read it. The notions of inheritance that were widely held in his day made it difficult to see how variant forms of organisms could exist for very many generations. Yet, individual organisms of the same species clearly did not look exactly alike and differences were heritable! This persistent variability was one of the foundations of Darwin's theory of evolution by natural selection. The laws of inheritance Mendel discovered would have explained for Darwin the observed persistence of variants across generations.

Mendel's principles were rediscovered in the early 1900s, and many scientists then took up the subject, which became known as genetics. There ensued two eras of research in this field. The first, called classical genetics, concentrated on the quantitative and abstract approach pioneered by Mendel; the second, called molecular genetics, became more important after 1940. Thomas Hunt Morgan (1866–1945) of Columbia University applied Mendelian ideas to the study of *Drosophila melanogaster*, the tiny fruit fly. Morgan's laboratory discovered the linkage of certain genes with a particular chromosome, the X chromosome, which in both *Drosophila* and humans is found in two copies in females and one copy in males. In humans, the Y chromosome that pairs with the X determines male development.

Scientists have since discovered, thanks to research based on organisms such as *Drosophila*, that human diseases with inheritance patterns like hemophilia are caused by recessive genes located on the X chromosomes, where one of a woman's two X chromosomes carries a recessive allele, so that her sons have a 50 percent chance of inheriting the disease. Diseases that are inherited like Huntington's are caused by dominant genes on the nonsex chromosomes, so that both males and females can inherit the disease with a probability of 50 percent if one of their parents has one copy of the dominant gene. Thus, genetic research on the fruit fly has had an illuminating effect on medicine.

Morgan and his students also discovered the linkage of genes to one another. This permitted the construction of *genetic maps*, depictions of the locations of many genes along the chromosomes (Morgan 1965). In Mendel's work, for example,

Mutant fly (right) with a normal fly (left). The mutant fly has four wings instead of the normal two. The mutant fly was constructed by putting together three mutations in *cis* regulators of the *ultrabithorax* gene. These mutations effectively transform the third thoracic segment into another second thoracic segment (i.e., halteres into wings). Pascal Goetgheluck/Photo Researchers, Inc.

round and yellow alleles were switched randomly with wrinkled and green ones, recombining with recessive alleles half the time. Morgan observed this behavior in *Drosophila* as well. Morgan, however, also found linkage between certain genes in *Drosophila*. For some pairs of different genes, the two alleles associated in each of the parents of hybrids tended to stay with each other. These appeared together in the offspring of the hybrids, so that the percentage of recombination was less than the 50 percent expected under Mendel's principle of independent assortment. Genes that were closer together would recombine less often than those far apart. By defining a *map unit* as equivalent to 1 percent recombination, Morgan and his students used the recombination data to construct a map showing one gene separated from another by say 10 map units, followed by another x map units further on, and so on. The physical basis of this behavior was discovered from careful examination of the behavior of chromosomes during egg or sperm cell formation. Each individual receives one copy of each chromosome from each parent, or a copy of the X and a copy of the Y in the case of humans and *Drosophila*. During egg or sperm cell formation (*meiosis*), the chromosomes derived from the two parents coil up and pair with one another after

replication of the DNA and then, remarkably, exchange segments or cross over with one another at roughly random positions along their length. For genes that are close together, such an exchange is less likely to fall between them and result in switching of alleles, whereas for genes that are farther apart, such a switching is more likely. Thus, the further apart two different genes are along the chromosome, the more likely they are to be switched or recombined. Thus the genetic map, based on recombination data, reflects the physical organization of the genes along the chromosome. These maps remain valid, but more detail has been added to them. They include DNA sequences that have no known function, in addition to the classical genes that control such things as wing shape or the presence of antennae on the head.

The structure and function of DNA

The structure of DNA (Watson and Crick 1953) provided the first clues to the mechanism of replication, a key requirement for the genetic material to perpetuate itself from one generation to the next. Each DNA molecule contains two strands, with four organic bases, adenine (A), cytosine (C), guanine (G), and thymine (T), in complementary sequences along their length. The complementarity arises because chemically A pairs easily only with T, and G pairs easily only with C, so that the sequence of one strand could be used to tell what the sequence of the opposite strand is. If the two strands were to separate, then adding new bases together, one at a time, using each strand as a template, would produce two DNA molecules identical to the first. Matthew Meselson and Franklin W. Stahl (1958) soon confirmed this mechanism.

Throughout the early twentieth century, genetics research concentrated on the transmission of hereditary information and the construction of genetic maps. Some geneticists, however, studied two other very profound questions: The first was the theory of population genetics, and the second was the problem of developmental genetics. Population genetics essentially began when G. H. Hardy (1908) and Wilhelm Weinberg (1908) independently published letters explaining why variant recessive traits almost never exist at the level of 25 percent in populations, as some naïve people thought would be true if there were such traits. They showed that theoretically in a large population with random breeding, no selection, and no migration, a recessive allele that was rare would remain equally rare from one generation to the next under Mendelian inheritance. This is the foundation of population genetics. For it could be shown, with great mathematical sophistication (Fisher 1930), that if population size were reduced, breeding became nonrandom, or natural selection or migration was permitted, the allele frequency could change, and very rapidly. This field, after some years, led around 1940 to the evolutionary synthesis, that is, the great theoretical linkup between genetics and Darwinism (Mayr 2001).

The second research trend, developmental genetics, stemmed directly from the kind of work Morgan had done. Indeed Morgan started to work on *Drosophila* because he thought this would be a good way to gain important insights into developmental phenomena. All mutations that geneticists were studying had a developmental connection, because all effects they produced emerged eventually *via* changes occurring in a fertilized egg. Darwin had proposed that evolution occurred by tiny steps (*gradualism*); mutations geneticists were studying were sometimes subtle, including those that made only partial or quantitative contributions to a trait, but by preference they studied somewhat more dramatic and distinct ones just as Mendel had done. Richard Gold-schmidt (1940) even thought that evolution might take place by the admittedly rare occurrence of extremely dramatic mutations rather than the gradual accumulation of small changes as proposed by Darwin. It was therefore of no small interest when mutations showed up that had profound effects on development. One of the most famous was the brachyury mutation in mice (Gluecksohn-Schoenheimer 1938), which produced individuals without tails. The richest source of such mutants, however, turns out to have been Morgan's first choice: *Drosophila melanogaster* (Jaim Etcheverry 1995). One of these, *antennapedia*, results in the production of a leg where the antenna is normally found. Another mutation, the recessive and lethal *bicoid*, causes the embryo to develop two tails without any head. About 100 genes of this type, exerting profound effects on development, have now been identified in

Drosophila alone. They code for various kinds of signaling molecules that normally cause cells to change their patterns of behavior, and molecules that they synthesize, in effect changing the fertilized egg stepwise into a fully formed organism with hundreds of different cell types arranged in a specific fashion. Many of these genes share sequences with genes in other, very distantly related species, which indicates a common evolutionary origin deep in the history of higher organisms (e.g., McGinnis et al. 1984). Mutations in genes such as this are believed to be responsible for many morphological changes that distinguish different species of animals, fungi, and plants.

Changes in the appearance of organisms were initially thought to be caused, at least in part, by effects on enzymes. This idea received great support from research in the 1940s and 1950s on the biochemical basis of mutations in the bread mold *Neurospora crassa* (Beadle 1964). This organism grows on a simple defined chemical medium. Mutations led to the inability to produce certain key biochemicals, which could be identified by supplementing the growth medium. By combining mutations in different ways, researchers figured out sequences of biochemical transformation. A principle emerged that each gene could code for a specific protein or enzyme that functioned as a catalyst to convert one biochemical to another.

Mutations that alter enzymes or noncoding sequences may have other, less obvious effects, or even no effect at all. The latter are nonetheless useful because they allow estimates to be made of the background rate of change in DNA and thus can help determine the pathway and rate of evolutionary change (Kimura 1968).

The study of genes that control activities of other genes gained tremendous momentum from experiments on the bacterium *Escherichia coli* (Jacob 1972). This organism grows on a supply of the six-carbon monosaccharide sugar glucose with a few mineral salts added. If the glucose is left out and cells are supplied with the disaccharide sugar lactose, in a very short time they start to make three enzymes that allow them to convert this into glucose and resume growth. François Jacob and his colleagues hit upon ways of studying mutations that affected this process, including mutations that interfered with the synthesis of all three enzymes at once. They mapped these *regulatory* mutations and found three different places for them—two quite close to the genes for the enzymes and the third a little further away on the single bacterial chromosome. Through some very clever experiments they showed that the product of one of these regulatory genes has to interact with another gene in a specific way to prevent the lactose-utilizing genes from being expressed. This *repressor molecule* was later isolated and shown to bind to a particular sequence of DNA (called the *operator*) near the beginning of the genes for lactose utilization. This discovery, that a gene product (*repressor*) could turn off other genes, and another finding—that this process could be controlled by environmental or nutritional factors—started a still-continuing avalanche of discovery of molecules that control the expression of genes in both bacteria and higher organisms. Many of the proteins that control development in mammals share features in common with the lactose-repressor protein from *E. coli*.

Color enhanced platinum-shadowed electron micrograph (freeze-fracture TEM) showing characteristic attachment of RNA polymerase molecules to DNA strands. Omikron/Photo Researchers, Inc.

The ability of lactose to trigger synthesis of the enzymes needed for its use in *E. coli* is so sensitive and rapid—and so dependent on continued supplies of lactose—that Jacob and Monod suspected that an unstable intermediate was produced by the DNA. Indeed, this intermediate was soon identified as a nucleic acid, this time a single-stranded form called *messenger RNA*. In the 1960s, many researchers worked out the manner in which the sequence of bases in the DNA was converted to RNA form and then used to specify the sequence of amino acids in proteins, which in turn fold and assemble into the active catalysts or enzymes that mediate the vast and complex network of chemical reactions that go on inside cells (Lewin 2008; Leder and Nirenberg 1964). All organisms use this same genetic code. If they were of independent origin there would be no reason to expect this, as the code itself appears to be arbitrary. The only explanation for its universality, therefore, is that all organisms are descended from a common ancestor that used this same code.

Within fifteen years of the discovery of the regulatory genes in *E. coli*, scientists began physically isolating genes and determining their structure. Like many discoveries in biology, the clue to how to do this came from research that was pointed in another direction: the restriction of susceptibility to infection by viruses in bacteria. Some bacterial virus strains cannot attack certain strains of bacteria. Scientists learned that this restriction of host range was caused by a remarkable adaptation on the part of the bacteria, which consisted of enzymes (now called restriction enzymes) that would attack DNA at short specific sequences, such as GGATCC. This particular sequence would be expected in just about any DNA of a certain length (about 4,100 base pairs long). So if a virus with this big a DNA molecule introduces its DNA into a cell that has the right enzyme, it is likely to be destroyed, permitting the cell to survive. Why does the enzyme not destroy bacterial DNA? The GGATCC sequences in the bacteria's own DNA molecules are modified beforehand by another enzyme (*host modification enzyme*), so that the restriction enzyme does not attack them.

Now these restriction sites have an interesting property: They are *palindromic*, a palindrome referring to something reading the same backward as forward. The sequence GGATCC on one strand is complementary to the sequence CCTAGG on the other. But as James D. Watson and Francis H. C. Crick (1953) showed, the chemical orientations of the two strands are opposite, so that chemically the sequence CCTAGG should be read in the reverse direction, or as GGATCC—which is exactly what the restriction enzyme recognizes. Hence the enzyme cuts both strands. Not only that, but usually the cut is not centered but occurs between the fifth and sixth base. This leaves a single-stranded whisker (*sticky end*) at the end of the cut DNA molecule, and this sticky end is complementary to that on the end of another DNA molecule cut with the same enzyme. In principle, these two sticky ends could bind back to one another by complementary base pairing.

This set the stage for a remarkable development: After cutting DNA molecules from different organisms with the same restriction enzymes, the complementary sticky ends would not be able to tell the difference, and there would be a lot of recombination—DNA from one organism linking up with DNA from another. By applying yet another enzyme, called DNA ligase, to the mixture one could seal up single-stranded gaps at junctions.

Further steps remained, including how to recover interesting recombined molecules from the mixture. This was accomplished by making sure that the interesting ones had genes in them that somehow could be picked out. This can be done if some of the genes code for antibiotic resistance: Just allow the recombinant DNA molecules that contain antibiotic resistance genes to be taken up by antibiotic-sensitive bacteria. These bacteria are the only ones that will grow when the antibiotic is added to the growth mixture. Thus, when the bacteria grew to a healthy sized population, they could be harvested and the recombinant DNA isolated from them (Berg 1993).

The ability to obtain large amounts of the DNA coding for a particular sequence made it possible to apply chemical techniques to determine the exact sequence of bases (Maxam and Gilbert 1977; see also Sanger 1993). Initially fairly laborious, DNA sequencing became increasingly automated, so that by 2008 sequences could be obtained very rapidly at reduced expense. Making this easier was the polymerase chain reaction (*PCR*) technique devised by Kary B. Mullis (1997). This enabled researchers to make many copies of DNA in test tubes without having to transform DNA into host organisms. By 2008, thanks to this and other technological advances, scientists have uncovered the sequence of the human genome, several higher animals and plants, and many bacteria and viruses, and the list is growing rapidly. Large databases, most of them freely accessible on the Internet, make it possible for investigators to determine relationships between newly discovered genes and previously established sequences; the National Center for Biotechnology Information maintains one such database, known as Entrez.

At about the same time that sequencing became possible, rapid progress began on uncovering the overall organization of the genomic DNA. Some of the information that emerged from this was quite startling—it was found for example that huge amounts of DNA in humans and other higher organisms did not code for proteins, and that many sequences were repeated many times within the DNA (Lewin 2008). Contrary to the earlier notion of a single gene for each character, it came to be understood that even the genes for common proteins such as hemoglobin were repeated, with slight variations, along the chromosome. Odd phenomena were uncovered, such as the existence of enzymes that could use RNA as a template to make DNA and sequences that could integrate into DNA and then be copied to make new ones that would integrate elsewhere (Lewin 2008). Some of this work reignited curiosity and respect for the work of Barbara McClintock (1993), who, using breeding techniques, had documented the odd behavior of mobile genetic elements in maize in the 1930s and 1940s. The genome itself was increasingly being recognized as an immense domain where Darwinian processes could operate at the level of molecules (Lewin 2008; Dawkins 2006). The enormous variation in sequences soon gave rise to the identification of sequence markers, stretches of DNA of varying length. A wide variety of such sequences have now been studied, and they go by such acronyms as VNTR (*variable number tandem repeat*), RFLP (*restriction fragment length polymorphism*), SINE (*short interspersed nuclear element*), and LINE (*long interspersed nuclear element*). These occur with differing frequencies in populations and can be used to identify different-sized populations or subpopulations of animals, plants, fungi, or bacteria. This is the basis for the famous DNA fingerprinting technique (Lewin 2008; Jeffreys, Wilson, and Thein 1985), which, among other uses, allows one to: identify samples from crime scenes, delineate taxonomic relationships among related species, track down disease genes in afflicted families, and follow migrations of ancient human groups that spread around the world (such as through the National Geographic Society's Genographic Project).

The growing importance of genetic research and development

Applications of genetic research have transformed biology from a strictly academic discipline into a powerhouse of industry. Recombinant DNA techniques are used to produce pharmaceuticals, such as human growth hormone, and crop plants with unique properties, such as insect or salt stress resistance. Regulations are in place to prevent misuse or unwanted dispersal of recombinant organisms and products derived from them. Of course, as with any regulated business, frequent controversies occur about regulations and disagreements arise between interested parties. For example, in France, growth of genetically engineered maize is more restricted than in the United States. In 2000 a genetically

engineered maize strain (StarLink), approved only for animal consumption, entered the human food chain in the United States (in the form of taco shells). This concerned government officials, because insecticidal protein engineered into this strain of corn shared properties with known food allergens. An extensive investigation was conducted to determine if any exposed people had a reaction to the protein. As it turns out, none did, but the story argues for regulation and for advanced testing of genetic engineering products (CDC 2001).

Given abilities such as those cited above, the claim that DNA is the blueprint of life gains considerable credibility. By introducing a new gene or even several genes (or even an entire chromosome) into an organism, one can significantly alter its properties.

This had led to speculation that a synthetic organism might eventually be produced. Synthesis of an organic compound from inorganic precursors was first accomplished in the nineteenth century by Friedrich Wöhler (1828). Philosophically this marked the end of the artificial distinction between living and nonliving matter. By 2008 the artificial production of a living organism seemed much closer with the publication of a remarkable feat of genetic engineering, the complete test tube synthesis of a genome for a very small microbe called *Mycoplasma*

John Craig Venter was the first to decipher the entire genome of an organism in 1995 and had a role in creating the first cell with a synthetic genome in 2010. © Ron Sachs/CNP/Sygma/Corbis.

genitalium (Gibson et al. 2008). Leaving no room for doubt about the origin of the DNA, the authors introduced *watermark* sequences based on their own names at specific nonfunctional locations. Although this in itself does not constitute a synthetic organism, the next step, which would be to introduce this sequence into a cell from which the DNA had been removed, would go in that direction. If this works, the case for DNA as the blueprint for life would be greatly augmented. The complete synthesis of a cell from chemical precursors would require a very large number of enzymes to be made and combined correctly within a synthetically created cell membrane, to elicit instructions from DNA in the form of messenger RNA, for example. Impressive though this might be, it would not necessarily explain how the first living organisms arose.

What were the first genes and biochemical systems, and how did the first cells arise? As Darwin proposed, and many studies have confirmed since, evolution of organisms generally does not proceed by sudden jumps. It seems reasonable to suppose that the same is true of molecular evolution. The famous experiment by Stanley L. Miller and Harold C. Urey (1959) and subsequent studies have shown that many organic compounds can be formed spontaneously under conditions thought to have existed on the primitive, prebiotic Earth. If such reactions took place in the wild today, living organisms would quickly consume the evidence. But such organisms by definition must have been absent in the prebiotic world, so a complex brew of organic compounds could have existed. DNA itself is not generally thought to be the first form in which genes appeared on earth, but rather RNA, which being single stranded and capable of making copies of itself in some instances (Spiegelman et al. 1965) is a more plausible candidate for the first genetic material. This conjures up the image of an *RNA world* in the prebiotic past (Woese 1967). Scientists do not know exactly how the first self-replicating molecules were formed, but research is continuing in this field and it would be naïve to suppose that it will never produce a plausible scenario, even a detailed and reproducible one, leading to the formation of more complex self-replicating systems. If that were to be achieved, the model of natural selection first conceived by Darwin to explain the diversity of whole organisms could be further tested with regard to self-replicating molecules.

Resources

Books

Dawkins, Richard. 2006. *The Selfish Gene*. 30th anniversary edition. Oxford: Oxford University Press.

Fisher, R. A. 1930. *The Genetical Theory of Natural Selection*. Oxford: Clarendon Press.

Goldschmidt, Richard. 1940. *The Material Basis of Evolution*. New Haven, CT: Yale University Press.

Jaim Etcheverry, Guillermo. 1995. "Nobel Prize of Physiology or Medicine, 1955: Edward B. Lewis, Christiane Nüsslein-Volhard, Eric Wieschaüs.

Lewin, Benjamin. 2008. *Genes IX*. Sudbury, MA: Jones and Bartlett.

Mayr, Ernst. 2001. *What Evolution Is*. New York: Basic Books.

Wöhler, Friedrich. 1828. "Ueber künstliche Bildung des Harnstoffs" [On the artificial production of urea]. *Annalen der Physik und Chemie* 12: 253–256. English translation in *Readings in Elementary Organic Chemistry*, ed. L. A. Goldblatt, New York: Appleton-Century, 1938.

Woese, Carl R. 1967. *The Genetic Code*. New York: Harper & Row.

Periodicals

Avery, Oswald T., Colin M. MacLeod, and Maclyn McCarty. 1944. "Studies on the Chemical Nature of the Substance Inducing Transformation of Pneumococcal Types: Induction of Transformation by a Desoxyribonucleic Acid Fraction Isolated from Pneumococcus Type III." *Journal of Experimental Medicine* 79(2): 137–158.

Fairbanks, Daniel J., and Bryce Rytting. 2001. "Mendelian Controversies: A Botanical and Historical Review." *American Journal of Botany* 88(5): 737–752.

Gibson, Daniel G., Gwynedd A. Benders, Cynthia Andrews-Pfannkoch, et al. 2008. "Complete Chemical Synthesis, Assembly, and Cloning of a *Mycoplasma genitalium* Genome." *Science* 319(5867): 1215–1220.

Gluecksohn-Schoenheimer, Salome. 1938. "The Development of Two Tailless Mutants in the House Mouse." *Genetics* 23(6): 573–584.

Hardy, G. H. 1908. "Mendelian Proportions in a Mixed Population." *Science* 28(706): 49–50.

Jeffreys, Alec J., Victoria Wilson, and Swee Lay Thein. 1985. "Hypervariable 'Minisatellite' Regions in Human DNA." *Nature* 314(6006): 67–73.

Kimura, Motoo. 1968. "Evolutionary Rate at the Molecular Level." *Nature* 217(5129): 624–626.

Leder, Philip, and Marshall W. Nirenberg. 1964. "RNA Code-words and Protein Synthesis, III. On the Nucleotide Sequence of a Cysteine and a Leucine RNA Codeword." *Proceedings of the National Academy of Sciences of the United States of America* 52 (6): 1521–1529.

Maxam, Allan M., and Walter Gilbert. 1977. "A New Method for Sequencing DNA." *Proceedings of the National Academy of Sciences of the United States of America* 74(2): 560–564.

McGinnis, W., M. S. Levine, E. Hafen, et al. 1984. "A Conserved DNA Sequence in Homoeotic Genes of the *Drosophila* Antennapedia and Bithorax Complexes." *Nature* 308(5958): 428–433.

Meselson, Matthew, and Franklin W. Stahl. 1958. "The Replication of DNA in *Escherichia coli*." *Proceedings of the National Academy of Sciences of the United States of America* 44(7): 671–682.

Miescher, Friedrich. 1871. "Ueber die chemische Zusammen-setzung der Eiterzellen" [On the chemical composition of pus cells]. *Hoppe-Seyler's medicinisch-chemische Untersuchungen* 4: 441–460.

Miller, Stanley L., and Harold C. Urey. 1959. "Organic Compound Synthesis on the Primitive Earth." *Science* 130 (3370): 245–251.

Spiegelman, S., I. Haruna, I. B. Holland, et al. 1965. "The Synthesis of a Self-Propagating and Infectious Nucleic Acid with a Purified Enzyme." *Proceedings of the National Academy of Sciences of the United States of America* 54(3): 919–927.

Watson, James D., and Francis H. C. Crick. 1953. "Molecular Structure of Nucleic Acids: A Structure for Deoxyribose Nucleic Acids." *Nature* 171(4356): 737–738.

Weinberg, Wilhelm. 1908. "Über den Nachweis der Vererbung beim Menschen" [On the demonstration of heredity in man]. *Jahreshefte des Vereins für vaterländische Naturkunde in Württemberg* 64: 368–382.

Other

Beadle, George. 1964. "Genes and Chemical Reactions in *Neurospora*." In *Nobel Lectures: Physiology or Medicine, 1942–1962*. Amsterdam: Elsevier. Also available from http://nobelprize.org/nobel_prizes/medicine/laureates/1958/beadle-lecture.html.

Berg, Paul. 1993. "Dissections and Reconstructions of Genes and Chromosomes." In *Nobel Lectures: Chemistry, 1971–1980*, ed. Tore Frängsmyr and Sture Forsén. Singapore: World Scientific Publishing. Also available from http://nobelprize.org/nobel_prizes/chemistry/laureates/1980/berg-lecture.html.

Centers for Disease Control and Prevention (CDC). 2001. "Investigation of Human Health Effects Associated with Potential Exposure to Genetically Modified Corn." Available from http://www.cdc.gov/nceh/ehhe/Cry9cReport/pdfs/cry9-creport.pdf.

Jacob, François. 1972. "Genetics of the Bacterial Cell." In *Nobel Lectures: Physiology or Medicine, 1963–1970*. Amsterdam: Elsevier. Also available from http://nobelprize.org/nobel_prizes/medicine/laureates/1965/jacob-lecture.html.

McClintock, Barbara. 1993. "The Significance of Responses of the Genome to Challenge." In *Nobel Lectures: Physiology or Medicine, 1981–1990*, ed. Tore Frängsmyr and Jan Lindsten. Singapore: World Scientific Publishing. Also available from http://nobelprize.org/nobel_prizes/medicine/laureates/1983/mcclintock-lecture.html.

Mendel, Gregor. 1866. "Versuche über Pflanzen-Hybriden" [Experiments in plant hybridization]. *Verhandlungen des naturforschenden Vereines in Brünn* 4: 3–47. English translation available from http://www.mendelweb.org/Mendel.html.

Morgan, Thomas Hunt. 1965. "The Relation of Genetics to Physiology and Medicine." In *Nobel Lectures: Physiology or Medicine, 1922–1941*. Amsterdam: Elsevier. Also available from http://nobelprize.org/nobel_prizes/medicine/laureates/1933/morgan-lecture.html.

Mullis, Kary B. 1997. "The Polymerase Chain Reaction." In *Nobel Lectures: Chemistry, 1991–1995*, ed. Bo G. Malmström. Singapore: World Scientific Publishing. Also available from http://nobelprize.org/nobel_prizes/chemistry/laureates/1993/mullis-lecture.html.

National Center for Biotechnology Information. "Entrez, the Life Sciences Search Engine." Available from http://www.ncbi.nlm.nih.gov/sites/gquery.

National Geographic Society. "The Genographic Project." Available from https://www3.nationalgeographic.com/genographic/.

Sanger, Frederick. 1993. "Determination of Nucleotide Sequences in DNA." In *Nobel Lectures: Chemistry, 1971–1980*, ed. Tore Frängsmyr and Sture Forsén. Singapore: World Scientific Publishing. Also available from http://nobelprize.org/nobel_prizes/chemistry/laureates/1980/sanger-lecture.html.

Harry Roy

● ● ● ● ●

Genes and development

With a little simplification, all forms of life on Earth can be viewed as nothing more than specially organized configurations of matter. But how are complex configurations of matter such as that in a shark, oak, ant, bat, giraffe, human, or any other multicellular organism achieved? What does it take to develop from a tiny single totipotent cell, such as a zygote (a fertilized egg), into a big adult creature composed of billions of specialized cells? And what does individual development, or ontogeny, have to do with the historical development of life and all its forms on Earth, that is, evolution?

Before answering these questions, it is necessary to summarize briefly the ontogeny of a multicellular organism. Surprisingly, this is not easy at all: "Given that Aristotle began the study of development over two millennia ago, we might expect the mechanisms of development to be well understood, but alas, they are not" (Hall 1999, p. 111).

Nevertheless, what seems to be clear enough is that ontogeny is a process that starts at a stage of a single cell and leads to the formation of a complex individual phenotype, which consists of billions of cells arranged into a functional system. This process runs under the control of something that can be metaphorically denoted as an "information system," which derives from two sources: a genetic code and the developmental (including cellular) environment. Until recently, biologists believed (and many still believe) that these two sources of information are not equal, that is, that the role of genes is primary and the role of the developmental environment is secondary. In other words, genes have been supposed to be "environmentally activated to produce the phenotype from what is thought to be latent in the genotype" (Robert 2004, p. 62). New insights have revealed, however, that developmental processes are much more complicated, specifically that genes themselves do not contain information sufficient to build up the phenotype of a multicellular metazoan organism and that the successful formation of the definite phenotype requires interactions not limited exclusively to gene activation but rather involving positive and negative feedback loops at a variety of levels (Robert 2004).

From Haeckel's biogenetic law to evo-devo

Individual phenotypes of metazoan organisms represent just a temporary existence of living forms, because they are mortal.

Nevertheless, these mortal phenotypes have the capability to reproduce, and thus to maintain the long-term existence of their own form of life (a species). If the temporary existence of an individual phenotype results from ontogeny, then the long-term existence of a species can be viewed as a sequence of individual ontogenies. On a long-term scale, however, species are not unchangeable, because all forms of life must be ready to answer the challenges posed permanently by altering environmental conditions, and as a result they evolve. This begs the question: How do developmental processes affect evolutionary changes? And, in turn, how has evolution affected ontogeny itself?

Ernst Haeckel came up with the idea that "ontogeny recapitulates phylogeny". © Library of Congress - digital ve/Science Faction/Corbis.

For centuries, these questions have been central to many philosophers and naturalists. Substantial progress in unraveling the links between ontogeny and evolution was achieved in the late 1800s by Ernst Haeckel (1834–1919), a prominent German biologist. His thinking initially derived from the Lamarckian notion of evolution—that evolution progressed via an accumulation of acquired characteristics. If true, then every descendant organism, during its development, would have to pass through the stages of its evolutionary ancestors first and could reach its adult appearance (the definite phenotype) only subsequently. "Ontogeny recapitulates phylogeny," asserted Haeckel and described this principle as his *biogenetic law*. Also known as the recapitulation theory, the biogenetic law promoted Charles Darwin's theory of evolution by means of natural selection, and as such it has had a strong influence on biology. It was, however, also extremely controversial.

The major problem with Haeckel's biogenetic law is its ambiguity. In their comprehensive review of the topic, Michael K. Richardson and Gerhard Keuck (2002) wrote: "It is not always clear whether he [Haeckel] was advocating the repetition [recapitulation] of entire developmental stages or of individual characters only. Furthermore, he was not consistent when describing the extent of parallelism between evolution and development" (p. 503). Haeckel's work is also associated with one of the biggest scandals in the history of science: He was charged (and in some cases also convicted) of plagiarizing, doctoring, and fabricating some of his famous/infamous drawings, which were to "prove" his recapitulation theory. Stephen Jay Gould (2000) even concluded that Haeckel had in some respects committed the "academic equivalent of murder."

In spite of this controversy, Haeckel's contribution to evolutionary and developmental biology was fundamental. Although there is no evidence from vertebrates that entire evolutionary stages are recapitulated during development, several modern studies support the biogenetic law in the case of single-character transformations. Haeckel recognized the evolutionary diversity in early embryonic stages, in line with modern thinking. His criticized embryo drawings are important as evidence for evolution. "Despite his obvious flaws, Haeckel can be seen as the father of a sequence-based phylogenetic embryology," Richardson and Keuck conclude (2002, p. 495).

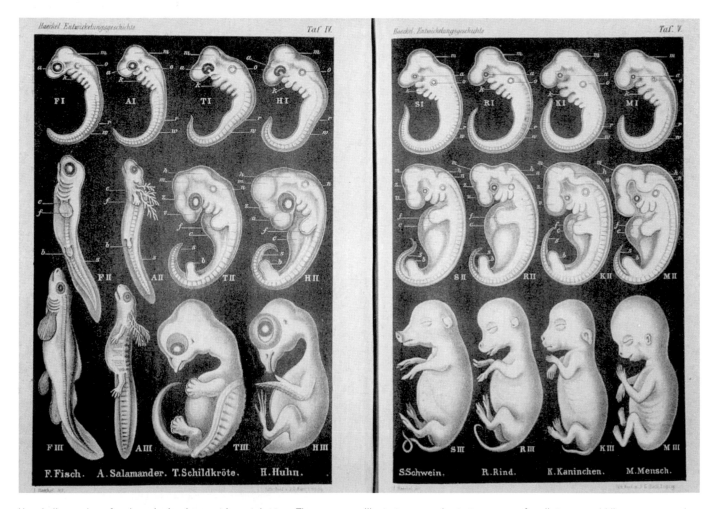

Haeckel's version of embryonic development in vertebrates. The upper raw illustrates an early stage common for all groups, middle raw represents a medium stage, and the lower raw an advanced stage of development. From left to right: fish, salamander, turtle, chick, pig, cow, rabbit and human. Haeckel, E., (1874): Anthropogenie oder Entwickelungsgeschichte des Menschen, 1. a 2. vyd. Engelmann, Leipzig; tab. IV a tab. V. Lithograph by J.G. Bach of Leipzig after drawings by Haeckel, from *Anthropogenie* published by Engelmann Public domain.

Nevertheless, in the 1930s, the rapid progress in genetics strongly affected developmental biology, and since that time only embryologists have taken embryology seriously. Most others have been fascinated by the elegance and simplicity of genetic models that reduced ontogeny to gene activity. "Thanks to its tremendously productive models, genetics established a monopoly position in both evolutionary biology and embryology in the first half of the twentieth century: embryology was redefined as the study of changes in gene expression, whereas the task of evolutionary biologists was recast as the study of changes in gene frequencies in a population" (Robert 2004, p. 58).

Serious doubts about the primacy of genetics in evolutionary processes emerged, however, after the surprising finding that one group of extremely important genes—homeotic genes, which play the regulatory role in the development of plants, fungi, and animals—have remained highly conserved throughout the entire evolution of all animals from sponges to mammals. But the mainstream reductionist view, which overestimates the role of genetic variation and natural selection in evolution, assumed that genetic and ontogenetic systems would evolve at exactly the same rate as the morphological characters they produce. Thus, the discovery of almost identical homeotic genes throughout the entire evolution of all animals, together with further advances in molecular biology, morphology, ecology, and developmental biology, contributed to the development of a new scientific discipline—evolutionary developmental biology (evo-devo). And it is just evo-devo that brings new insights into evolutionary processes, often as an alternative to mainstream evolutionary biology, that is, neo-Darwinism or the modern synthesis. Indeed:

> to evolutionary biologists, the emerging understanding of developmental mechanisms is an opportunity to understand the origins of variation not just in the selective milieu but also in the variability of the developmental process, the substrate for morphological change. Ultimately, evolutionary developmental biology ... expects to articulate how the diversity of organic form results from adaptive variation in development. (Dassow and Munro 1999, p. 307)

Genecentrism versus epigeneticism

Modern biology views the ontogeny of multicellular organisms as a process with modular structure and hierarchical arrangement. This generates very important implications for the scientific approach to developmental processes, because the hierarchical nature of development and evolution necessitates the study of emergent properties inexplicable from lower (or higher) hierarchical levels (Robert 2004). Most developmental biologists also agree that to understand ontogenetic processes it is necessary to focus on the interactions between genotype and developing phenotype, which draws attention to the fundamental ontogenetic importance of epigenetics and epigenesis.

Indeed, epigeneticism has become an influential branch within evolutionary developmental biology. Epigenetics, however, is not the same as epigenesis. In fact, even the term

epigenetics may have different meanings for different scientists. For molecular biologists it is mainly the study of heritable changes in gene function that cannot be explained by changes in DNA sequence, whereas for functional morphologists epigenetics means the entire series of interactions among cells and cell products that leads to morphogenesis and differentiation (Haig 2004). The latter notion of epigenetics is closer to the original meaning of this term, coined by the British embryologist Conrad H. Waddington in 1942, and it refers to the central phenomenon in developmental biology—epigenesis.

Epigenesis is the process responsible for the individual development of metazoan (multicellular) organisms. It is an intrinsic element of ontogeny, and it occurs out of the genetic code. Developmental biologists assume that epigenesis may also play an important role in evolution. For example, Eugene K. Balon has developed a theoretical concept of alternative ontogenies and evolution (2004). This concept follows Waddington's famous model of the epigenetic landscape, which illustrates a ball rolling over a landscape with branching valleys (see Figure 1). The model represents a metaphor for possible developmental pathways of embryos or their cells, for example, the possible developmental pathways of stem cells and their further differentiation into specialized cells, such as muscle cells, neurons, and bone cells. The cells are canalized along the developmental pathways represented by the types of gene activities. At each bifurcation, the cell is directed into one of the two possible pathways ("canals") by the cues from the intracellular, extracellular, and/or external environment, which alternatively switch different genes on or off.

Balon's theory of alternative ontogenies and evolution (2004) derives from the fact that the process of building a definite phenotype of a metazoan organism from a single nonspecialized cell requires two sources of information: the so-called programmatic source (genetic code) and the developmental epigenetic source (environment, including cellular). If the programmatic information comes from the genotype (but see below), then the epigenetic information is supplied

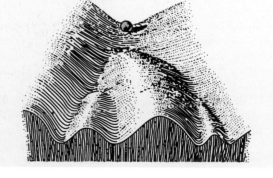

Figure 1. Waddington's Epigenetic Landscape. The ball represents cell fate. The valleys are the different fates the cell might roll into. At the beginning of its journey, development is plastic, and a cell can become many fates. However, as development proceeds, certain decisions cannot be reversed. From Waddington, C. H. (1940): Organisers & Genes. Cambridge: Cambridge University Press. Reproduced by permission of Gale, a part of Cengage Learning.

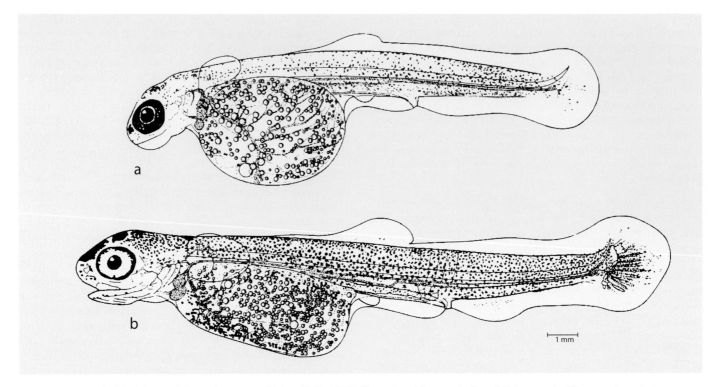

Precocial (top) and altrical (bottom) form of sunapee. (Balon, E. K. 1980). Reproduced by permission of Gale, a part of Cengage Learning.

by the developing phenotype—this is exactly about the cues that direct the ball in Waddington's model into one of the two possible canals (see Figure 1). At the same time, both of these sources of information are also sources of variation, which enters the process at each transition from one developmental interval (module) to another. As a result, the life history of each population and/or species can vary back and forth, from generation to generation, along a continuum between the most generalized (altricial) and the most specialized (precocial) extremes. Epigenesis, as the mechanism of ontogenies, creates in every generation alternative variations that enable the organisms to survive in the changing environments as either altricial or precocial forms (Balon 2004).

One of the best demonstrations of the phenomenon of alternative ontogenies can be seen in fishes. Experimental comparative studies of early development in chars (salmonid fishes of the genus *Salvelinus*) indicated that some females produced smaller eggs or eggs with denser yolk than other females. The smaller eggs resulted in more generalized progeny (more tolerant to various ecological factors) in comparison with the more specialized progeny (less tolerant to various ecological factors) from larger eggs, even if the rearing conditions were identical. When smaller eggs were kept under two different temperature regimes, however, the warm incubated progeny was more specialized than that from colder water, and the same results were obtained with larger eggs. Subsequently, these laboratory experiments indicating the existence of alternative ontogenies were supported by studies conducted in the field (both reviewed in Balon 2004).

Two tiny, closely related North American killifishes (*Lucania goodei* and *Lucania parva*), which inhabit very different environments, provided material for this "natural experiment." The former lives in a stream with stable conditions and little diel, or seasonal fluctuations, whereas the latter has to cope with highly unpredictable conditions—altering expositions to seawater and freshwater, as well as large diel and seasonal temperature and oxygen fluctuations. Even though the distance between these two populations is only 10 kilometers, the life history characteristics exhibited by *L. goodei* were found to be more specialized than those of *L. parva*. Females of *L. goodei* produced significantly fewer eggs with significantly more yolk. Consequently, their offspring developed more rapidly than those of *L. parva* and reached the definitive phenotype at an earlier age. "All these differences clearly agreed with the expected differences caused by epigenetic processes ultimately responsible for the true species-pair divergence" (Balon 2004, p. 58).

Ironically, another good example supporting the theory of alternative ontogenies comes from an unwanted recent global phenomenon—biological invasions. Despite the huge negative impact that the invaders may pose to native ecosystems, biological invasions also offer the potential for positive use, in particular the expansion of scientific knowledge of biological processes, because the invaded areas also represent sophisticated, in situ laboratories. The round goby, *Neogobius melanostomus*, is a relatively small gobiid fish originating from the Ponto-Caspian region (the species' native range covers the Black Sea, Sea of Azov, Caspian Sea, and surrounding waters). Since the early 1980s, this fish—originally rather inconspicuous—has

Comparative life-history studies between native and nonnative populations of round goby (*Neogobius melanostomus*) (both European and American) reveal that this species demonstrates great flexibility in its biological traits resulting from alternative ontogenies. Edward Kinsman/ Photo Researchers, Inc.

managed to colonize the largest European rivers from the Volga and Dnieper to the Danube and Rhine, and thus to expand its area of distribution thousands of kilometers away from its native range, reaching the shores of the Baltic Sea and North Sea. Moreover, it was also introduced unintentionally into the Great Lakes in North America, where it has become a nuisance nonnative aquatic animal.

Nevertheless, comparative life-history studies between native and nonnative populations of round goby (both European and American) reveal that this species demonstrates great flexibility in its biological traits resulting from alternative ontogenies. Indeed, the same shift from the highly specialized life history of the indigenous round goby back toward a more generalized life history (especially earlier maturation at a smaller size) in the invasive populations occurred independently on two continents (Balážová-Lavrinčíková and Kováč 2007). The same pattern was found in the bighead goby, *Neogobius kessleri*—a close relative of the round goby that is also invasive, though in Europe only. In the bighead goby, the shift toward the alternative, more generalized ontogeny is even more apparent, as the females of an invasive Danubian population not only mature at a smaller size but also produce a higher number of smaller eggs.

Thus, similar to the alternative ontogenies and life-history strategies in the two killifish species, both the round and bighead gobies follow the same scenario. Their ontogenies produce specialized forms under stable conditions (represented by native areas in the case of gobies), but when the conditions become unpredictable (as is the case of the unknown environment in the invaded areas), they shift toward more generalized alternatives.

Many biologists believe that DNA is a structure that itself contains a predefined genetic program that controls the development of the organism. This idea, however, is simply not correct. In fact, without the highly structured cellular environment, which itself is not constructed by DNA, the DNA molecule is inert, relatively unstructured, nonfunctional, and therefore not ontogenetically meaningful. "DNA is a dead molecule, among the most non-reactive, chemically inert molecules in the world—not only is DNA not capable of making copies of itself—but it is incapable of making anything else" (Lewontin 2000, p. 141). Indeed, both the structure and the function (the ontogenetic role) of genes derive from spatial and temporal aspects of the state of the cell-organism. "In turn, genes so produced help to regulate ontogenetic processes as participants in nonlinear feedback and feedforward networks generating and being generated by the developing organism. Consequently, the usual idea of genetic primacy is rendered incoherent" (Robert 2004, p. 75).

In other words, the programmatic information (as a whole) is not simply deposited in DNA but is also always generated de novo (anew) in coordination with the three-dimensional structure of the cell. If the genome comes from the information recorded as the "memory" of past environments, developments, and their genetic assimilations, then the phenotype is formed by an interaction with the present environment. In terms of the theory of alternative ontogenies and evolution, the living systems work on a principle similar to that known from personal computers—a principle of bifurcation. Each operation in such a computer is based on a series of steps that have passed through the bifurcation of two options (0 or 1). The same applies for both the short-term development (ontogeny) and the long-term development (evolution) of living forms. Every organism has the potential to respond during its ontogeny to cues from both external and internal environments in two ways: 0 or 1. Thanks to this principle, various forms (individuals) of the same species, as well as various forms of organisms (species), can arise.

The theory of saltatory ontogeny

The resemblance of this concept of ontogeny with Waddington's epigenetic landscape (see Figure 1) is not accidental. It is based on the theory of saltatory ontogeny, which considers the individual development of a multicellular phenotype to be a sequence of stabilized states, that is, developmental steps (or modules). Central to this theory is the idea that a developing individual cannot remain stabilized during the constant additions and subtractions of structures and functions, and during the constantly changing multitude of environmental, cellular, structural, and endocrine interrelations. During a stabilized state (developmental step), cells and tissues differentiate, and structures grow at various rates, as if accumulated and canalized in preparation for the next, more specialized, stabilized state. The developing system "resists" destabilization for as long as possible, enabling structures to be completed and functions to progress without interfering with stabilized life activities. When ready for new or additional integrative actions, the organism shifts via a far-from-stable developmental threshold into the next stabilized state, that is, the next step of ontogeny (Balon 2004). The shift into the subsequent developmental step is possible, however, only through the bifurcation of the two options (0 or 1) that represent either an altricial (more generalized) or precocial (more specialized) direction for further development.

The cue that influences which of the two directions will finally take place comes from the external or internal environment of the developing organism. Thus, it appears that metazoans are more than sheer products of epigenetically triggered, preformed genetic programs. Ontogeny cannot be reduced to differential gene expression. Alternatively, it is a creative process, nonlinear, or emergently epigenetic (Robert 2004). Inasmuch, when two genetically identical individuals develop in different environments, they finally create two different phenotypes. Their development has been affected by their environments and thus canalized through the bifurcations into different developmental trajectories. At the population level, this results in a variety of individual phenotypes. This variety, as a whole, represents a discrete continuum with "extremely" generalized forms at one end and "extremely" specialized forms at the other end.

Of course, there are limits to this flexibility, and such intraspecific differences between the general (altricial) and the specialized (precocial) forms are usually very small. Nonetheless, the ontogeny of a metazoan appears to be an extraordinarily complex process that supplies organisms with an enormous potential to generate a variety of phenotypes, an emergent property that is essential for species to survive in various environments and consequently very useful for a species to establish self-sustaining populations in novel environments. The phenotypic plasticity of a species, therefore, appears to be a function of epigenetic mechanisms, which are usually expressed through the creation of both altricial and precocial forms within and/or among populations.

Beyond the timescale of generations, different ontogenetic trajectories (from generalized to specialized) in generation lineages, given an appropriate environment and/or isolation, may result in formation of new taxa (Balon 2004). The theory of alternative ontogenies and evolution, together with the theory of saltatory ontogeny, also has another implication for evolutionary processes: Both predict that the processes of evolution can be very rapid rather than gradual. This is at odds with the traditional mainstream neo-Darwinian gradualist view of evolution, though the idea of nongradual events in evolution is neither new nor exclusive.

The theory of punctuated equilibrium

Certainly the most famous concept of rapid (but not saltational) evolution is the theory of punctuated equilibrium developed by Niles Eldridge and Stephen Jay Gould (1972). Central to this theory is the idea that evolution consists of long periods of relatively little change (i.e., stasis) that are punctuated occasionally by short bursts of rapid evolution associated with the emergence of new species. In other words, "speciation is a rare and difficult event that punctuates a system in homeostatic equilibrium" (Eldridge and Gould 1972, p. 115). The theory of punctuated equilibrium was originally proposed based on an evaluation of fossil records and as such has become a subject of controversy among paleontologists and other evolutionary biologists, especially population geneticists. Unexpected support for punctuated equilibrium, however, has recently been provided from molecular biology (Pagel, Venditti, and Meade 2006). If rapid evolution used to

be considered an exception rather than a common phenomenon, this study suggests that as much as 22 percent of substitutional changes at the DNA level can be attributed to punctuational evolution, with the remainder accumulating from background gradual divergence. Thus, it appears that "punctuational episodes of evolution may play a larger role in promoting evolutionary divergence than has previously been appreciated" (Pagel, Venditti, and Meade 2006, p. 119).

The theory of punctuated equilibrium is now widely recognized as a valid model for evolutionary change that largely depends on environmental circumstances (Gould 2002), as well as developmental plasticity: "there is often a very good correspondence between patterns of developmental plasticity and major patterns of evolutionary change. This supports the view that the flexible phenotype, not the genome, is the leader of the evolutionary parade, and it is a strong argument in favor of a multilevel approach to understanding macroevolutionary change" (West-Eberhard 2003, p. 616).

The theory of synchrony and heterochrony in ontogeny

In terms of space and time, developmental plasticity can be viewed as a result of variation in timing—a mechanism known as heterochrony. The role of heterochrony in ontogeny is explained within the theory of synchrony and heterochrony in ontogeny (Kováč 2002). This theory speculates that ontogeny is a process that strongly depends on time and that two timing mechanisms are essential for the development of a metazoan— synchrony (coordinating) and heterochrony (implementing).

Indeed, during a developmental step, not all the developing structures are synchronized all the time, for the development of these structures will not progress at exactly the same rate. As such, they will not all be, at any given moment, at the same "distance" from their definitive developmental state. Therefore, within the steps, those structures still under development cannot form a fully functional system. For the living system to maintain its functionality, however, a higher level of synchronization must be reached at a certain point. This higher level of synchronization can be attained only during a transition to a new step (a higher stabilized state). Such a transition ensures that the organism acquires new functions and skills, and thus enhances its chances of survival.

Thus, if developing structures cannot be synchronized constantly throughout the entire ontogeny, but their synchronization must be achieved during a transition to a new step, then it appears that two timing mechanisms working in concert (synchrony and heterochrony) enter ontogenetic processes. Heterochrony operates at the level of individual structures (cellular organs, cells, tissues, organs), speeding up or slowing down their development so that a relevant group of structures is ready at the same moment. Synchrony operates at the level of groups or systems of structures (e.g., organ systems) to harmonize them so that the various groups of structures are ready simultaneously, providing the organism with new functions (abilities) and interactions associated with the next (higher) stabilized state. Both of these mechanisms, however, are integral parts of the

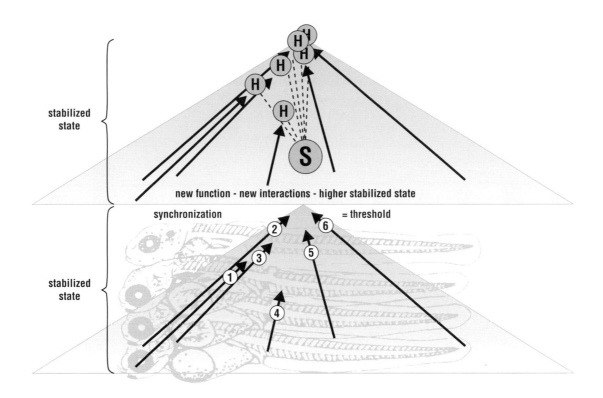

Hypothetical model of how synchrony and heterochrony act in ontogeny of fish. In Zingel streber, E7 is characterized by preparations for exogenous feeding (Kov!a$c, 2000). The onset of exogenous (i.e. active) feeding is an important threshold, both ontogenetically and ecologically. When endogenous supplies (yolk, oil globule) are nearly depleted, embryos must begin active feeding. However, this requires that all necessary organs are not only well developed but also synchronized in a manner that ensures efficient cooperation and functionality (prey detection, capture, handling, digestion). This requires synchronization of the jaw apparatus (1), eyes (2), respiratory and digestive systems (3, 4), muscles (5), finfold (6), circulation, etc., to form a functional system. To achieve this, some of these organs require accelerated development (4, 5), others decelerated formation (1, 2, 3, 6), i.e. heterochronies (H), and the whole process is coordinated by synchrony (S). (Kovac, 2002). Reproduced by permission of Gale, a part of Cengage Learning.

same entity—synchrony acts as a coordinating mechanism, heterochrony as an implementing mechanism.

The theory of synchrony and heterochrony itself may not seem to be so exciting, but it has important implications for evolution. Overall, at least three levels of heterochrony should be distinguished: interspecific, intraspecific and intra-individual. Widely accepted heterochrony, which appears as evolutionary changes in morphology that result from shifts in the rate or timing of ancestral developmental patterns, is the same heterochrony that is the source of individual intraspecific variation in a population, and the same heterochrony that is essential to each individual phenotype during its ontogeny. "The difference between these three types of heterochrony is not in the phenomenon itself but in the way we perceive and classify it" (Kováč 2002, p. 506).

Inheritable epigenetics

Although the heritability of epigenetic features has been discussed for decades, most biologists still view the potential role of epigenetics in evolution with caution. There have been an increasing number of scientists, however, who concede that epigenetic mechanisms could affect evolutionary processes. One of them is Wallace Arthur: "phenotypic plasticity should not just be thrown out of the window by evolutionary theorists because it is not inherited. If you take one step back from the non-inheritance of 'plastic' variants, you will probably find that the pattern of plasticity is itself inherited" (Arthur 2004, p. 151). Arthur maintains that "biases in the ways that the embryos can be altered are just as important as natural selection in determining the directions that evolution has taken, including the one that had led to the origin of humans" (2004, back cover).

And the biases in the embryos are nothing else but epigenetics. Indeed, more and more evidence has been emerging from various scientific studies that epigenetic features are heritable. In fact, changes in a phenotype do not require any change in the DNA sequence: A change in gene activities, which has been proven heritable, is enough. Therefore, biologists coined the term *epigenetic inheritance systems*. Molecular biologists have described at least four types of epigenetic inheritance: memories of gene activity (so-called self-sustaining loops), architectural memories (structural inheritance), chromosomal memories (chromatin-marking systems), and silencing of the genes (RNA interference) (Jablonka and Lamb 2005).

Nevertheless, before the role of epigenetics in evolution can be acknowledged, one more fundamental question must be resolved: How can the epigenetically acquired attributes of a metazoan phenotype pass through the powerful barrier between the somatic and reproductive cells, and thus from parents to their offspring? In fact, one of the possible answers is already known—RNA interference (RNAi), a well-described molecular mechanism that inhibits gene expression (for the discovery of RNAi, Andrew Z. Fire and Craig C. Mello were awarded the 2006 Nobel Prize in Physiology or Medicine).

Be that as it may, the debate about the details of evolutionary mechanisms continues. One thing is nevertheless sure: "Ideas about heredity and evolution are undergoing a revolutionary change. New findings in molecular biology challenge the gene-centered version of Darwinian theory according to which adaptation occurs only through natural selection of chance DNA variations" (Jablonka and Lamb 2005, front flap). The latest insights into evolutionary biology, taken from a developmental view, make clear that induced and acquired changes also play a role in evolution.

Genes and environment

Nevertheless, considering other than genetic aspects seriously, both in development and evolution, does not mean one must neglect the role of genes in these processes. Metazoans are functionally integrated wholes, and "functional integration arises from temporal chains of stable interacting genes" (Sinervo and Svensson 2004, p. 171). These temporal chains of interacting genes fit well the theory of synchrony and heterochrony in ontogeny (see above), and the interactions within them are known as epistasis. Epistasis occurs when the activity of one gene is modified by one or several other genes.

A highly illustrative example of how epistasis affects development, and thus also phenotypic plasticity and evolution, can be found for instance in primitive amphibians—the axolotls. The ontogeny of a metazoan is a process regulated by the products of endocrine systems—hormones. In the Mexican axolotl, *Ambystoma mexicanum*, the genetically based deletion of thyroxine regulation leads to a loss of the terrestrial adult period of life (Sinervo and Svensson 2004). Thus, this species spends its whole life in the water. It is evolutionary derived from the tiger salamander, *A. tigrinum*, which has retained larval metamorphosis. So what happened with the Mexican axolotl? It simply reached sexual maturation in its juvenile period of life, which is a classical example of heterochrony called juvenilization or paedomorphosis. And the mechanism, how it happened, is just the alteration in the endocrine regulation that resulted from the alteration in gene interactions, that is, epistasis.

However, in the mole salamander, *A. talpoideum*, a close relative of the two above species, the same phenomenon—heterochrony—has taken a different shape and resulted in different effects. Under certain conditions, some individuals of the same population exhibit temporal plasticity (delay) in the onset of metamorphosis relative to the normal ontogenetic

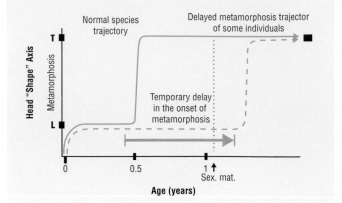

Intraspecific heterochrony in the mole salamander. Some individuals of *Ambystoma talpoideum* (dashed line) exhibit temporal plasticity (delay) in the onset of metamorphosis (large arrow) relative to the normal species trajectory of metamorphosis in the first year of life (solid line). Metamorphosis is delayed for one or more years but virtually all individuals eventually proceed through normal transformation to the terminal shape for this species (T on y-axis). (Reilly, S. M., Wiley, E. O., Meinhardt, D. J. 1997). Reproduced by permission of Gale, a part of Cengage Learning.

trajectory of the species. Their metamorphosis is delayed for one or more years, but virtually all individuals eventually proceed through normal transformation to the definitive shape for this species (Reilly, Wiley, and Meinhardt 1997). Thus, heterochrony provides the population alternative ways of development, which increases its chance to survive under unpredictable conditions.

Thus, in axolotls, two ontogenetic phenomena—epistasis and heterochrony—may lead not only to phenotypic plasticity within a population of the same species but also to the most advanced step in evolution—the origin of a new species (speciation). Nevertheless, perhaps the most illustrative example of phenotypic plasticity can be found in a completely different group of animals: the social insects. Colonies of social insects, for example, ants, thrive thanks to the division of their members into distinct castes, such as workers and soldiers. Depending on the environmental cues (temperature, photoperiod, or nutrition) received during development, an egg develops into either a winged queen or a wingless worker. Therefore, although the members of the castes differ dramatically in appearance and behavior, the differences among them are not genetic. The completely different phenotypes emerge during their ontogeny, which takes one of the possible developmental trajectories that depends on the treatment by the queen and the workers, who manipulate developmental factors, for example, larval diet and/or incubation temperature.

Of course, the destiny of each individual ant is not completely "gene-free," for the range of its potential developmental trajectories is restricted genetically. In other words, the genome of each individual contains all the instructions needed to develop into any one of several ant phenotypes, but only a subset of genes necessary for one developmental trajectory is activated

In the Mexican axolotl (*Ambystoma mexicanum*), the genetically based deletion of thyroxine regulation leads to a loss of the terrestrial adult period of life. Kenneth W. Fink/Photo Researchers, Inc.

epigenetically. Nevertheless, the definite developmental "program" for each individual is not predefined in the genes but emerges during ontogeny.

Genes, phenotype plasticity, and humans

If ontogeny and evolution of multicellular organisms is basically an extremely complex game between genes and environments, then the development of *Homo sapiens* is not an exception. Humans show the mammalian type of phenotype plasticity and evolution, though extended up to four dimensions: genetic, epigenetic, behavioral, and symbolic. Molecular biology has shown that cells transmit information to daughter cells not only through the genes but also through epigenetic inheritance, which means that all organisms have at least two systems of heredity. Many animals can transmit information to their offspring by their behavior (observational learning), which represents the third heredity system. "And we humans have a fourth, because symbol-based inheritance, particularly language, plays a substantial role in our evolution" (Jablonka and Lamb 2005, p. 1).

Both human development and evolution in these four dimensions have various implications, from biological (including those associated with health) up to cultural and social. For example, "in humans, career decisions can be strongly influenced by body size [a subject of phenotype plasticity], as shown by the size differences between the average racing jockey or college professor and the average basketball player" (West-Eberhard 2003, p. 390). For the existence of humans, however, the so-called grandmother effect is much more important. The theory of saltatory

ontogeny mentioned above is closely associated with the hierarchical model of individual development that is comprised of five basic intervals (periods) of life: embryonic, larval, juvenile, adult, and senescent (Balon 2004). Because ontogeny is also modular, not all metazoans must follow this model completely—many species lack some of the periods. For example, some fishes do not have a larval period but some do. Similarly, many animal species do not have a senescent period, but many—including fishes—do. The senescent period is the interval of life that follows after the reproductive activity of an adult has stopped. In females, including female humans, it starts with the onset of menopause.

The grandmother hypothesis, developed by Kristen Hawkes (2004), suggests that grandmothers play a very important role in both human development and evolution. *Homo sapiens* is a species with an extremely large portion of parental investment in a low number of offspring. This includes a long period of pregnancy as well as a very long period of parental care. Both pregnancy and childbirth, however, can have detrimental consequences for the health and longevity of women. Therefore, the capability of older mothers, those who have lost their fertility, to spend more of their time helping, protecting, and teaching their grandchildren appears extremely important for optimizing the survival chances of subsequent generations.

Epigenetic phenomena may also have serious effects on human health. For instance, several human disorders result from genomic imprinting. This phenomenon, typical for mammals, may occur when the epigenetic patterns, transmitted by father and mother into specific genes in their germ cells, differ from each other.

Much worse, epigenetics also appears to be associated with one of the most common civilization disease—cancer. Not only are some compounds considered epigenetic carcinogens (associated with tumors but without mutagenesis), but they also may have detrimental effects on human fetuses and can be thus passed to the subsequent generation. Interestingly, the epigenetic roots of cancer may be associated with the above-mentioned developmental mechanism—heterochrony. In the early 1980s Roy Douglas Pearson coined the term *cellular heterochrony*. In terms of oncogeny (oncological development), the theory of cellular heterochrony represents an analogy between the malignant cell and metamorphic tissue changes seen in amphibians, and this is in turn associated with hormonal regulation (Pearson 2004). And, hormonal regulation serves to control the "temporal chains of stable interacting genes" (as referenced above, from Sinervo and Svensson 2004, p. 171), which brings the whole process back to epistasis and thus to the genes. In this way, the circle set up by the genes, environment, and epigenetic mechanisms is completed.

Resources

Books

Arthur, Wallace. 2004. *Biased Embryos and Evolution*. Cambridge, UK: Cambridge University Press.

Balážová-Ľavrinčíková, Mária, and Vladimír Kováč. 2007. "Epigenetic Context in the Life History of the Round Goby, *Neogobius melanostomus*." In *Biological Invaders in Inland Waters: Profiles, Distribution, and Threats*, ed. Francesca Gherardi. Dordrecht, Netherlands: Springer.

Balon, Eugene K. 2004. "Alternative Ontogenies and Evolution: A Farewell to Gradualism." In *Environment, Development, and Evolution: Toward a Synthesis*, ed. Brian K. Hall, Roy Douglas Pearson, and Gerd B. Muller. Cambridge, MA: MIT Press.

Eldredge, Niles, and Stephen Jay Gould. 1972. "Punctuated Equilibria: An Alternative to Phyletic Gradualism." In *Models in Paleobiology*, ed. Thomas J. M. Schopf. San Francisco: Freeman, Cooper.

Gould, Stephen Jay. 2002. *The Structure of Evolutionary Theory*. Cambridge, MA: Harvard University Press, Belknap Press.

Hall, Brian K. 1999. *Evolutionary Developmental Biology*. 2nd edition. Dordrecht, The Netherlands: Kluwer Academic.

Lewontin, Richard. 2000. *It Ain't Necessarily So: The Dream of the Human Genome and Other Illusions*. New York: New York Review of Books.

Jablonka, Eva, and Marion J. Lamb. 2005. *Evolution in Four Dimensions: Genetic, Epigenetic, Behavioral, and Symbolic Variation in the History of Life*. Cambridge, MA: MIT Press.

Pearson, Roy Douglas. 2004. "The Determined Embryo: Homeodynamics, Hormones, and Heredity." In *Environment, Development, and Evolution: Toward a Synthesis*, ed. Brian K. Hall, Roy Douglas Pearson, and Gerd B. Muller. Cambridge, MA: MIT Press.

Robert, Jason Scott. 2004. *Embryology, Epigenesis, and Evolution: Taking Development Seriously*. Cambridge, UK: Cambridge University Press.

Sinervo, Barry, and Erik I. Svensson. 2004. "The Origin of Novel Phenotypes: Correlational Selection, Epistasis, and Speciation." In *Environment, Development, and Evolution: Toward a Synthesis*, ed. Brian K. Hall, Roy Douglas Pearson, and Gerd B. Muller. Cambridge, MA: MIT Press.

West-Eberhard, Mary Jane. 2003. *Developmental Plasticity and Evolution*. Oxford: Oxford University Press.

Periodicals

Dassow, George von, and Ed Munro. 1999. "Modularity in Animal Development and Evolution: Elements of a Conceptual Framework for EvoDevo." *Journal of Experimental Zoology* B 285(4): 307–325.

Gould, Stephen Jay. 2000. "Abscheulich! (Atrocious!) The Precursor to the Theory of Natural Selection." *Natural History* 109(2): 42–49.

Haig, D. 2004. "The (Dual) Origin of Epigenetics." *Cold Spring Harbor Symposia on Quantitative Biology* 69: 67–70.

Hawkes, Kristen. 2004. "Human Longevity: The Grandmother Effect." *Nature* 428(6979): 128–129.

Kováč, Vladimír. 2002. "Synchrony and Heterochrony in Ontogeny (of Fish)." *Journal of Theoretical Biology* 217(4): 499–507.

Pagel, Mark; Chris Venditti; and Andrew Meade. 2006. "Large Punctuational Contribution of Speciation to Evolutionary Divergence at the Molecular Level." *Science* 314(5796): 119–121.

Reilly, Stephen M.; E. O. Wiley; and Daniel J. Meinhardt. 1997. "An Integrative Approach to Heterochrony: The Distinction between Interspecific and Intraspecific Phenomena." *Biological Journal of the Linnean Society* 60(1): 119–143.

Richardson, Michael K., and Gerhard Keuck. 2002. "Haeckel's ABC of Evolution and Development." *Biological Reviews* 77(4): 495–528.

Vladimír Kováč

Evolutionary ecology

Ecology is the study of the relationship between organisms and their environment. The environment can be abiotic (physical, such as rainfall, temperature, and wave action) as well as biotic (other organisms). Evolutionary ecology emphasizes those aspects of ecology affected most strongly by evolutionary processes. Biologists define *evolution* as a change in gene frequency within a population over time (a *population* is a set of organisms of the same species in some place). Gene frequencies are manifested as distributions of phenotypic traits—morphological, physiological, and behavioral—for the individuals within a population. Such distributions give the frequency of individuals with various values of the traits, for example, having various body sizes. This article begins with aspects of individual ecology then extends upward, asking how interactions between species lead to properties of an ecological *community* (defined as a set of species' populations in some place). A main focus is the major species interactions—competition, predation, and mutualism—and the article explores how such interactions affect community characteristics, including species diversity.

Natural selection

Charles Darwin's concept of *natural selection* has three components, according to Richard C. Lewontin (1970). First, populations contain individuals with different phenotypes, so that, for example, moths of certain species have both dark and light forms. Second, different phenotypic variants have different survival and reproductive output over time, so that their lifetime reproduction differs as well. The term *individual fitness* is defined as the number of offspring produced over an individual's life that themselves reach reproductive age. Third, the traits that affect fitness are inherited from parents to offspring, that is, are *heritable*. Because of the emphasis on individuals, the terms *natural selection* and *individual selection* have become synonymous.

Hundreds if not thousands of examples of natural selection in action have been observed. One such example is industrial melanism, in which a number of moth species have increased in frequency of dark (melanic) forms following the blackening of trees from pollution, and then reversed this trend after the adoption of pollution controls. The color pattern is a trait that apparently increases camouflage and thereby reduces predation. Two other examples especially interesting to animal biologists follow.

Beak size in Galápagos Island finches

The finches first studied by Darwin during the famous voyage of the *Beagle* in the 1830s have now been investigated by many scientists. On the island of Daphne Major, finches during dry years eat relatively larger seeds, which are more available then (Boag and Grant 1981). Larger seeds are harder to crack, so more massive beaks are favored. During such dry years, most finches die (in 1977, 85% died). Finches surviving have mostly big beaks, and beak size is heritable. During wet years, selection reverses and favors smaller beaks, better at handling smaller seeds, and the advantage oscillates between smaller and larger beaks with changing weather conditions over time.

Life history characteristics of fish

Because of the collapse of the North Atlantic cod fishery, a moratorium on cod fishing was enacted in the late twentieth century. In the past, however, the very intense harvest pressure gradually favored over time individuals maturing at earlier ages and smaller sizes (Olsen et al. 2004; see "Population Structure and Reproductive Strategies" section below). David N. Reznick, Helen Rodd, and Leonard Nunney (2004) did experiments with guppies, introducing those from predator-free environments into environments that had large predators and vice versa (here the predators were other fish, not fishermen). Guppies with predators evolved the same changes as the cod. When the researchers changed back the environment, the guppy traits shifted back. These results show that natural selection can occur quickly, sometimes during the short time of ecological experiments.

Predator avoidance

The industrial melanism example illustrates how certain traits of animals can allow them to avoid being eaten by predators. Such traits are *adaptations*, in the sense that they fit or suit the animal to its environment; in this case the cryptic color/pattern trait reduces predation (and therefore increases fitness) by increasing camouflage. Indeed, nearly all traits of animals are believed to be the result of natural selection, even though of course scientists could not observe the evolution of most of them as they could in melanic moths. Two kinds of antipredator adaptations—morphological and behavioral—can be distinguished.

Some animals, such as porcupines, have spines or hairs that make consumption by a predator difficult to impossible. Linda Freshwaters Arndt/Photo Researchers, Inc.

Six categories of morphological antipredator adaptations are as follows:

1. *Protective structures*. Some animals have spines or hairs that make consumption by a predator difficult to impossible. Examples are the spiny sea urchin *Diadema*, mammals such as porcupines and echidnas, and certain bristly caterpillars.

2. *Protective resemblance*. Some animals, such as the melanic moths discussed earlier, have colors and patterns that match their background, providing camouflage. In some cases, such as *Anolis* lizards, the female is more likely to be camouflaged than males, probably because males engage in fierce territorial defense for which conspicuousness is an advantage.

3. *Aposematic coloration and pattern*. This condition is the opposite of the previous condition; here the animal is often gaudily colored and patterned. Such organisms are typically poisonous or distasteful, and their appearance is as if to say "don't eat me." A famous example is the monarch butterfly, whose larvae obtain a harmful chemical from their milkweed food plant. A second example are species of the genus *Dendrobates*,

the very colorful poison dart frogs of the Amazon. Scientists have wondered how natural selection can produce aposematic traits (Guilford 1990): If the prey dies while educating the predator, how does that help the prey's individual fitness? If the predator dies, how does that teach other predators not to eat the aposematic prey? It is interesting that aposematic caterpillars are often aggregated (see "Levels of Selection" section below).

4. *Frightening coloration and pattern*. Some moths have a startle display, whereby for example, they flash eyespots or other structures at an approaching predator. These eyespots are on the hind wings and are normally concealed by the forewings, but when the latter are quickly extended, the eyespots suddenly appear, frightening the would-be predator.

5. *Batesian mimicry*. This involves several species, in which certain species mimic, that is, come to resemble, model species (Wickler 1968). Mimics are harmless, but models are harmful and often aposematic. For example, harmless flies can mimic stinging bees and wasps. A more complex example involves North American butterfly species: A single model

species (the pipevine swallowtail) is mimicked by a number of species, including some in which only females, or only some females, resemble the model. One mimic species looks like the model only where it overlaps the model geographically and is completely different elsewhere.

6. *Mullerian mimicry.* In this kind of mimicry, all species are harmful and have a common, aposematic pattern. Sometimes this convergence in pattern occurs between closely related species, such as certain butterflies occurring at particular vegetation heights in tropical rain forests. In other cases, convergence can occur between rather unrelated organisms, such as beetles and moths.

Five categories of behavioral antipredator adaptations are as follows:

1. *Passive defense.* In this case the animal tries to remain as still as possible, hoping the predator will not find it (cryptic behavior). This obviously is the best strategy, at least in the initial portions of an encounter, for camouflaged animals. It is also the strategy pursued when an animal cannot escape rapidly: In the lizard *anolis lineatopus*, the warmer the body temperature,

the closer a simulated predator was able to approach before the lizard fled, because the cooler the lizard, the less quickly it can run (Rand 1964).

2. *Active defense.* This is most likely when the prey are large and/or in a group, such as the eland (an African antelope) defending against hyenas.

3. *Active escape.* This is perhaps the most common form of defense. An interesting variant is that certain moths will drop out of the sky suddenly upon the approach of an insectivorous bat; apparently they can detect the bat's sonar. A twist on this involves tiger moths emitting an ultrasonic reply, to which bats respond negatively only for those moth species containing harmful chemicals—resulting in aposematic sounds (Hristov and Conner 2005).

4. *Patterns of grouping/spacing.* Animals can be either spaced out or clumped, depending on whether they are camouflaged or aposematic, respectively. The assumption is that predators learn to find prey by reinforcement, so the secret of the camouflage is more likely to be forgotten when the next prey is encountered in the distant future. Just the opposite spacing pattern is beneficial to aposematic animals;

The monarch butterfly, whose larvae obtain a harmful chemical from their milkweed food plant, is a famous example of Aposematic coloration and pattern. Scott Camazine/Photo Researchers, Inc.

The tiger moth (*Arctia caja*) emits an ultrasonic noise to which bats respond negatively. Thus the moth avoids being eaten by the bat. © phodopus. Image from BigStockPhoto.com.

here the harmfulness of the pattern should not be forgotten between encounters, and that is more likely if prey are clumped.

5. *Alarm calls*. This is a special category of defense in which certain members of a group vocalize an alarm in the presence of a predator. But as for an aposematic trait, how can this be selected for, if the caller dies (i.e., is at a selective disadvantage for displaying the trait)? To explain fully, one needs to examine selection at levels higher than the individual level.

Levels of selection

The alarm call of certain animals is an example of *true altruism*, defined as a morphological or behavioral trait that reduces the individual fitness of the bearer/performer while increasing the individual fitness of the recipient(s). *Kin selection* is one mechanism that can produce true altruism. Kin selection works for individuals related by descent. It operates to cause an increase in the frequency of certain genes, especially those that help siblings or other relatives. This will increase the siblings' fitness. If an individual helps a sibling enough, then the likelihood that those genes will increase in the next generation increases even if it costs the individual something in terms of his or her own fitness. Fitness that includes the benefits to relatives is sometimes called *inclusive fitness*. William D. Hamilton (1964) proposed the following inequality giving when a kin-selected trait, such as performing an alarm call, is favored: (B)(COR)>C. Benefit to relative (B) and cost to performer (C) are measured in fitness units and are the results of the behavior being performed. *Coefficient of relatedness* (COR) is a measure of the fraction of genes on average shared with a relative. The more closely related, the more genes shared on average. For example, a human shares more genes with a sibling (one-half) than with a first cousin (one-eighth). If the benefit to a relative is sufficiently high relative to the cost to the performer, kin selection will operate to increase the frequency of the altruistic trait. Interestingly, unlike most organisms such as humans, which are diploid, members of the insect order Hymenoptera (ants, bees, and wasps) are haplodiploid, that is, males are haploid and develop from unfertilized eggs, and females are diploid and develop from fertilized eggs. This results in sisters having three-quarters of their genes in common. Workers (sisters) are more likely to help one another because the COR is so high. In particular, raising younger sisters rather than one's own offspring (for which the COR is one-half) is better. In fact, the female workers are sterile (have zero individual fitness), and this can be argued to have evolved by kin selection, although other explanations exist (Queller and Strassmann 1998).

Some excellent tests of kin selection have been performed. In his 1977 study of Belding's ground squirrels, Paul W. Sherman found that females with young offspring call more often than those without—but this can be explained as parental care, consistent with individual selection because offspring that are raised to reproductive maturity are included in the calculation of individual fitness. But when one focuses only on females without young offspring, those with female relatives (sister, mother) call more often than those without. This trait *is* kin selected. A second example is the white-fronted bee-eater, studied by Stephen T. Emlen, Peter H. Wrege, and Natalie J. Demong (1995) in Kenya. These birds live in cliffside colonies of up to twenty-five families. Sometimes young birds are fed and cared for by individuals other than their parents; such helpers at the nest increase the survival of the young. The more closely related a bird is to the parent, the more likely it is to help, so for example sons commonly help fathers raise the latter's offspring rather than go off and breed themselves. A third example is courtship in some birds, in which groups of males display together, and this attracts females more than a single male would. In black grouse, peacocks, and turkeys, these courting males are related. But in the tropical long-tailed manakin, a bird studied by David B. McDonald and Wayne K. Potts (1994), two males display and one gets 98 percent of the matings, yet the two are unrelated. This final example cannot be explained by kin selection, and it is thought that the subordinate male might eventually inherit the site if the dominant male dies.

Although not as well tested, it is hypothesized that kin selection might also be involved in the evolution of aposematic traits (see "Predator Avoidance" section above). A broad survey found that two-thirds of gregarious moth caterpillars are aposematic (Järvi, Sillén-Tullberg, and Wiklund 1981). If the caterpillars are related, which is especially likely when a mother lays her eggs in a single clump, they benefit by kin selection if one dies from a predator and "teaches" the predator to avoid prey, including the relatives of the dead individual, that have the particular aposematic appearance.

The second mechanism that can result in true altruism is *group selection*. Here, individuals are not necessarily related by descent. The key structure is Richard Levins's *metapopulation* (1970). This is a group of subpopulations close enough in space to interchange individuals. Each subpopulation inhabits its own site, such as an island. Suppose two genotypes exist, "selfish" and "altruistic," according to individual fitness. By definition, selfish individuals outreproduce the altruists *within a site*. They will eventually replace them there. But too many selfish individuals mean extinction of the subpopulation at the site—because they overeat food, or predators are attracted (because, for example, of no alarm calls) and wipe them out. This causes selfish individuals, relative to altruistic, to be lost from the entire metapopulation. Empty sites are sometimes recolonized by altruists alone—such sites thrive until they are "contaminated" by selfish individuals—then they eventually become extinct. These processes can balance so that altruists are preserved at some constant frequency over the entire metapopulation (Slatkin and Wade 1978).

Although possible in theory, a group-selection mechanism is rarely needed to explain particular traits, because individual or kin selection is adequate. But has group selection ever been observed in nature? Lewontin (1970) argued that a host–pathogen system shows group selection in action. A Мухома virus was introduced to Australia to kill off introduced rabbits that were considered pests—it killed 99 percent. Then rabbits became more abundant. While the evolution of resistance in rabbits was part of the story, rabbits with no history of exposure to the virus were also tested and were not as affected as when the virus was first introduced. The virus had become less virulent. In this conceptualization, rabbits are "sites," infected rabbits are "occupied sites," and uninfected rabbits are "empty sites" that are "colonized" from infected rabbits.

Physiological ecology

Physiological ecology is the study of the relation of an individual organism to its physical environment. The two major features of the physical environment are temperature and moisture. Only temperature will be considered here.

First, two pairs of terms need to be defined. *Regulators* (or *homeotherms*) maintain a constant body temperature in the face of varying environmental temperatures, whereas *conformers* (or *poikilotherms*) vary body temperature in proportion to the environmental temperature. *Endotherms* have an internal metabolic heat source, allowing them to maintain their warm body temperatures by metabolizing energy stores, whereas *ectotherms* must obtain heat from the external

environment. Birds and mammals are endotherms, whereas fish, reptiles, amphibians, and invertebrates are ecotherms. Endotherms are regulators much more than they are conformers. But some kinds of ectotherms can be both conformers and regulators. How would an ectotherm be a regulator? The main method is by behavioral thermoregulation, which operates in three major ways. First, the animal can shuttle to and from sunny (warm) and shady (cool) places, depending on whether it wants to heat or cool. Thus lizards in early morning are in the sun, and they later move into the shade when it becomes hot. Second, the animal can change its color; objects heat up more quickly when they are dark. Third, the animal can vary the percent of its surface area exposed to the sun by changing its orientation and raising or lowering crests.

The *metabolic rate* of an animal is a measure of how much energy per time is needed to run the animal's bodily functions. Ectotherms and endotherms differ fundamentally in the relation of metabolic rate to environmental temperature (see Figure 1). Ectotherms show a geometrically increasing metabolic rate with environmental temperature; the hotter it is, the more energy is used (Gordon et al. 1968). Endotherms show a U-shaped relation: Over a certain range of temperatures (the thermal-neutral zone), metabolic rate is constant; in this zone a constant body temperature is maintained by minor adjustments such as shifting the coat of hairs in mammals or ruffling the feathers in birds. Above *and* below that zone, metabolic rate increases with more extreme temperatures. Thus endotherms consume more energy when the ambient temperature is more extreme (either very hot *or* very cold).

The metabolic rate increases with the body mass of an animal, but to its three-quarters power: metabolic rate = $a(body\ mass)^{0.75}$, which implies a straight line on a log-log plot: log metabolic rate = $\log(a) + 0.75(\log\ body\ mass)$ (where a is a constant when the equation is fitted to the data; the bigger the

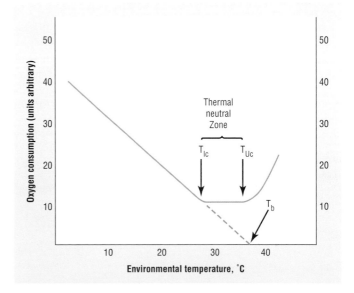

Figure 1. Relation of oxygen consumption (a measure of metabolic rate) to environmental temperature in a hypothetical mammal. T_{lc} = lower critical temperature; T_{uc} = upper critical temperature; T_b = core body temperature. Source: Gordon et al. (1969). Reproduced by permission of Gale, a part of Cengage Learning.

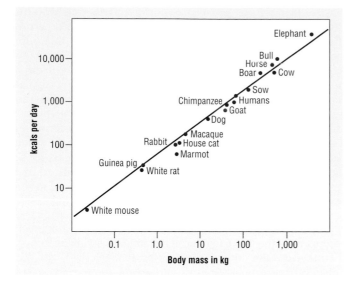

Figure 2. The mouse-to-elephant curve, showing a linear relation of the logarithm of energy metabolism to the logarithm of body mass. (Kleiber, 1932). Reproduced by permission of Gale, a part of Cengage Learning.

a, the higher is the curve or line relative to the *x* axis). These equations give the total metabolic cost of running the animal per time. Because of the fractional power, however, the per-mass cost is negatively related to body mass (divide both sides of the first equation by (*body mass*)$^{1.0}$). This means it costs more to feed a small animal (mouse) per unit of its own mass than to feed a large animal (elephant) per unit of its own mass, and indeed small mammals and birds consume a very high fraction of their own weight in food per day.

Endothermy would seem to be a great evolutionary advance, as it allows a large degree of independence from environmental temperature. In particular, endotherms can inhabit cold locations where ectotherms would be unable to operate. Why then did endotherms not replace ectotherms everywhere? The main reason is because endotherms are more expensive to run: For the same body mass, endotherms require about an order of magnitude more energy than an ectotherm. So in places where it is warm enough—for example, most deserts, equatorial latitudes—ectotherms are quite common.

Finally, the standard metabolic-rate-to-body-mass curves are for resting (or basal) metabolism. The curves for active metabolism often have about the same slope but are five to ten times higher: For example, territorial defense in humming-birds is about five times resting cost. But some animals can use up much more energy in their activities: Sphinx moths, for example, when flying very fast, then metabolize at 172 times resting rate.

Feeding ecology

The study of *feeding ecology* can be separated into two parts, diet—what to eat—and habitat—where to eat (Schoener 1971b; Stephens and Krebs 1986). These are discussed in turn.

What to eat

Imagine an animal such as a jay foraging over the forest floor looking for food. When it encounters an item, should it eat the item or go on and look for other items? The answer might seem to be that the jay should stop and eat the item so long as its energy gain per consumption time (time once found) is positive. But this is neither true in theory nor in reality. Animals are often finicky, skipping certain items whose energy yields per consumption time are too low. The reason is that during the time it would take to pursue, capture, handle, and eat the item, the animal might be able to find a substantially better item, one that averages a higher energy yield per unit of time even when search time and energy are included.

This kind of thinking is embodied in models of optimal diet, which specify from a set of kinds of food (such as different food sizes) which food kinds should be selected and which skipped. Two predictions emerge from the models. First, the greater the density (number per area) of all food kinds combined, the more selective the feeding animal should be (fewer kinds of food in diet). Second, whether a food kind should be included or not depends only on the abundance of better kinds, not on its own abundance. Earl E. Werner and Donald J. Hall (1974) tested the first prediction using bluegill sunfish, which eat aquatic invertebrates. The researchers showed that the sunfish broadened their diet from one to two kinds of invertebrates, and from two to three kinds of invertebrates, as total density of all kinds of invertebrates was lowered. (The model also accurately predicted the threshold densities at which these broadenings actually took place.) John R. Krebs and colleagues (1977) tested the first and second predictions using great tits. These birds were allowed to select food kinds from a conveyor belt. When large and small foods were low in total density, the birds ate both. As total density increased, the birds at some threshold ate only the large items, confirming the first prediction. Then the investigators radically reduced the density of the better (larger) foods. Yet no matter how rare, the larger foods were still eaten whenever they passed by, confirming the second prediction.

Where to eat

The question of where an animal should feed is studied with models of optimal habitat patch use. Imagine an animal going through the environment. When it encounters a habitat patch (perhaps a tree), should it stop and feed in that patch, and if so, for how long before leaving? Eric L. Charnov (1976) developed a model that made two predictions. First, for a particular environment (say, a group of trees), the better the patch type the longer it should be fed in before leaving. In fact, the animal should leave a particular patch when it has reduced the food therein (by its own consumption) to the point that its rate of energy gain per time spent feeding is equal to the average rate it would get across the environment as a whole. Second, for a particular patch quality, the better the overall environment for feeding, the sooner that type should be left (an environment is good because it has lots of high-quality patches, or because patches are close together so little time and energy is expended getting from one patch to another). Both predictions were confirmed in experiments on black-capped chickadees (Krebs, Ryan, and Charnov 1974).

Spacing ecology and territoriality

Organisms space themselves over the landscape in three ways. First, they can be randomly distributed, essentially exhibiting no spacing pattern. Second, they can be clumped, perhaps because their habitat is restricted, or because of benefits of grouping such as coordinated predator defense (see "Predator Avoidance" section above). Third, they can be spaced out (technically *overdispersed*), in which they are widely and regularly spaced. While the last pattern in particular might seem less appropriate for animals than for plants (especially in orchards but also in deserts and other natural habitats), many animals defend areas of space that result in their being overdispersed. This defense of an area is called *territoriality*, and it is found mostly in vertebrates (especially certain birds, lizards, and fish) but can also occur in invertebrates (e.g., damselflies, members of the insect order Odonata; Córdoba-Aguilar 2008).

Territoriality can be viewed as an adaptation and as such has a variety of functions for the individual animal, including securing a regular food supply, facilitating mating and parental care, and even preventing disease by reducing contagion. There is much evidence for the first function, food supply. In birds, larger species, which need more food energy, have larger territories than smaller species, and species feeding on less dense food (less food per unit area) have larger territories than species feeding on more dense food. In certain species, territory size varies with food density. For example, the tundra territories of snowy owls shrink during peaks in their lemming food, and ovenbirds defend smaller territories, the denser their forest-floor food. Carol A. Simon (1975) did an intriguing experiment with a species of fence lizard, in which she artificially increased then decreased food density by varying the number of mealworms provided; the lizards correspondingly shrank, then expanded, their territory sizes. John P. Ebersole (1980) obtained a more complicated result with damselfish: When he provided artificial food (in this case, bits of coral covered with the fishes' algal food), males shrank their territories whereas females expanded them. The difference stems from the fixed energy requirement of males (so the greater the density, the less food needed), versus the flexible energy requirement of females, who are able to make more eggs the more food consumed. A final, more complicated case involves the Jamaican lizard *Anolis lineatopus*, studied by A. Stanley Rand (1967). In this species adult males defend large territories against other adult males of the same species only. Females, by contrast, defend against any similarly sized lizard (lizards of the same size are more likely to be food competitors) regardless of species. Male territory sizes seem more adjusted to acquiring mates (their territories can encompass those of several females and are much too large for their food requirements), whereas female territory sizes are more adjusted to acquiring food.

Mating ecology

Mating systems can be classified in several ways. *Monogamy* has a single male and female as the mating unit, whereas *polygamy* involves more individuals. Within polygamy, *polygyny* has one male and several females as the mating unit, whereas *polyandry* has one female and several males as the mating unit.

Of the two types of polygamy, polygyny is much commoner than polyandry. The reason probably has to do with the basic biological nature of males versus females. By definition, the female is that sex with the larger gamete. This usually implies that the female has the developing zygote within her body, and that in turn implies that the individual female is not able to have potentially as many offspring as a male. Hence, males in some species mate as often as possible and leave the females to foster the (perhaps numerous) offspring. Females, having fewer offspring, must pay more attention to each one, and so they are more concerned about the quality of their offspring's male parent (in terms of, for example, life expectancy, see "Population Structure and Reproductive Strategies" section below). This argument can break down when parental care is involved, because then the male helps raise the offspring and so cannot spread himself too thinly, devoting too little time and energy to each offspring and thereby reducing his fitness.

The distribution of mating systems in terrestrial (land-dwelling) vertebrates tends to support the preceding arguments (Orians 1969). Among passerine (perching) birds, which show parental care by feeding their nestlings, 95 percent are monogamous. Exceptions occur in highly productive and patchy environments; for example, a red-winged blackbird male can support a number of females plus her offspring in his territory if his territory is good enough, and it can be shown that females do better in such large mating units than in smaller groups that are in less food-rich territories (Pianka 2000, Figure 10.6). In contrast to passerine birds, lizards have essentially no parental care, and they tend to be highly polygynous, an example being the Jamaican lizard *Anolis lineatopus* (see "Spacing Ecology and Territoriality" section above). Mammals are in between: The females have mammary glands to feed the young, and this implies (all other things being equal) that females would invest more in parental care. This predisposes mammals toward polygyny, but not always; in the yellow-bellied marmot (a relative of the groundhog), female fitness decreases the more females are in the male's group (this is in contrast to red-winged blackbirds), but male fitness first increases then decreases as the number of females in his group increases (Downhower and Armitage 1971). So males prefer mild polygyny, but females prefer monogamy—a conflict in terms of their fitness benefits.

What about polyandry? This occurs rarely among terrestrial vertebrates, but several examples are intriguing. In the chickenlike tropical bird called a tinamou, the male incubates the eggs, while the female goes off to pair with another male (Handford and Mares 1985). In this case, the polyandry is sequential, not simultaneous as in the polygyny examples. A second example of polyandry is in the Tasmanian hen (a kind of rail): This bird does have simultaneous polyandry, but the several males are brothers, so the polyandry likely results from kin selection rather than individual selection (Emlen 1978; see "Levels of Selection" section above). But in a third example, the Galápagos hawk, polyandry is simultaneous—the several males share feeding of the young—yet the males are no more related than random (Faaborg et al. 1995). Polyandry is thus an area showing that particular ecological phenomena can have multiple explanations.

The various mating systems sometimes favor alternative strategies within one sex or the other. For males, alternate strategies are especially likely in polygynous systems, in which dominant males can appropriate most matings. Rather than play the dominant male's game, such as noisily and aggressively displaying to females, the alternative in the ruff (a bird similar to a snipe) is to be a satellite male, sneaking copulations when the displaying male is busy (Davies 1978). Similarly, in the Gila topminnow that employs external fertilization, nonterritorial males rush in and release their sperm when the territorial male is releasing his (Constanz 1975). In both examples, the resemblance of the satellite males to the females facilitates the alternative strategy. For females, alternative strategies are less blatant, but surprising data, thanks to recent genetic advances, have shown how imperfect monogamy is among passerine birds: On average 15 percent of the nestlings raised by a particular pair have a father other than the male feeding the offspring (Petrie and Kempenaers 1998). This is thought to result from female choice: In the blue tit, Bart Kempenaers, Geert R. Verheyen, and André A. Dhondt (1997) found that females having many of their offspring by outside males are paired with a male that has on average offspring that are smaller, shorter lived, and poorer at raising young.

Perhaps the ultimate mating strategy is changing sex. In hogfishes (Hoffman 1983), a male will have a harem of females, and when he dies or is removed, the dominant female of the harem will transform into a male in a week or so. Thus in this highly polygynous system, only when an individual is going to be able to control a harem is it adaptive for that individual to be male. In contrast, *Crepidula* mollusks change from male to female as they age (Proestou, Goldsmith, and Twombly 2008): These limpets occur in stacks, with the oldest and therefore largest at the bottom; it is advantageous in this nonpolygynous species for larger individuals to be female, as they produce the larger gametes and so require a larger body to do so.

Population structure and reproductive strategies

Recall that a population is a set of individuals of a given species in some place. Populations are not composed of identical individuals, particularly in higher animals. Instead, populations have variation between individuals called the *population structure*, and the two major ways individuals can vary is by sex and age.

Variation by sex

The *sex ratio* describes the frequency of the two sexes, and it is usually measured as the number of males divided by the number of females. The primary sex ratio is the ratio at birth, but because of differential mortality of the sexes (usually males survive less well) it can change with age.

Ronald A. Fisher (1930) formulated the now famous argument for why the sex ratio should be one (equal numbers of males and females), according to natural selection. In diploid sexually reproducing species, males and females contribute equally to the next generation: Each individual has a mother and a father. If the sex ratio is unbalanced, individuals of the underrepresented sex leave more descendents on average and are thereby more valuable. Selection, therefore, favors shifting the allocation toward the rarer sex. This will continue until the sex ratio reaches one. Fisher's argument may imply a primary sex ratio of one, but if mortality occurs during parental care (before the offspring can be counted toward fitness; see "Natural Selection" section above), higher or lower values may be favored. David O. Conover and David A. Van Voorhees (1990) tested Fisher's argument with the Atlantic silverside. This fish species has a temperature-dependent sex ratio (though no one knows why). Early in the year, when the temperature is low, the fish mostly produce females; later in the year during high temperatures, this reverses. The average sex ratio over the entire year is one, agreeing with Fisher. The ability to modify the sex ratio is genetically based, so the investigators were interested in seeing if they could induce selection for a new sex ratio. Five populations from different localities were kept at a constant temperature, low or high all the time, so their average sex ratio over the year was no longer one. In five to six years, each population went back to a sex ratio of one, some by increasing male production, others by increasing female production.

Exceptions to Fisher's argument are known in both directions and can be explained by an extended argument using natural selection. Robert L. Trivers (1976) studied a polygynous *Anolis* lizard in Jamaica and found that males had an S-shaped relation of reproductive success (in this case mating frequency) to their size and condition, whereby small males had almost no reproductive success and large males had much more (see Figure 3, top). In females, this same relation was much more gradual. The differing relations predict the following: Producing males is better if one is producing larger, better-conditioned males, and vice versa for producing females (Trivers and Willard 1973). In fact, while studies with other animals have not all agreed with these predictions, Polley Ann McClure's experiments with wood rats (1981) were quite consistent: When food was reduced, female parents produced more females by rejecting male offspring. In contrast to a situation expected to be favored by polygyny, the reverse sex-ratio bias was argued to occur for parasitic wasps by Charnov (1982). A female wasp lays a single egg on a weevil (beetle) larva. The beetle larva inhabits a wheat grain, which it hollows out while feeding. Observations of the parasitic wasp that eventually emerges (which has eaten the beetle larva alive) show that if the beetle larva was small, a female wasp tends to hatch out and vice versa. In this case the curves for reproductive success are reversed from the polygynous situation (see Figure 3, bottom); similar to limpets (see "Mating Ecology" section above), larger females produce more eggs.

How could an animal adjust its sex ratio? In haplodiploid animals such as wasps, this can be relatively easy; recall (from the "Levels of Selection" section) that whether an egg is fertilized or not determines whether the individual is female or male. In diploid organisms a mechanism is more difficult to imagine, but a fascinating study by Ben C. Sheldon and colleagues (1999) on the blue tit shows that the female can still

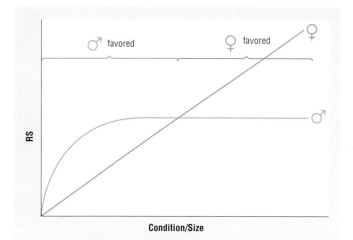

Figure 3, top. Reproductive success as a function of size/condition of parent and offspring (the two are assumed closely correlated) for a species with male-male competition for mates, in which larger body size is more advantageous. Reproduced by permission of Gale, a part of Cengage Learning.

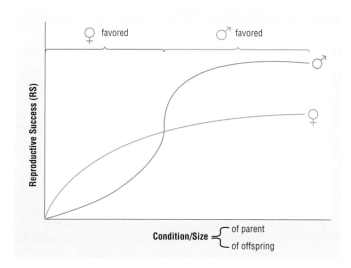

Figure 3. Bottom. Reproductive Success with No Size-Affected Male-Male Competition. Reproductive success as a function of size/condition of parent and offspring (the two are assumed closely correlated) for a species with no size-affected male-male competition for mates and in which larger females have higher have higher reproductive success because they produce more offspring (e.g., have more eggs in a clutch). Reproduced by permission of Gale, a part of Cengage Learning.

control the sex ratio. Males that give off higher ultraviolet (UV) radiation survive better, and the stronger the UV given off by the male mate, the more male offspring are produced. Investigators used sunblock to diminish a male's UV, and more females were produced.

Variation by age

Variation by age within a population is described by the *age distribution*, giving the fractions of individuals of different ages in a single *cohort*, that is, all individuals born during a particular time interval (e.g., during the year 1990).

A *survivorship curve* gives the fraction of individuals surviving to age x (this is the same as the probability of surviving to age x). Three types of survivorship curves (log survival versus age) are distinguished (see Figure 4). Type I shows most mortality in later rather than earlier years; humans, some other mammals, and a few lizards belong to this type. Type II shows an equal probability of dying at any age; most lizards and birds and some mammals belong to this type. Type III shows most mortality among younger individuals; fish and most invertebrates belong to this type (Pianka 2000).

The other major way to characterize animals of different ages within a population is by their *fecundity*, the average number of offspring produced by an individual of a certain age while at that age. The study of *reproductive strategies* is concerned with how natural selection might favor different patterns of fecundity and their relation to survivorship. There are two major questions. First, how should fecundity be scheduled through the lifetime, and particularly should there be one or a series of reproductive bouts? Second, if there is a series, how many offspring should be produced per reproductive bout?

The two major scheduling patterns are *iteroparous*, meaning more than one reproductive bout per lifetime, and *semelparous*, meaning one reproductive bout per lifetime. Most vertebrates, including humans, are iteroparous. Many insects are semelparous, unsurprising in view of their short, often seasonally restricted, lifetimes. Yet, the much longer-lived Pacific salmon are also semelparous: After maturing in the ocean they migrate upstream to engage in their single reproductive bout, called spawning, then die. The interesting question is why some long-lived species such as salmon are semelparous. While not completely understood, it is thought that if the effort of reproduction is necessarily huge and/or risky, it is adaptive to perform it only once; in the case of salmon, for example, the chance of two successful spawning journeys, which are so perilous, is slim.

When there is more than one reproductive bout (iteroparous), what is the number of offspring per bout favored by selection? Because reproduction is energetically costly, the key idea here is a trade-off between survival and fecundity. For example, if fecundity is high one year, less energy is available for the following year, so little perhaps that the animal dies before reproducing then. Lizards show an inverse relation between fecundity and survival—the more eggs produced, the lower the probability of living afterward (Pianka 2000). Such considerations lead to the issue of the optimal size of the brood or clutch, where optimal here means conferring the maximum fitness and thereby being favored by selection. It has been shown in some birds that the larger the clutch (number of eggs) the less each young weighs, a factor that can affect survival. From such considerations, David Lack (1954) inferred that the clutch size giving the highest production of fledged offspring (those able to leave the nest successfully) is optimal and so should be commonest. Detailed studies of great tits (Boyce and Perrins 1987), however, showed that during a single reproductive bout, a clutch larger than the commonest gave more fledged offspring. Adult mortality is apparently a key factor giving this outcome: Fitness is lifetime number of offspring, and the lower survivorship of birds with larger clutches balances out that advantage during a single reproductive bout.

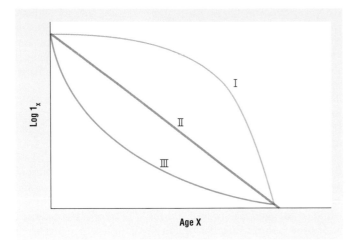

Figure 4. Types of Survivorship Curves. 1x is the fraction of individuals surviving to Age x, or the probability of surviving to Age x. See text for explanation. Reproduced by permission of Gale, a part of Cengage Learning.

Clutch size in birds is known to vary with a variety of geographic factors, being greater in higher latitudes, at higher altitudes, and on mainlands as opposed to islands (Cody 1966). Three hypotheses for the latitudinal variation are prominent (Pianka 2000). First, Lack's day-length hypothesis posits that there is more time to feed when temperate birds are breeding (spring and summer have longer days) than when tropical birds are breeding (in the tropics, day length is about the same throughout the year). The latitudinal trend, however, also holds for nocturnal birds, which argues against this hypothesis. Second, the Skutch clutch hypothesis (originated by the tropical ornithologist Alexander Skutch) postulates that the tropics have more nest predators: Smaller clutches are favored there because the greater the number of young, the more times parents must visit their nest and possibly give away its location. This hypothesis has strong support. Gordon H. Orians (1969) found lower nestling survivorship among red-winged blackbirds in tropical Costa Rica than in Washington State. Moreover, hole-nesting birds, which are relatively protected from predators, show little latitudinal variation in clutch size. Third, the spring bloom hypothesis states that food is maximally more abundant in temperate areas than in the tropics, certainly during the time birds are breeding; again, data from the great tit show that clutch size is higher, the higher the density of food. An explanation for all three trends involves r-selection versus K-selection (MacArthur and Wilson 1967). These are kinds of individual selection favored at different stages of population expansion. When a population is near K, its carrying capacity (defined as the maximum number the environment can support), K-selection predominates, whereas when a population is just getting going, r-selection predominates (where r is the intrinsic rate of natural increase found in the exponential equation of population growth [Pianka 2000]). Competition is highest when the population is crowded, so K-selection favors individuals with traits good in competition, such as larger size and delayed reproduction (because larger and older animals are better at reproduction); r-selection favors the reverse. For higher latitudes (temperate and arctic), higher altitudes, and mainlands (where the weather is more severe), r-selection should operate more than K-selection, implying more offspring per reproductive bout.

Competition and ecological niche

Competition occurs when species populations affect one another negatively; population size and/or population growth decrease. Competition is one of the three major ecological interactions, the others being *predation*, in which one species (the predator) is affected positively and the other species (the prey) is affected negatively, and *mutualism*, in which each species population affects one another positively. Two general mechanisms of competition are recognized. *Exploitative competition* occurs when resources that both species use are limited; each species indirectly reduces the other by consuming resources that the other would have obtained. *Interference competition* occurs when there is direct interaction between individuals. Examples are encounters that lead to physical harm or even death, as well as poisoning by metabolic products or adaptive release of chemicals that harm the other species. Competition has both a between-species (interspecific) and within-species (intraspecific) manifestation. When one species' population is sufficiently superior because of a competition mechanism, it will eliminate the other in a process called competitive exclusion. When individuals of each species have a sufficiently greater (per-capita) effect on the other species than on their own species, one or the other species will win depending on their numbers at the start of the competitive interaction; the species having numerical superiority is favored, all other things being equal. Finally, when individuals of each species have a sufficiently greater (per-capita) effect on their own species than on the other species, both will coexist, each at a population size smaller than would be attained if the other species were absent. These various outcomes of competition have been demonstrated in the laboratory using protozoa, beetles, and flies, among other animals.

If species are different enough in their resource use, they will fall under the third outcome, coexistence. The study of how species differ in resource use, and thereby potentially avoid competitive exclusion, is called the study of *resource partitioning* (Schoener 1986). Species partition resources in three general ways. First, they may differ in food type, such as food size, food hardness, or food taxon. Second, they may differ in where the food is obtained—that is, have different habitat preferences. Differences occurring over narrow spatial scales are microhabitat differences, whereas those over broad spatial scales are macrohabitat differences. Third, they may differ in the time when food is obtained, and this may be daily activity time or seasonal occurrence.

The *ecological niche* is a concept that allows ecologists to describe resource partitioning. This concept has a number of interpretations (Schoener), and the one used here is Robert H. MacArthur and Richard Levins's multidimensional utilization distribution (1967). Imagine a single axis along which food resources can be characterized, such as by food size. Each species population's use can be represented as a frequency histogram, giving the fraction of that population's diet in each food-size class. A second axis could pertain to microhabitat, such as feeding height. Other dimensions could be added, and the

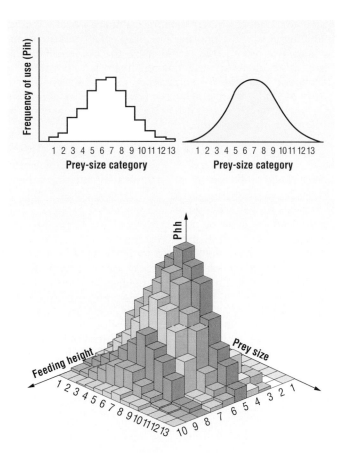

Figure 5. (a) The utilization for Species i as a frequency histogram. This utilization is one-dimensional, where the dimension is prey size. Numbers refer to prey-size categories, indexed by h. (b) The same utilization smoothed. (c) A utilization for two resource dimensions, prey size and feeding height. (Schoener, 1986). Reproduced by permission of Gale, a part of Cengage Learning.

totality of the joint utilization over all axes (the axes are called *niche dimensions*) is the ecological niche. Species in laboratory experiments often did not coexist because they were too similar in resource use, an example being experiments of Georgyi Frantsevitch Gause (1969 [1934]) on protozoa, and this led to the Gause Principle: Species whose niches are too similar cannot coexist. Ecologists now routinely document resource partitioning between co-occurring species. One example are the five species of terns studied on Christmas Island by N. Philip Ashmole (1968), who found that differences in beak size correlate with differences in food size. A second example, studied by MacArthur (1958) in the northeastern United States, involves five species of warblers (small birds), in which each species specializes on feeding in a different part of a tree. A more complicated case involves *Anolis* lizards studied by Thomas W. Schoener and Amy Schoener (1971) in Puerto Rico: Nine widespread species differ in macrohabitat (rain forest, intermediate forest, or dry forest), microhabitat (where they perch and feed in the vegetation), and to a lesser extent food size.

There are two major mechanisms whereby resource partitioning resulting from competition can develop. First,

species can compete to the point at which one or more is eliminated, leaving only those species sufficiently different in resource use. Second, *coevolution* (the sequential development over time of genetically based adaptations in a set of interacting species) can occur, in which species diverge from one another via evolutionary change; if the divergence is rapid and extreme enough, competitive exclusion is averted. Sometimes scientists can infer or observe this divergence in *character displacement*, defined as the situation in which a species is more different from a second species in sympatry (where they co-occur) than allopatry (where only one or the other occurs). One such example is among stickleback fishes, studied by Dolph Schluter (2000): Where two species co-occur in a lake, they diverge in a trait related to the resources they consume; in lakes having only one species, that species is intermediate in the trait. A second example, from an investigation by Tom Fenchel (1975), is two species of *Hydrobia* gastropods, which ingest diatoms and inorganic particles covered with microorganisms; the larger the gastropod, the larger the diatoms and particles consumed. These species did not overlap in geographic range until 1825 when a seawall broke open and they slowly came together. One of the species became smaller and the other larger, and this apparently happened independently a number of times. The difference in the niches of the two species where they co-occur is about that expected from *limiting-similarity theory*, which attempts to predict—using mathematical models—the degree of difference at which species can just coexist (in the sense that if their niches were any more similar, one would competitively exclude the other).

Two other kinds of data are consistent with the idea that resource partitioning developed via the two mechanisms just discussed. First, separation of species can be complementary with respect to several niche dimensions. For example, *Anolis* lizards of Bimini Island differ either in food size, in microhabitat, or in both, but in no case are they similar in both dimensions (Schoener 1986). A second example involves ducks of California's Central Valley studied by Paul J. DuBowy (1988): The species differ in food type, habitat, or both, but only in winter when food is scarce; during summer when food is abundant, the pattern breaks down and species often overlap in both niche dimensions. A second kind of data involves comparison with what is called a *null model*. In this case, the null model specifies the degree of difference between species that would be expected were the species to co-occur at random. For example, bird-eating hawks co-occur at localities around the world in communities of two to five species; their body-size differences (indicating food-size differences) are greater than expected were they to co-occur randomly (Schoener 1984).

Predation

The second major kind of ecological interaction, predation, covers a variety of mechanisms in which one species is affected positively and the other negatively. First is *predation proper*, in which the predator kills animal prey relatively quickly, such as a lion killing and eating a gazelle. Second is *herbivory*, in which an animal consumes plant food (the plants are essentially the prey). Third is *parasitism*, in which the victim or host is not killed quickly but loses health, weight, and/or offspring and may eventually die.

Laboratory experiments using protozoans or arthropods have produced cycles in which populations of the predator species and the prey species fluctuate out-of-phase. These cycles are well reproduced by mathematical models in which the only process included is the predator–prey interaction (Harrison 1995). These models, as well as the laboratory data, show that the cycles tend to persist (and thereby allow coexistence of the predator and prey) under two conditions. First, the prey must be relatively restricted by its own food supply or other factors that cause it to have a small carrying capacity K (see "Population Structure and Reproductive Strategies" section above). Second, the predator must not be too proficient at searching for and finding prey.

An open question is the degree to which such models and laboratory demonstrations pertain to predator–prey cycles in the field. A famous such cycle is that between lynx and hare, which show a cycle period of between ten and eleven years. This cannot be purely a lynx–hare cycle, however, because the hare cycle on an island without the lynx. Tree-ring studies showed that the plant food the hares consume cycle with the same period as hares do; thus the hare is the predator and the plants the prey, with the lynx brought along where it occurs. Overlaying all is an extrinsic factor, sunspot cycles, also having the same period and perhaps pacing the cycle, as would a metronome, even though not causing it per se (Sinclair et al. 1993). A second kind of cycle occurs in forest moths, which eat leaves. The period here is more variable, although averaging about the same as the lynx–hare cycle (Myers 1998). Again, sunspots may set the pace, but the root cause of the cycle is suspected mostly to involve parasitic insects such as wasps whose larvae consume the moth larvae (Turchin et al. 2003); cycling of plant nutritional quality, of plant defenses, or of disease may also be involved. A completely different period—three to four years—is found in small rodents such as lemmings and voles. Lemmings apparently set up a cycle between their food supply (mosses of the tundra) and their own populations, thereby acting as predators in the same sense as the hares are predators on their plant food (Turchin et al. 2000). Voles, by contrast, have their populations controlled by higher predators such as weasels; their cycles are closer to the classical concept of predator–prey interactions (Turchin and Hanski 1997).

Mutualism

Mutualism is the ecological interaction in which species populations affect one another positively. Two types are recognized. In *obligatory mutualism*, a given species requires those species or group of similar species that provide the benefit in order to exist—without the latter, the given species would decline to extinction. In *facultative mutualism*, the given species does not require the benefiting species in order to exist, although it is still helped by them in the sense of having a higher population size or population growth rate.

Perhaps the most common type of mutualism involves dispersal over space. Pollen is commonly spread from one plant to another by pollinating animals—bees, moths, hummingbirds, and even bats. The plant offers nectar, pollen, or both as a reward (the nectar itself has no benefit to the plant other than in attracting pollinators). Sometimes a pollination relationship is obligatory, as in the famous star orchid of Madagascar, which has a flower tube 30 centimeters (11.5 inches) long, and whose sole pollinator, a sphinx moth, has a tongue of equal length.

A second type of mutualism involves trophic (food) relations between the mutualist species. Mutualists can be integrated into the body of the partner and help with essential substances, in which case they are called *endosymbionts*. An example is termites and their endosymbionts, flagellated protozoa. The termites eat wood but cannot digest cellulose; the protozoa do the latter (in part with their own bacterial endosymbionts) and make glucose that is usable by the termites. A second example is an even more complete integration: Corals have protozoa called zooxanthellae living in their tissues. The zooxanthellae have chloroplasts and so can photosynthesize, producing glucose and oxygen. In both these tightly integrated examples, the mutualism is obligatory.

Sometimes both trophic and dispersive relationships are involved in mutualisms. The yucca plant is eaten by the yucca moth; the former requires the latter for pollination, whereas the latter has only one food plant, the yucca (Ricklefs and Miller 2000). Again, this mutualism is obligatory.

A final kind of mutualism involves defense (Heil and McKey 2003). Daniel H. Janzen (1966) studied ants and acacias in Central America. The acacias have thorny stems sheltering ant colonies; in addition, the acacias provide food for the ants using nectaries (a plant structure that produces nectar) and nutritious areas on the tips of their leaves. In return, the ants eat the herbivorous insects that would otherwise consume the acacias. The mutualism is obligatory for the ants but facultative for the acacias; thus some mutualisms are a combination of the two types.

A final sort of interaction, called *commensalism*, has one species benefiting but the other scarcely affected. For example, cattle egrets follow herds of cattle to eat the insects the latter stir up, but the cattle are apparently unaffected by the egrets.

Food webs

Food webs are a representation of who eats whom. If one species eats another, this is indicated by a *link* in the food web (generally an arrow pointing from the prey to the predator). Each such link represents a *direct effect* in the food web. In addition, *indirect effects* can occur whereby species affect one another via one or more intermediate species. When a food web is very complex, there are many indirect pathways, and this can sometimes make understanding how the food web works quite difficult.

Trophic cascades are among the most important effects in a food web. These refer to the downward propagation of the actions of predator species at the top of the food web. Predators affect not only their prey (directly, by definition) but also the prey of their prey, and such an indirect effect

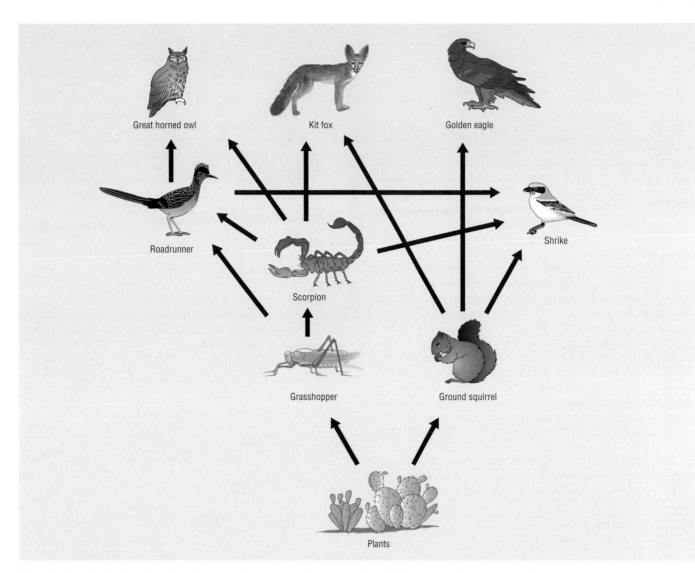

Figure 6. A food web for a North American desert. Reproduced by permission of Gale, a part of Cengage Learning.

might even extend to the base of the food web, the green plants (called *producers*). Numerous examples of trophic cascades have been described. A fascinating one is from Christmas Island (Terborgh 2009). Red crabs are a major herbivore on the vegetation, with a square kilometer (0.4 square miles) supporting 145,000 kilograms (320,000 pounds) of crabs. In 1989 multi-queened supercolonies of the yellow crazy ant, an introduced species, were noticed to be spreading across the island, reaching densities in the thousands per square meter. These ants wiped out the defenseless crabs, and in so doing caused a massive top-down effect on the vegetation. Areas whose crab density is yet unaffected are largely open, with few seedlings and little leaf litter. Areas with ants and many fewer crabs showed a spectacular increase in plant biomass, because of the saplings that were now able to survive, and the number of plant species increased threefold.

The exact way a top predator affects plants depends on the number of intermediate species between them. In the Christmas Island example, there is one intermediate species,

the crab, between the ant top predators and the plants. In another ecological system showing strong cascades, small islands of the Bahamas, lizards are at the top and generally have a positive effect on the plants by consuming herbivores. But one kind of herbivore, a midge (a kind of fly) whose larvae produce galls, is also consumed by spiders, and lizards consume spiders as well, so this kind of herbivore is helped by the lizards, making its leaf damage more common where lizards occur (Spiller and Schoener 1996). A four-level or three-link system is also found in certain rivers (Power 1990) and lakes: Piscivorous (fish-eating) fish are at the top, and two levels (smaller predators, herbivores) separate them from plants, so where piscivorous fish occur, plants are reduced.

Predator removal can affect the species diversity of food webs in various ways. When Robert T. Paine (1966) removed a giant predatory starfish from intertidal areas of the northwestern United States, the number of animal species dropped in half: The predator was preventing the other species from competitively exterminating one another, so when it was removed, the rest of the species competed vigorously, producing numerous extinctions. In contrast, in the Bahamas system, the top predatory lizards decrease the number of species of spiders by both consuming the latter and competing with them for certain food: Lizard predator removal resulted in an increase in the number of species (Schoener and Spiller 1996).

Species diversity

The previous section described how top predators affected numbers of species, one measure of species diversity. This section discusses species diversity more generally and shows how its study is paramount for conservation.

Measures of diversity

The simplest way to represent the species diversity of an area is to count the number of species, S. A more complicated way is to use the following measure from information theory:

$$H' = \sum_{i=1}^{S} p_i \log p_i$$

where p_i equals the fraction of individuals (of the total number of individuals of all species combined) belonging to species i. H' increases with both the number of species S and the evenness (degree of equality) of the abundances of the species. Consider two communities, each with two bird species. The first has fifty individuals of each species, and the second has ninety-nine of one species and one individual of the second. The second community is less diverse by H', even though it has the same S as the first. Moreover, for perfectly even abundances, the more species, the higher H' will be.

Latitudinal variation in diversity

One of the most striking differences in species diversity over the face of the earth is between tropical (low latitude) regions compared to temperate and polar (high latitude) regions. This huge latitudinal diversity gradient first became apparent during the age of the great nineteenth-century naturalists: Darwin, Alfred Russel Wallace, and Henry Walter Bates. Bates (responsible for the Batesian mimicry concept—see "Predator Avoidance" section above) spent eleven years on the Amazon and its tributaries living as a tramp, sending back specimens of everything for little pay. He collected 14,712 species of animals (Bates 1962, Introduction). Near Pará (Belém) he collected about 700 species of butterflies alone (Bates 1962, p. 62), more than the number known for the entire continent of Europe. Modern research has documented more precisely this huge tropical diversity. For example, the American entomologist Terry L. Erwin meticulously studied a single tree species in Panama (Wilson 1988). He found 1,100 beetle species in its canopy, of which 160 were specific to that tree alone. These data can be used to calculate how many insect species exist on tropical trees. Given that 40 percent of all insects are beetles, and that there are 50,000 tree species in the tropics, it follows that the number of tropical insect species is estimated at roughly 20 million (from the calculation (160 x 50,000)/0.4). Yet as of 2007 scientists had described only about 1.5 million species of animals, plants, and algae from the entire

earth, suggesting that much more work remained to be done (World Conservation Union 2007, Table 1). Six hypotheses have been advanced to explain why there are more species in the tropics. The first two are evolutionary, and the other four are ecological.

First, more generations per year are possible in the tropics because there is no harsh season. This leads to a greater rate of evolution and speciation (generation of new species) in the tropics.

Second, there is greater reproductive isolation in the tropics, and this leads to a greater speciation rate. Individuals of tropical species are more sedentary; birds, for example, do not migrate as much. Moreover, Daniel H. Janzen has suggested that organisms are more likely to cross mountain barriers in regions with seasonal climatic variation (Ghalambor et al. 2006); thus in higher latitudes winter may confine species to lowlands, but in summer the conditions they are used to in winter are found at higher altitudes, and so they can go up there and eventually get across the mountain barrier. The more effective the barrier (as in the tropics), the greater is the isolation, which contributes to a high speciation rate.

Third, tropical species have a greater niche specialization—that is, their utilization distributions (see "Competition and Ecological Niche" section above) are narrower (smaller standard deviation). A greater specialization would result from climatic stability, both within a single year and over many years. The narrower each niche, the more species can be packed along the available range of a resource dimension. An example is J. Mark Scriber's work (1973) on swallowtail butterflies. Swallowtails have many more species in the tropics, and tropical species tend to be more specialized, in the sense of eating only a single species of plant, than are swallowtail species of higher latitudes.

Fourth, a greater range of resource kinds occurs in the tropics. Bird species diversity, for example, shows an explosive increase toward the tropics. H' for bird species diversity increases with the diversity of the forest vegetation structure (e.g., more layers), and tropical forests have a more diverse vegetation structure (MacArthur and Wilson 1967). For the same vegetation-structure diversity, however, the tropics have more bird species. Hence, an explanation beyond greater foliage diversity is needed for high tropical bird diversity. One might be that tropical birds have more kinds of resources they can reliably eat during the year; fruit, for example, is available year-round in the tropics but is available only during a restricted time in higher latitudes. A second explanation involves insects. Compared to higher latitudes, there are more large insects in the tropics but about the same number of small insects. The great increase in insectivorous bird species in the tropics is entirely among large-billed species, the ones that eat large insects (Schoener 1971a).

Fifth, the tropics have a higher productivity (Cain, Bowman, and Hacker 2008); that is, a greater amount of energy flows from the plant base to the higher trophic levels in the tropics than for high-latitude systems (certainly over the entire year). Indeed, the most productive vegetation type in the world is tropical rain forest. Greater productivity might imply more energy for more species, in the sense that if there is a minimum population size below which a species is likely to become extinct, the higher the productivity the greater the number of minimum-sized populations that can occur in an ecological community. Recall (from the "Predation" section), however, that the more food a prey species has (which increases with productivity), the more likely is the predator–prey system to crash to extinction, leading to lower species diversity.

Sixth, the tropics have a greater amount of predation, implying that more species of competitors can coexist there. Recall (from the "Population Structure and Reproductive Strategies" section) that red-winged blackbirds have a higher rate of nest predation in the tropics. Also recall (from the "Food Webs" section) Paine's experiment with starfish showing that when this predator was removed, species diversity decreased. The same effect, however, does not occur in terrestrial arthropods, as shown by Bahamas experiments, and because most arthropod taxa show a huge increase in species diversity in the tropics, this argues against the predation hypothesis.

Diversity on islands

It has long been noted that islands tend to have fewer species than the same sized area on comparable mainland habitats. Specifically, small islands are especially low in diversity, and isolated islands (those far from sources of colonization) are especially low in diversity. In 1963 MacArthur and Edward O. Wilson devised a model of island species diversity that predicted these area and isolation trends (MacArthur and Wilson 1967). Their model proposed two major properties for the species of a given island. First, after sufficient time, an equilibrium is attained, whereby the number of species stays roughly constant. Second, the equilibrium is dynamic, in the sense that species turnover occurs, whereby those species becoming extinct are replaced by species immigrating to the island (see Figure 7). Both equilibrium and turnover have received major testing. Equilibrium for a set of species is more likely the better its members are at dispersal (e.g., more likely for birds than lizards) and the less disturbed the island is by outside factors (e.g., if devastating hurricanes are frequent enough, species are wiped out before reaching equilibrium, so the process has to constantly restart). In an elaborate test of the model, Daniel S. Simberloff and Wilson (1969) hired a pest-extermination company to erect tents over tiny mangrove islands off mainland Florida and kill all the arthropods therein. Subsequent monitoring agreed with the main predictions of the model, including species turnover and the regaining of equilibrium rather quickly in these good dispersers.

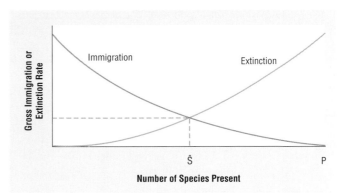

Figure 7. The model is for a particular island. Ŝ is the number of species at equilibrium (when gross immigration equals gross extinction), and P is the number of species in the source pool. Rate curves are monotonic but nonlinear. The intercept of the dashed line on the ordinate is the turnover rate at equilibrium. Reproduced by permission of Gale, a part of Cengage Learning.

Hotspots

Twenty-five areas of the earth with many unique species have been designated conservation *hotspots* (Mittermeier, Myers, and Mittermeier 1999), according to two criteria. First, an area must have a large percentage of endemic species—those found nowhere else; this criterion was originally evaluated using vascular plants (ferns and seed plants) but was later broadened to higher vertebrates—mammals, birds, reptiles, and amphibians. Second, the area must have 25 percent or *less* of its land area in intact vegetation, meaning vegetation relatively undisturbed by humans. Fifteen of the twenty-five hotspots have tropical rain forest, that vegetation type with the highest species diversity, and another eleven have tropical dry forest. Eleven hotspots include large islands, which tend to generate unique species by speciation. What is necessary to save the intact areas of all the hotspots, and how many species will thereby be preserved? Out of the combined area of all hotspots, 13 percent is intact vegetation. This represents only 1.44 percent of the total earth's surface. Yet that relatively small percentage contains 43.8 percent of all known vascular plant species and 35.5 percent of all known higher vertebrate species (estimates for known plus unknown species run about 70 percent in both cases). Unfortunately, only 0.6 percent of the intact vegetation was being protected in the early twenty-first century, so humankind had a long way to go to avoid major reductions in animal and plant diversity.

Resources

Books

Bates, Henry Walter. 1962. *The Naturalist on the River Amazons.* Foreword by Robert L. Usinger. Berkeley: University of California Press.

Cain, Michael L.; William D. Bowman; and Sally D. Hacker. 2008. *Ecology.* Sunderland, MA: Sinauer.

Charnov, Eric L. 1982. *The Theory of Sex Allocation.* Princeton, NJ: Princeton University Press.

Emlen, Stephen T. 1978. "The Evolution of Cooperative Breeding in Birds." In *Behavioural Ecology,* ed. J. R. Krebs and N. B. Davies. Oxford: Blackwell Scientific.

Córdoba-Aguilar, Alex, ed. 2008. *Dragonflies and Damselflies: Model Organisms for Ecological and Evolutionary Research.* Oxford: Oxford University Press.

Fisher, Ronald A. 1930. *The Genetical Theory of Natural Selection.* Oxford: Clarendon Press.

Gause, Georgyi Frantsevitch. (1969). *The Struggle for Existence* (1934). New York: Hafner.

Gordon, Malcolm S., George A. Bartholomew, Alan D. Grinnell, et al. 1968. *Animal Function: Principles and Adaptations.* London: Macmillan.

Guilford, Tim. 1990. "The Evolution of Aposematism." In *Insect Defenses: Adaptive Mechanisms and Strategies of Prey and Predators*, ed. David L. Evans and Justin O. Schmidt. Albany: State University of New York Press.

Lomolino, Mark V., Brett R. Riddle, and James H. Brown. 2006. *Biogeography.* 3rd edition. Sunderland, MA: Sinauer.

MacArthur, Robert H., and Edward O. Wilson. 1967. *The Theory of Island Biogeography.* Princeton, NJ: Princeton University Press.

Mittermeier, Russell A., Norman Myers, and Cristina Goettsch Mittermeier. 1999. *Hotspots: Earth's Biologically Richest and Most Endangered Terrestrial Ecoregions.* Mexico City: CEMEX.

Pianka, Eric R. 2000. *Evolutionary Ecology.* 6th edition. San Francisco: Benjamin Cummings.

Reznick, David N., Helen Rodd, and Leonard Nunney. 2004. "Empirical Evidence for Rapid Evolution." In *Evolutionary Conservation Biology*, ed. Régis Ferrière, Ulf Dieckmann, and Denis Couvet. Cambridge, UK: Cambridge University Press.

Ricklefs, Robert E., and Gary L. Miller. 2000. *Ecology.* 4th edition. New York: Freeman.

Schoener, Thomas W. 1986. "Resource Partitioning." In *Community Ecology: Pattern and Process*, ed. Jiro Kikkawa and Derek J. Anderson. Melbourne, Australia: Blackwell Scientific.

Schoener, Thomas W. 1989. "The Ecological Niche." In *Ecological Concepts*, ed. J. M. Cherrett. (British Ecological Society Symposium volume). Oxford: Blackwell Scientific.

Spiller, David A., and Thomas W. Schoener. 1996. "Food-Web Dynamics on Some Small Subtropical Islands: Effects of Top and Intermediate Predators." In *Food Webs: Integration of Pattern and Dynamics*, ed. Gary A. Polis and Kurt O. Winemiller. New York: Chapman and Hall.

Stephens, David W., and John R. Krebs. 1986. *Foraging Theory.* Princeton, NJ: Princeton University Press.

Terborgh, John. 2009. "The Trophic Cascade on Islands." In *The Theory of Island Biogeography at 40: Impacts and Prospects*, ed. Jonathan B. Losos and Robert E. Ricklefs. Princeton, NJ: Princeton University Press.

Wickler, Wolfgang. 1968. *Mimicry in Plants and Animals*, trans. R. D. Martin. New York: McGraw-Hill.

Wilson, Edward O., ed. 1988. *Biodiversity.* Washington, DC: National Academy Press.

Periodicals

Ashmole, N. Philip. 1968. "Body Size, Prey Size, and Ecological Segregation in Five Sympatric Tropical Terns (Aves: Laridae)." *Systematic Zoology* 17: 292–304.

Boag, Peter T., and Peter R. Grant. 1981. "Intense Natural Selection in a Population of Darwin's Finches (Geospizinae) in the Galápagos." *Science* 214(4516): 82–85.

Boyce, Mark S., and C. M. Perrins. 1987. "Optimizing Great Tit Clutch Size in a Fluctuating Environment." *Ecology* 68(1): 142–153.

Charnov, Eric L. 1976. "Optimal Foraging: The Marginal Value Theorem." *Theoretical Population Biology* 9(2): 129–136.

Cody, Martin L. 1966. "A General Theory of Clutch Size." *Evolution* 20(2): 174–184.

Conover, David O., and David A. Van Voorhees. 1990. "Evolution of a Balanced Sex Ratio by Frequency-Dependent Selection in a Fish." *Science* 250(4987): 1556–1558.

Constanz, George D. 1975. "Behavioral Ecology of Mating in the Male Gila Topminnow, *Poeciliopsis occidentalis* (Cyprinodontiformes: Poeciliidae)." *Ecology* 56(4): 966–973.

Davies, N. B. 1978. "Ecological Questions about Territorial Behavior." In *Behavioural Ecology*, ed. J. R. Krebs and N. B. Davies. Oxford: Blackwell Scientific.

Downhower, Jerry F., and Kenneth B. Armitage. 1971. "The Yellow-Bellied Marmot and the Evolution of Polygamy." *American Naturalist* 105(944): 355–370.

DuBowy, Paul J. 1988. "Waterfowl Communities and Seasonal Environments: Temporal Variability in Interspecific Competition." *Ecology* 69(5): 1439–1453.

Ebersole, John P. 1980. "Food Density and Territory Size: An Alternative Model and a Test on the Reef Fish *Eupomacentrus leucostictus*." *American Naturalist* 115(4): 492–509.

Emlen, Stephen T., Peter H. Wrege, and Natalie J. Demong. 1995. "Making Decisions in the Family: An Evolutionary Perspective." *American Scientist* 83(2): 148–157.

Faaborg, J., P. G. Parker, L. DeLay, et al. 1995. "Confirmation of Cooperative Polyandry in the Galápagos Hawk (*Buteo galapagoensis*)." *Behavioral Ecology and Sociobiology* 36(2): 83–90.

Fenchel, Tom. 1975. "Character Displacement and Coexistence in Mud Snails (Hydrobiidae)." *Oecologia* 20(1): 19–32.

Ghalambor, Cameron K., Raymond B. Huey, Paul R. Martin, et al. 2006. "Are Mountain Passes Higher in the Tropics? Janzen's Hypothesis Revisited." *Integrative and Comparative Biology* 46(1): 5–17.

Hamilton, William D. 1964. "The Genetical Evolution of Social Behaviour," pts. I and II. *Journal of Theoretical Biology* 7(1): 1–16, 17–52.

Handford, Paul, and Michael A. Mares. 1985. "The Mating Systems of Ratites and Tinamous: An Evolutionary Perspective." *Biological Journal of the Linnean Society* 25(1): 77–104.

Harrison, Gary W. 1995. "Comparing Predator–Prey Models to Luckenbill's Experiment with *Didinium* and *Paramecium.*" *Ecology* 76(2): 357–374.

Heil, Martin, and Doyle McKey. 2003. "Protective Ant–Plant Interactions as Model Systems in Ecological and Evolutionary Research." *Annual Review of Ecology, Evolution, and Systematics* 34: 425–453.

Hoffman, Steven G. 1983. "Sex-Related Foraging Behavior in Sequentially Hermaphroditic Hogfishes (*Bodianus* Spp.)." *Ecology* 64(4): 798–808.

Hristov, Nickolay I., and William E. Conner. 2005. "Sound Strategy: Acoustic Aposematism in the Bat-Tiger Moth Arms Race." *Naturwissenschaften* 92(4): 164–169.

Janzen, Daniel H. 1966. "Coevolution of Mutualism between Ants and Acacias in Central America." *Evolution* 20(3): 249–275.

Järvi, Torbjörn, Birgitta Sillén-Tullberg, and Christer Wiklund. 1981. "Individual versus Kin Selection for Aposematic Coloration: A Reply to Harvey and Paxton." *Oikos* 37(3): 393–395.

Kempenaers, Bart, Geert R. Verheyen, and André A. Dhondt. 1997. "Extrapair Paternity in the Blue Tit (*Parus caeruleus*): Female Choice, Male Characteristics, and Offspring Quality." *Behavioral Ecology* 8(5): 481–492.

Kleiber, M. 1932. *Hilgardia* 6: 315–353.

Krebs, John R., Jonathan T. Erichsen, Michael I. Webber, and Eric L. Charnov. 1977. "Optimal Prey Selection in the Great Tit (*Parus major*)." *Animal Behaviour* 25(1): 30–38.

Krebs, John R., John C. Ryan, and Eric L. Charnov. 1974. "Hunting by Expectation or Optimal Foraging? A Study of Patch Use by Chickadees." *Animal Behaviour* 22(4): 953–964.

Lack, David. 1954. *The Natural Regulation of Animal Numbers*. Oxford: Clarendon Press.

Levins, Richard. 1970. "Extinction." In *Some Mathematical Questions in Biology*, Vol. 2., ed. Murray Gerstenhaber. Providence, RI: American Mathematical Society.

Lewontin, Richard C. 1970. "The Units of Selection." *Annual Review of Ecology and Systematics* 1: 1–18.

MacArthur, Robert H. 1958. "Population Ecology of Some Warblers of Northeastern Coniferous Forests." *Ecology* 39(4): 599–619.

MacArthur, Robert H., and Richard Levins. 1967. "The Limiting Similarity, Convergence, and Divergence of Coexisting Species." *American Naturalist* 101(921): 377–385.

McClure, Polley Ann. 1981. "Sex-Biased Litter Reduction in Food-Restricted Wood Rats (*Neotoma floridana*)." *Science* 211 (4486): 1058–1060.

McDonald, David B., and Wayne K. Potts. 1994. "Cooperative Display and Relatedness among Males in a Lek-Mating Bird." *Science* 266(5187): 1030–1032.

Myers, Judith H. 1998. "Synchrony in Outbreaks of Forest Lepidoptera: A Possible Example of the Moran Effect." *Ecology* 79(3): 1111–1117.

Olsen, Esben M., Mikko Heino, George R. Lilly, et al. 2004. "Maturation Trends Indicative of Rapid Evolution Preceded the Collapse of Northern Cod." *Nature* 428(6986): 932–935.

Orians, Gordon H. 1969. "On the Evolution of Mating Systems in Birds and Mammals." *American Naturalist* 103(934): 589–603.

Paine, Robert T. 1966. "Food-Web Complexity and Species Diversity." *American Naturalist* 100(910): 65–75.

Petrie, Marion, and Bart Kempenaers. 1998. "Extra-Pair Paternity in Birds: Explaining Variation between Species and Populations." *Trends in Ecology and Evolution* 13(2): 52–58.

Power, Mary E. 1990. "Effects of Fish in River Food Webs." *Science* 250(4982): 811–814.

Proestou, Dina A., Marian R. Goldsmith, and Saran Twombly. 2008. "Patterns of Male Reproductive Success in *Crepidula fornicata* Provide New Insight for Sex Allocation and Optimal Sex Change." *Biological Bulletin* 214(2): 194–202.

Queller, David C., and Joan E. Strassmann. 1998. "Kin Selection and Social Insects." *BioScience* 48(3): 165–175.

Rand, A. Stanley. 1964. "Inverse Relation Between Temperature and Shyness in the Lizard *Anolis lineatopus*." *Ecology* 45(4): 863–864.

Rand, A. Stanley. 1967. "Ecology and Social Organization in the Iguanid Lizard *Anolis lineatopus*." *Proceedings of the United States National Museum* 122(3595): 1–79.

Schluter, Dolph. 2000. "Ecological Character Displacement in Adaptive Radiation." *American Naturalist* 156(S4): S4–S16.

Schoener, Thomas W. 1971a. "Large-Billed Insectivorous Birds: A Precipitous Diversity Gradient." *Condor* 73(2): 154–161.

Schoener, Thomas W. 1971b. "Theory of Feeding Strategies." *Annual Review of Ecology and Systematics* 2: 369–404.

Schoener, Thomas W. 1984. "Size Differences among Sympatric, Bird-Eating Hawks: A Worldwide Survey." In *Ecological Communities: Conceptual Issues and the Evidence*, ed. Donald R. Strong Jr., Daniel Simberloff, Lawrence G. Abele, and Anne B. Thistle. Princeton, NJ: Princeton University Press.

Schoener, Thomas W., and Amy Schoener. 1971. "Structural Habitats of West Indian *Anolis* Lizards. II. Puerto Rican Uplands." *Breviora* 375:1–39.

Schoener, Thomas W., and David A. Spiller. 1996. "Devastation of Prey Diversity by Experimentally Introduced Predators in the Field." *Nature* 381(6584): 691–694.

Scriber, J. Mark. 1973. "Latitudinal Gradients in Larval Feeding Specialization of the World Papilionidae (Lepidoptera)." *Psyche* 80:355–373.

Sheldon, Ben C., Staffan Andersson, Simon C. Griffith, et al. 1999. "Ultraviolet Colour Variation Influences Blue Tit Sex Ratios." *Nature* 402(6764): 874–877.

Sherman, Paul W. 1977. "Nepotism and the Evolution of Alarm Calls." *Science* 197(4310): 1246–1253.

Simberloff, Daniel S., and Edward O. Wilson. 1969. "Experimental Zoogeography of Islands: The Colonization of Empty Islands." *Ecology* 50(2): 278–296.

Simon, Carol A. 1975. "The Influence of Food Abundance on Territory Size in the Iguanid Lizard *Sceloporus jarrovi*." *Ecology* 56(4): 993–998.

Sinclair, Anthony R. E., J. M. Gosline, G. Holdsworth, et al. 1993. "Can the Solar Cycle and Climate Synchronize the Snowshoe Hare Cycle in Canada? Evidence from Tree Rings and Ice Cores." *American Naturalist* 141(2): 173–198.

Slatkin, M., and M. J. Wade. 1978. "Group Selection on a Quantitative Character." *Proceedings of the National Academy of Sciences of the United States of America* 75(7): 3531–3534.

Stachowicz, John J. 2001. "Mutualism, Facilitation, and the Structure of Ecological Communities." *BioScience* 51(3): 235–246.

Trivers, Robert L. 1976. "Sexual Selection and Resource-Accruing Abilities in *Anolis garmani*." *Evolution* 30(2): 253–269.

Trivers, Robert L., and Dan E. Willard. 1973. "Natural Selection of Parental Ability to Vary the Sex Ratio of Offspring." *Science* 179(4068): 90–92.

Turchin, Peter, and Ilkka Hanski. 1997. "An Empirically Based Model for Latitudinal Gradient in Vole Population Dynamics." *American Naturalist* 149(5): 842–874.

Turchin, Peter, Lauri Oksanen, Per Ekerholm, et al. 2000. "Are Lemmings Prey or Predators?" *Nature* 405(6786): 562–565.

Turchin, Peter, Simon N. Wood, Stephen P. Ellner, et al. 2003. "Dynamical Effects of Plant Quality and Parasitism on Population Cycles of Larch Budworm." *Ecology* 84(5): 1207–1214.

Werner, Earl E., and Donald J. Hall. 1974. "Optimal Foraging and the Size Selection of Prey by the Bluegill Sunfish (*Lepomis macrochirus*)." *Ecology* 55(5): 1042–1052.

Other

World Conservation Union. 2007. 2007 IUCN Red List of Threatened Species. Summary Statistics for Globally Threatened Species. Table 1: Numbers of threatened species by major groups of organisms (1996–2007). Available from http://www.iucnredlist.org.

Thomas W. Schoener

Adaptation and evolutionary change

The field of evolutionary biology addresses the genetic changes that occur in living organisms over short as well as very long timescales. Evolutionary biologists often evaluate change at several different levels as well, such as among individuals within a population, between subpopulations, within a species, or even among species. Genetic changes may occur as a result of both genetic and environmental factors. For example, one genetic morph, or type (scientifically termed a genotype), in a population may be more fit (i.e., produce more surviving offspring and thus copies of its genes) than another for a multitude of reasons. Perhaps the relatively fit genotype carries genes that provide for being a faster runner than other conspecifics (members of the same species), which may allow this individual to capture more prey than less-fit genotypes. Or the more-fit genotype's color pattern may be better matched to its background and therefore make it better at evading predators. Differential survival occurs when some individuals survive at higher rates than others. Additionally, some individuals have higher reproductive success than others. This may result, for example, from being more colorful during the breeding season and attracting more mates.

Differential survival and reproduction, as described above, are the core drivers of natural selection, which is a primary force behind evolutionary change. Reproductive fitness reflects the number of viable offspring individuals produce, and differential reproduction occurs when one individual has more young than other, conspecific individuals, which often occurs as a result of adaptations. Such adaptations have typically evolved over long periods of time because they proved beneficial and allowed their carriers to fare better than others with respect to food or mate acquisition or predator avoidance, and to have more viable offspring. Beneficial, adaptive traits allow a genotype that is better at survival and reproduction to increase in frequency in subsequent generations, relative to other, less-fit genotypes.

In addition to genetic constitution, environmental factors may affect the relative survival of different genotypes. Consider two broods of young that are produced by the same parents and therefore closely related genetically. If one brood is born into a food-rich environment and the other into a food-depauperate one, even though the two broods have the same genes the individuals in the first environment may be healthier, survive at a higher rate, and bear more offspring as a result of their high-quality diet. In this example, an ability to survive and reproduce is driven primarily by the environmental circumstances the individuals were born into, rather than genetic factors. Environmental factors are not heritable.

While not a requirement of evolution by definition, evolutionary changes often and typically involve adaptations that result from natural selection. There are many specific examples of heritable, beneficial traits that allow organisms to be better suited to live in particular environments. Adaptations include a wide array of interesting changes that result in response to natural selection, or differential survival of different genotypes. A simple example of an adaptation involves background matching. This occurs when, for example, octopi, squid, frogs, and fishes contract (or expand) melanin-pigment-containing cells to lighten (or darken) overall coloration so as to blend better with their light (or dark) background. Depending on the taxa, this adaptive response may serve for predator evasion or as a means of surprising prey, prior to capture. Many adaptations occur in the sensory systems of animals, primarily because the senses are essential for so many basic functions related to survival and reproduction.

Sensory adaptations

Color vision is adaptive for a multitude of purposes including detecting and recognizing predators, prey, and potential mates. Color vision is based on a divergent family of genes, called opsins, which have arisen as a result of gene duplication (see Horth 2007). Opsins produce the visual pigments found in the eyes' retinal cones. These pigments work with light-absorbing chromophores to provide color vision (Terai et al. 2002). Different opsin genes produce particular proteins that absorb light at specific wavelengths. In vertebrates, opsin classes vary somewhat among taxa, but for color vision there are basically one or two classes of short-wavelength- (or ultraviolet-) sensitive proteins (SWS1, SWS2) produced by specifically associated opsin genes, as well as long-wavelength-sensitive (LWS) and middle-wavelength-sensitive (MWS) proteins. Some rhodopsin gene proteins are also MWS in fish (e.g., RH2) but these opsins have not been found in mammalian genomes (Yokoyama 2002). Broadly, SWS1 and SWS2 are sensitive to ultraviolet and/or shades of blue-violet (van Hazel et al. 2006), whereas LWS/MWS (and RH2) opsins are red/green sensitive (Trezise and Collin 2005). In most organisms, rhodopsins tend to be useful for dim-light vision and contrast detection.

Divergent evolution occurs when organisms that share a recent common ancestor experience a form of selection that requires them to diverge, or adapt to different circumstances. For example, the hundreds of colorful African cichlid fish species of Lake Victoria and Lake Malawi are predicted to be undergoing rapid radiation and divergent selection (Allender et al. 2003). These fishes actually have eight opsin genes that arose from multiple, rapid gene-duplication events (Terai et al. 2002; Trezise and Collin 2005; Carleton, Hárosi, and Kocher 2000; Spady et al. 2006). Positive, Darwinian selection (or natural selection favoring particular types, or alleles, allowing them to increase in relative frequency) occurs on all but two of these cichlid opsins (Sugawara, Terai, and Okada 2002; Spady et al. 2005). Thus, color vision is crucial for cichlids.

In cichlids, the LWS (red-sensitive) opsin has a lot of variability (fourteen different alleles) in the coding region compared to other opsin genes (SWS2-B) (Terai et al. 2002). Some of the mutations comprising this variability are predicted to shift the absorbance spectrum of this gene, which may be adaptive for visual acuity in a turbid environment. Directional, sexual selection has also been demonstrated for female mate choice on the red nuptial coloration in males of the cichlid species *Pundamilia nyererei* (Maan et al. 2004). Lake Victoria is a turbid lake where red wavelengths are likely transmitted better than many other colors. For several *Pundamilia* species, the wavelength of maximum absorption differs for the LWS pigment, but not three other pigments. These absorption differences are associated with LWS opsin sequence differences and male coloration differences while water transparency is associated with the ratio of expression of different cone types (red/red vs. red/green) (Carleton et al. 2005).

In contrast, Lake Malawi is a clear-water environment, and therefore more colors can be successfully transmitted through the water column (Terai et al. 2002). In fish from this lake, the LWS genes display very little genetic variability in their coding region. Yet, two species with vibrant blue coloration, *Metriaclima zebra* and *Labeotropheus fuelleborni*, have a higher frequency of RH2 and SWS2-B opsins than they do other cone opsins (Carleton and Kocher 2001). Further, blue markings can also be ultraviolet reflective, and *M. zebra* has been identified as also having ultraviolet-sensitive cones (Carleton, H&árosi, and Kocher 2000). Finally, body coloration is broadly similar among the lakes for several cichlid species. Yet, consistent with an adaptive scenario that addresses an association between environment and mate-choice colors, markings that are typically yellow on fish in Lake Malawi are red in Lake Victoria species (Terai et al. 2002).

An interesting visual adaptation is one of very deepwater organisms, such as the Comoran coelacanth (*Latimeria chalumnae*), which lives some 200 meters (650 feet) below the ocean's surface. In very deep water, sunlight penetration is drastically reduced and only a narrow band of light in the blue-range (with a wavelength of about 480 nanometers) is perceptible (Yokoyama et al. 1999). The coelacanth appears to use only two visual genes (similar to rhodopsin RH1 and RH2), which are shifted 20 nanometers toward blue, thus allowing for maximum light absorbance of the only visible wavelength that penetrates this unusual habitat.

A similar issue arises in southern Siberia in Lake Baikal, because it is the deepest freshwater lake on Earth. Here, the Cottoid fishes demonstrate a maximum light wavelength absorbance for the rhodopsin (RH1) gene at 516 nanometers, in the species residing in the shallow water. However, in deeper water species, RH1 is actually blue-shifted and demonstrates a maximum absorbance of 484 nanometers at greater depths. This absorbance shift allows for greater visual acuity in limited, blue-light conditions (Hunt et al. 1996), just as in the case of the Comoran coelacanth.

Lastly, in the Antarctic Notothenioids, there is no functional LWS visual (opsin) gene at all. This is entirely consistent with the fact that red light does not penetrate this deep-sea environment, making the protein that this gene typically produces not useful. Not surprisingly, this particular opsin is no longer functional or no longer extant in other deep-sea species as well (Pointer et al. 2005).

In contrast to refining color vision as an adaptation, nocturnal organisms often rely on other sensory systems as a primary means of survival. For example, insectivorous (insect-eating) bats use sound, or more specifically echolocation (as well as other nocturnal adaptations), to locate and identify their prey in the dark. Bats that tend to fly through large open spaces emit calls that are well suited to facilitate catching prey in this environment. This is evidenced in the specific traits associated with these calls, including long duration, long pulse intervals, and low frequencies (Holderied et al. 2008). Insectivorous bat calls tend to range between 20 kHz and 60 kHz (Fenton et al. 1998) but typically do not fall lower than this because echoes are weak when the wavelength transmitted exceeds the prey insect's wing length (see Jones and Holderied 2007).

The use of echolocation to catch prey is posited to have resulted in the coevolution of ultrasonic hearing in, for example, many lepidopteran butterflies (Rydell 2004). Earless moths appear to have coevolved to the bats' sensory

Musk ox (*Ovibos moschatus*) have developed a thick, two-layered coat to protect them from harsh winters. © Nilanjan Bhattacharya. Image from BigStockPhoto.com.

The desert iguana (*Dipsosaurus dorsalis*) has adapted to extreme temperatures and practices behavioral thermoregulation. © Gert Hochmuth. Image from BigStockPhoto.com.

adaptation as well, by flying less in open spaces and more in vegetation, where it is harder for bats to detect them via echolocation (Lewis, Fullard, and Morrill 1993). Interestingly, human engineers have mimicked sonar and the echolocation function of bats and dolphins in robots. The objective of this work was to evaluate the value of sound waves in lieu of vision for habitat exploration. Sound waves allow for an immense amount of detail to be detected (Kuc 1997).

Adaptations to extreme environments

While echolocation serves as an excellent adaptation for nonvisual hunting, organisms that live in extreme temperatures also have a unique set of adaptations to survive. Musk oxen (*Ovibos moschatus*) are large, hooved mammals that live in Arctic regions. They have adapted to severe winter temperatures that fall well below freezing, via several adaptations. These include practicing energy conservation during the extreme winter months, gaining the ability to digest low-quality forage, retaining high body fat, and having a very thick, warm coat that is comprised of two layers (Reynolds, Wilson, and Klein 2002). The outercoat is comprised of long, tough, water-resistant guard hairs, which grow for several years and are never shed. The undercoat, or qiviut, is comprised of down and serves as insulation during the winter but can be shed each spring (Rowell et al. 2001).

Animals that do not have fur coats have adapted to frigid habitats in other ways. Some species have evolved antifreeze glycoproteins as an adaptation to cold environments. In fact, several Antarctic notothenioid fishes, as well as northern cod, have converged on the same solution to freezing water temperatures by making these unique antifreeze proteins, even though these fishes are not at all closely related and are found in different orders and superorders. The antifreeze glycoproteins that these fishes synthesize are very similar to one another and are encoded by a family of polyprotein genes. Each gene produces multiple glycoprotein molecules in tandem (Chen, DeVries, and Cheng 1997). The gene sequences and substructures of the proteins, however, demonstrate that they have evolved independently in the different fishes (Chen, DeVries, and Cheng 1997). Thus, these unrelated fishes have evolved the same adaptation that allows for survival in a harsh ecological environment. When relatively unrelated organisms independently evolve the same, or very similar, adaptations to a particular circumstance, this is called convergent evolution.

In stark contrast, many plants and animals have adapted to the extreme temperatures, the high variance in hot and

To maintain optimum temperature, garter snakes (*Thamnophis elegans*) can shield themselves from heat and cold by settling in burrows or under rocks, or they can opt for sunny or shady surface locations. © David Stevenson/Lexington Herald-Leader/MCT.

cold temperatures, and the dry, hot habitat of the desert. Consider the desert iguana (*Dipsosaurus dorsalis*), an ectotherm (where body temperature is controlled by the temperature outside) that can actually function well in temperatures ranging from 15°C to 45°C (and can even survive short bouts at temperatures as low as 0°C and as high as 47°C). Moreover, iguanas practice behavioral thermoregulation: They actively select a warmer location that is more optimal for their survival. Similarly, when it is too hot, they adaptively select shadier locales, which are cooler (see Huey and Kingsolver 1989).

A similar scenario holds true for garter snakes (*Thamnophis elegans*). To maintain optimum temperature, garter snakes can shield themselves from heat and cold by settling in burrows or under rocks, or they can opt for sunny and shady surface locations. After identifying the preferred body temperatures of these snakes in scientific laboratories by evaluating the temperature at various times of the day and night in different habitats, researchers were able to determine that at night the optimal location for a snake is under rocks of medium thickness. By examining the frequency of available rocky sites and comparing this with the sites that were actually used by snakes, the researchers ascertained that over half of the snakes surveyed were found under medium-sized rocks, even though the environment contained roughly one-third each of thin, medium, and thick rocks (Huey et al. 1989). Thus, snakes also demonstrate behavioral thermoregulation, actively seeking the optimum rock thickness and protection from harsh temperatures.

Human adaptation and the fossil record

The adaptive properties of the human pelvis are slowly being revealed by the fossil record. Childbirth is tricky in humans, as compared to some other primates, because the size of the human newborn's head, which matches the size of the

birth canal (Weaver and Hublin 2009). The relatively large size of the human pelvis is thought to be adaptive to account for birthing large-brained individuals. At 3 million years ago (mya), early hominids had relatively small brains (McHenry 1975). Hominids from 2 mya displayed larger variance in brain size. At the same time, the pelvis appears to have begun to be reconfigured (McHenry 1975).

A fossil pelvis from an adult female *Homo erectus*, dated at about 1.2 mya, provides further evidence, via specific adaptations, that an adaptive pelvis shape (shorter and wider) emerges in response to increasing fetal brain size (Simpson et al. 2008). This conclusion is based on a large number of different measurements of the pelvis; the presence/absence, size, and shape of specific traits on the pelvis; and a comparison of these measurements with those of additional fossils and modern humans. Sacrum shape and the size of the sciatic nerve notch were also evaluated (Simpson et al. 2008).

Previously, *Homo erectus* brains were thought to have been no more than 230 cubic centimeters. This fossil pelvis demonstrates, however, that a baby with a brain of up to 315 cubic centimeters in size, or about the same size as a modern-day thirty-six-week-old, could have been delivered. For comparison, the brain size of infants delivered today is about 380 cubic centimeters.

Mimicry among species

Mimicry is a form of evolution that occurs between species. Here, one species evolves to resemble a model species because the model has a particular warning signal, which is typically used to ward off predators. There are two major types of mimicry, namely Müllerian and Batesian. In Müllerian mimicry (Müller 1879), both the model and the mimic share an aposematic trait, one that warns predators they are harmful or distasteful. Aposematic warning traits are conventionally bright, often red or yellow, visual warning coloration, but mimics do exist that exploit other sensory systems, too. In Batesian mimicry (Bates 1862), a non-noxious organism mimics the appearance of a dangerous or toxic one. In this way, the mimic benefits from the appearance of the model, without actually having to incur the cost of producing a noxious substance. The mechanism works on the theory that if a predator has tasted a noxious insect and recognizes that insect's color pattern, then the predator will avoid attempting to consume the mimic because of its similar appearance to the noxious model.

A classic example of Müllerian mimicry is found in the bright and starkly colored, distasteful, lepidopteran *Heliconius* butterflies of the neotropics. The butterflies in this genus display a warning coloration on their wings to deter predators. There are many species in the genus, which has radiated into hundreds of mimics (Turner and Mallet 1996). Two species, *Heliconius erato* and *H. melpomene*, which are only distantly related to one another, have extremely similar wing color patterns. Moreover, these species are sympatric, meaning their distributions overlap geographically. What is most interesting is that while the species have about thirty different wing patterns across their range, in any given region the

The small postman butterfly (*Heliconius erato*) is an example of Müllerian mimicry. Adam Jones/Photo Researchers, Inc.

species' wing patterns are similar to one another (see Kronforst and Gilbert 2008).

Similarly, in *Heliconius cydno* there are also several different color-pattern morphs. When tested against a common morph in nature, it has been shown that the rare patterned individuals do actually have lower survival rates as a result of predation than the common morph, as expected with this type of mimicry. Further, when tested in nature, locally abundant morphs have higher survival rates than rarer morphs (Kapan 2001).

Numerous Batesian mimics occur that resemble stinging wasp species. Red-headed borers (*Neoclytus acuminatus*) and lilac ash borers (*Podosesia syringae*) are a few of the many ash-boring moths in the family Sesiidae, or clearwing wasps, that mimic stinging paper wasps (e.g., *Polistes fuscatus*). The basis for this mimicry is that predators who have experienced the impact of the stinging wasp will not attempt to consume the ash borer, which has no stinger. These moth species are interesting because they mimic the wasp in many physical characteristics, including their banded brown-and-yellow color,

their long and slender size and shape, and even their basic flight pattern.

Genetic drift

Not all changes are adaptive. Evolution may also result from chance events that can change genotype or allele frequencies and can rapidly drive alleles either extinct or to fixation (a frequency of 1.0) in a population. When populations are small in size, they are particularly susceptible to genetic drift, or chance events. For example, assume that a population of the bladder campion (*Silene vulgaris*), a roadside weed, is comprised of ten plants. An individual plant may be hermaphroditic and have both male (pollen) and female (ovules) mating structures in the flowers on a single plant. Some plants, however, are comprised of only female flowers. If this population of ten plants is comprised of nine female plants and one hermaphrodite and chance events, such as drought, annihilate the hermaphrodite, the female-only population will go extinct, unless pollen comes in from another population. Thus, a dramatic consequence arises from this chance event occurring in a small population. Similarly, it can be seen that migration can have a dramatic effect on this population. In fact, migration is also capable of homogenizing subpopulations' genes quite rapidly.

In the example above, 1 in 10, or 10 percent, of the plants were hermaphrodites. For comparison, if the population size is larger, for example, 1,000 plants, and 10 percent are hermaphrodites, then 100 plants would have to go extinct during the drought for genetic drift to have the same catastrophic effect on the large population. If one individual hermaphrodite goes extinct during the drought, this might not affect the fitness of the population at all. Hence it can be seen that genetic drift is more likely to affect smaller populations than larger ones.

Levels of natural selection

Generally, Darwinian selection, where the fittest individuals have the highest survival or produce the most surviving

The largest female mosquitofish (*Gambusia affinis*) produces the most young. © blickwinkel/Alamy.

Cannibalistic tiger salamanders (*Ambystoma tigrinum*) are more likely to eat unrelated individuals than they are to eat siblings or cousins. Suzanne L. & Joseph T. Collins/Photo Researchers, Inc.

young, is considered to operate at the individual level. Thus, the fastest cheetah gets the gazelle, or the largest female mosquitofish produces the most young. There are now a substantial number of examples of natural selection operating at different levels, from the gene to the group, some of the most interesting of which occur for kin (genetically related individuals). A few explicit examples are considered here.

The ability of individuals to recognize conspecific individuals is useful for a variety purposes, including identifying potential mates, benefiting from protective shoaling (loose aggregations of fish) and avoiding unnecessary competition. The ability of an individual to discriminate between other individuals and recognize the degree of relatedness between oneself and another is the basis for the concept of kin selection. William D. Hamilton (1964) proposed the theory of inclusive fitness to explain how altruism (or unselfish, costly behavior) might evolve as a result of the ability of individuals to increase their own fitness by aiding in the propagation of genes in highly related individuals. The theory basically states that the benefit of aiding a conspecific multiplied by its relatedness to oneself must be greater than the cost of aiding an individual for kin selection to evolve.

Carnivorous spadefoot toad tadpoles (*Spea bombifrons*, *S. multiplicata*) will cannibalize nearby individuals. In situations where there is a choice, however, these tadpoles tend to elect to eat unrelated tadpoles over siblings, even when they have never previously been exposed to either potential prey individual (Pfennig 1999). In this situation the benefit of relatedness to the offspring potentially produced by siblings appears to outweigh the cost of additional future offspring production resulting from cannibalism.

Tiger salamanders (*Ambystoma tigrinum*) display a similar pattern (Pfennig, Collins, and Ziemba 1999). Here, however, distinguishing between relatives of different relatedness occurs as well. In research studies, cannibalistic salamanders were more likely to eat relatively unrelated individuals than they were to eat siblings or cousins. Cannibals, however, took less time to discriminate between siblings and relatively unrelated individuals than they did to discriminate between cousins and relatively unrelated individuals (Pfennig, Sherman, and Collins 1994).

In addition to higher levels of selection such as kin selection, there are also lower levels of selection. These include some examples of peculiar chromosomes that are often responsible for nothing other than self-replication.

Sex-ratio distortion, meiotic drive, and microchromosomes

Sex-ratio theory predicts that species will exhibit a 50:50 sex ratio of males to females (Fisher 1930). In cases in which this is not true and the sex ratio is biased, the rare sex will have greater reproductive success. In subsequent generations, however, as the rare sex becomes more common, the opposite sex will take on a fitness advantage. This is a form of frequency-dependent selection, where the rare morph (or sex) has an advantage when rare, but that same morph (or sex) then has a disadvantage when common. Ultimately, this conflict should result in a return of the sex ratio to 50:50 (Fisher 1930).

In some cases, a bias in sex ratio occurs during the production of young, as a result of meiotic drive. Meiotic drive elements are considered selfish genetic elements that disproportionately increase the rate of their own transmission during meiosis. This can affect the sex ratio at the level of individual broods as well as populations. Theoretically this, or any sex-ratio distorter, could actually drive a species to extinction (Fisher 1930; Hamilton 1967).

Some of the organisms in which meiotic drive occurs include plants, fungi, insects, mammals, and fish. In many taxa, including mice, there are also meiotic drive enhancer and suppressor genes (Pomiankowski 1999). Meiotic drive occurs when there is a bias in the transmission of the chromosomes controlling for one of the sexes. For example, in diploid heterogametic species (with two difference sex chromosomes), either the X- or the Y-bearing sperm may be successfully used in the production of young at a higher rate than the alternate sperm, resulting in a sex bias in the progeny.

Similarly, B chromosomes are often called supernumerary or extra chromosomes, because they occur in addition to the typical chromosome repertoire of a species. Sometimes such chromosomes are costly to the genome, some appear neutral, and a few provide a benefit. B chromosomes may derive from autosomes or sex chromosomes and have been identified in more than 1,300 plant and 500 animal species (Camacho, Sharbel, and Beukeboom 2000). They tend to have few functions, though a primary one across species is self-replication. B chromosomes, transposons (replicating DNA segments, which can move in the genome independent of homologous recombination), segregation distorters (causing deviation of normal segregation and hence overrepresentation of self/chromosome), and cytoplasmic genetic elements are all considered selfish, mobile genetic elements, because they defy typical Mendelian inheritance patterns and instead primarily self-replicate.

One particularly interesting example of a microchromosome with more than a neutral function appears in fish. The Amazon molly (*Poecilia formosa*) is a parthenogenetic species composed of only females. The eggs of this species develop without fertilization and are therefore clones of the mother. Additionally, this species is considered a sexual parasite because it appears to use sperm from the males of closely related species merely to trigger embryogenesis. Given that Amazon mollies are clonal and that meiosis does not occur to form eggs, males have been considered to make no genetic contribution to this all-female lineage. The total number of chromosomes these mollies have is forty-six. The wild-type or common coloration of Amazon mollies is silver, or olivaceous-gray.

In nature, Amazon mollies mate with *P. Mexicana* or *P. latipinna*. In a laboratory setting, Amazon mollies can also be mated to a lab strain known as the black molly (*Poecilia* sp.). In one study, about 1 in 1,000 offspring expressed black pigmentation that ranged from light-black speckling or blotches to completely black patterning (Schartl et al. 1995). Offspring of both color patterns show the same DNA fingerprint, indicating that if there is a genetic difference in them, it must be smaller than this technology can detect. Consistent with this, cytological evidence shows that in addition to the traditional forty-six chromosomes, microchromosomes are also found in Amazon mollies in nature and in these lab fish (Schartl et al. 1995). Wild-type pigmented fish were found to have one microchromosome whereas black fish had multiple microchromosomes. Many species within this fish family have rare, black spotted morphs, and mating-behavior differences are found between black spotted and wild-type (silver) pigmentation morphs of poeciliid species (Horth 2003).

Mutation

From an evolutionary perspective, time may elapse on a scale that is measured in a few days or in millions of years, depending on the amount of time it takes for a generation to elapse for a given organism and on the type of evolutionary changes being evaluated. Some organisms, such as the beneficial bacteria in the human intestine (*Escherichia coli*), complete one generation in less than an hour (in cell culture; Meselson and Stahl 1958), whereas other organisms, such as the largest terrestrial species, the African elephant (*Loxodonta africana*), requires about twenty-five years or more for a single generation to elapse (Blanc 2008).

Why does the amount of time a generation takes make a difference in the evolution of organisms? In brief, this is because the number of generations that elapse, not the amount of absolute time, dictate the amount of novel, genetic variation that can be generated in the genomes of the organisms comprising a population. There are two basic categories of genetic variation—standing genetic variation, or that which exists at any given time in the individuals within a species or population, and novel genetic variation, or that which arises from spontaneous (also called de novo) mutations.

Spontaneous mutations that can be inherited by subsequent generations occur during the cell division and replication that is associated with reproduction. Though estimates of actual mutation rates vary (Lynch and Walsh 1998), the pervasive rarity of novel, or de novo, mutations is unquestionable. Mutation rates on the order of one in a million or less (or about 10^{-6} per gene, or per nucleotide) per generation are documented for many genes such as the five different coat-color loci in mice, where spontaneous mutations from the wild-type or common coat-color morph accrue at the rate of 8.9×10^{-6} (and reversions back to that wild-type color morph state occur at a rate of 2.7×10^{-6}; Schlager and Dickie 1967). Both higher and lower mutation rates have been documented as well. While the rates of mutation for different situations (e.g., transitions, transversions, insertions/deletions, CpG sites) in humans vary, the accepted rate in humans tends to be on the order of 10^{-8} for humans (Nachman and Crowell 2000). For hemophilia B it is documented as 2.14×10^{-8} per nucleotide base per generation (Giannelli, Anagnostopoulos, and Green 1999).

Mutation is important because simple or small mutational changes, such as substituting one amino acid for another in the protein (or polypeptide chain) that is being coded for by a gene, may result in visible (or phenotypic) changes in, for example, an organism's overall color, body shape, or in the ability to metabolize a particular substance for nutrition. In humans, for example, just a few nucleotide changes in a single gene (in this example, tyrosinase) can cause the disruption of typical pigment production and result in the lack of, or reduced, pigmentation known as albinism (King, Mentink, and Oetting 1991).

Biomechanics

The field of biomechanics is found at the intersection of biology, mathematics, and engineering. Through the use of principles derived from the latter two fields a greater understanding of how an organism navigates the world can emerge. For example, different lizard species obtain their prey using different foraging strategies. Essential functional and mechanical differences are found between lizards that sprint and those that move slowly and steadily, to track down and catch their prey.

Through mathematical and biological comparative analyses, several traits have been evaluated for differences in the mechanics of how lizards move when foraging. Lizards basically have two major strategies to catch prey. Some species sit motionless and wait, visually inspecting their surroundings for prey, then sprint fast to ambush their prey (called sit and wait). Others move about slowly over larger distances, probing with their tongues for chemosensory cues from their prey (called wide foraging). Sit-and-wait lizards sprint faster, but wide-foragers have greater endurance (McElroy, Hickey, and Reilly 2008).

Large-scale analyses of fifteen different lizard species demonstrated that slow locomotion mechanics did not evolve just once for all slow lizards but instead have evolved several different times, independently, and in a few different ways. Thus, wide-foraging lizards use several different walking gaits and vaulting mechanics. In contrast, sit-and-wait species use only fast speeds and trotting gaits plus bouncing mechanics (McElroy, Hickey, and Reilly 2008). Foraging mode and biomechanical foraging pattern match extremely well, indicating that lizards use adaptive locomotor functioning to catch their prey (McElroy, Hickey, and Reilly 2008).

Another example of the elucidation of biomechanics can be found in the manner in which boxfish navigate currents.

Unlike many fishes, these marine species have hard exterior carapaces that vary in shape by species and make conventional swimming difficult. Through biomechanics—in this case, applying engineering and mathematical principles to study the flow of water around the boxfishes—it has been shown that at least two species are able to maintain a stable position by using a method somewhat akin to a sailor navigating winds and currents by continually making small adjustments to maintain course (Bartol et al. 2008). When strong ocean currents pull a boxfish off course (perhaps pointing upward instead of forward), unique water-flow patterns develop around particular locations along the body of the fish because of its boxy shape. For example, small vortices form around the keels, and this allows what have been termed self-correcting trimming forces to operate (Bartol et al. 2008). This means that the boxfish can right themselves efficiently, despite their seemingly nonstreamlined shape.

What is fascinating is that this self-correction mechanism occurs automatically, without any neural processing by the fish (Bartol et al. 2008). As a rule there is a trade-off between being stable and being maneuverable, because shifting course is counteracted by stabilizing mechanisms. Boxfishes are unique, however, and have adapted to be both stable and maneuverable. Some of the nonscientific paybacks that result from understanding such evolutionary adaptations as these biomechanical processes are the potential financial gains that result from such knowledge. For example, this boxfish research has been applied to the development of automobiles, such as a concept car from Mercedes-Benz known as the bionic car.

A plethora of examples in nature serve as demonstrations of adaptation and provide evidence of natural selection and evolution. Background matching, which results in camouflage, as well as colorful patterning, which solicits mating, are classical examples of adaptive phenotypes. The timescale for adaptations to arise can be anywhere from a few generations to hundreds, thousands, or more generations. The necessary requirement is merely that a trait evolve that benefits its bearer relative to other individuals in a population. Through

It has been shown that at least two species of boxfish are able to maintain a stable position by using a method similar to a sailor navigating winds and currents by continually making all adjustments to maintain course. © McClatchy-Tribune.

natural selection such a trait would be predicted to increase in frequency over time. While chance events such as genetic drift can drive populations to extinction, especially numerically small populations, many examples persist of genetic material at multiple scales increasing in frequency over time as a result of their adaptive properties.

Resources

Books

Fisher, Ronald A. 1930. *The Genetical Theory of Natural Selection.* Oxford: Clarendon Press.

Lynch, Michael, and Bruce Walsh. 1998. *Genetics and Analysis of Quantitative Traits.* Sunderland, MA: Sinauer.

Pomiankowski, Andrew. 1999. "Intragenomic Conflict." In *Levels of Selection in Evolution,* ed. Laurent Keller. Princeton, NJ: Princeton University Press.

Rydell, Jens. 2004. "Evolution of Bat Defense in Lepidoptera: Alternatives and Complements to Ultrasonic Hearing." In *Echolocation in Bats and Dolphins,* ed. Jeanette A. Thomas, Cynthia F. Moss, and Marianne Vater. Chicago: University of Chicago Press.

Periodicals

Allender, Charlotte J., Ole Seehausen, Mairi E. Knight, et al. 2003. "Divergent Selection during Speciation of Lake Malawi Cichlid Fishes Inferred from Parallel Radiations in Nuptial Coloration." *Proceedings of the National Academy of Sciences of the United States of America* 100(24): 14074–14079.

Bates, Henry Walter. 1862. "Contributions to an Insect Fauna of the Amazon Valley: Lepidoptera: Heliconidae." *Transactions of the Linnean Society of London* 23(3): 495–566.

Camacho, Juan Pedro M., Timothy F. Sharbel, and Leo W. Beukeboom. 2000. "B-Chromosome Evolution." *Philosophical Transactions of the Royal Society* B 355(1394): 163–178.

Carleton, Karen L., Ferenc I. Hárosi, and Thomas D. Kocher. 2000. "Visual Pigments of African Cichlid Fishes: Evidence for Ultraviolet Vision from Microspectrophotometry and DNA Sequences." *Vision Research* 40(8): 879–890.

Carleton, Karen L., and Thomas D. Kocher. 2001. "Cone Opsin Genes of African Cichlid Fishes: Tuning Spectral Sensitivity by Differential Gene Expression." *Molecular Biology and Evolution* 18(8): 1540–1550.

Carleton, Karen L., Juliet W. L. Parry, James K. Bowmaker, et al. 2005. "Colour Vision and Speciation in Lake Victoria Cichlids of the Genus *Pundamilia.*" *Molecular Ecology* 14(14): 4341–4353.

Chen, Liangbiao, Arthur L. DeVries, and Chi-Hing C. Cheng. 1997. "Convergent Evolution of Antifreeze Glycoproteins in Antarctic Notothenioid Fish and Arctic Cod." *Proceedings of the National Academy of Sciences of the United States of America* 94 (8): 3817–3822.

Fenton, M. B., C. V. Portfors, I. L. Rautenbach, and J. M. Waterman. 1998. "Compromises: Sound Frequencies Used in Echolocation by Aerial-Feeding Bats." *Canadian Journal of Zoology* 76(6): 1174–1182.

Giannelli, F., T. Anagnostopoulos, and P. M. Green. 1999. "Mutation Rates in Humans," Pt. II: "Sporadic Mutation-Specific Rates and Rate of Detrimental Human Mutations Inferred from Hemophilia B." *American Journal of Human Genetics* 65(6): 1580–1587.

Hamilton, William D. 1964. "The Genetical Evolution of Social Behaviour," Pts. I and II. *Journal of Theoretical Biology* 7(1): 1–16, 17–52.

Hamilton, William D. 1967. "Extraordinary Sex Ratios." *Science* 156(3774): 477–488.

Horth, Lisa. 2003. "Melanic Body Colour and Aggressive Mating Behavior Are Correlated Traits in Male Mosquitofish (*Gambusia holbrooki*)." *Proceedings of the Royal Society* B. 270:1033–1040.

Horth, Lisa. 2007. "Sensory Genes and Mate Choice: Evidence That Duplications, Mutations, and Adaptive Evolution Alter Variation in Mating Cue Genes and Their Receptors." *Genomics* 90(2): 159–175.

Huey, Raymond B., and Joel G. Kingsolver. 1989. "Evolution of Thermal Sensitivity of Ectotherm Performances." *Trends in Ecology and Evolution* 4(5): 131–135.

Huey, Raymond B., Charles R. Peterson, Stevan J. Arnold, and Warren P. Porter. 1989. "Hot Rocks and Not-So-Hot Rocks: Retreat-Site Selection by Garter Snakes and Its Thermal Consequence." *Ecology* 70(4): 931–944.

Hunt, David M., Jude Fitzgibbon, Sergey J. Slobodyanyuk, and James K. Bowmaker. 1996. "Spectral Tuning and Molecular Evolution of Rod Visual Pigments in the Species Flock of Cottoid Fish in Lake Baikal." *Vision Research* 36(9): 1217–1224.

Jones, Gareth, and Marc W. Holderied. 2007. "Bat Echolocation Calls: Adaptation and Convergent Evolution." *Proceedings of the Royal Society* B 274(1612): 905–912.

Kapan, Durrell D. 2001. "Three-Butterfly System Provides a Field Test of Müllerian Mimicry." *Nature* 409(6818): 338–340.

King, Richard A., Margaret M. Mentink, and William S. Oetting. 1991. "Non-random Distribution of Missense Mutations within the Human Tyrosinase Gene in Type I (Tyrosinase-Related) Oculocutaneous Albinism." *Molecular Biology and Medicine* 8(1): 19–29.

Kronforst, Marcus R., and Lawrence E. Gilbert. 2008. "The Population Genetics of Mimetic Diversity in *Heliconius* Butterflies." *Proceedings of the Royal Society* B 275(1634): 493–500.

Kuc, Roman. 1997. "Biomimetic Sonar Locates and Recognizes Objects." *IEEE Journal of Oceanic Engineering* 22(4): 616–624.

Lewis, Francine P., James H. Fullard, and Scott B. Morrill. 1993. "Auditory Influences on the Flight Behaviour of Moths in a Nearctic Site," Pt. II: "Flight Times, Heights, and Erraticism." *Canadian Journal of Zoology* 71(8): 1562–1568.

Maan, Martine E., Ole Seehausen, Linda Söderberg, et al. 2004. "Intraspecific Sexual Selection on a Speciation Trait, Male Coloration, in the Lake Victoria Cichlid, *Pundamilia nyererei.*" *Proceedings of the Royal Society* B 271(1556): 2445–2452.

McElroy, Eric J., Kristen L. Hickey, and Stephen M. Reilly. 2008. "The Correlated Evolution of Biomechanics, Gait, and Foraging Mode in Lizards." *Journal of Experimental Biology* 211 (7): 1029–1040.

McHenry, Henry M. 1975. "Fossils and the Mosaic Nature of Human Evolution." *Science* 190(4213): 425–431.

Meselson, Matthew, and Franklin W. Stahl. 1958. "The Replication of DNA in *Escherichia coli.*" *Proceedings of the National Academy of Sciences of the United States of America* 44(7): 671–682.

Müller, Fritz. 1879. "*Ituna* and *Thyridia*: A Remarkable Case of Mimicry in Butterflies," trans. Raphael Meldola. *Proceedings of the Entomological Society of London* 1879: 20–29.

Nachman, Michael W., and Susan L. Crowell. 2000. "Estimate of the Mutation Rate Per Nucleotide in Humans." *Genetics* 156 (1): 297–304.

Pfennig, David W. 1999. "Cannibalistic Tadpoles That Pose the Greatest Threat to Kin Are Most Likely to Discriminate Kin." *Proceedings of the Royal Society* B 266(1414): 57–61.

Pfennig, David W., James P. Collins, and Robert E. Ziemba. 1999. "A Test of Alternative Hypotheses for Kin Recognition in Cannibalistic Tiger Salamanders." *Behavioral Ecology* 10(4): 436–443.

Pfennig, David W., Paul W. Sherman, and James P. Collins. 1994. "Kin Recognition and Cannibalism in Polyphenic Salamanders." *Behavioral Ecology* 5(2): 225–232.

Pointer, Marie A., Chi-Hing Christina Cheng, James K. Bowmaker, et al. 2005. "Adaptations to an Extreme Environment: Retinal Organisation and Spectral Properties of Photoreceptors in Antarctic Notothenioid Fish." *Journal of Experimental Biology* 208(12): 2363–2376.

Rowell, J. E., C. J. Lupton, M. A. Robertson, et al. 2001. "Fiber Characteristics of Qiviut and Guard Hair from Wild Muskoxen (*Ovibos moschatus*)." *Journal of Animal Science* 79(7): 1670–1674.

Schartl, Manfred, Indrajit Nanda, Ingo Schlupp, et al. 1995. "Incorporation of Subgenomic Amounts of DNA as Compensation for Mutational Load in Gynogenetic Fish." *Nature* 373(6509): 68–71.

Schlager, Gunther, and Margaret M. Dickie. 1967. "Spontaneous Mutations and Mutation Rates in the House Mouse." *Genetics* 57(2): 319–330.

Simpson, Scott W., Jay Quade, Naomi E. Levin, et al. 2008. "A Female *Homo erectus* Pelvis from Gona, Ethiopia." *Science* 322 (5904): 1089–1092.

Spady, Tyrone C., Juliet W. L. Parry, Phyllis R. Robinson, et al. 2006. "Evolution of the Cichlid Visual Palette through Ontogenetic Subfunctionalization of the Opsin Gene Arrays." *Molecular Biology and Evolution* 23(8): 1538–1547.

Spady, Tyrone C., Ole Seehausen, Ellis R. Loew, et al. 2005. "Apative Molecular Evolution in the Opsin Genes of Rapidly Speciating Cichlid Species." *Molecular Biology and Evolution* 22 (6): 1412–1422.

Sugawara, Tohru, Yohey Terai, and Norihiro Okada. 2002. "Natural Selection of the Rhodopsin Gene during the Adaptive Radiation of East African Great Lakes Cichlid Fishes." *Molecular Biology and Evolution* 19(10): 1807–1811.

Terai, Yohey, Werner E. Mayer, Jan Klein, et al. 2002. "The Effect of Selection on a Long Wavelength-Sensitive (LWS)

Opsin Gene of Lake Victoria Cichlid Fishes." *Proceedings of the National Academy of Sciences of the United States of America* 99 (24): 15501–15506.

Trezise, Ann E. O., and Shaun P. Collin. 2005. "Opsins: Evolution in Waiting." *Current Biology* 15(19): R794–R796.

Turner, John R. G., and James L. B. Mallet. 1996. "Did Forest Islands Drive the Diversity of Warningly Colored Butterflies? Biotic Drift and the Shifting Balance." *Philosophical Transactions of the Royal Society* 351(1341): 835–845.

Weaver, Timothy D., and Jean-Jacques Hublin. 2009. "Neandertal Birth Canal Shape and the Evolution of Human Childbirth." *Proceedings of the National Academy of Sciences of the United States of America* 106(20): 8151–8156.

Yokoyama, Shozo. 2002. "Molecular Evolution of Color Vision in Vertebrates." *Gene* 300(1–2): 69–78.

Yokoyama, Shozo, F. Huan Zhang, Bhernhard Radlwimmer, and Nathan S. Blow. 1999. "Adaptive Evolution of Color Vision of the Comoran coelacanth (*Latimeria chalumnae*)." *Proceedings of the National Academy of Sciences of the United States of America.* 96: 6279–6284.

Other

Bartol, I. K., M. S. Gordon, P. Webb, et al. 2008. "Evidence of Self-Correcting Spiral Flows in Swimming Boxfishes." *Bioinspiration and Biomimetics* 3(1): doi:10.1088/1748-3182/3/1/014001. Available from http://www.iop.org/EJ/article/1748-3190/3/1/014001/bb8_1_014001.pdf.

Blanc, Julian. 2008. "*Loxodonta africana.*" In IUCN Red List of Threatened Species. Version 2009.1. Available from http://www.iucnredlist.org/details/12392/0/full.

Holderied, Marc W., Chris J. Baker, Michele Vespe, and Gareth Jones. 2008. "Understanding Signal Design during the Pursuit of Aerial Insects by Echolocating Bats: Tools and Applications." *Integrative and Comparative Biology* 48(1): 74–84. Available from http://icb.oxfordjournals.org/cgi/reprint/48/1/74.

Karren, Jay B. 1993. "Ash/Lilac Borer." Utah State University Extension Fact Sheet No. 36. Logan: Utah State University Cooperative Extension. Available from http://extension.usu.edu/files/publications/factsheet/ash-lilac-borers93.pdf.

Reynolds, Patricia E., Kenneth J. Wilson, and David R. Klein. 2002. "Muskoxen." In *Arctic Refuge Coastal Plain Terrestrial Wildlife Research Summaries*, ed. David C. Douglas, Patricia E. Reynolds, and E. B. Rhode. Biological Science Report USGS/BRD/BSR-2002-0001. Reston, VA: U.S. Geological Survey. Available from http://www.absc.usgs.gov/1002/section7part1.htm.

Tao, Yun, Luciana Araripe, Sarah B. Kingan, et al. 2007. "A Sex-Ratio Meiotic Drive System in *Drosophila simulans*," Pt. II: "An X-Linked Distorter." *PLoS Biology* 5(11): e293. Available from http://www.plosbiology.org/article/info:doi/10.1371/journal.pbio.0050293.

van Hazel, Ilke, Francesco Santini, Johannes Müller, and Belinda S. W. Chang. 2006. "Short-Wavelength Sensitive Opsin (SWS1) as a New Marker for Vertebrate Phylogenetics." *BMC Evolutionary Biology* 6: 97. Available from http://www.biomedcentral.com/1471-2148/6/97.

Lisa Horth

Sexual selection

Any consideration of the diversity of animal life has to take into account the power of sexual selection. Many of the most elaborate traits and behaviors of animals are associated with courtship and evolved under the influence of sexual selection: the massive antlers of the extinct Irish elk and its living relatives, the elongated tail (actually trains) of a peacock, the elaborate melodies of songbirds, the incessant chirping of many frogs and insects, and the brilliant colors of many butterflies and fish. To understand sexual selection is to understand the cause of a major component of biodiversity.

Darwin's theories

Charles Darwin's main contribution to biology did not come from convincing others that evolution took place, a fact that was not doubted by most of Darwin's contemporaries. It was, instead, his argument about the process that resulted in evolution of adaptations for survival—natural selection—that caused the most controversy. In *On the Origin of Species* (1859), Darwin explained: "if variations useful to any organic being do occur, assuredly individuals thus characterized will have the best chance of being preserved in the struggle for life; and from the strong principles of inheritance, these will tend to produce offspring similarly characterized" (p. 127).

Thus if variation in certain traits causes variation in survivorship, and if variation in those traits is caused by underlying variation in genes, then natural selection results in the evolution of these traits. Natural selection brings about adaptations for survival.

Counter to Darwin's theory of natural selection, however, are the host of traits that hinder survivorship. These traits all share some similarities. They usually are present or more elaborated in males compared to females, and if they are expressed only part of the time, it is usually during the breeding season. Darwin suggested an alternative theory to natural selection to explain the evolution of these exaggerated traits. He called it sexual selection and explained it thusly: "This form of selection depends not on a struggle for existence in relation to other organic beings or the external conditions, but on the struggle between individuals of one sex, generally the males, for the possession of the other sex" (1872, p. 69).

Many sexual traits, such as antlers, large tails, elaborate vocalizations, and bright colors, evolved for the struggle to acquire mates, although they often hinder male survivorship. Thus, if variation in certain traits causes variation in mating success, and if variation in those traits is caused by underlying variation in genes, then sexual selection results in the evolution of these traits. Sexual selection brings about adaptations for acquiring mates, not adaptations for survival.

Sexual selection versus natural selection

Many biologists now consider sexual selection as a form of natural selection. Both forms of selection favor traits that enhance an individual's Darwinian fitness—the number of genes an individual transmits to the next generation relative to other individuals in its population. Fitness has two major components, survival and reproduction. An individual must do both to transmit its genes into the next generation.

There can be a tradeoff between survivorship and mating success. Thus natural selection and sexual selection can exert counter selection pressures on the same traits. As mentioned above, many traits that males use in courtship seem to hinder survivorship. Developing display traits requires an expenditure of energy. When nutritional resources are low, males often divert energy from display traits and focus on survival. Also, many sexually selected traits develop under the influence of testosterone, and this hormone has detrimental effects on the immune system, which can further decrease a male's prospects for survival. Once a male has a courtship trait, such as a long tail or bright coloration, displaying it can also be energetically costly. Singing in birds, frogs, and insects, for example, can increase the metabolic rate several hundred percent. Furthermore, the primary function of a display trait is to attract the attention of females, but these traits also attract the attention of "eavesdroppers." Bright colors of male guppies attract predators as well as mates (Endler 1978). Complex calls of male túngara frogs are more attractive not only to females but also to predators and parasites such as frog-eating bats and blood-sucking flies (Tuttle and Ryan 1981).

Natural selection can constrain the degree to which sexually selected traits evolve. The longer a male bird's tail the more attractive it might be to females, but it will not evolve to the extent that the male's survivorship is so low (e.g., because it reduces the individual's ability to evade predators) that it negates the benefits the male achieves from attracting

The great frigatebird (*Fregata minor*) adult male extends his goular pouch in courtship display. © Krystyna Szulecka Photography/Alamy.

kingdom in the form of horns, antlers, canines, and claws. Darwin saw evidence for sexual selection by female mate choice in the beauty of animal ornaments—colors, odors, dances, and songs. Female animals, he conjectured, seem to have aesthetic senses similar to humans. Females must prefer more elaborately ornamented males, but it was not clear to him why. Sexual selection by female choice is perhaps the best studied and the most controversial aspect of sexual selection theory.

Sexual difference and sexual conflict

In 1948 A. J. Bateman conducted a simple and insightful experiment with fruit flies. He mated males and females several times. Males showed a steady increase in offspring number with number of matings, but once a female mated, additional matings had no influence on her reproductive success. These results suggested that sexual selection should favor males to increase their number of mates, but that females would not be under such selection.

Geoffrey A. Parker (1970) and Robert L. Trivers (1972) uncovered the importance of Bateman's principle in the 1970s in their theories of sexual conflict and parental investment, respectively. Their notion is that males and females invest their energy very differently as parents, especially in their respective gamete production. Males produce many small gametes, whereas females produce fewer and larger gametes. Because of these differences in gametic investment, and in order to maximize their reproductive fitness, males should tend toward promiscuity and mate with as many females as possible because they are unlikely to exhaust their abundant and comparatively "cheap" sperm supply. Conversely, females should be more circumspect in their mate choice because they produce relatively few, large, energetically expensive eggs, and should therefore choose to mate with better quality males rather than more males (i.e., they should be more picky in their choice of mates). Thus, the sexes are inherently in conflict as to how to maximize their reproductive success, and in their willingness to mate. One outcome is that all females will tend to be mated, but only a fraction of the males will ever reproduce. This introduces greater variance in mating success in males than in females. Whenever there is more variance in a trait, there is greater opportunity for selection to act as a powerful agent of evolutionary change.

The variance in mating between the sexes can be skewed further by some realities of the mating system. Females often do not mate after their eggs are fertilized, whereas males continue to do so. Thus at any point in time more males are available for mating than females. This skew in the operational sex ratio results in keen competition among males for access to females and provides females with an abundance of males from which to choose. This also promotes higher variance in mating success in males than in females, and thus sexual selection will act more strongly on male traits than on female traits. This explains why most sexually dimorphic traits are more elaborate in males than in females.

The differences between the sexes can result in sexual conflict and pit them against one another in an evolutionary "arms race." In many species, multiple matings decrease

more mates. But natural selection and sexual selection need not always be in conflict. Foraging efficiently, for example, enhances survivorship and provides nutrients critical for the growth of sexual ornaments.

How sexual selection happens

Just as an individual can increase its fitness in two general ways, survivorship and reproduction, an animal can increase its ability to acquire mates in two general ways: mate competition and mate choice:

> …in the one it is between the individuals of one sex, generally the male sex, in order to drive away or kill their rivals, the female remaining passive; whilst in the other, the struggle is likewise between the individuals of the same sex, in order to excite or charm those of the opposite sex, generally the females, which no longer remain passive but select more agreeable partners. (Darwin 1871, vol. 2, p. 398)

The evidence for sexual selection through male competition and female choice was clear to Darwin. Weapons of offense used in male competition abound in the animal

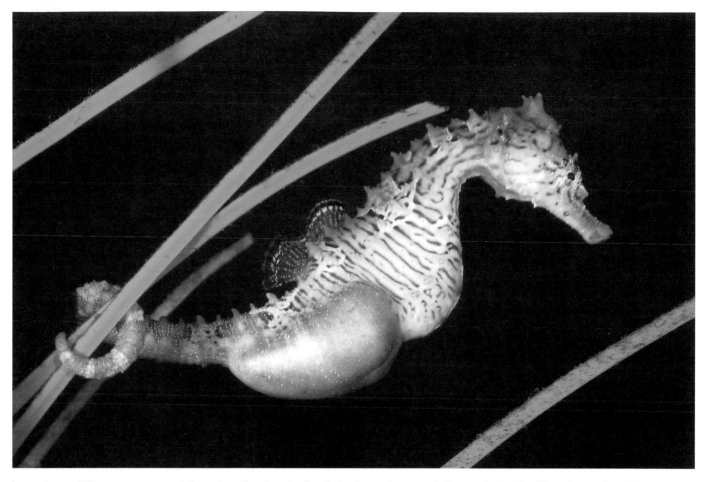

In sea horses (*Hippocampus erectus*) the male rather than the female incubates the eggs. © Gregory G. Dimijian/Photo Researchers, Inc.

female survivorship because of the increased probability of injury or disease transmission. When a male fruit fly mates a female, he deposits toxins, along with his seminal fluid, that cause a female to delay remating. This benefits a male because it protects his paternity by killing the sperm of other males. These compounds also have the incidental consequence of increasing female mortality. Artificial selection experiments show that females can evolve resistance to these male compounds. Thus, conflict between the sexes initiates a cycle of coevolution in which males evolve more toxic sperm and females counter by evolving resistance; this is called chase-away selection (Rice 1996).

Sex reversal and mating system variation

Bateman's results with fruit flies should not be over-interpreted. Göran Arnqvist and Tina Nilsson (2000) reviewed over one hundred experimental studies of insects showing that females' reproductive success increases with multiple matings. Thus sexual selection will have the opportunity to act on females as well as males, even if it will act stronger on the latter in many cases.

Sexual selection acts strongest on females when they are competing for mates rather than choosing them. Extreme

cases of female polyandry offer incontrovertible support for the rule of sexual conflict and parental investment. In these cases, not only do females mate with multiple males, but in addition the more common roles assumed by males and females in courtship and parental care are reversed. In sea horses and pipefish, for example, the male rather than the female incubates the eggs. In these cases, the pattern elucidated by Bateman also tends to be reversed: The number of mates has a greater influence on female than on male reproductive success (Jones et al. 2000).

Competition and choice can also interact. In elephant seals, for example, a female's initial choice of a male can incite a round of male–male competition, after which the female mates with the eventual winner (Cox and Le Boeuf 1977). Thus, males and females both can compete, both actively and passively, and choose their mates, and choice and competition can interact.

Mechanism of sexual selection: male competition

There is substantial evidence of males competing for access to females. In some species, males forcefully copulate with females. In others, males defend areas where females gather for resources, or they gather females in groups and exclude other males from these groups. In mixed male groups, dominant males often have priority access to ovulating females.

Another arena of mate competition can take place inside the female (Parker 1970). As females often mate multiply, the sperm of multiple males have the potential to interact. In some cases, the sperm mix and the outcome is similar to a lottery. The male's probability of paternity is dependent on the quantity (and quality) of sperm deposited in the female. In other cases timing is important; in the phenomenon of sperm precedence, either the first or the last male to inseminate the female fertilizes most of the eggs. Males can evolve adaptations to sperm competition. For example, a male will often guard a mate after he inseminates her to prevent her from mating with rivals. Males can also include chemicals in their seminal fluids, as do some fruit flies, to influence the female's reproductive physiology and to delay her time to remating. If a female does mate again, the males of some species will deposit mating plugs in the female's reproductive tract to block access to her eggs by the sperm of competing males.

Mechanism of sexual selection: Female choice

The most controversial aspect of Darwin's theory was that females attend to differences in male courtship traits when they choose a mate from among conspecific males. Definitive proof became available only when researchers could manipulate male traits experimentally and demonstrate that these actually influenced a female's mating preferences and that a female's preferences were correlated with patterns of male mating success in nature.

The first such study was by Michael J. Ryan in a 1980 study of túngara frogs. Males produce a mating call consisting of a whine that can be followed by zero to seven chucks. Females prefer calls with chucks, and all males add chucks in choruses. In nature, females can swim through the chorusing males and exercise unimpeded choice of mates. They are more likely to choose larger males than smaller ones, and larger ones also tended to fertilize more of the female's eggs. Larger males produce chucks with lower frequencies because they have a larger larynx. In mate choice experiments, two calls were broadcast to females from speakers. Both had the same whine, but the chucks differed in frequency. Females were preferentially attracted to the lower-pitch chucks characteristic of larger males. Although it had been known for some time that females use calls to identify members of their own species, this experiment showed that females also evaluate differences in calls to choose larger males who, in turn, increase the female's

In Southern elephant seals (*Mirounga leonina*), a female's initial choice of a male can incite a round of male–male competition, after which the female mates with the eventual winner. Rod Planck/Photo Researchers, Inc.

The calling male Túngara frog (*Physalaemus pustulosus*) has an unusually large external vocal sac to attract a mate with his call. Kentwood D. Wells.

reproductive success by passing these traits along to her male offspring, who, in turn, will be more successful.

Many demonstrations of female mate choice have followed. Vocalizations in birds, frogs, and crickets; colors in birds, lizards, fish, and butterflies; and odors in mammals, fish, and many insects are among the many traits that have been shown to influence mate choice. The question is why?

The evolution of female preferences

Mate choice exerts sexual selection on male traits and is responsible for most of the diversity in sexually dimorphic traits that abound in nature. An important question is why selection forces favor the evolution of female preferences.

Traits can evolve under direct selection or indirect selection (Kirkpatrick and Ryan 1991). Direct selection occurs when the focal trait has an immediate effect on reproductive success. Indirect selection occurs when the focal trait does not directly influence reproductive success but is correlated with other traits that do. Female preferences can evolve under both scenarios.

Direct selection on female mating preferences is common as males often influence the number of offspring a female produces. One example is when females choose males who can fertilize more eggs. Selection will favor females who prefer the more virile males, and that preference will evolve to become fixed in the population.

Direct selection also occurs when a female's preference for more conspicuous male traits enhances her survivorship. This will occur when such preferences reduce the time and effort spent searching for a mate. Reducing the time spent searching for mates reduces both a female's energy expenditure and exposure to predation risk. Conspicuous males are easier to find than cryptically colored ones.

Females use their sensory systems—their eyes or ears or sense of smell—to choose mates. Those sensory systems, however, are used for other tasks as well. Direct selection in other contexts could influence the evolution of sensory systems. For example, in surfperch, prey detection and selection promotes the evolution of photoreceptor tuning, which is unique to each species' habitat. This, in turn, has resulted in the evolution of different visual biases for

Kin selection

In his book *On the Origin of Species*, Charles Darwin (1809–1882) postulated that evolution occurred via the process of natural selection, whereby individuals that possessed traits well suited to the environment would pass those successful traits on to their progeny. Darwin bolstered his hypothesis by attempting to account for biological features that appeared inconsistent with the action of natural selection. He was particularly troubled with "one special difficulty, which at first appeared to me insuperable, and actually fatal to my whole theory. I allude to the neuters or sterile females in insect-communities: for these neuters often differ widely in instinct and in structure from both the males and fertile females, and yet, from being sterile, they cannot propagate their kind" (Darwin 2006 [1859], p. 600). Darwin was thus very concerned with understanding how distinctive traits displayed by sterile social insects could be passed on to future generations if the bearers of these traits failed to reproduce. More broadly, he wondered if the presence of these individuals provided evidence against his theory of evolution by natural selection.

Darwin ultimately realized that the presence of sterile social insects did not invalidate his theory. Instead, the existence of these individuals can be explained by *kin selection*. Kin selection is a type of natural selection that depends on social interactions between individuals that share *alleles* (alternate forms of a gene). Kin selection can be formally defined as the process by which a focal allele changes frequency within a population, specifically because the allele influences social behaviors of the bearer that lead to the differential reproductive success of other individuals that possess the allele. Fundamentally, the idea underlying kin selection theory is that an allele can be transmitted through personal offspring production *or* through offspring production of relatives.

Darwin's deep insight into biology gave him a basic understanding of kin selection. Indeed, he noted that "insects and other articulate animals in a state of nature occasionally become sterile; and if such insects had been social, and it had been profitable to the community that a number should have been annually born capable of work, but incapable of procreation, I can see no very great difficulty in this being effected by natural selection" (Darwin 2006 [1859], p. 600). Thus, Darwin realized that an individual's effects on relatives could strongly influence trait evolution and even allow for the production of sterile individuals that could not transmit genes by themselves.

The scope of kin selection

Kin selection is an extremely important natural process that affects the fitness of interacting individuals. The term *fitness* is used in the evolutionary sense and refers to the ability of an individual to reproduce. An individual's total *inclusive fitness* is composed of *direct fitness*, which arises from the personal offspring production of the individual that would be attained without the influence of the social actions of others, plus *indirect fitness*, which arises from changes in offspring production of relatives caused by the social actions of the focal individual.

Kin selection can lead to the evolution of a wide array of social behaviors. Here, social behaviors are defined broadly to include any actions that affect the fitness of both the actor and the recipient of the behavior. From an evolutionary standpoint, social behaviors fall into four different categories (West, Griffin, and Gardner 2007). A behavior that (1) increases the direct fitness of both the actor and recipient is *mutually beneficial*; (2) increases the direct fitness of the actor and decreases the direct fitness of the recipient is *selfish*; (3) decreases the direct fitness of the actor and increases the direct fitness of the recipient is *altruistic*; and (4) decreases the

A carnivore morph Mexican spadefoot (*Spea multiplicata*) cannibalizing an omnivore. Courtesy of David Pfennig.

direct fitness of both the actor and recipient is *spiteful*. It is worth noting that words such as spiteful, altruistic, and selfish carry specific connotations when used nonscientifically to describe human actions and attitudes. These words should be divorced, however, from any emotional, moral, motivational, or ethical implications when discussing the evolution of social behaviors. They simply describe how particular behaviors affect the direct fitness of interacting individuals.

Kin selection is useful in understanding the evolution of all four types of social behaviors. Mutually beneficial and selfish behaviors, however, are often best understood in terms of classical natural selection because they both provide direct fitness benefits for the actor. Therefore, the idea of kin selection is most often used to explain altruistic or spiteful behaviors, which decrease the direct fitness of the actor.

The process of kin selection

So how does kin selection work? Consider a hypothetical population of birds. A fraction of these birds possess an allele that sometimes causes them to share food with other birds. The action of sharing food is altruistic because food is a source of energy required for reproduction, and so the donating bird will reproduce less and the recipient bird will reproduce more. This leads to the following question: Will the allele and, consequently, the altruistic behavior of food sharing increase in frequency in this population? It turns out that the answer depends on with whom the altruistic bird shares food.

First, consider the case in which the altruistic bird shares food without regard to the genetic makeup of the recipient. In this case, the altruistic bird helps individuals both with and without the altruistic allele. Consequently, the altruistic behavior will not increase in frequency because the birds that receive help do not possess the altruistic allele at a frequency above the population average. In fact, the altruistic behavior will likely decrease in frequency because birds with the altruistic allele, and thus share food, will have relatively low fitness. Now, consider a second situation in which the altruistic bird tends to share food with other birds that possess the altruistic allele. In this case, the altruistic allele and behavior might increase in frequency because the sharing behavior is specifically directed at other birds with the altruistic allele.

The question of whether the allele actually increases in frequency ultimately depends on three interacting factors: (1) the cost of food sharing that is borne by the altruistic bird, (2) the benefit of food sharing that the recipient bird obtains, and (3) the probability that an altruistic bird shares food with another altruistic bird. This probability depends on the *relatedness* between the interacting individuals. Relatedness is a measure of genetic similarity and can be defined as the overall proportion of alleles that individuals share relative to the proportion shared by individuals randomly selected from the population.

From this discussion, it can be seen that it is fundamentally important for individuals that possess the altruistic alleles to interact for kin selection to operate. Such interactions can

occur in three nonexclusive ways (Sachs et al. 2004). First, interactions might occur among individuals sharing the allele if genetic relatives (i.e., kin) recognize and interact with each other. There is, in fact, evidence that some mammals make use of physical cues (e.g., appearance, odor) or familiarity to discriminate kin from nonkin. In addition, kin discrimination may be indirect and arise from environmental cues learned early in development, as seems to be the case in many social insects, which learn to recognize kin because they share common odors obtained from the colony and environment. Certain mammals (including ground squirrels) are also known to use odor to discriminate kin.

The second way that individuals sharing alleles might interact is if they simply fail to disperse far from their natal site and tend to interact with neighbors. In this case, relatives interact as a by-product of limited dispersal. In fact, almost all organisms show limited dispersal to some degree (Avise 2004). Thus, in many cases, this mechanism leads to interactions among relatives.

Finally, individuals sharing the altruistic allele may interact if others that possess it recognize the focal allele, and these individuals behave differently toward those that bear the allele and those that do not. This unusual phenomenon is known as the *green beard* phenomenon. The term was coined to describe the existence of a hypothetical allele that causes its bearers to develop a green beard and to act beneficially toward other individuals with green beards. Surprisingly, some examples of green beard recognition have been identified in natural populations. For example, in flour beetles, the presence of the *medea* allele on the maternal side of a cross leads to the

Ants, such as the red imported fire ant (*Solenopsis invicta*) shown here, display the most advanced levels of sociality. The queen is much larger than her sister workers and is capable of flight. In contrast, fire ant workers never develop wings, nor can they ever reproduce. However, workers still obtain fitness indirectly by helping rear fertile sisters who develop into queens. Courtesy of Michael Goodisman.

Yellow jackets are highly social wasps. Southern yellow jacket (*Vespula squamosa*) workers, shown here, forage for food, defend the colony, and construct the nest. The helping and cooperative behaviors displayed by workers evolved through the process of kin selection. Courtesy of Michael Goodisman.

death of embryos that do not carry the allele (the allele is named after the tragic mythological character medea who killed her own children). Similarly, fire ant workers possessing the '*b*' allele preferentially kill queens that do not possess this allele. Thus the *medea* and *b* alleles conform to green beard expectations because bearers of the allele distinguish individuals that also possess the allele and treat these individuals in a relatively beneficial way.

The history of kin selection

Many important scholars of evolutionary biology have contributed to the elaboration of kin selection theory (Dugatkin 2007). Darwin was the first to draw attention to the problem, and solution, of how altruistic traits evolved. But he lacked the knowledge of modern genetics to fully formulate the underlying causes of the evolution of social actions. The issue was considered again in the 1930s by two of the major architects of the modern evolutionary synthesis, J. B. S. Haldane (1892–1964) and Ronald A. Fisher (1890–1962). Haldane, for example, is believed to have told friends that he would risk his life by jumping into a river to save two brothers or eight cousins. This amusing anecdote showed that Haldane

understood how behaviors that affected close versus distant relatives might influence the transmission of alleles. Fisher pointed out that a noxious insect might lose its life if attacked by a predator. But the nauseated predator might depart the area thereby sparing relatives of the unlucky prey. Fisher thus correctly suspected that the trait of distastefulness could increase in frequency through benefits to relatives. Despite these insights, however, neither Haldane nor Fisher developed kin selection ideology into a coherent body of work.

William D. Hamilton (1936–2000) is credited with fully constructing a mathematical framework describing kin selection. Hamilton's key contribution was to quantify exactly how and when social traits would be subject to kin selection (Hamilton 1964). Hamilton found that the conditions under which a social action was favored by selection could be summarized by a simple equation, $r\,b > c$. This formula, which came to be known as *Hamilton's rule*, indicates that a social behavior increases in frequency if the relatedness between the actor and recipient, r, times the benefit to the recipient of the action, b, exceeds the cost of the action to the actor, c.

Hamilton called his set of ideas *inclusive fitness theory*. The term kin selection was actually coined by John Maynard Smith

The bumblebee (*Bombus* sp.) is another highly social insect. This worker bee forages for pollen on a flower. Courtesy of Michael Goodisman.

in 1964. Kin selection theory and inclusive fitness theory are now used more or less interchangeably. Since Hamilton's 1964 contribution, mathematical theory surrounding the evolution of social behaviors has been refined and expanded (Frank 1998), but Hamilton's key insight that alleles can be transmitted via collateral reproduction by relatives remains.

It should be noted that kin selection theory has not been without controversy. In particular, there has been considerable debate regarding the relationship between kin selection and group selection, which is the differential reproduction or survival of groups based on heritable characteristics of individuals within those groups. A full review of the meaning and importance of group selection is beyond the scope of this entry, but it is sufficient to say that certain group selection models represent different ways to conceptualize the evolution of social actions. It has been shown, however, that kin selection and group selection models are essentially equivalent mathematically and that they rely on the same factors to explain trait evolution. Kin selection formulations are generally viewed as more easily interpretable, readily constructed, and testable than group selection models (West, Griffin, and Gardner 2007), but the subject is not beyond debate.

Natural examples of kin selection

Kin selection has been extremely important in explaining the evolution of social behaviors in a wide variety of species. However, social insects have played a particularly important role in testing kin selection theory (Bourke and Franks 1995; Queller and Strassmann 1998). For example, as discussed above, kin selection explains the evolution of the extremely unusual trait of sterility found in many social insects. The genes conferring sterility clearly cannot be passed on directly but instead are transmitted by fertile relatives who carry, but do not express, the focal genes for this trait.

In addition, hymenopteran social insects (ants, social bees, and social wasps) have been models for understanding social behaviors because they possess a genetic system known as haplodiploidy (Crozier and Pamilo 1996). Haplodiploidy leads to unusual relationships among kin. Specifically, a haplodiploid female is related to her sister by 0.75 but to her brother by only 0.25, as opposed to standard genetic systems in which these relatedness values are both 0.5. The asymmetries in relatedness allow for important tests of kin selection theory. For instance, kin selection might influence how hymenopteran social insect workers, which are always female, allocate resources into producing fertile individuals. In particular, the unusual relatedness patterns between the sexes suggest that workers should bias sex allocation toward their sisters and away from their brothers. Overall, this prediction has been supported, as sex allocation in hymenopteran social insects is generally female-biased. In addition, kin selection theory predicts that variation in relatedness among individuals from different colonies might lead to situations in which some colonies produce exclusively males or reproductive females. This prediction has also been upheld in several species, thereby providing strong support for kin selection operating on sex ratio in hymenopteran social insects.

Kin selection theory has also been successfully used to explain a variety of behaviors in vertebrates. For example, helping behavior in cooperative breeding birds and mammals is preferentially directed to close relatives, and helping of relatives is more pronounced in species in which help is more valuable (Griffin and West 2003). In addition, kin selection has been instrumental in understanding conflict among siblings or between parents and offspring. Interestingly, such conflict is expected to occur under many conditions because an individual is more closely related to itself than to its relatives. Kin selection theory can explain the genetic mechanisms and mating systems that affect struggles among relatives (Parker, Royle, and Hartley 2002).

Dispersal also appears to be subject to kin selection. Dispersal has important social implications, because individuals that remain in their natal territory potentially compete with their relatives for resources or mates. Such selective pressures are believed to be common in promoting the evolution of dispersal in a variety of mammalian species (Lawson Handley and Perrin 2007).

There has been considerable interest in understanding whether nonhuman primates display kin-selected traits, because the study of primates may yield insight into the evolution of humans. Available data indicate that almost all primates live in highly social kin groups (Chapais and Berman 2004). Kin groups form because individuals of one sex tend to remain associated with their natal group. Group members, which are often matrilineal kin, engage in a variety of supportive behaviors including grooming, cooperative breeding, food sharing, and coalition forming (Silk 2006). In addition, increasing evidence suggests that male kin groups behave nepotistically as well (Widdig 2007). Thus there is strong evidence that kin selection has played an important role in the evolution of nonhuman primate behavior.

Group interaction, such as these chimpanzees (*Pan troglodytes*) grooming each other, plays an important role in kin selection. © Simon Allen/Alamy.

The finding of kin-selected behaviors in nonhuman primates has naturally led to the investigation of such behaviors in humans. Current evidence does indeed suggest that some human behaviors have evolved via kin selection. For example, violence within family groups tends to be directed toward nonkin more frequently than kin. The sharing of resources, such as food and money, is also sometimes dependent on relatedness. And individuals providing child care are often related to the individuals they help. Nevertheless, culture also plays an important role in shaping human societies and behaviors. Moreover, many human behaviors may not have a genetic basis or lead to variation in fitness among individuals, in which case they would not be subject to natural selection. Thus, culture may affect the process of evolution within human societies in unique ways (Fehr and Fischbacher 2003).

The process of kin selection was originally conceived to explain animal behaviors, but kin selection can operate in any biological system. For instance, kin selection theory is important to understanding the evolution of social behaviors of microbes. In particular, kin selection thinking has been used to understand how microbes use resources within their environment. There is considerable interest in this social trait, because resource use may be related to virulence in pathogenic microorganisms (Buckling and Brockhurst 2008). Kin selection theory predicts that microbes should limit resource use if the population consists of kin. In contrast, genetically mixed populations of microbes are expected to lead to competition among nonkin and, potentially, host damage. Despite this important theoretical prediction, however, insufficient data are available to determine if kin selection affects levels of virulence in microbes in this manner.

Another example of the operation of kin selection in microbes concerns the production of public goods. Public goods are products, such as antibiotics or digestive enzymes, made by some individuals that can be used by others in the environment. The production of these public goods is potentially subject to the process of kin selection. For example, kin selection theory predicts that selfish behaviors, whereby individuals use public goods without producing them, should evolve when populations are composed of mixed strains. In fact, such predictions have been upheld in a variety of bacterial and viral systems. Specifically, it has been shown that the evolution of selfish behaviors in microbes is affected by variation in relatedness among interactants (West, Diggle, et al. 2007).

A final example of the effect of kin selection on the evolution of a "social action" underscores the scope and importance of the process. Plants compete for resources with each other by varying how they produce roots and leaves. Studies now show that some plants allocate less mass to roots involved in underground competition when they are grown with kin than when they are grown with nonkin. This result is consistent with kin selection theory because greater allocation to roots increases competitive ability, which is expected to occur most fiercely when nonkin interact.

Overall, evidence suggests that kin selection is a fundamental evolutionary mechanism leading to the elaboration of both cooperative and competitive behaviors. Moreover, kin selection operates in diverse groups of organisms, including microbes, animals, and plants. Consequently, kin selection appears to be a critically important process affecting the evolution of social actions among virtually all biological organisms.

Resources

Books

Avise, John C. 2004. *Molecular Markers, Natural History, and Evolution*. 2nd edition. Sunderland, MA: Sinauer Associates.

Bourke, Andrew F. G., and Nigel R. Franks. 1995. *Social Evolution in Ants*. Princeton, NJ: Princeton University Press.

Chapais, Bernard, and Carol M. Berman, eds. 2004. *Kinship and Behavior in Primates*. Oxford: Oxford University Press.

Crozier, Ross H., and Pekka Pamilo. 1996. *Evolution of Social Insect Colonies: Sex Allocation and Kin Selection*. Oxford: Oxford University Press.

Darwin, Charles. 2006 (1859). *On the Origin of Species*. In *From So Simple a Beginning: The Four Great Books of Charles Darwin*, ed. Edward O. Wilson. New York: Norton.

Frank, Steven A. 1998. *Foundations of Social Evolution*. Princeton, NJ: Princeton University Press.

Silk, Joan B. 2006. "Practicing Hamilton's Rule: Kin Selection in Primate Groups." In *Cooperation in Primates and Humans: Mechanisms and Evolution*, ed. Peter M. Kappeler and Carel P. van Schaik. Berlin: Springer.

Periodicals

Dugatkin, Lee Alan. 2007. "Inclusive Fitness Theory from Darwin to Hamilton." *Genetics* 176(3): 1375–1380.

Fehr, Ernst, and Urs Fischbacher. 2003. "The Nature of Human Altruism." *Nature* 425(6960): 785–791.

Griffin, Ashleigh S., and Stuart A. West. 2003. "Kin Discrimination and the Benefit of Helping in Cooperatively Breeding Vertebrates." *Science* 302(5645): 634–636.

Hamilton, William D. 1964. "The Genetical Evolution of Social Behaviour." *Journal of Theoretical Biology* 7(1): 1–52.

Lawson Handley, Lori J., and Nicolas Perrin. 2007. "Advances in Our Understanding of Mammalian Sex-Biased Dispersal." *Molecular Ecology* 16(8): 1559–1578.

Parker, Geoff A., Nick J. Royle, and Ian R. Hartley. 2002. "Intrafamilial Conflict and Parental Investment: A Synthesis." *Philosophical Transactions of the Royal Society* B 357(1419): 295–307.

Queller, David C., and Joan E. Strassmann. 1998. "Kin Selection and Social Insects." *Bioscience* 48(3): 165–175.

Sachs, Joel L., Ulrich G. Mueller, Thomas P. Wilcox, and James J. Bull. 2004. "The Evolution of Cooperation." *Quarterly Review of Biology* 79(2): 135–160.

West, Stuart A., Stephen P. Diggle, Angus Buckling, et al. 2007. "The Social Lives of Microbes." *Annual Review of Ecology, Evolution, and Systematics* 38: 53–77.

West, Stuart A., Ashleigh S. Griffin, and Andy Gardner. 2007. "Social Semantics: Altruism, Cooperation, Mutualism, Strong Reciprocity, and Group Selection." *Journal of Evolutionary Biology* 20(2): 415–432.

Widdig, Anja. 2007. "Paternal Kin Discrimination: The Evidence and Likely Mechanisms." *Biological Reviews* 82(2): 319–334.

Buckling, Angus, and Michael A. Brockhurst. 2008. "Kin Selection and the Evolution of Virulence." *Heredity* 100(5): 484–488.

Michael A. D. Goodisman

Coevolution

Coevolution, the process of reciprocal evolutionary change between interacting species, is thought to have played a significant role in shaping almost all life on Earth (Thompson 2005). Some of the most important steps in the diversification of life have probably been made through coevolutionary processes. For example, the fundamental leap from prokaryote to eukaryote cell design is thought to have evolved through symbiotic relationships. The endosymbiotic theory of eukaryote origins, which was largely developed by Lynn Margulis, postulates that cell organelles (particularly mitochondria and chloroplasts) are derived from the engulfment of one prokaryote (bacterial) cell by another (Margulis and Sagan 2002). Chloroplasts are thought to have once been free-living photosynthetic bacteria, and mitochondria were once aerobic heterotrophs, oxygen-dependent organisms incapable of producing their own food. How they gained entry to larger cells is uncertain, but they could have been engulfed as prey or even through evolution of a parasitic relationship. But once inside, the possible benefits to both the engulfer and engulfee are not difficult to envisage: Photosynthetic symbionts could supply nourishment, and as the world became increasingly oxygenated, anaerobic cells would benefit from aerobic symbionts, which used oxygen to their advantage. In turn, residing within a larger cell may provide a safe haven from other would-be predators. Presumably, the relationship between hosts and symbionts eventually became so interdependent that it was impossible to separate one from another, finally resulting in a single organism. Lines of evidence for the endosymbiotic theory are many and include the fact that many enzymes and transport systems in the inner membranes of chloroplasts and mitochondria are most similar to those found in prokaryote cells. Chloroplasts and mitochondria also divide through a binary fission process (not mitosis), which is most like bacteria; similar to prokaryotes, the DNA found within them is also in the form of circular molecules and is not associated with histones or other proteins as it is in the nucleus of eukaryotes. Finally, mitochondrial genes are more similar to those of free-living bacteria than they are to genes in the nucleus of the cell in which they reside.

Symbiotic relationships continued to play significant roles in the diversification of life. For example, colonization of land by early plants, which lacked roots, may not have been possible without the aid of mycorrhizal fungi. Fossil evidence indicates that even the earliest plants were associated with mycorrhizae, which help with the absorption of nutrients. In addition, many chemical defense compounds produced in plants may not be produced by the plants themselves but by the fungi often obligately associated with them (Herre et al. 2005). Tropical seas, which are so poor in nutrients, would probably be comparatively lifeless were it not for the extraordinary relationship between coral polyps and dinoflagellate algae. The polyps provide "housing" and waste products that dinoflagellate algae use in photosynthesis. In return, polyps supplement their daily food intake with photosynthate produced by resident algae. The enormous success of the corals, which are among the largest non-human-made structures on Earth, has allowed them to provide most of the structural heterogeneity and nutrients on which many tropical marine species have become so dependent. Even the process of digestion, which most people take for granted in their daily lives, cannot take place without the aid of endosymbiotic bacteria. Although it may be hard to believe that coevolution did not play an important role in some of the most important biological leaps in the explosion of Earth's biodiversity, there have been very few explicit tests of its putative role. Part of the reason for this is that coevolution is notoriously difficult to demonstrate, especially for events that took place millions of years ago or that involved the interactions of numerous species.

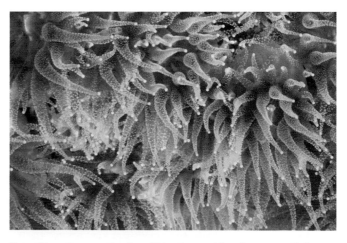

The colony of cup coral polyps (*Tubastrea* sp.) functions essentially as a single organism by sharing nutrients via a well-developed gastrovascular network, and are dependent upon unicellular algae called zooxanthellae. Polyps are clones, each having the same genetic structure. © Stuart Westmorland/Science Faction/Corbis.

As cheetahs (*Acinonyx jubatus*) evolve to become faster, it forces their prey to become more efficient runners and vice versa. © Winifried Wisnlewski/ Corbis.

What is coevolution?

To show that organisms have coevolved, scientists must demonstrate the process of reciprocal evolutionary change between interacting species, driven by natural selection (Thompson 2005). This definition of coevolution is a modification of Daniel H. Janzen's (1980) definition, which Janzen put forward to halt the injudicious use of the term, which had often been used to describe a host of relationships that had probably not coevolved at all. Prior to Janzen's definition, the word *coevolution* had been used to explain all manner of symbiotic relationships in which organisms were closely associated but for which there was no direct evidence for reciprocally driven evolutionary change. For example, Janzen questioned the use of the term when it was assumed that the traits of certain mammal-dispersed fruit had coevolved with a particular mammal's dietary requirements. He postulated that a mammal could enter a new habitat with its dietary preferences already established and would begin to feed on the fruits of plants that fulfilled those requirements. Seeds of those plants could evolve adaptations to the gut of the new herbivore, but unless the herbivore evolved adaptations to the new seeds, the relationship would not be coevolutionary.

The English naturalist Charles Darwin (1809–1882) was the first to envisage the reciprocity of coevolution, where changes in one species drive changes in another species, which in turn drive changes in the first species. He wrote about his ideas in *On the Origin of Species* (1859), where he envisaged "how a flower and a bee might slowly become, either simultaneously or one after the other, modified and adapted in the most perfect manner, by the continued preservation of individuals which presented slight deviations of structure mutually favorable to each other" (pp. 94–95). Darwin was referring to red clover flowers, which are normally visited by bumblebees with long proboscises, whose nectar was mostly inaccessible to honeybees with shorter tongues. He imagined the scenario of bumblebees becoming locally rare and changes in honeybee morphology, such as proboscis length, that would enable them to take advantage of the new resource by better matching the floral morphology of red clovers. But he also envisaged that floral morphology might also change to make honeybee pollination more efficient.

Darwin was a masterful natural historian who had an excellent understanding of how intricately adapted flowers influenced the evolutionary path of their pollinators and vice versa. This was best demonstrated in his first book after *Origin of Species*, a detailed account of how floral parts are sculpted by their pollinators over evolutionary time. He described how minute floral folds and convolutions are in fact adaptations to pollinator behavior and morphology. In particular he planted the seed of how the extremely long nectaries of flowers evolved and how these could often be the result of coevolution. For an example of this, he used the Madagascan star orchid (*Angraecum sesquipedale*), an orchid he had never seen in the wild.

This orchid is unusual because it has an absurdly long spur (in excess of 30 centimeters [12 inches]), at the bottom of which a few drops of nectar can be found. Orchid pollen is found in sacks called pollinaria, and these sacks have sticky attachments that adhere to various parts of their pollinator's anatomy when the pollinator contacts the flower's reproductive parts. Darwin postulated that only a large moth with its head pushed hard up against the flower, in its effort to drain the last drop of nectar, could remove the pollinaria. His example suggested that very long-tongued moths would not have to push their heads up against shorter-tubed flowers to get the nectar, and as a result the pollinaria of such flowers would not be removed and the moths would not deposit pollen on their stigmas. Thus, long-tongued moths exert a selective pressure on flowers to evolve corolla tubes or spurs that are longer than the tongues of the moths. But at the same time, moths that have short tongues

Comet orchid (*Angraecum sesquipedale*), also known as the Madagascar star or Darwin orchid. This species of orchid was discovered by Darwin during his journey on the HMS *Beagle*. Sinclair Stammers/Photo Researchers, Inc.

cannot reach all the nectar found at the base of very deep flowers. To access all the floral nectar, selective pressures on moths would favor tongues longer than the tubes of flowers visited. Darwin recognized that "there has been a race in gaining length between the nectary of *Angraecum* and the proboscis of certain moths" (1877, p. 116), where as one species evolves greater length, it forces the other species to evolve greater length and vice versa. This line of logic led Darwin to make his bold, and at the time frequently ridiculed, prediction that the Madagascan star orchid was pollinated by an enormous hawkmoth with a tongue that matched the orchid spur in length. Unfortunately he never lived to see his prediction tested because the hawkmoth was only found forty years later, long after his death.

Scientists have conducted several tests of the prediction that selection favors increased tube length in flowers pollinated by long-tongued insects, but perhaps the most famous is the experiment by L. Anders Nilsson (1988). Nilsson artificially shortened the spurs of some plants in a natural orchid population and found that plants with shortened spurs had fewer pollinaria removed and less pollen deposited on their stigmas. Although no conclusive tests have yet shown that long-tubed plants drive selection of long-tongued pollinators, coevolutionary arms races have been used to describe many other directional trends, such as the observation that brains of predators and their prey get larger over time, or that their bodies become increasingly adapted to increased speed over time. Even mollusk shells thicken over evolutionary time in response to evolutionary advances made by their predators. In a similar way to the nuclear arms races of the cold war, these so-called evolutionary arms races can produce an escalation in traits, but the fitness of one organism in relation to the other remains more or less the same.

The relationship between plants and insects is often not mutualistic. Every plant on the planet is probably consumed by insects, but many plants evolve mechanical (e.g., spines) or often chemical defenses to protect themselves from herbivores. To eat plants with chemical defenses, insects must evolve resistance to the plants' natural insecticides. In one of the most important works on coevolution, Paul R. Ehrlich and Peter H. Raven (1964) examined dietary patterns of butterfly larvae. They found that the insects' diets were restricted to a particular subset of often-unrelated plants, which had similar chemical defense mechanisms. They interpreted this pattern as evidence for the manner in which plant defenses and insect resistance coevolve. When an insect evolves a new defense against a particular plant chemical, all those plants possessing that chemical suddenly become available as a new food resource. This can open a host of new niches for the insect to exploit, allowing the insect to diversify and perhaps even speciate. Plants, at the same time, are continually evolving new chemicals to counter the evolution of insect resistance, which may have also resulted in diversification and speciation in plants.

The close associations of angiosperms with their pollinators and herbivores have led some biologists to postulate that diversification through coevolutionary relationships between plants and animals may have been responsible for the enormous species diversity in both of these groups. If these two groups played major roles in each other's speciation rates,

Morgan's sphinx moth (*Xanthopan morgani*) uses its long tongue specifically adapted to pollinate and feed at orchids, whose nectar is at bottom of 10 inch tube. © iStockPhoto.com/BobME.

they would have rapidly diversified at the same time. The fossil record suggests that major diversification started in the angiosperms during the Cretaceous. Conrad C. Labandeira and John J. Sepkoski Jr. (1993) calculated the proliferation of insect families from about 25 million years ago to the recent past and found that the number of insect families increased logarithmically with time, with no apparent acceleration in the Cretaceous. Although this can be taken as evidence against coevolutionarily induced radiations of these taxa, the methods used by Labandeira and Sepkoski have been severely criticized. Rapid radiations are often characterized by diversification at higher taxonomic levels, such as within genera. Thus looking for a diversification at higher taxonomic levels, such as insect families, may not be appropriate if insect diversity had radiated at lower taxonomic levels. Other tests of this hypothesis have examined whether insect-pollinated plant taxa are more diverse than plant taxa pollinated by abiotic means (e.g., wind). Michael E. Dodd, Jonathan Silvertown, and Mark W. Chase (1999) compared related branches of the angiosperm phylogeny, where one branch was insect pollinated but the other was abiotically pollinated. In contrast to that of Labandeira and Sepkoski, the results for Dodd and colleagues showed that insect-pollinated branches of the angiosperm phylogeny are more diverse than abiotically pollinated branches, suggesting that associations with insects

did increase angiosperm diversity. This is not necessarily coevolution, however, because, in return, scientists also need to test whether plant diversity played a role in the diversification of insect taxa. So the jury is still out on whether reciprocal associations between insects and plants were responsible for each other's diversity.

Although scientists are still unsure whether coevolution was responsible for reciprocal insect and plant radiations, many believe that coevolution is implicated in the evolution of sex itself. And sexual reproduction has been shown to speed up rates of evolution and adaptation, which may in turn have consequences for speciation rates and ultimately the diversity of life on Earth. Evolution and maintenance of sex is one of the thorniest biological debates in science, but one of the main contending hypotheses is the red queen hypothesis, which is based on host–parasite coevolution. Here, parasites evolve virulence toward the most common host genotype in a population. In the next generation, however, a previously rare genotype will be the most resistant to parasite virulence, and hosts with this genotype should increase in frequency until it becomes beneficial for parasites to evolve virulence toward them instead. Thus, the environment for both hosts and parasites is constantly changing every generation, so that parasites and hosts, like Lewis Carroll's Red Queen, may

The thorny acacia tree has evolved a defense against herbivory. © Chris Fourie. Image from BigStockPhoto.com.

By isolating both rabbit and virus strains before the virus was introduced (pre-introduction rabbits and pre-introduction viruses), scientists showed that rabbit recovery stemmed from both a reduction of virus virulence and an increase in host resistance. This was demonstrated after viruses taken from Australian rabbits, several years after the virus introduction, failed to kill as many pre-introduction rabbits as the pre-introduction viruses killed. Similarly the evolution of rabbit resistance was demonstrated by "evolved" rabbits having a lower mortality rate than pre-introduction rabbits after both were exposed to pre-introduction virus strains.

The geographic mosaic of coevolution

Despite the examples of coevolution given above, most coevolution probably does not involve just a single species pair. In any population, several species may exert selective forces upon one another. Over time, the composition and abundance of communities may fluctuate, and at the same time, the strength of selection imposed by each species will fluctuate as well. Similarly, a single species may interact with

A Yucca moth (*Tegeticula yuccasella*) pollinating Yucca flower (*Yucca torreyi*). © Michael & Patricia Fogden/Minden Pictures/Getty Images.

continually run a cyclical arms race instead of a directional one. Because sex allows recombination of genotypes, it can recreate genotypes that were lost in the past because they were so disadvantageous. These recreated genotypes may be needed in future bouts of coevolution to combat parasite virulence or alternatively to combat host resistance. The same cannot be said for asexually reproducing organisms, because when their disadvantageous genotypes are lost they may take hundreds of generations to rebuild through mutational processes. Strong evidence for the red queen hypothesis comes from Mark F. Dybdahl and Curtis M. Lively (1998), who used snails and their trematode parasites to show that parasites were adapting to the most common host genotype in a population and that, as a result, hosts underwent cycles of their genotype frequencies.

Hosts and their parasites provide some of the best examples of coevolution (not only red queen coevolution), such as the classic work on the myxoma virus and their rabbit hosts, which were introduced without their parasites to Australia where they rapidly became pests. After the virus was introduced, the rabbit population was nearly eliminated but then recovered.

A South Hills crossbill (*Loxia sinesciuris*) is foraging on a lodgepole pine cone. The crossbill laterally abducts its lower mandible to spread apart the cone scales and access the seeds. Thickening of the cone scales would impede crossbills and drive the evolution of larger more powerful bills. Courtesy of Craig W. Benkman.

several species across its geographic range so that the outcomes of interactions can vary geographically. Since the 1980s, these ideas have been carefully formulated in a series of books and papers, mostly written by John N. Thompson and colleagues (see Thompson 1994 and 2005, and the references therein). One of Thompson's model study systems is yucca plants and their close relatives, which are visited by moths that frequently pollinate as well as parasitize them. In a landmark paper, Thompson and Bradley M. Cunningham (2002) showed that in certain parts of the plant's range the relationship between plant and moth was mutualistic. In other parts of the range, however, the relationship was antagonistic, and in still others the relationship was commensalistic. The divergent outcomes of this relationship appear to be the result of species composition, as well as the presence, absence, and abundance of copollinators at each site.

Yet another example of how geography and community context can play a decisive role in structuring coevolutionary outcomes can be seen in the three-way relationship between crossbills, lodgepole pines, and red squirrels. In the central and northern Rocky Mountains, red squirrels (*Tamiasciurus hudsonicus*) are the main seed predators on lodgepole pines (*Pinus contorta*), and consequently these pines show many defensive character traits adapted primarily to squirrel predation (Benkman, Holimon, and Smith 2001). Because crossbills (*Loxia* spp.) exert comparatively little selective pressure on the pines here, pines show no adaptations to specifically prevent predation by crossbills. Crossbills have nevertheless evolved bill adaptations specifically to pry apart the scales of pinecones. In contrast, some areas peripheral to the Rocky Mountains have no red squirrels, and so the major seed predators in these systems are crossbills (Benkman et al. 2003). In these systems, the plants have adapted to crossbill predation by evolving thicker scales where most of the seeds are located. In turn, the crossbills here have evolved, through coevolutionary processes, even larger and more decurved bills to pry apart the thickened scales. Thus, geographically variable selective pressures have caused both matches and mismatches of phenotypically complementary traits in interacting organisms.

Using examples such as these, Thompson has constructed what he calls the geographic mosaic theory of coevolution. Here, the outcomes of interspecific interactions vary across the geographic landscape, in some places forming coevolutionary hot spots and in other places coevolutionary cold spots. Mediated by geneflow, trait remixing, and community context, this selection mosaic of coevolutionary hot spots and cold spots will pulse and shift across the landscape. These divergent and morphous patterns of selection should have enormous consequences for how biological communities evolve and function and ultimately on how Earth's biodiversity is organized. Only since the formulation of the geographic mosaic of coevolution hypothesis have scientists really begun to appreciate the pervasiveness of coevolution on everyday life and the ramifications it has on applied sciences such as biological control, agriculture, human epidemiology, and conservation.

Resources

Books

Darwin, Charles. 1859. *On the Origin of Species by Means of Natural Selection; or, The Preservation of Favoured Races in the Struggle for Life*. London: John Murray.

Darwin, Charles. 1877. *On the Various Contrivances by Which British and Foreign Orchids Are Fertilised by Insects*. 2nd edition. London: John Murray.

Herre, Edward Allen, Sunshine A. Van Bael, Zuleyka Maynard, et al. 2005. "Tropical Plants as Chimera: Some Implications of Foliar Endophytic Fungi for the Study of Host Plant Defense, Physiology, and Genetics." In *Biotic Interactions in the Tropics: Their Role in the Maintenance of Species Diversity*, ed. David F. R. P. Burslem, Michelle A. Pinard, and Sue E. Hartley. Cambridge, UK: Cambridge University Press.

Margulis, Lynn, and Dorion Sagan. 2002. *Acquiring Genomes: A Theory of the Origins of Species*. New York: Basic Books.

Thompson, John N. 1994. *The Coevolutionary Process*. Chicago: University of Chicago Press.

Thompson, John N. 2005. *The Geographic Mosaic of Coevolution*. Chicago: University of Chicago Press.

Periodicals

Benkman, Craig W., William C. Holimon, and Julie W. Smith. 2001. "The Influence of a Competitor on the Geographic Mosaic of Coevolution between Crossbills and Lodgepole Pine." *Evolution* 55(2): 282–294.

Benkman, Craig W., Thomas L. Parchman, Amanda Favis, and Adam M. Siepielski. 2003. "Reciprocal Selection Causes a

Coevolutionary Arms Race between Crossbills and Lodgepole Pine." *American Naturalist* 162(2): 182–194.

Dodd, Michael E., Jonathan Silvertown, and Mark W. Chase. 1999. "Phylogenetic Analysis of Trait Evolution and Species Diversity Variation among Angiosperm Families." *Evolution* 53(3): 732–744.

Dybdahl, Mark F., and Curtis M. Lively. 1998. "Host–Parasite Coevolution: Evidence for Rare Advantage and Time-Lagged Selection in a Natural Population." *Evolution* 52(4): 1057–1066.

Ehrlich, Paul R., and Peter H. Raven. 1964. "Butterflies and Plants: A Study in Coevolution." *Evolution* 18(4): 586–608.

Janzen, Daniel H. 1980. "When Is It Coevolution?" *Evolution* 34 (3): 611–612.

Labandeira, Conrad C., and John J. Sepkoski Jr. 1993. "Insect Diversity in the Fossil Record." *Science* 261(5119): 310–315.

Nilsson, L. Anders. 1988. "The Evolution of Flowers with Deep Corolla Tubes." *Nature* 334(6178): 147–149.

Thompson, John N., and Bradley M. Cunningham. 2002. "Geographic Structure and Dynamics of Coevolutionary Selection." *Nature* 417(6890): 735–738.

Bruce Anderson

Optimal reproductive tactics and life history theory

Life on Earth is diverse, showing many different modes of lifestyle. Life history theory takes this diversity as its focus and seeks to explain how and why so many different ways of making a living and procreating exist. The fundamental assumption of life history theory, and of evolutionary theory in general, is that life histories are constrained by trade-offs among traits; that is, devoting time or energy into one activity will detract from another activity. Thus, for example, on the one hand, attention paid to searching for food could detract from vigilance in spotting predators and hence lead to increased mortality. On the other hand, spending too much time watching out for predators will diminish food intake, which could reduce both survival and reproduction.

Life history theory encompasses many disciplines, including ecology, behavior, reproduction, population biology, and genetics. Because genetics and trade-offs are so central to life history theory, one necessarily inquires about the action of genes both from the point of view of their direct expression and also from a statistical perspective, which is the only way to describe how thousands of genes interact to produce the coordinated series of traits that comprise a life history. A study of trade-offs involves not only genetic studies but also physiological investigations of the immediate causal components of a trade-off. The importance of circumstance in determining the variation in phenotypes (phenotypic plasticity) leads one to investigate ecological circumstances. Thus, life history theory requires its practitioners to move from the minute (the gene) to the megascale (the ecosystem) in order to fully understand how the diversity of life histories has come about. R. A. Fisher anticipated such a need in 1930.

Darwinian fitness

The central components of life history theory are survival and reproduction, both of which contribute to Darwinian fitness, which is assumed will be maximized by natural selection. In the simplest case, Darwinian fitness can be defined as the number of offspring a female produces, or a male sires, that themselves survive to reproduce. This is the most frequently given definition, and while it is correct under some, if not most circumstances, it is not universally true because it does not take into account variation in generation times, which results in different rates of increase. Suppose there are two "types" of individuals in a hypothetical population, one reaching maturity in one day, producing

three offspring and then dying, and another that takes two days to reach maturity but produces five offspring and then dies (this type of reproduction, in which there is a single bout of reproduction, is known as a *semelparous* life history, whereas a life history that consists of multiple bouts of reproduction is termed *iteroparous*). Starting on day one with one of each type of organism in the hypothetical population, the following occurs: On the second day the first type reproduces, producing 3 offspring, which grow and produce in total 9 offspring on the third day, which then give rise to 27 offspring on the fourth day, which produce 81 offspring

Colored historical artwork of Thomas Robert Malthus, British economist and clergyman. SPL/Photo Researchers, Inc.

Elephants (*Loxodonta africana*) are an example of iteroparity. ©Chris Fourie. Image from BigStockPhoto.com.

on the fifth day. In contrast, over this five-day period the second type reproduces only twice, on days three and five, first producing 5 offspring, which then produce 25 offspring. Thus, although the second type produces more offspring per bout of reproduction it has a lower rate of increase in descendants than the first type. Both Charles Darwin and Fisher recognized the importance of considering the rate of increase of a type, rather than simply the number of surviving offspring. In his landmark 1930 book, *The Genetical Theory of Natural Selection*, Fisher called this rate the *Malthusian parameter*, in recognition of the importance that the English economist Thomas Robert Malthus (1766–1834) put on the "law of geometric increase." This name appears to have largely fallen into disuse, and the parameter is frequently called the *innate rate of increase* and is denoted with the symbol r_m or simply r, though, as discussed below, the latter can cause confusion.

There is a precise mathematical definition for r_m, represented by an equation called the *characteristic equation* or sometimes the *Euler equation*. The important components of the characteristic equation are the age-specific rates of survival and reproduction (more precisely, assuming an equal sex ratio, the number of female births per time increment). For any population, these two life history characteristics can be brought together in a life table, which gives at any age the number surviving to that age and the number of female offspring produced at that age. Life history theory is thus concerned with the life table and those factors that alter age-specific rates of survival and reproduction. At each age one can use the characteristic equation to compute the *reproductive value* of an individual, which is a measure of the extent to which an individual contributes to future generations. The innate capacity to increase is a measure of fitness provided that the population is in a stable age distribution (that is, although the population as a whole might be increasing [$r_m > 0$] or decreasing [$r_m < 0$], the proportional representation of each age class in the population remains constant) and that the population growth rate is independent of density. Under these conditions natural selection maximizes r_m and also maximizes reproductive value at each age relative to reproductive effort at that age (though not necessarily with respect to other ages, an important caveat that has led to some confusion in the literature—see Caswell 1982).

At first glance the assumption of a stable age distribution may seem to preclude r_m as a useful measure of fitness, because variations in vital rates are likely to be ubiquitous and hence populations will rarely be in a stable age distribution. However, populations disturbed from a stable age distribution

Parry agave (*Agave parryi*) is a form of semelparity. The plant dies after blooming because all of its resources are put into the stalk, flowers, and seeds. © Tom Bean/Alamy.

very quickly return to this distribution, and even in nonstable distributions r_m will in many cases still be a useful index of fitness even if it does not represent fitness itself.

When the population growth rate is density dependent one can still calculate a population rate of increase (or decrease), but it is then referred to simply as the *rate of increase*, symbolized by r. Density dependence undoubtedly occurs in populations, but its effects are frequently difficult to measure and their importance in shaping life history variation has been rather neglected. Because no population can increase without limit, in the long run the average rate of increase of a population must be zero or the population becomes extinct. Given this, how can one justify the use of r_m as a fitness measure? Simply put, if the factor that prevents the population from increasing indefinitely acts upon all types equally, then the Malthusian parameter of each type gives the rate at which each type increases in frequency within the population, and hence the relative fitness of each type.

The rationale for using r_m as a measure of fitness assumes that within a population there exists "types" and implicitly

these types remain distinct. For organisms, such as aphids, that produce clones (genetic copies of themselves) this assumption is biologically reasonable, although even in the case of clonal species, sexual reproduction generally occurs periodically. Sexual reproduction obviously mixes up genomes and hence mixes up "types." Thus, while the above example has heuristic merit, it actually provides insufficient grounds for taking r_m as a measure of fitness. The validity of using r_m as a measure of fitness was formally demonstrated by Brian Charlesworth (1994) who showed mathematically that a mutation that increased r_m would increase in frequency in the population.

If generation time does not vary among genotypes then the first measure of fitness, the lifetime reproductive success of a female (or male), is adequate, provided either that traits of interest are not themselves directly influenced by density-dependent factors or that such factors affect all genotypes equally. If density dependence affects genotypes differentially, then fitness is measured as the relative number of individuals that pass successfully through the stage during which density dependence acts. For example, suppose that density dependence occurs during a particular phase of development and that a trait whose evolution one is interested in predicting influences survival during this period. In this case, relative fitness is a function of survival in relation to the value of the trait and the density experienced during this critical period (the trait is not affected during other phases and does not itself affect relative fitness at other ages). A general method of assessing fitness in models with or without density dependence and with constant or variable environments is known as *invasibility analysis*.

Frequency-dependent selection

The above definitions of fitness are appropriate provided that selection is not frequency dependent. With frequency dependence, in which fitness is a function of the frequency of a particular genotype in the population, multiple phenotypes (a phenotype is the manifested expression of the genotype) may persist at equilibrium. A classic example of frequency-dependent selection is that of dimorphism in some male salmon and beetle species. In these species, two types of males occur, one equipped by virtue of its size or armaments (hook jaws in salmon and horns in beetles) to hold a territory and another that is smaller and/or less well armed that acts as a satellite to the territorial males and attempts to intercept females attracted to the territorial males.

In Atlantic salmon, territorial males measure about 1 meter (39 inches) in length, whereas satellite males, known as jacks, attain less than 30 centimeters (12 inches). Territorial males are more successful at obtaining mates than satellite males, but satellite males have the advantage of developing in less time (one year versus two years), and as a consequence both morphs persist in the population. A population of only territorial males is stable but, clearly, if territorial males became extinct the satellite strategy could not persist, and hence as the frequency of the satellite morph increases in the population its relative fitness at some point must decrease. The satellite strategy can also arise not because of any intrinsic advantage of the morph (such as reduced development time) but because environmental factors prevent a male from successfully attaining territorial status.

In Atlantic salmon (*Salmo salar*), territorial males measure about 1 meter (39 inches) in length, whereas satellite males, known as jacks, attain less than 30 centimeters (12 inches). Territorial males are more successful at obtaining mates than satellite males, but satellite males have the advantage of developing in less time (one year versus two years), and as a consequence both morphs persist in the population. Herve Berthoule/Jacana/Photo Researchers, Inc.

If, for example, food is limited, some males cannot grow to a size that allows them to successfully defend a territory, in which case adopting a satellite strategy may be making "the best of a bad situation." This phenomenon is a case of phenotypic plasticity, in which a genotype is able to express several different phenotypes depending on conditions experienced during development. Phenotypic plasticity can also be expressed at a behavioral level: In some species of frogs, the strategy a male frog adopts depends on the quality of its neighbor. If the neighbor has a deep croak, signifying that it is large and hence dominant, the neighbor will adopt a satellite strategy.

Trade-offs

As noted earlier, trade-offs are a central concept of life history theory. Trade-offs have been demonstrated on many occasions, an example of which is the inverse relationships between size and number of offspring. The demonstration of a trade-off is not by itself sufficient evidence that the trade-off is evolutionarily significant; that is, one cannot conclude from the simple observation of a trade-off that the trade-off will influence evolutionary trajectories. The reason for this is that evolutionary

changes in traits require a change in gene frequencies, which means that both the traits and their interaction must be inherited.

The most likely manner in which two traits are genetically coupled is that some of the genes that influence one trait also influence the second trait (a phenomenon known as *pleiotropy*). Thus selection acting on one trait will alter the gene frequencies of that trait and hence also, by virtue of the shared genes, the gene frequencies of the second trait. A simple example of pleiotropy is in the correlation of morphological components such as arms and legs. Trade-offs that affect life history components are likely to include allocation trade-offs in which a limited supply of resources must be divided among several traits that affect fitness. An excellent example is the trade-off between the number and size of offspring: In many organisms, particularly vertebrates, a female is constrained in the number and size of offspring, either because of geometrical constraints (a fixed space in the body cavity) or because the resources that can be supplied to the eggs or embryos are limited.

In some cases a trade-off is expected but not observed. The foregoing example of a trade-off between the size and number of offspring is a case in point. Although trade-offs are frequently observed, there are also cases in which no trade-off is discernable.

Female peacocks prefer to mate with males with elaborate plumage.
© iStockPhoto.com/Rainlady.

The general reason for a failure to observe a trade-off between two traits is that there is an unmeasured third trait that is interfering with the correlation. Females are likely to vary in the quantity and quality of resources (energy) that they can acquire, and females that can acquire a large quantity can invest in both many and large offspring. The consequence of this variation is that the expected trade-off is obscured: If resource acquisition is taken into account the trade-off becomes evident.

In other cases, a correlation between fitness-related traits is observed, but this does not result in any changes in the traits. In some European birds, for example, those individuals that breed early in the spring are the most successful, primarily because their offspring have an increased period of time in which to accumulate resources before the onset of inclement winter conditions. Over time, the onset of breeding should progress to an earlier date. But this is not the case. The explanation for this lack of an evolutionary response lies in the relationship between condition and onset of breeding. Only birds in good condition have the resources necessary to commence breeding early in the year. Yet, variation in condition among birds was shown to be only phenotypic; that is, birds in good condition just happened to be those that managed to acquire resources, and this acquisition, in this instance, did not have a genetic basis. Thus, although early onset of breeding was favored, there was not the underlying genetic variation allowing it to proceed through natural selection.

An interesting case in which trade-offs are important in reproductive strategies is that of sexual selection. Consider, for example, the peacock's tail: There is no doubt that such an ornament must impede movement and lead to increased mortality, relative to the female that both lacks the monster tail and the vibrant plumage of the male. Given that male plumage reduces survival one must wonder, as did Darwin, why selection favors its development and retention. The

answer is that females prefer to mate with such males, and hence these males produce more offspring than their less well-endowed congeners. Provided that the benefits (number of matings achieved) exceed the costs (reduced male survival), the elaborate ornamentation will increase in the population. Because of female preference for particular types of males there will develop a genetic correlation between female preference and the favored male traits.

Unlike trade-offs discussed thus far, the genetic correlation between female preference and male traits is not likely to be due to shared genes, but rather that genes that produce elaborate ornamentation and genes that generate female preference for them become statistically linked, a phenomenon called *linkage disequilibrium*. With this type of correlation the covariation between the traits is transitory in that it will disappear under random mating. This actually makes it difficult to work on under laboratory conditions, because stocks are generally not kept in such circumstances where mate selection can be expressed. In fact, this could be disastrous in some circumstances, such as the breeding of an endangered animal. Suppose, for example, a zoo were attempting to breed an antelope species in which females strongly preferred males with large horns (or that large horns gave the males an advantage in combat for territory and breeding rights): In the relatively close confines of the paddock and with a small number of males, only a single male may sire all the offspring. The result could be significant inbreeding and loss of vigor. From the point of view of genetic variation it would be preferable to arrange conditions such that multiple males sired offspring, even if in the long run it might affect female preference and horn length (worrying about the long run is meaningless if the species goes extinct in the short run). This is precisely why scientifically managed captive-breeding programs seek to equalize the genetic contributions of each founder animal in a population (Ballou and Foose 1996).

Although the introduction of organic pesticides and antibiotics was heralded as the beginning of an era free of many pest and health problems, resistance to pesticides, herbicides, rodenticides, and antibiotics develops rapidly. Because such resistance was not originally found at high frequency in the population, life history theory predicts that such resistance is likely to bear a cost and thus be maintained at particular frequencies as a consequence of a trade-off between resistance and reproductive success. This has been demonstrated in a number of insect pest species in which resistance comes about because of the massive overexpression of a detoxifying enzyme. Such production is metabolically expensive and reduces growth rate and fecundity. In the absence of the pesticide the faster growth and higher fecundity of the susceptible genotype is strongly favored, and within five to ten generations resistance frequently disappears or is reduced to a very low frequency. By careful manipulation of the spraying regime it is theoretically possible to strike a balance between the two extremes and while not eradicating the pest at least keeping it below a level at which significant economic damage will occur.

Resources

Books

Ballou, Jonathan D., and Thomas J. Foose. 1996. "Demographic and Genetic Management of Captive Populations." In *Wild Mammals in Captivity: Principles and Techniques*, ed. Devra G. Kleiman, Mary E. Allen, Katerina V. Thompson, and Susan Lumpkin. Chicago: University of Chicago Press.

Charlesworth, Brian. 1994. *Evolution in Age-Structured Populations*. 2nd edition. Cambridge, UK: Cambridge University Press.

Charnov, Eric L. 1993. *Life History Invariants: Some Explorations of Symmetry in Evolutionary Ecology*. Oxford: Oxford University Press.

Fisher, R. A. 1930. *The Genetical Theory of Natural Selection*. Oxford: Clarendon Press.

Fox, Charles W., Derek A. Roff, and Daphne J. Fairbairn, eds. 2001. *Evolutionary Ecology: Concepts and Case Studies*. Oxford: Oxford University Press.

Grant, B. Rosemary, and Peter R. Grant. 1989. *Evolutionary Dynamics of a Natural Population: The Large Cactus Finch of the Galápagos*. Chicago: University of Chicago Press.

Maynard Smith, John. 1982. *Evolution and the Theory of Games*. Cambridge, UK: Cambridge University Press.

Roff, Derek A. 1992. *The Evolution of Life Histories: Theory and Analysis*. New York: Chapman and Hall.

Roff, Derek A. 2002. *Life History Evolution*. Sunderland, MA: Sinauer Associates.

Stearns, Stephen C. 1992. *The Evolution of Life Histories*. Oxford: Oxford University Press.

Stearns, Stephen C., and Jacob C. Koella, eds. 2008. *Evolution in Health and Disease*. 2nd edition. Oxford: Oxford University Press.

Periodicals

Caswell, Hal. 1982. "Optimal Life Histories and the Maximization of Reproductive Value: A General Theorem for Complex Life Cycles." *Ecology* 63(5): 1218–1222.

Derek A. Roff

Evolutionary physiology

Natural populations of organisms are adapted to their habitats in a variety of ways. Plants and animals time their reproduction to coincide with the most ideal conditions, hibernate or become dormant when conditions become too cold or food becomes scarce, and have extra features and functions to cope with unique problems encountered in their habitat, such as glands that excrete excess salt and antifreeze that builds up in cells to prevent ice from damaging their structure. Because environmental conditions fluctuate on a daily, seasonal, and annual basis, individuals must often survive and reproduce across a wide range of conditions. Ecological physiologists are interested in understanding how this occurs—how the interaction between internal processes and the environment allows organisms to survive where they occur. This requires an understanding of mechanisms underlying a wide variety of physiological traits, such as resistance to climatic stresses including heat, cold, and aridity, as well as the development of hibernation and dormancy, which allows organisms to successfully evade the stressful conditions that can occur in extreme seasonal conditions.

Evolutionary physiologists are interested in how mechanisms and processes of physiological adaptation have evolved over time. As all organisms share a common ancestor, many similarities are expected between taxa both at the DNA and cellular level (Garland and Carter 1994). Despite this, species vary in physiological responses, and this may restrict them to a particular habitat. For example, two lizard species may appear very similar in appearance, but one may be found in a cool, moist rain forest and the other in a very hot and dry habitat because of a difference in their ability to tolerate or avoid desiccation stress. These differences have evolved through a process of adaptation. Evolutionary physiology aims to understand how physiological function evolves to generate differences between organisms and why particular lineages have successfully adapted to a wide range of habitats, whereas others have failed to evolve and as a consequence have become restricted in their distribution.

By combining an understanding obtained from general physiological studies with ideas of evolutionary biology, hypotheses about the evolution of physiological function can be tested. For instance, physiological differences among species might reflect phylogeny (historical patterns of physiological evolution) and/or the evolutionary processes influencing physiological processes (i.e., natural selection) (Feder, Bennett, and Huey 2000). The importance of these processes can be examined through comparative studies (i.e., comparing closely related species or populations that show differences in physiology) or laboratory-based manipulation experiments.

Comparative studies and the evolution of physiological traits

Comparative studies at the species level are particularly useful for understanding historical processes influencing physiological traits. Traditionally, inferences about evolution of traits were made on the basis of comparisons of two or a few species. If a species occupying one type of environment was found to have a particular value for a physiological trait while a related species from a contrasting environment had a different value for the same trait, trait values were assumed to have evolved to allow organisms to live in particular environments. These types of inferences are weak, however, because different trait values might have nothing to do with the environmental conditions being compared; just because a snail with a dark shell lives in a cold environment and a related species with a light shell lives in a warmer environment does not mean that shell color has evolved to allow adaptation to different climatic conditions. Instead, more powerful evolutionary inferences can be made by considering more than two species within a phylogenetic framework. The need to consider many species in any comparison is obvious—if ten species of dark snails and ten species of light snails show the above association, it increases confidence about an association between shell color and climatic conditions.

The importance of phylogenetic relationships among species is illustrated in Figure 1. On the left side of this figure, all ten light snails form one closely related group and all ten dark snails form the other closely related group. While shell color and environment are strongly related at the species level, this could have arisen through a single evolutionary shift producing light shells from a dark ancestor (or vice versa). Therefore any physiological association between environmental conditions and shell color might occur by chance because of a single event. However, the phylogenetic relationship on the right of Figure 1 shows a much more complex association between shell color and environmental conditions. In this case, shell color has evolved repeatedly within the phylogeny, and one can be much more confident that shell color represents an adaptation to climatic conditions because it has evolved repeatedly and independently under similar conditions.

Figure 1. * represents snail species from a cold climate; # from a warm climate. The image on the right depicts a scenario with a strong influence of phylogeny on shell color. In this instance, all snails that live in a cold climate have darker colored shells, but this may be a result of ancestral color and not an example of evolution to the environment. The image on the left however, shows a scenario of phylogenetically independent evolution of shell color in response to the environment. All snails that live in cold climates have darker shells regardless of the color of their ancestors. Reproduced by permission of Gale, a part of Cengage Learning.

Evolutionary inferences about species patterns are made through the application of phylogenetic contrast analysis (Felsenstein 1985) that allows common ancestry and species differentiation to be taken into account when testing for adaptive shifts. To achieve this, a phylogenetic tree (such as the ones in Figure 1) is created with branch lengths assigned based on the level of difference between the species. These branch lengths are then used to weight traits so that comparisons between species become independent of phylogeny and to determine with some certainty the ancestral form of the trait (Mongold, Bennett, and Lenski 1996). From such comparisons, one can gain an understanding of how a trait has changed throughout evolutionary history and postulate factors that may have influenced its evolution.

An example of this is Enrico L. Rezende, Francisco Bozinovic, and Theodore Garland Jr.'s 2004 study of climatic adaptation and basal metabolic rate (BMR) and maximum metabolic rate (MMR) in rodents. The authors were interested in whether specific environmental factors—and in particular mean annual maximum temperature (Tmax) and minimum temperature (Tmin), altitude, latitude, and/or diet (herbivore, granivore, etc.)—could explain diversification among rodents

for basal and maximum metabolic rates. When phylogeny was ignored, Tmax, Tmin, and latitude best explained variation in BMR between species. When phylogenetic effects were included, however, only latitude was correlated with BMR, suggesting that BMR had evolved independently from temperature and diet. In contrast, for MMR, Tmin explained differences among species, regardless of whether species' phylogenetic relationships had been considered. Therefore for BMR, associations with Tmax and Tmin were actually artifacts of phylogeny and not historical selection pressures, whereas for MMR, Tmin directly influenced the evolution of this trait. A combination of temperature and latitude might therefore explain differentiation in measures of metabolic rate between rodent species, with diet showing no particular influence.

Population comparisons and population selection

Species have separate evolutionary histories, often over millions of years, leading to divergence in numerous traits, and thus also making it difficult to use species comparisons to establish causal links between environmental conditions and the evolution of specific genes or traits. Because evolution occurs over such a long time frame, this often cannot be avoided. In some natural situations, however, one can examine evolution at the population level. Populations are much less diverged genetically than species and often continue to exchange genes, making it more likely that trait and gene differences between populations are adaptive.

Evolutionary changes in populations are evident in cases in which native predators have evolved rapidly in response to an introduced species population. An example of this involves the rapid evolution of the snake *Pseudechis porphyriacus* in response to the toxic South American cane toad (*Bufo marinus*). These toads were purposely introduced into northern Australia in 1935 to control insect pests in cane fields. As Australia has no native *Bufo* species, native predator populations could not tolerate toxins produced by glands in the toad's skin, and populations were severely affected by the introduction of this novel prey source.

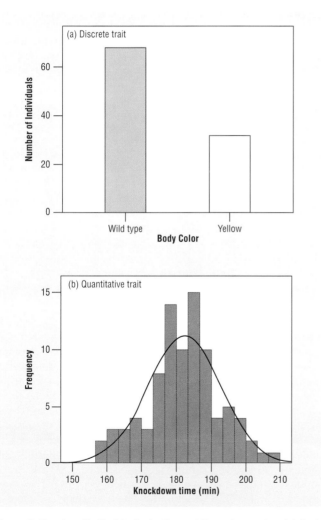

Figure 2. Two hypothetical traits in *Drosophila melanogaster*. (a) Body color, a discrete trait where individuals depict either a wild (normal, brown body) or mutant (yellow body) phenotype. (b) Heat resistance, measured as time to knockdown (i.e. inability to stand) after being subjected to a high temperature (38° C). This is a quantitative trait that varies in the population and follows a standard bell curve. Reproduced by permission of Gale, a part of Cengage Learning.

Ben L. Phillips and Richard Shine (2006) used comparisons between naive and exposed populations of the snake to determine if the strong selection pressure imposed by cane toads had caused an adaptive shift in toxin resistance. Snakes from naive and exposed populations were subjected to equal doses of toxin and assessed for locomotory ability (swimming speed) before and after injection. Snakes from naïve populations exhibited a far greater reduction in swimming speed (potentially associated with a loss of fitness) compared to those from exposed populations, indicating that in a matter of approximately twenty-three snake generations, *P. porphyriacus* had rapidly evolved increased toxin resistance to *B. marinus*.

Witnessing evolution in the field is rarely possible, but conditions necessary for selection and physiological evolution can be simulated in the laboratory. When organisms can be reared easily and quickly in the laboratory, many variable environmental conditions experienced in nature can be manipulated, allowing researchers to control minute details of the environment and monitor evolutionary responses to specific conditions. When organisms respond to stimuli above or below a set threshold, these individuals can be selected to propagate the next generation. Species with short generation times commonly used in such artificial selection experiments include bacteria (particularly *Escherichia coli*), mice, and the fruit fly *Drosophila melanogaster*. By "choosing" individuals with a particular phenotype to continue the next generation (i.e., selectively breeding), strong selection pressure can be enforced and trait changes can be observed. The intensity of selection induces a change in the mean phenotype of the population, and because humans are doing the selecting, this process is termed *artificial selection*. When populations are exposed repeatedly to conditions that impose selection, such as through having toxins in an environment that decrease survival and/or reproduction, this process is termed *experimental evolution*.

Selection experiments can allow researchers to test specific hypotheses about how natural selection is acting and to test the repeatability of the evolutionary process. For example, according to a 1998 study by John G. Swallow, Patrick A. Carter, and Theodore Garland Jr., voluntary running in an exercise wheel (wheel running) is a behavioral trait linked to aerobic fitness and can be selected for in house mice. Swallow and colleagues achieved this by providing individual mice with access to a running wheel for six days and measuring the amount of activity (total number of wheel revolutions) on the fifth and sixth days, that is, after allowing time for acclimation (see below). Individuals with the highest voluntary wheel running were then used to continue the next generation. After ten generations of selection, wheel-running activity increased 75 percent when compared to control mice that had been exposed but not selected. Thus, wheel running in mice has a genetic basis that can be readily selected and may be related to increased aerobic capacity.

Artificial selection and experimental evolution can be used to study evolutionary processes in any trait, but they have proved particularly useful for the study of evolution of physiological traits. The inherent complexity in these traits often makes it difficult to identify underlying mechanisms and genes, unless populations have diverged substantially and specifically for traits.

Quantitative genetics

Natural selection acts only on heritable traits and requires genetic variability (Falconer and Mackay 1996). Many genetic studies, such as the classic work on peas undertaken by the Austrian botanist Gregor Mendel (1822–1884), focus on understanding the genes controlling distinct phenotypes such as tall or short plants, pink or purple flowers, and so on (called *discrete traits*). Physiological traits, however, tend not to be discrete but instead tend to vary continuously between individuals and in response to the environment. When traits vary continuously, multiple genes act together to influence variation for a trait, making it difficult to isolate effects of individual genes. Such traits can be analyzed using quantitative genetics, instead of the more traditional Mendelian genetics in which discrete forms are recognized.

Heritability

The main trait parameters measured in quantitative genetics are the trait's heritability and its genetic correlation to other traits. The narrow-sense heritability (h^2) of a trait is an estimate of the amount of phenotypic variation within the trait that is genetically based so that selection can act upon, and it is provided as a proportion between 0 and 1. Essentially, it is an estimate of the evolutionary potential of a trait given a particular level of variation in a trait. By measuring a trait in multiple generations at the population level, heritability of physiological traits can be estimated (Falconer and Mackay 1996). While this can be achieved in many ways, a simple method is to regress values of the offspring against the mean value of their parents (see Figure 2). Other relationships among relatives can also be used. In these cases, the slope of the line for the population gives an indication of the similarity of generations, and therefore how much of the variation in a trait is passed on across generations because of heritable factors. The slope of the regression with a mid-parental value varies from 0 to 1, with a slope of 1 indicating that all phenotypic variation is passed on (see Figure 2). The scatter about the regression line can also be informative as it indicates how accurate the heritability has been estimated.

Estimates of heritability can also be obtained from selection experiments. In selection experiments, the aim is to shift the population mean of a trait in a particular direction. The extent to which the trait mean can be shifted depends on the strength of selection (*S*) (i.e., how far selected individuals deviate in phenotype from the population) and the trait's heritability (Falconer and Mackay 1996). The heritability of a trait can be calculated as a ratio of the amount of change between the mean values of parents and offspring (response to selection, *R*) and the strength of selection, that is, $h^2 = R/S$. This is termed the *realized heritability* because it is normally calculated from response to selection across several generations. Swallow and colleagues (1998) used this method to determine the heritability of wheel running in mice in the experiment described above. The realized heritability estimate calculated across ten generations for the increased running lines was $h^2 = 0.19$.

The ideal method for calculating heritability uses an animal model analysis and works with large numbers of individuals (e. g., Kruuk 2004). By setting up matings between individuals so that relationships of all offspring and their parents are known, this method compares traits between all known related individuals, such as father, mother, and siblings, rather than just parents to offspring. This analysis is thus described as an "animal model" because it is calculated at the individual or animal level. The animal model offers a more powerful analysis and therefore results in more accurate estimates of genetic parameters. This is particularly important when examining traits with low levels of heritability as is often the case for physiological traits. A trait with a low level of heritability will evolve more slowly than one with a high level of heritability as long as the traits exhibit a similar level of variation; a trait with a low level of variation will also show a low rate of evolution.

Current theory suggests that additive genetic variation (i.e., new genetic variation available for selection, V_A) is maintained within a population by the continuous accumulation of mutations within the genome balanced by stabilizing selection, which removes deleterious mutations. Based on this theory and assuming that a substantial number of genes contribute to variation in a trait, the availability of heritable variation in a trait is unlikely to limit evolutionary responses, but this does occur. Some traits might have very low levels of heritability/low V_A because selection on trait variation in one direction has been strong or because genes that control variation are no longer under selection.

A 2003 study showed limited heritability for desiccation resistance in rain-forest specialist *Drosophila* flies using realized heritability estimates (Hoffmann, Hallas, et al. 2003). This was later confirmed as being due to an absence of additive genetic variation in *D. birchii* and another rain-forest specialist, *D. bunnanda*, using the animal model (Kellermann et al. 2006). Heritability estimates are very important in understanding the prior strength of selection on these traits and for predicting a species' ability to change in the future. They can also provide insight into limits to species distributions (e.g., an absence of desiccation resistance restricting *D. birchii* and *D. bunnanda* to rain-forest environments).

Genetic correlation

As stated previously, variation in quantitative physiological traits may be quite complex because of multiple genes contributing to this variation. These genes might influence enzyme activity in a metabolic pathway, control the levels or effectiveness of hormones, act as regulatory elements on the expression of other genes, regulate the movement of ions and proteins across cell membranes, and so on. Because pathways and control mechanisms are typically interconnected and influence multiple outcomes, the same genes will often influence more than one trait (termed *pleiotropy*) (Falconer and Mackay 1996). If selection for one trait causes a change in allele frequencies at a locus with pleiotropic effects, then other traits will also be affected, either beneficially or adversely. The interdependence between traits in quantitative genetics because of gene action is termed the *genetic correlation* between traits. This can be quantified by examining the genetic covariance between traits (i.e., variation shared between traits) and partitioned in a way similar to the variance components of heritability (Falconer and Mackay 1996). This information can provide some evidence of the mechanisms underlying physiological traits; if two traits are genetically correlated, a mechanism influenced by underlying pleiotropic genes may be suggested. Genetic correlations can also influence responses to selection in a trait and even limit selection responses. For instance, increased resistance to aridity stress in an insect might be attained through preventing water loss via spiracles, favoring genes that regulate spiracle control. Yet, the same genetic changes that restrict water loss rate will also limit gaseous exchange, potentially limiting metabolism and growth. Such pleiotropic effects could set limits on the extent to which aridity resistance can be selected and evolve.

Repeatability

Estimates of narrow-sense heritability and other genetic parameters such as V_A are quite difficult to obtain, and they normally require cross-breeding under controlled conditions or information on relatedness of individuals from field populations. This is possible for organisms that can be reared in the laboratory or where tagged individuals are available in closed populations across multiple generations, but it is almost impossible to obtain genetic estimates for most free-ranging

species. Although physiological traits are complex and prone to environmental effects (such as circadian rhythm and seasonal rearing conditions influencing heat tolerance in *Drosophila*), some useful information about genetic variation in traits can nevertheless be obtained for studies of evolutionary physiology from the measurement of *repeatability*.

Whereas other estimates from quantitative genetics focus on partitioning the variance of traits into the additive, genetic, and phenotypic portions, repeatability involves taking repeated measures of the same trait from the same individual. This type of estimate can be obtained for almost any individual, and estimates of variation from within one individual can then be compared to variation between individuals to determine the upper limit of heritability. While this does not give an exact estimate for the heritable variation of a trait, it is much more practical to apply in many situations particularly when organisms have a long generation time. Nevertheless, using repeatability to estimate heritability involves certain assumptions. M. R. Dohm (2002) outlined situations in which repeatability may not accurately reflect narrow-sense heritability, including genotype *x* environment interactions (where environmental conditions can directly affect the genes being expressed) and maternal or paternal effects (i.e., where conditions experienced by either

Rainforest specialist, *Drosophila birchii*, found only in the Wet Tropics rainforest of northern Australia and Papua New Guinea. This species may be restricted to the moist rainforest due to lack of narrow sense heritability that limits its ability to improve tolerance of dry climates. Courtesy of Ary Hoffman.

parent may have flow-on effects to their offspring). These factors can be estimated only through complex breeding designs and ideally through animal models.

Phenotypic plasticity

Phenotypic plasticity is the extent to which the same genotype can express different phenotypes in response to environmental variation. Physiological traits tend to be particularly susceptible to environmental conditions; for instance, levels of resistance to thermal extremes are influenced by the conditions under which organisms are reared and held (through *acclimation*), and they are affected by rapid changes in thermal conditions over short periods (*hardening*) (Chown 2001). Short-term beneficial changes are particularly important in physiological traits as they may allow species to persist in suboptimal conditions and adjust to fluctuating environments without necessarily requiring genetic changes in a trait. For example, phenotypic plasticity allows tadpoles to speed up their development time when the pool they inhabit begins to dry up, therefore allowing a greater proportion of the cohort to continue on to the next generation. Such plastic effects influence phenotypes across generations as well as within generations. An example is seasonal variation in color exhibited by *Bicyclus* butterflies in Malawi (Brakefield and Reitsma 1991). These butterflies exhibit distinct wet and dry season color forms that switch from one generation to the next and are induced when butterflies are reared at particular temperatures and relative humidities.

Although phenotypic plasticity represents an environmental response, the extent to which traits are phenotypically plastic (i.e., the range of possible variation) is ultimately determined by genetic factors. For this reason, levels of plasticity in a trait are expected to be determined by selection; traits that exhibit high levels of plasticity are expected to show fitness benefits that derive from this high level of plasticity. The fitness benefits of acclimation have been tested specifically under the beneficial acclimation hypothesis (BAH) (Leroi, Bennett, and Lenski 1994). When organisms are raised in an environment and become acclimated to it, the BAH states that they should then have a relatively higher fitness in that environment compared to other environments. If the plastic acclimation response provides a fitness advantage in the environment in which it is induced, it is an adaptive response. Little evidence supports the BAH, however, and many hypotheses exist for why this may occur (Ghalambor et al. 2007). One quite simple explanation is that if all individuals within a habitat experience the same stressful event and the subsequent plastic change associated with it, the plastic response has no selective advantage. Plasticity can inhibit selection in certain situations by reducing the effects of selection on the genotype, but it can also facilitate selection by providing a ready source of variation for selection to act upon, assuming it has a (variable) genetic basis.

Phenotypic plasticity is an important component of evolutionary responses to changing environmental conditions that is not well understood mechanistically. Perhaps a greater understanding of the BAH will emerge once mechanisms and specific genetic limits to plastic responses emerge.

Combining approaches: Thermal tolerance in *Drosophila*

Since the mid-1980s, evolutionary analysis of thermal physiology has progressed significantly, most particularly in ectotherms (animals whose internal body temperature varies with external temperature), which are susceptible to temperature fluctuations (see Angilletta, Niewiarowski, and Navas 2002 for a review). Insects, and in particular *Drosophila* species, have become the organisms of choice for this work. *Drosophila* species differ markedly in *thermal sensitivity*, defined as the strength of a species' physiological or behavioral response to changing thermal conditions (Gilchrist 2000). They also differ in levels of *thermal tolerance* and *performance* (i.e., a species ability to survive and function in a particular environment).

Climatic conditions are usually warmer at the equator and steadily decrease toward the poles (i.e., with increasing latitude). This latitudinal thermal gradient can provide natural conditions in which to examine how species and populations adapt to variation in temperature. In insects, many studies have found that thermal resistance (in particular cold resistance) has evolved to change with thermal conditions and latitude. For instance, in the early twentieth century the

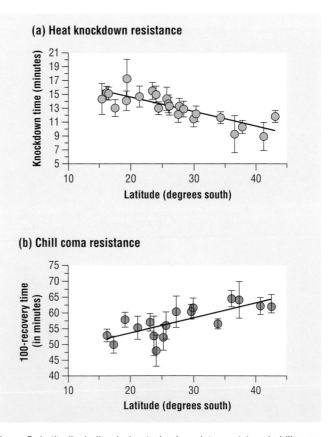

Figure 3. Latitudinal cline in heat shock resistance (a) and chill coma recovery (b) in *Drosophila melanogaster*. These graphs provide evidence of rapid local adaptation to the gradient in temperature that exists down the east coast of Australia, with more heat-resistant flies occurring in populations closer to the equator, and more cold-resistant flies in populations toward the South Pole. (Adapted from Figure 1 in Hoffmann, Anderson, and Hallas [2002].) Reproduced by permission of Gale, a part of Cengage Learning.

fruit fly *Drosophila melanogaster* was introduced into northern Australia. Since then, this generalist species has become widespread down the east coast of Australia, expanding its range to include a latitudinal gradient of 28 degrees (Hoffmann, Anderson, and Hallas 2002). Flies collected along this gradient differ in heat resistance (measured as knockdown time, i.e., time to physical incapacitation) and cold resistance (measured as the time it takes adult flies to recover from a comatose state after exposure to chilling). By comparing these traits across different populations, a strong opposing clinal trend was found in both heat and cold resistance (see Figure 3). Within a century of exposure to these conditions, flies have become locally adapted, reflecting relatively rapid physiological adaptation.

Laboratory selection methods have been used to investigate the selection potential of heat resistance in *Drosophila melanogaster* (Gilchrist and Huey 1999). Replicate populations were selected for increased heat resistance by breeding only the flies with heat resistance in the top 25 percent of the population. The control population exhibited a bimodal distribution of resistance to increasing temperatures, indicating the existence of individuals with relatively high or low heat resistance. In contrast, after twenty generations of selection, this bimodality was lost from both selection regimes. The change in the mean (see Figure 4) is indicative of the potential to evolve away from the current mean in this population of *D. melanogaster*. In the lines selected for increased resistance, however, this selection response appears to reach a plateau after approximately twenty generations. Once selection was halted, populations did not return to the original mean, suggesting that no adverse effects are associated with increased resistance that would otherwise favor genes with a lower tolerance to heat stress. This, together with the limit in selection for increased heat resistance, may indicate a limit to adaptation that cannot be attributed to decreased fitness associated with high resistance (i.e., these traits are not genetically correlated). Limits to adaptation have been found for other traits in *Drosophila*, as indicated by the lack of heritability for desiccation resistance in the rain-forest specialists *D. birchii* and *D. bunnanda* mentioned earlier (Hoffmann, Hallas, et al. 2003; Kellermann et al. 2006).

Drosophila studies have also identified one of the few regulatory mechanisms influencing the evolution of phenotypic plasticity. In *D. melanogaster*, pre-exposure (hardening) to high temperatures can provide a 40 percent increase in survival for females when exposed to 40°C (Krebs and Loeschcke 1994). The mechanisms inferring this response were discovered only by chance when a "puff" (expanded region of the chromosome; evidence of transcriptional up-regulation) was identified after exposure to heat. These puffs generated during the response could be induced within a few minutes after exposure to heat, and they receded with removal of heat stress. They were associated with the production of a new RNA product and subsequent proteins. This response is now known as the *heat shock response*. The

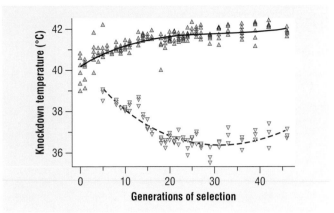

Figure 4. Change in mean knockdown temperature of up-selected (filled triangles) and down-selected (open triangles) lines of *Drosophila melanogaster* after each generation of selection. For up-selected lines, mean resistance after each generation of selection increases significantly up to generation twenty, where it plateaus, indicating a limit to the selection potential in this trait. Down-selected lines show a strong influence of selection through every generation of selection, with a slight increase toward the end, an indication of sufficient selection potential. (Adapted from Figure 3(d) in Gilchrist and Huey [1999].) Reproduced by permission of Gale, a part of Cengage Learning.

protein product produced during heat stress was hypothesized to inhibit regulation of other proteins during heat stress, reducing the influence of heat-induced protein denaturation and therefore the stress associated with heat shock. This was later confirmed by an RNA expression analysis that identified vast quantities of one particular RNA product (named Hsp70) in cells of *D. melanogaster* reared at 37°C, in stark contrast to the numerous others expressed at 25°C. Further work has uncovered a large family of genes responsive to heat shock conserved throughout the animal kingdom.

As has been emphasized in this entry, an understanding of the evolution of physiological traits can be obtained by combining a variety of approaches. In general, physiological traits are complex and likely to be controlled by multiple genes and influenced by many factors such as environmental conditions. Nevertheless, artificial selection experiments allow many environmental factors to be controlled and evolution of physiological traits to be observed within a laboratory setting. Comparisons among different species using phylogenetically independent contrasts as well as the examination of traits between populations along an environmental gradient can offer insights about how evolution shapes physiological traits. Quantitative genetic techniques provide an understanding of genetic interactions influencing physiological traits as well as limits to their evolution. Genes involved in certain traits have been identified, such as Hsp70 and the heat shock response in plastic changes that alter heat resistance. With the advancement of molecular techniques and the development of next-generation sequencing technology, many more genes affecting the evolution of physiological traits are likely to be identified.

Resources

Books

Falconer, D. S., and Trudy F. C. Mackay. 1996. *Introduction to Quantitative Genetics*. 4th edition. Harlow, UK: Longman.

Gilchrist, George W. 2000. "The Evolution of Thermal Sensitivity in Changing Environments." In *Environmental Stressors and Gene Responses*, ed. K. B. Storey and J. M. Storey. Amsterdam: Elsevier.

Mongold, Judith A., Albert F. Bennett, and Richard E. Lenski. 1996. "Experimental Investigations of Evolutionary Adaptation to Temperature." In *Animals and Temperature: Phenotypic and Evolutionary Adaptation*, ed. Ian A. Johnston and Albert F. Bennett. Cambridge, UK: Cambridge University Press.

Periodicals

Angilletta, Michael J., Jr., Peter H. Niewiarowski, and Carlos A. Navas. 2002. "The Evolution of Thermal Physiology in Ectotherms." *Journal of Thermal Biology* 27(4): 249–268.

Brakefield, Paul M., and Nico Reitsma. 1991. "Phenotypic Plasticity, Seasonal Climate, and the Population Biology of *Bicyclus* Butterflies (Satyridae) in Malawi." *Ecological Entomology* 16(3): 291–303.

Chown, Steven L. 2001. "Physiological Variation in Insects: Hierarchical Levels and Implications." *Journal of Insect Physiology* 47(7): 649–660.

Dohm, M. R. 2002. "Repeatability Estimates Do Not Always Set an Upper Limit to Heritability." *Functional Ecology* 16(2): 273–280.

Feder, Martin E., Albert F. Bennett, and Raymond B. Huey. 2000. "Evolutionary Physiology." *Annual Review of Ecology and Systematics* 31: 315–341.

Felsenstein, Joseph. 1985. "Phylogenies and the Comparative Method." *American Naturalist* 125(1): 1–15.

Garland, Theodore, Jr., and Patrick A. Carter. 1994. "Evolutionary Physiology." *Annual Review of Physiology* 56: 579–621.

Ghalambor, C. K., J. K. McKay, S. P. Carroll, and D. N. Reznick. 2007. "Adaptive versus Non-adaptive Phenotypic Plasticity and the Potential for Contemporary Adaptation in New Environments." *Functional Ecology* 21(3): 394–407.

Gilchrist, George W., and Raymond B. Huey. 1999. "The Direct Response of *Drosophila melanogaster* to Selection on Knockdown Temperature." *Heredity* 83(1): 15–29.

Hoffmann, Ary A., Alisha Anderson, and Rebecca Hallas. 2002. "Opposing Clines for High and Low Temperature Resistance in *Drosophila melanogaster*." *Ecology Letters* 5(5): 614–618.

Hoffmann, Ary A., Rebecca J. Hallas, J. A. Dean, and Michele Schiffer. 2003. "Low Potential for Climatic Stress Adaptation in a Rainforest *Drosophila* Species." *Science* 301(5629): 100–102.

Hoffmann, Ary A., Jesper G. Sørensen, and Volker Loeschcke. 2003. "Adaptation of *Drosophila* to Temperature Extremes: Bringing Together Quantitative and Molecular Approaches." *Journal of Thermal Biology* 28(3): 175–216.

Kellermann, Vanessa M., Belinda Van Heerwaarden, Ary A. Hoffmann, and Carla M. Sgro. 2006. "Very Low Additive Genetic Variance and Evolutionary Potential in Multiple Populations of Two Rainforest *Drosophila* Species." *Evolution* 60(5): 1104–1108.

Krebs, Robert A., and Volker Loeschcke. 1994. "Effects of Exposure to Short-Term Heat Stress on Fitness Components in *Drosophila melanogaster*." *Journal of Evolutionary Biology* 7(1): 39–49.

Kruuk, Loeske E. B. 2004. "Estimating Genetic Parameters in Natural Populations Using the 'Animal Model.'" *Philosophical Transactions of the Royal Society* B 359(1446): 873–890.

Leroi, Armand M., Albert F. Bennett, and Richard E. Lenski. 1994. "Temperature Acclimation and Competitive Fitness: An Experimental Test of the Beneficial Acclimation Assumption." *Proceedings of the National Academy of Sciences of the United States of America* 91(5): 1917–1921.

Lindquist, Susan. 1986. "The Heat-Shock Response." *Annual Review of Biochemistry* 55: 1151–1191.

Phillips, Ben L., and Richard Shine. 2006. "An Invasive Species Induces Rapid Adaptive Change in a Native Predator: Cane Toads and Black Snakes in Australia." *Proceedings of the Royal Society* B 273(1593): 1545–1550.

Rezende, Enrico L., Francisco Bozinovic, and Theodore Garland Jr. 2004. "Climatic Adaptation and the Evolution of Basal and Maximum Rates of Metabolism in Rodents." *Evolution* 58(6): 1361–1374.

Swallow, John G., Patrick A. Carter, and Theodore Garland Jr. 1998. "Artificial Selection for Increased Wheel-Running Behavior in House Mice." *Behavior Genetics* 28(3): 227–237.

Other

Maddison, Wayne P., and David R. Maddison. 2009. "Mesquite: A Modular System for Evolutionary Analysis." Version 2.6. Available from http://mesquiteproject.org.

Katherine A. Mitchell
Ary A. Hoffmann

Industrial melanism in moths

The evolution of dark, or melanic, forms of insects and other animals, in industrial regions of Britain and elsewhere, is known as industrial melanism. Industrial melanism in the peppered moth, *Biston betularia*, is one of the most quoted examples of an evolutionary change that has occurred through the process of natural selection. The reasons for this are that the changes are visually obvious, that the factors that caused changes in the frequencies of the melanic and nonmelanic forms are easy to understand and supported by a wealth of experimental evidence, and that the changes in the frequencies of the forms have been relatively recent, and continue to this day. Moreover, the peppered moth case is of additional interest because it has engendered considerable controversy since the late 1990s.

The rise of the black peppered moth

The peppered moth is typically a white moth, liberally speckled with black scales. It flies at night, resting by day on the bark of trees, where it relies on crypsis (camouflage) for defense against predators. In 1848 an almost completely black form of this species was recorded in Manchester. This form spread and rapidly increased in frequency in industrial parts of Britain, but not elsewhere. By 1895, 98 percent of the peppered moths in Manchester were of the black, *carbonaria* form. In 1896 the Victorian lepidopterist James William Tutt put forward a hypothesis to explain the rapid increase in *carbonaria* frequency in industrial regions. He proposed that two pollutants, sulfur dioxide, which kills lichens, and airborne soot particles, which blackened lichen-denuded surfaces, had substantially changed the environment in industrial regions of Britain. On such surfaces, the black form of the peppered moth, and indeed melanic forms of other moth species, would be better camouflaged than the pale *typica* form, which would thus be selected against by predators, such as birds, that hunt by sight. This hypothesis was not accepted initially because ornithologists and entomologists of the time did not believe that birds were major predators of cryptic moths.

In the early twentieth century, breeding experiments showed that the *carbonaria* form of the peppered moth resulted from a mutation in a single gene, and that the *carbonaria* allele was genetically dominant to the *typica* allele. In 1924 J. B. S. Haldane calculated the selection coefficient required to explain the increase in the *carbonaria* form in Manchester in the second half of the nineteenth century, concluding that *carbonaria* would have had to be 1.5 times as fit as (reproduce successfully 1.5 times more often than) the *typica* form to explain its increase. This showed that the differences in fitness between forms could be considerable.

The lack of belief in Tutt's hypothesis led to a number of other explanations being suggested to explain the rise of melanic forms of moths in industrial areas. These included that

Peppered moth (*Biston betularia*) showing three forms on lichen. © Andrew Darrington/Alamy.

A melanic version of the peppered moth (*Biston betularia*). © Steve McBill. Image from BigStockPhoto.com.

some pollutants were mutagenic agents, that the nonmelanic/ melanic heterozygotes were fitter than homozygous forms, or that melanic forms had greater physiological hardiness, particularly in adverse environmental conditions. None of these hypotheses was supported by empirical evidence.

Bernard Kettlewell's experiments

In the 1950s Bernard Kettlewell, at Oxford University, began work to test Tutt's differential bird predation hypothesis, using the peppered moth as subject material. There were three major components to Kettlewell's work: field predation experiments, mark-release-recapture experiments, and survey work to assess whether *carbonaria* frequencies were correlated to pollution levels.

In one set of field predation experiments, Kettlewell released three live *carbonaria* and three live *typica* moths onto tree trunks, and watched them from a bird hide. Once three of one form had been taken, the set of six was restored. This procedure was first carried out in a highly polluted woodland in Birmingham in 1953, where the trees and lichens had been discolored. The experiments were repeated in the same woodland and an unpolluted woodland in Dorset in 1955 where the bark and lichens remained unchanged. A variety of species of bird were observed to find and eat the peppered moths that were released. The results showed that the *typica* form was predated more often than *carbonaria* in the polluted woodland, the reverse being the case in the unpolluted wood (reviewed in Kettlewell 1973).

In the mark-release-recapture experiments, Kettlewell set out a network of moth traps in the two woodlands. He released peppered moths of both forms that he had marked with a spot of paint on the underside of the wings (where the spot would be hidden when the moth was at rest). Any marked moths subsequently recaptured in his traps were recorded. He then calculated the recapture rates of the *carbonaria* and *typica* forms for each wood. He found that the recapture rate for *carbonaria*

was approximately twice that of *typica* in the polluted wood, in both 1953 and 1955, the reverse being the case for the unpolluted wood in 1955 (reviewed in Kettlewell 1973).

To determine whether the frequency of *carbonaria* was correlated to sulfur dioxide and soot pollution levels, Kettlewell persuaded lepidopterists across Britain to record the frequencies of the forms of the peppered moth that they encountered in moth traps. The results showed that the frequency of *carbonaria* was positively correlated to both pollutants, with highest *carbonaria* frequencies being recorded in highly industrial urban areas and regions downwind of such localities.

Kettlewell's results, therefore, supported Tutt's differential bird predation hypothesis. It was the reciprocal nature of Kettlewell's results from the two woodlands, allied to the correlation between *carbonaria* frequencies and pollution, that made the case so persuasive and easy to understand. Here then was Charles Darwin's missing evidence: an observed change in a conspicuous inherited trait of an organism over time, together with the agent of selective change. As Robert H. MacArthur and Joseph H. Connell (1966) wrote, "It used to be argued that natural selection was only a conjecture, because it had not actually been witnessed". But in the peppered moth, the change and the agent of selection were witnessed by Kettlewell, and, moreover, with the help of the well-known ethologist Niko Tinbergen, some of the experiments were filmed so that others could also see natural selection in action.

The post-Kettlewell period

In the ensuing years, many other researchers, in Britain and elsewhere, collected additional evidence in support of Tutt's hypothesis. For example, the ecological geneticist J. A. Bishop (1972) collected evidence along a transect from Liverpool into rural north Wales showing that the frequency of *carbonaria* was inversely related to the number of lichen species growing on trees. Moreover, mapping first records of *carbonaria* across Britain produced data that was consistent with all the *carbonaria* in Britain arising from a single mutation event.

During this period, many details of the ecology and behavior of the peppered moth were uncovered. Work in Finland, using large cages, showed that peppered moths usually rest on the underside of lateral branches, rather than on tree trunks, as Kettlewell had previously suggested. This view was endorsed by similar experiments by Tony G. Liebert and Paul M. Brakefield (1987). By detailed observation, Liebert and Brakefield further found that the dispersal of the two sexes of peppered moth was very different. The adult moths hatch from subterranean pupae in the late afternoon. Once their wings have expanded, females climb a tree trunk and "call" to males using scent pheromones. Most females mate on the evening that they hatch. The pair will stay *in copula* for the rest of the night and through the next day. Once parted the following evening, males fly to seek other females. Females also fly from the pairing site in a "dispersal flight" (Liebert and Brakefield 1987), and once they land on an appropriate tree, they commence laying eggs. On subsequent nights, females do not fly again, but crawl around the tree they are on, laying batches of eggs on several different branches.

Other studies highlighted areas that needed to be taken into consideration if a full understanding of the case was to be gained. For example, a Cambridge University group showed that the apparent levels of crypsis of the forms of the moth on bark or lichen backgrounds, as judged by humans, differs from those apparent to birds (Majerus, Brunton, and Stalker 2000). This is because most bird species have an additional type of cone cell in their eyes that is sensitive to the ultraviolet spectrum, to which humans are almost blind.

The work on the peppered moth in the four decades following Kettlewell's classical experiments in general endorsed Tutt's differential bird predation hypothesis and added considerable detail to the story. Although some empirical studies questioned details of the "textbook" story of industrial melanism in the peppered moth, none seriously undermined the qualitative accuracy of the story.

Less detailed work has been conducted on some of the many other species of moth that exhibit industrial melanism. These researches include studies of full industrial melanic polymorphisms, in which melanic forms became established only after the Industrial Revolution, and partial industrial

Peppered moths (*Biston betularia*). Black and speckled forms on soot-blackened tree trunk. Michael Willmer Forbes Tweedie/Photo Researchers, Inc.

melanic polymorphisms, in which melanic forms that were known prior to the Industrial Revolution increased subsequently in polluted parts of Britain. Examples of the first category include *Stauropus fagi*, *Tethea octogesima*, and *Gonodontis bidentata*. Species showing partial industrial melanic polymorphism include *Apocheima pilosaria*, *Allophyes oxyacanthae*, and *Apamea monoglypha*. In general, research on other species that exhibit industrial melanism supports the major influence of selective predation by birds on these polymorphisms (reviewed in Majerus 1998). In addition, comparative analysis of dispersal rates and spatial changes in the frequencies of three species, *B. betularia*, *Apocheima pilosaria*, and *G. bidentata*, have shown the important influence of migration on melanic polymorphism. *B. betularia* is the most mobile of these species, whereas *G. bidentata* is the most sedentary. Theoretically, the lower the rate of migration, the more the frequencies of forms will be the product of adaptation to local conditions. This leads to the prediction that spatial changes in melanic and nonmelanic form frequencies of *G. bidentata* will be more abrupt than those in *Apocheima pilosaria*, which will in turn be more abrupt than those in *B. betularia*. Field surveys have verified this prediction.

The decline of the black moth

Starting around 1960, a further environmental change led the case of the peppered moth as an example of Darwinian evolution in action to be strengthened further. In the 1950s and subsequently, enactment of antipollution laws in Britain brought about significant reductions in both sulfur dioxide and soot pollution. On the basis of Tutt's hypothesis, the cleaning up of the environment should, after a lag period, result in reduced pollution on tree bark, the regrowth of lichens on trees, and a reduction in the frequency of *carbonaria*. A number of surveys have shown that both of these predictions have been fulfilled (e.g., Mani and Majerus 1993; Grant, Owen, and Clarke 1996). Similar declines in industrial melanic form frequencies have been reported in many other species (reviewed in Majerus 1998). Importantly, such declines have been reported not only from Britain but also from Holland and the United States. In essence, these independent declines represent replicate natural experiments. Moreover, they show that evolution is not a one-way process but can be reversed if the direction of selection changes.

The peppered moth story comes under attack

The Cambridge University evolutionary geneticist Michael E. N. Majerus (1998), in reviewing the peppered moth case, concluded that the cause of "the rise and fall of the melanic peppered moth" was "differential bird predation in more or less polluted regions, together with migration... almost to the exclusion of other factors". Despite this unequivocal conclusion, Jerry A. Coyne (1998), from the University of Chicago, argued that Majerus's review revealed serious flaws in the story, writing that the book shows the peppered moth story "to be in bad shape" and that "for the time being we must discard *Biston* as a well-understood example of natural selection in action." Coyne's review was itself criticized by an American scientific commentator, Donald Frack (1999), who wrote, "There is essentially no resemblance between Majerus' book and Coyne's review of it.... If I hadn't known differently, I would have thought that the review was of some other book." Despite Frack's view, many creationists and other antievolution lobbyists seized upon Coyne's review. Articles with such titles as "Scientists Pick Holes in Darwin Moth Theory," Darwinism in a Flutter," "Moth-Eaten Statistics," and "The Piltdown Moth" appeared in newspapers, on creationist Web sites, and in books. Moreover, in 2002 Judith Hooper, a journalist, published the book *Of Moths and Men: An Evolutionary Tale; Intrigue, Tragedy, and the Peppered Moth*, in which she all but accuses Kettlewell of scientific fraud asking, "who arranges the scientific data to arrive at the desired result?" Notably, none of these criticisms are published in peer-reviewed scientific journals, and none of their authors have ever worked on the peppered moth in the field.

The peppered moth resurrected

The issues raised by these critics of the peppered moth story have been answered in a number of papers in peer-reviewed journals (e.g., reviewed in Majerus 2005). These authors based their defense of the peppered moth story on previously published work. They note that the critics of the story cite evidence selectively and focus strongly on Kettlewell's predation experiments, ignoring the numerous independent predation experiments conducted subsequently, each of which also supports Tutt's differential predation hypothesis (reviewed by Cook 2003). In addition, they entirely ignore the many other species of moth that exhibit industrial melanism in Britain and elsewhere (reviewed by Kettlewell 1973; Lees 1981; Majerus 1998). Moreover, the American scientific historian David Wÿss Rudge (2005), in analyzing Hooper's evidence that Kettlewell was guilty of scientific fraud, concluded, "that Hooper (2002) does not provide one shred of evidence to support this serious allegation."

Majerus took a different approach. Rather than reappraise previously published data, he embarked on a suite of experiments to address some of the problems in the story. This work was designed to determine whether bird predation is sufficient to explain observed changes in peppered moth form frequencies over a number of years, to discover precisely where peppered moths do spend the day in the field, and to consider the effect of bat predation on peppered moths (an issue raised by Hooper 2002). The results of the main predation experiment, which was designed to avoid the problems identified in Kettlewell's methodologies (Majerus 2005), showed that over a six-year period, bird predation of *carbonaria* was greater than that of *typica* to a degree sufficient to explain the 11 percent decline in the frequency of *carbonaria* over this period (Majerus 2007). In addition, he showed that bats do not prey on *carbonaria* and *typica* to different extents. Finally, a data set of the resting positions of 135 peppered moths found in the wild revealed that the majority (50.4 percent) rest on lateral branches, with smaller proportions resting on vertical trunks (37 percent) and foliate twigs (12.6 percent) (Majerus 2007). These results fully support Tutt's differential bird predation hypothesis (1896), and they endorse Kettlewell's view (1973) of the factors involved in the evolution of industrial melanism in the peppered moth.

Conclusion

After first envisaging his hypothesis of evolution by natural selection, Darwin spent two decades collecting evidence before publishing *On the Origin of Species* (1859). But he missed the peppered moth. It was during this period that *carbonaria* first appeared in favorable selective circumstances and began to increase in frequency and spread. Had the case of this melanic moth been drawn to Darwin's attention, one wonders what Darwin's reaction would have been. Possibly, he would have simply ignored it, because he did not view polymorphic species as showing signs of selection. Indeed, in defining natural selection, he specifically dismisses polymorphisms, writing, "This preservation of favourable variations, and rejection of injurious variations I call natural selection. Variations neither useful nor injurious would not be affected by selection and would be left a fluctuating element, as perhaps we see in the species called polymorphic." Alternatively, he may have latched onto the case and found in the peppered moth evidence for observable natural selection in action.

Industrial melanisms in moths, such as the peppered moth, are the most cited examples of Darwinian selection in action. Despite the peppered moth story's somewhat checkered career, it has stood the tests of both time and criticism, and because of its visual impact and the ease of understanding of the selective agent responsible for the evolutionary changes, it remains the clearest and most accessible teaching example of Darwin's central mechanism of evolution.

Now, 150 years after the publication of *On the Origin of Species*, the melanic peppered moth is fading. Analysis of the rate of decline in the frequency of the forms of the peppered moth have led to the prediction that by 2019, *carbonaria* will not comprise more than 1 percent of any population in Britain (Mani and Majerus 1993). More recent data suggests that this prediction will be fulfilled. If, as seems likely, the melanic peppered moth does disappear from the country in which it has been most heavily researched, this will not be the end of the story of industrial melanism in moths. Work on industrial melanism will simply need to move to parts of the world where industrial pollution is less well controlled than in

Two forms of the peppered moth (*Biston betularia*) occur, a pale form (seen here) and a dark form. John Devries/Photo Researchers, Inc.

Europe and North America. Moreover, the case of the black peppered moth will doubtless still engender attack from the antievolution lobby, yet will remain, as the great American evolutionary geneticist Sewall Wright (1978) described it, "the clearest case in which a conspicuous evolutionary process has been actually observed."

Resources

Books

Darwin, Charles. 1859. *On the Origin of Species by Means of Natural Selection; or, The Preservation of Favoured Races in the Struggle for Life*. London: John Murray.

Hooper, Judith. 2002. *Of Moths and Men: An Evolutionary Tale; Intrigue, Tragedy, and the Peppered Moth*. London: Fourth Estate.

Kettlewell, Bernard. 1973. *The Evolution of Melanism*. Oxford: Clarendon Press.

Lees, David R. 1981. "Industrial Melanism: Genetic Adaptation of Animals to Air Pollution." In *Genetic Consequences of Man Made Change*, ed. J. A. Bishop and L. M. Cook. London: Academic Press.

MacArthur, Robert H., and Joseph H. Connell. 1966. *The Biology of Populations*. New York: Wiley.

Majerus, Michael E. N. 1998. *Melanism: Evolution in Action*. Oxford: Oxford University Press.

Majerus, Michael E. N. 2002. *Moths*. London: HarperCollins.

Majerus, Michael E. N. 2005. "The Peppered Moth: Decline of a Darwinian Disciple." In *Insect Evolutionary Ecology: Proceedings of the Royal Entomological Society's 22nd Symposium*, ed. M. D. E. Fellowes, G. J. Holloway, and J. Rolff. Wallingford, UK: CABI Publishing.

Tutt, James William. 1896. *British Moths*. London: George Routledge.

Wright, Sewall. 1978. *Evolution and the Genetics of Population*, Vol. 4: *Variability within and among Natural Populations*. Chicago: University of Chicago Press.

Periodicals

Bishop, J. A. 1972. "An Experimental Study of the Cline of Industrial Melanism in *Biston betularia* (L.) (Lepidoptera) between Urban Liverpool and Rural North Wales." *Journal of Animal Ecology* 41(1): 209–243.

Cook, Laurence M. 2003. "The Rise and Fall of the *Carbonaria* Form of the Peppered Moth." *Quarterly Review of Biology* 78 (4): 399–417.

Coyne, Jerry A. 1998. "Not Black and White." *Nature* 396(6706): 35–36.

Grant, B. S., D. F. Owen, and C. A. Clarke. 1996. "Parallel Rise and Fall of Melanic Peppered Moths in America and Britain." *Journal of Heredity* 87(5): 351–357.

Haldane, J. B. S. 1924. "A Mathematical Theory of Natural and Artificial Selection." *Transactions of the Cambridge Philosophical Society* 23: 19–41.

Liebert, Tony G., and Paul M. Brakefield. 1987. "Behavioural Studies on the Peppered Moth *Biston betularia* and a Discussion of the Role of Pollution and Lichens in Industrial Melanism." *Biological Journal of the Linnean Society* 31(2): 129–150.

Majerus, Michael E. N., Clair F. A. Brunton, and James Stalker. 2000. "A Bird's Eye View of the Peppered Moth." *Journal of Evolutionary Biology* 13(2): 155–159.

Mani, G. S., and Michael E. N. Majerus. 1993. "Peppered Moth Revisited: Analysis of Recent Decreases in Melanic Frequency and Predictions for the Future." *Biological Journal of the Linnaean Society* 48(2): 157–165.

Rudge, David Wÿss. 2005. "Did Kettlewell Commit Fraud? Re-examining the Evidence." *Public Understanding of Science* 14(3): 249–268.

Other

Frack, Donald. 1999. "Peppered Moths—in Black and White." E-mail to Anticreation mailing list, March 30, 1999.

Majerus, Michael E. N. 2007. "The Peppered Moth: The Proof of Darwinian Evolution." Available from http://www.gen.cam.ac.uk/Research/Majerus/Swedentalk220807.pdf.

Michael E. N. Majerus

Galápagos finches

Several adaptive radiations have occurred in the isolated Galápagos Archipelago in the Pacific Ocean off the coast of Ecuador. Perhaps the most famous occurred with the island's endemic finches, which were discovered by the English naturalist Charles Darwin (1809–1882) and his shipmates in 1835, during the voyage of the HMS *Beagle*. Darwin's finches have since become one of the most noteworthy examples of natural selection and evolution. They continue to challenge scientific notions about how species are formed and classified, just as they offer clues to novel processes of species formation and the underlying basis of traits that are shaped by natural selection.

Basic description

The Galápagos finches, also known as Darwin's finches, comprise thirteen classically recognized species that can be found on all major islands and many minor islands of the archipelago. The Galápagos are approximately 900 kilometers (560 miles) west of the coast of Ecuador. Darwin's finches now include a recently discovered species of warbler finch and the Cocos finch, which occupies isolated Cocos Island, about 500 kilometers (310 miles) off the western coast of Costa Rica. Galápagos finches are traditionally divided into three main groups: six species of ground finches, six species of tree finches, and the warbler and Cocos finches.

Large ground finch (*Geospiza magnirostris*) feeding. Image by Mark Moffett.

The small, medium, and large ground finches differ mainly in body size, beak size, and the size of seeds they eat. The small and large cactus finches specialize on the seeds, flowers, and fruit of the endemic prickly pear (*Opuntia*) cactus. Strong grasping beaks allow the small, medium, and large tree finches to tear vegetation, revealing concealed insects in the process. The woodpecker finch probes holes and uses tools (cactus spines) to extract insects and larvae. The sharp-beaked ground finch is a "jack-of-all-trades," as it consumes seeds, insects, and nectar. The vegetarian finch eats vegetation and flower buds with its crushing bill, while the much smaller warbler finches glean insects from vegetation and sip nectar with their slim, delicate bills. The only habitat specialist of the group is the mangrove finch, a tree finch that is currently restricted to Isabela, the largest island.

Adaptive radiations

The geologist and paleontologist George Gaylord Simpson (1902–1984) characterized adaptive radiations as groups of organisms that have diverged rapidly to exploit novel opportunities. Few places offer such abundant opportunities as isolated oceanic islands. As rich volcanic soil forms, rare colonists find opportunity in the form of resources (space, food, shelter) that are not being used by other organisms. Adaptations arise as solutions to physical, environmental, and biotic challenges. Adaptive traits often arise incrementally and by chance, but they are perpetuated because of the increased reproductive output of individuals who possess the advantageous trait. The fourteen major islands of the Galápagos (see Figure 1) were formed by volcanoes emerging periodically through a sliding plate comprising the seafloor, much like Hawaii and other isolated archipelagos. More than single isolated islands, clusters of islands appear to provide greater opportunities for speciation, diversification, and generation of novel biodiversity. Adaptive radiations can occur anywhere suitable habitat remains fragmented, such as on mountaintops surrounded by lowlands, in valleys separated by uplands, and in lakes whose margins are broken by different types of rocky or sandy habitat. Geography is thus considered an important component of adaptive radiations, but more difficult to assess is the notion of which species may be predisposed to undergo adaptive radiation or not. Recent adaptive radiations such as Darwin's finches offer some of the best opportunities to study the evolutionary process in action.

Discovery of Darwin's finches

The Galápagos finches bear his name, but Darwin's attention was not immediately drawn to these generally nondescript birds, nor did he ever fully grasp the extent and patterns of their diversification. During the voyage of the *Beagle*, Darwin was clearly intrigued by the amazing geological formations and fossils he encountered during the journey. His naturalist training shone through as he noticed how different the birds of Galápagos were compared to those on the South American mainland. It was not the finches but the Galápagos mockingbirds that he noticed most during the voyage, and it was about them that he made a very important observation: The mockingbirds looked different on islands only a few dozen miles apart. Similarly, he was told by the local governor of the then transiently occupied archipelago that the locals could tell which island a tortoise came from merely by looking at the shape of its shell. These observations questioned the notion that species were immutable, which was a prevailing notion of the time. Darwin later wrote that this observation—species varying in appearance over short geographical distances—was a key component of his theory of natural selection.

Darwin's writings indicate that he wanted to include the Galápagos finches as an example of geographical differentiation in his later writings. He was convinced they could be used to support this pattern of differentiation, but, unfortunately, the island of origin was not recorded for several of the specimens he collected. Perhaps it is not surprising that Darwin did not record the source locations, given that species were considered immutable. At the time of his voyage, Darwin suspected that the drab little finches belonged to several different families of birds that had colonized the Galápagos

Woodpecker finch (*Camarhynchus pallidus*) using a cactus spine to coax insects out of a hole in a tree. Image by Tierbild Okapia.

independently. It was not until months after he returned to England that the ornithologist John Gould (1804–1881) convinced Darwin that all of the finches shared a recent common ancestor. Unlike the tortoises and mockingbirds, and contrary to what Darwin suspected, species of Darwin's finches are generally not segregated on different islands. They occur together on most of the larger islands of the archipelago. The notion that species evolved in response to local conditions in different locations had a profound effect on Darwin and his ideas, yet this particular story of adaptive radiation had to be explained by additional factors leading to their coexistence in the same locations.

Early studies

Early studies of Darwin's finches were accelerated by a flurry of collecting activity in the late nineteenth and early twentieth centuries. The California Academy of Sciences expedition led by the American ornithologist and explorer Rollo Beck (1870–1950) in 1905 and 1906 was especially important for providing easy access to thousands of Darwin's finch specimens still being studied in the early twenty-first century. Early taxonomic classifications by Robert E. Snodgrass, Harry S. Swarth, and others culminated with David Lack's comprehensive study of morphological divergence among island populations of every species, published in 1947. His monograph was one of the first quantitative studies of adaptive radiation and an early masterpiece connecting morphology and biogeographical patterns to ecological and evolutionary processes. In the 1960s Robert I. Bowman extended Lack's work on

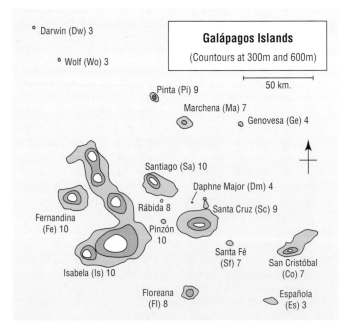

Figure 1. A map of the Galápagos Islands showing numbers of finch species breeding on each island. Reproduced by permission of Gale, a part of Cengage Learning.

Large cactus ground finch (*Geospiza conirostris*) on Espanola in the Galápagos Islands. Image by David Hosking.

morphology, while also compiling a most impressive study quantifying song variation between species, populations, and habitats.

In the 1970s, a group led by Peter R. and B. Rosemary Grant began to amass what eventually became one of the most impressive bodies of scientific work on any natural system. In one of the earliest studies to unequivocally show the process of natural selection in action, their student Peter T. Boag showed that during an extended drought, medium ground finches with larger beaks were more likely to survive than birds with smaller beaks, resulting in a beak size shift in the small population on the island of Daphne Major. Several key aspects of natural selection were demonstrated. Toward the end of the drought, most of the small seeds had been consumed, leaving mostly large seeds. Finches that could more efficiently process and consume larger seeds tended to survive better, and because beak size is highly heritable, the brief shift in size was sure to be perpetuated across later generations.

Environmental fluctuations

During El Niño (El Niño Southern Oscillation) events, the trade winds recede and warm water collects around the Galápagos Archipelago, resulting in abundant rain and explosive plant growth. In turn, insect populations grow and consume the flush of vegetation that also replenishes the seed bank for future years. Darwin's finches breed vigorously by feeding the insects to their young. The El Niño of 1983 triggered this cascade of effects, and the spectrum of available seeds was changed for years to come. A subsequent excess of small seeds and relative lack of larger seeds caused an evolutionary shift in the medium ground finch, but in the opposite direction to that observed in the previous drought. This time, in subsequent dry seasons, smaller finches had the advantage because they could more efficiently find enough small seeds to consume and sustain themselves, whereas the larger finches with larger beaks were less likely to survive. Establishing the connection between natural selection, beak size, and seed size allowed the Grants and student H. Lisle Gibbs to make sense of what appeared to be a contradictory pattern. The finches responded differently to drought because the seed spectrum had changed, favoring natural selection in the opposite direction.

Breeding and interbreeding

El Niño events not only promote breeding, but they also lead to more successful interbreeding between species. Studies of the medium ground finch suggest that Darwin's finches choose mates primarily by song. Passerines generally imprint on the song sung at the nest while nestlings. Males eventually sing the imprinted song, and females tend to choose mates that sing their father's song. Body size and beak size are also important factors in mate choice, but song is important enough to override other cues. Occasionally, a nestling imprints on a song sung by a different species, presumably a close neighbor. Males who grow up singing the song of a different species attract mates of the other species, which has led to several documented cases of hybridization. Interspecific hybrids tend not to survive as well as purebred birds, but hybrids survived at least as well as purebreds after the 1983 El Niño event.

Hybridization leads to genetic exchange and has the potential to erode the distinctive character of species. In Darwin's finches, genetic exchange is limited because of song (a behavioral barrier to interbreeding). Copying mistakes are uncommon, and hybrids and their subsequent offspring tend to mate with the species whose males sing the song of the original hybrid's father. Several species of ground finches are known to hybridize, and even this limited genetic exchange may have important consequences. Small populations that occur on many Galápagos islands may have limited ability to respond to natural selection if they possess little genetic variation. Hybridization can infuse such populations with genetic variation that may allow them the flexibility to adapt to future environmental changes. Genetic exchange between species may also cause molecular genetic markers, which are commonly used to reconstruct evolutionary relationships, to yield false or unclear patterns that can obscure species origins.

Evolutionary history

David Lack published an evolutionary tree of Darwin's finches based on morphology in 1947 (see Figure 2). His tree largely reflected ecological similarities and current nomenclature, with six ground finches and six tree finches forming two distinct groups, or clades, and with the Cocos island finch and warbler finch as distinct branches. Studies based on DNA variation have had success in reconstructing the deeper branches of the tree. The DNA tree (see Figure 3) resembles Lack's tree, but differs in several respects.

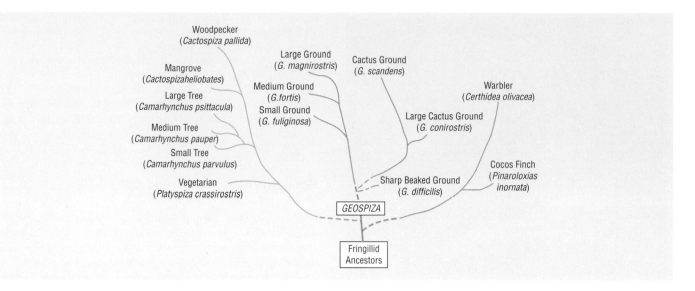

Figure 2. The evolutionary tree of Darwin's finches based on morphology. Reproduced by permission of Gale, a part of Cengage Learning.

The vegetarian finch, for instance, does not cluster with the other tree finches but forms a distinct lineage, which is perhaps not surprising given the dietary distinctness of this species. Also, Lack showed island populations of the sharp-beaked ground finch differed markedly, with each possessing traits reflecting other, more recently evolved species. Populations of sharp-beaked ground finches are indeed genetically distinct, with some populations forming a branch off the main trunk of the tree before other tree and ground finches and even before the vegetarian finch diverged. A most surprising pattern revealed by DNA analysis was that there are two distinct lineages of warbler finches. No prior study suspected that two groups of warbler finches were actually the most genetically distinct of Darwin's finches. The gray warbler finch generally occupies smaller, lower, drier islands, whereas the green warbler finch occupies moist, upland habitat of the larger central islands. These marked differences in habitat between islands may play a general role in the process of speciation and divergence.

Using DNA to resolve the sequence of evolution for the ten species remaining within the tree and ground finch groups has proved to be challenging. Evolutionary trees based on genetic distances from nuclear allelic variation closely parallel trees based on morphology and ecology, but DNA sequence variation, the most commonly used tool for reconstructing evolutionary history, cannot resolve species relationships within these groups. The difficulty is likely attributable to three factors. First, even the low levels of hybridization mentioned previously can obscure evolutionary history over longer periods of time. Second, population fragmentation within species complicates the task of resolving relationships between species, and most Darwin's finches occur on several islands. Finally, the adaptive radiation of Darwin's finches occurred so recently that there has not been much time for mutations to accumulate.

Darwin's finches are most closely related to the grassquits (genus *Tiaris*) and their relatives, which occur on the South

and Central American mainland and the Caribbean. The degree of DNA sequence divergence between Darwin's finches and their mainland relatives suggests that Darwin's finches first colonized the Galápagos less than three million years ago. The entire adaptive radiation took place in less than two million years based on a survey of existing species, although earlier forms may have gone extinct. The grassquits are small seed-eating birds much like the small ground finch (*Geospiza fuliginosa*). It is a challenging task to reconstruct the evolutionary sequence of beak evolution, but the tree topologies suggest that the earliest ancestors of Darwin's finches may have initially evolved from a seedeater into a warbler finch or sharp-beaked ground finch before completing the rest of the adaptive radiation.

Ecological species interactions

The Galápagos finches are a textbook example of the fundamental process of competitive character displacement. On islands that have both small and medium ground finches, the beak sizes, and thus the size of seeds consumed by birds, differ markedly. Species tend to partition resources to reduce competition. On islands that have only one species, the resident has an intermediate beak size. In the absence of a biological competitor for food resources, natural selection favors individuals with beak sizes that allow them to consume the greatest range of seeds.

A compelling case of a strong interaction between species was observed by the Grants after the 2003–2004 drought on Daphne Major. The presence of a newly founded population of large ground finches on Daphne Major changed the course of evolution of the medium ground finch. An earlier drought provided a famous case of natural selection as larger forms of the medium ground finch were favored because they could eat larger seeds. This niche was not available to them in 2004, and in response to the larger competitor, they instead evolved smaller beaks. This is compelling evidence for the role of

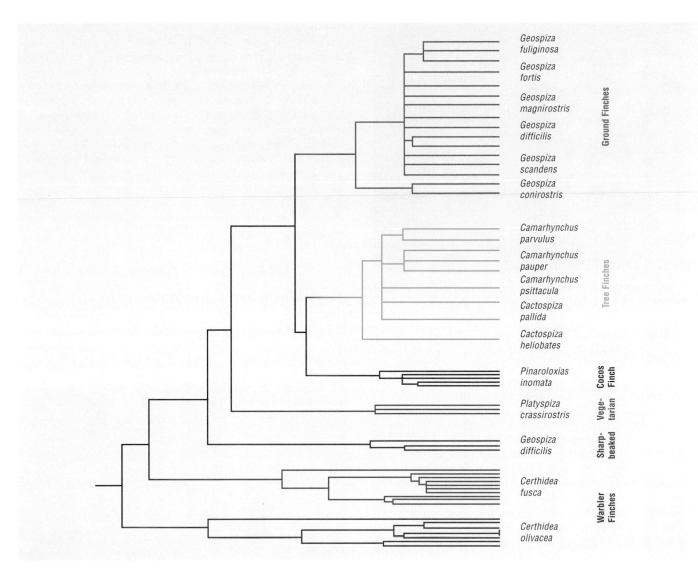

Figure 3. An evolutionary tree for Darwin's finches based on recent genetic sequence comparisons. Time proceeds from left to right. Reproduced by permission of Gale, a part of Cengage Learning.

competition and ecological character displacement in structuring communities.

With adaptive radiation there is typically a trend toward increasing specialization to maximize efficiency. This can be seen among the recently diverged small, medium, and large forms of both ground and tree finches. When the entire adaptive radiation is considered, however, it is difficult to imagine a finch that can consume all of the resources that are currently consumed by Galápagos finches. No finch can efficiently glean the undersides of leaves like a warbler finch and also crack a large hard seed like a large ground finch. Thus, along with a trend toward ecological specialization there is also a trend to produce species that occupy distinct ecological niches that were previously unoccupied. The evolution of novel structures to exploit new resources is a common pattern seen in the fossil record across longer timescales. Remaining a topic of some debate is how the underlying processes creating novelty differ from the kind of

size shifts observed in the process of ecological character displacement.

Species and speciation

How are species of Darwin's finches defined, if they do not all show consistent mutational differences in their DNA? The answer stems from the historical recognition of different forms living largely together in the same locations. All Darwin's finches co-occur in the same habitat with other species, and they generally do not interbreed. Each species has a distinct song, morphology, and ecology that distinguish it from other species, and, as alluded to previously, differences are most pronounced among populations that co-occur on the same island. Furthermore, on several occasions species on the same island have responded differently to environmental change. The ability to respond differently to environmental change is at the core of the concept of biodiversity. Populations that respond differently to environmental change

The warbler finch (*Cerhidea olivacea*) is an endemic small insectivore. Image by Tui De Roy.

should be recognized as independent units of biodiversity, or species. Thus, although studies of DNA variation offer a fast and clear way to distinguish many species, one cannot assume that convenient measures of overall genetic divergence capture all of the important aspects of biodiversity.

In the Galápagos finches, different island populations of the same species frequently exhibit morphological differences that are attributable to local adaptation. Accordingly, there are several instances of incipient speciation in the Galápagos. Nevertheless, few cases of speciation in Darwin's finches can be wholly attributed to differences evolved in geographic isolation. Darwin's finches thus differ from several other Galápagos vertebrates, such as the tortoises, lizards, and mockingbirds, whose different species show little geographical overlap.

To account for the buildup of finch species into a community of finches that occupy the same locations, other processes have been invoked. David Lack described a process in which initial differentiation occurs in isolation on different islands. At some later point, a chance colonization event brings the two incipient species together on the same island. If they do not interbreed—for instance, because their songs have changed over time—then one would expect that further divergence will take place through competitive character displacement. Each new species would then slowly colonize the rest of the archipelago, and the process would be repeated until the current community of species was established. This general model of community buildup has been applied to several other systems, but it is very difficult to devise a way to test such historical models.

Genetic studies of population migration have shed light about the speciation process in Darwin's finches. Several lines of evidence suggest that birds move throughout the archipelago much more frequently than previously assumed. There are several anecdotes of different looking birds arriving on an island, and more recently, genetic analysis has confirmed the occurrence of substantial levels of interisland movement. The

notion that interisland movement is not rare alters the way scientists perceive speciation. Contrary to Lack's model, interisland movement does not appear to be rare, but it usually results in interbreeding and genetic exchange among populations. On some occasions, immigrants choose to breed among themselves, as has been observed on Daphne Major. In rare instances, these newly founded populations persist to complete the second stage of speciation according to Lack's model. Other evidence suggests that founders of new populations and immigrants show nonrandom patterns of settlement that may be attributable to habitat choice, mate choice, or natural selection. In light of evidence for elevated levels of genetic exchange, the observed differences between island populations implies that natural selection that must be quite strong to counter the homogenizing effects of gene flow.

The genetic basis of trait differences

Scientific understanding of how natural selection operates in nature is incomplete without knowledge of the genetic basis of the traits under selection. Studies of gene expression have

Common cactus finch (*Geospiza scandens*) feeding on cactus in the Galápagos Islands. Image by Frans Lanting.

made significant inroads to this problem. Beak size and shape differences in Darwin's finches have been linked to increased expression of two proteins at a critical stage of development. BMP4 and calmodulin are known components of craniofacial development in a wide variety of organisms, and they clearly account for a significant amount of the observed beak size and shape differences among Darwin's finches. Such advances are a critical step toward revealing the causal chain leading from genetic mutations to the actual trait that is honed by natural selection. The ultimate genetic causes of beak shape differences may range from a simple mutation in the regions flanking these genes, to very complex interactions among several different loci that all pitch in to affect protein expression at the end of a cascade of interactions. One exciting aspect of this research is that it places the genetic basis of natural selection within scientists' reach.

Conservation

The Galápagos finches are unusual partly because they all still exist. Endemic island birds, such as the Hawaiian honeycreepers, are undergoing extinction at an alarming rate. No known species of Darwin's finch has been lost to extinction, which is one reason why these birds offer a unique and effective model for the study of evolutionary processes. Several populations, however, are known or suspected to be extinct, including several on the islands with the most extended record of human occupation, Floreana and San Cristóbal. Extinction is not limited to these islands. The mangrove finch has disappeared from Fernandina and is now confined to a single declining population on Isabela, the largest Galápagos island. It is considered critically endangered and is the only species of Galápagos finch in immediate peril. The causes of mangrove finch decline are unclear. Fernandina has never had a human settlement, but human disturbance (wood gathering) and introduced predators (rats) and parasites are past and present threats to the survival of this species. Efforts are underway to mitigate the effects of introduced predators, parasites, and pathogens on mangrove finches. One valuable tool for assessing recent population decline is genetic analysis of the vast collection of finches available in museums,

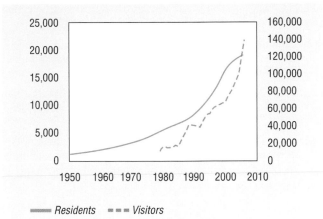

Figure 4. Numbers of Galápagos residents and tourists have increased dramatically over the last thirty years. Reproduced by permission of Gale, a part of Cengage Learning.

including the specimens collected by Darwin himself during the voyage of the HMS *Beagle*. Cross-temporal comparisons can yield valuable insight about how species have changed since the arrival of humans.

Compared to other islands, the Galápagos have enjoyed a much more limited period of human disturbance that is quickly coming to an end. Numbers of tourists and residents are increasing at an alarming rate (see Figure 4). Along with increased human traffic comes more introduced species carrying their associated parasites and pathogens. Efforts are underway to remove introduced species, such as goats and rats, that threaten several native species. The Galápagos are in a delicate balance between human disturbance and human efforts toward preservation. Careful monitoring and intervention are necessary if the Galápagos and its unique birds and other organisms, which took millions of years to evolve, will remain intact to the end of the twenty-first century.

Resources

Books

Bowman, Robert I. 1983. "The Evolution of Song in Darwin's Finches." In *Patterns of Evolution in Galápagos Organisms*, ed. Robert I. Bowman, Margaret Berson, and Alan E. Leviton. San Francisco: American Association for the Advancement of Science, Pacific Division.

Darwin, Charles R. 1839. *Journal of Researches into the Geology and Natural History of the Various Countries Visited by H.M.S. Beagle, under the Command of Captain FitzRoy, R.N. from 1832 to 1836*. London: Henry Colburn.

Grant, B. Rosemary, and Peter R. Grant. 1989. *Evolutionary Dynamics of a Natural Population: The Large Cactus Finch of the Galápagos*. Chicago: University of Chicago Press.

Grant, Peter R. 1999. *Ecology and Evolution of Darwin's Finches*. Princeton, NJ: Princeton University Press.

Lack, David. 1947. *Darwin's Finches*. Cambridge, UK: Cambridge University Press.

Schluter, Dolph. 2000. *The Ecology of Adaptive Radiation*. Oxford: Oxford University Press.

Weiner, Jonathan. 1994. *The Beak of the Finch: A Story of Evolution in Our Time*. New York: Knopf.

Periodicals

Boag, Peter T., and Peter R. Grant. 1981. "Intense Natural Selection in a Population of Darwin's Finches (Geospizinae) in the Galápagos." *Science* 214(4516): 82–85.

Gibbs, H. Lisle, and Peter R. Grant. 1987. "Oscillating Selection on Darwin's Finches." *Nature* 327(6122): 511–513.

Grant, Peter R., and B. Rosemary Grant. 2006. "Evolution of Character Displacement in Darwin's Finches." *Science* 313 (5784): 224–226.

Grant, Peter R., B. Rosemary Grant, and Arkhat Abzhanov. 2006. "A Developing Paradigm for the Development of Bird Beaks." *Biological Journal of the Linnean Society* 88(1): 17–22.

Petren, Kenneth, B. Rosemary Grant, and Peter R. Grant. 1999. "A Phylogeny of Darwin's Finches Based on Microsatellite DNA Length Variation." *Proceedings of the Royal Society* B 266 (1417): 321–329.

Petren, Kenneth, Peter R. Grant, B. Rosemary Grant, and Lukas F. Keller. 2005. "Comparative Landscape Genetics and the Adaptive Radiation of Darwin's Finches: The Role of Peripheral Isolation." *Molecular Ecology* 14(10): 2943–2957.

Sato, Akie, Herbert Tichy, Colm O'hUigin, et al. 2001. "On the Origin of Darwin's Finches." *Molecular Biology and Evolution* 18(3): 299–311.

Sulloway, Frank J. 1982. "Darwin and His Finches: The Evolution of a Legend." *Journal of the History of Biology* 15(1): 1–53.

Tonnis, Brandon, Peter R. Grant, B. Rosemary Grant, and Kenneth Petren. 2005. "Habitat Selection and Ecological Speciation in Galápagos Warbler Finches (*Certhidea olivacea* and *Certhidea fusca*)." *Proceedings of the Royal Society* B 272 (1565): 819–826.

Other

Watkins, Graham, and Felipe Cruz. 2007. "Galápagos at Risk: A Socioeconomic Analysis of the Situation in the Archipelago." Puerto Ayora, Province of Galápagos, Ecuador: Charles Darwin Foundation. Available from http://www.darwinfoundation.org/files/library/pdf/2007/Galapagos_at_Risk_7-4-07-EN.pdf.

Kenneth Petren

Artificial selection

The close association between widely different species of organisms is common throughout the animal kingdom, and if at least one partner benefits from the relationship it is known as symbiosis. There are different kinds of symbiosis: If one partner in the relationship suffers in some way, it is known as parasitic symbiosis. If one partner benefits and the other is unharmed, it is known as commensal symbiosis. And if both partners benefit, the relationship is known as mutual symbiosis. Most examples of symbiotic relationships have evolved over millions of years, including internal parasites and their hosts, barnacles that live as commensals on jaws of whales, and ants that live in a mutual relationship with aphids, which they protect from predators and in return "milk" for their sugary excreta. The relationship between humans and their domestic animals is also symbiotic but unique in that the behavior, the morphology, and in some species the physiology of animals (and plants) have been purposefully altered by the human partner. This form of symbiosis is known as domestication, and unlike evolution of symbiotic relationships in the natural world, those between humans and living animals are believed to have begun only around fifteen thousand years ago.

The process of domestication

The process by which species of animals have been changed into domestic forms was first published as the principle of selection by Charles Darwin in 1868, and little new can be added to his description of the three kinds of selection:

The principle of selection may be conveniently divided into three kinds. *Methodical selection* is that which guides a man who systematically endeavours to modify a breed according to some predetermined standard. *Unconscious selection* is that which follows from men naturally preserving the most valued and destroying the less valued individuals, without any thought of altering the breed; and undoubtedly this process slowly works great changes. Unconscious selection graduates into methodical, and only extreme cases can be distinctly separated; for he who preserves a useful or perfect animal will generally breed from it with the hope of getting offspring of the same character; but as long as he has not a predetermined purpose to improve the breed, he may be said to be selecting unconsciously. Lastly, we have *Natural selection* which implies that the individuals which are best

fitted for the complex, and in the course of ages changing conditions to which they are exposed, generally survive and procreate their kind. With domestic productions, natural selection comes to a certain extent into action, independently of, and even in opposition to, the will of man. (Darwin 1890, vol. 2, p. 177)

Darwin's term, *methodical selection*, has since become known as *artificial selection*, and it is the basis by which the thousands of breeds of domestic animals and plants all over the world have been produced and continue to be developed. To understand how artificial selection works, it is necessary to define domestication and describe the biological process that has enabled, for example, the wolf to become humankind's most highly valued animal companion, the domestic dog.

An animal may be said to be domesticated or "tame" when it has lost its fear of humans and will breed readily in captivity, but true domestication involves much more than this and can be defined as the keeping of animals in captivity by a human community that maintains total control over their breeding, organization of territory, and food supply. True domestication takes place from the combination of a biological and a cultural process. The biological process begins when a few animals are separated from the wild species and become habituated to

Two common remoras (*Remora remora*) cling to this green sea turtle (*Chelonia mydas*) and feed on parasitic copepods attached to their host's back. © Roman & Olexandra, 2010/Shutterstock.com.

Wild species	Domestic form
Perissodactyla	
Equus africanus (Heuglin and Fitzinger 1866) North African wild ass	*Equus asinus* (Linnaeus) Donkey
Equus ferus (Boddaert 1785) Russian extinct wild horse, tarpan	*Equus caballus* (Linnaeus) Domestic horse
Artiodactyla	
Camelus ferus (Przewalski 1878) Wild Bactrian camel, now restricted to the western Gobi desert	*Camelus bactrianus* (Linnaeus) Domestic Bactrian camel
Lama guanicoe (Müller 1776) South American guanaco	*Lama glama* (Linnaeus) Llama
Vicugna vicugna (Molina 1782, p. 313) South American vicuña	*Vicugna pacos* (Linnaeus) Alpaca
Bos primigenius (Bojanus 1827) Aurochs of Europe, Asia, and North Africa, extinct since 1627	*Bos taurus* (Linnaeus) Common cattle
Bos namadicus (Falconer 1853) Indian aurochs, extinct	*Bos indicus* (Linnaeus) Indian humped cattle or zebu
Bos gaurus (H. Smith 1827, p. 399) Gaur of India, Burma, and Malaya	*Bos frontalis* (Lambert) Gaur, mithan
Bubalus arnee (Kerr 1792) Indian water buffalo, arni	*Bubalus bubalis* (Linnaeus) Domestic water buffalo
Bos mutus (Przewalski 1883) Yak of mountains of Tibet, Nepal, and the Himalayas	*Bos grunniens* (Linnaeus) Domestic yak
Capra aegagrus (Erxleben 1777) Bezoar of the Middle East	*Capra hircus* (Linnaeus) Domestic goat
Ovis orientalis (Gmelin 1774) Mouflon of Western Asia	*Ovis aries* (Linnaeus) Domestic sheep
Sus scrofa (Linnaeus) Wild boar of Europe, Asia, and North Africa	*Sus domesticus* (Erxleben 1777) Domestic pig
Rodentia	
Cavia aperea (Erxleben 1777) South American cavy	*Cavia porcellus* (Linnaeus) Domestic guinea pig
Carnivora	
Canis lupus (Linnaeus) Wolf of the Palaearctic, India, and North America	*Canis familiaris* (Linnaeus) Dog (including dingo)
Mustela putorius (Linnaeus) Polecat of Europe, Middle East, and Morocco	*Mustelo furo* (Linnaeus) Ferret
Felis silvestris (Schreber 1777) Wildcat of Western Europe to Western China and Central India, much of Africa	*Felis catus* (Linnaeus 1758) Domestic cat
Osteichthyes	
Carassius gibelio (Bloch 1782) Prussian or gibel carp of Central Europe to East Asia	*Carassius auratus* (Linnaeus) Goldfish
Lepidoptera	
Bombyx mandarina (Moore 1862) Mulberry silk moth of China, Korea, and Japan	*Bombyx mori* (Linnaeus) Silkworm

Wild species and their domestic derivatives that traditionally have separate names. Reproduced by permission of Gale, a part of Cengage Learning.

humans. A few of these animals may breed within the human community, and if their offspring survive they will form a small isolated or founder group. If these animals then interbreed and increase in numbers over many successive generations, they will respond by means of small genetic changes to natural selection under the new regime of the human community and its environment, and later they will respond for economic, cultural, or aesthetic reasons first to Darwin's unconscious selection and then to artificial selection until the domestic breed is created. A breed may be defined as a group of animals that has been bred by humans to possess uniform characters that are passed down through the generations and distinguish the group from other animals within the same species.

The progress of domestication can be seen as an evolutionary process that mimics the sequence of events when a small group of wild animals becomes isolated from the main group, say on an island. Subsequent evolution of this tiny founder group will depend on the unique genetic composition of the individuals. The principle of how this works is known as the founder effect, and it will cause the original and the new isolated populations to evolve along different paths. Initially the founder effect will lead to a population bottleneck, this being caused by rapid fluctuations in the gene frequencies, known as genetic drift. The population contracts and then expands again by breeding, but with an altered genetic composition. Genetic drift and natural selection act together on the new population of wild animals, and after thousands or even millions of years a new species will evolve, as did Darwin's finches on the Galapagos Islands.

Before true speciation occurs in an isolated group of wild animals, there may be an intermediate stage, which is termed the subspecies. A subspecies is a geographically and reproductively isolated group of animals or plants that differs in hereditary characteristics from the main species but is not distinctive enough to be treated as a separate species. Geographic isolation is the important factor because it prevents interbreeding and hence a flow of genes between the populations. The subspecies is often a stepping-stone to true speciation.

If the isolated group consists of tamed or habituated animals that are under human control, then unconscious and artificial selection will also be powerful forces in the progeny, and the process of change will be speeded up. The question then has to be asked whether the new isolated group of humanly controlled animals can be considered to be a new species or subspecies. So what constitutes a species of animals? Until the past fifty years or so the definition of a species held that if a population of

Orchids are known for their easy hybridization. © Dwight Smith. Image from BigStockPhoto.com.

Feral breeds of dog, such as the dingo of Australia, inhabit lands where their ancestor the wolf has never lived. ANT Photo Library/Photo Researchers, Inc.

animals breeds freely and produces fertile offspring, then it is a species. If two populations do not interbreed or if the hybrid offspring are infertile, then they are separate species; in an example of the latter, if a horse and a donkey mate, they will produce an infertile mule. On its own, however, the state of fertility of hybrid offspring is an inadequate means of defining a species. Many mammals that are normally considered to be good species will interbreed, although, because of a behavioral barrier, they may not usually do so in the wild, and their offspring will be fertile—for example, the dog, wolf, jackal, and coyote, all of which will breed and produce fertile offspring, as will the European and American bison.

A new species definition, the biological species concept, avoids the hybridization issue. This has had many different wordings from taxonomists but its essence was defined by Ernst Mayr as long ago as 1940. His definition holds that "a species is a group of actually or potentially interbreeding natural populations that is reproductively isolated from other such groups" (Mayr 1966, p. 19). The next question is whether domesticated animals in general should be considered to be separate species from their wild progenitors, and this has been argued over since the time of Darwin. All domestic animals will interbreed with their wild progenitors so in that sense they are not separate species, but they are reproductively isolated from

them in the same way that the African jackal is isolated from the American coyote. A biologically more important factor is that molecular evidence is now proving that a number of domestic forms, such as domestic sheep, are descended from several wild lineages so they should not be aligned with one particular ancestor and should be treated as separate species in their own right. In this entry, domestic animals, with a few exceptions, are considered as belonging to species that are distinct from those of their wild progenitors.

This leads to how these domestic species should be named and whether they should be included in the Latin nomenclature that is in use for all wild species. A dog is certainly different from a wolf so it should not be called by the Latin name of the wolf, which is *Canis lupus*. The most parsimonious arrangement is to continue to call the dog and all other fully domestic animals by the Latin names they were first given by the Swedish botanist Carolus Linnaeus (1707–1778) in the eighteenth century, which for the dog was *Canis familiaris*. Table 1 lists the Latin names that are now accepted by the International Commission on Zoological Nomenclature (ICZN, Opinion 2027 Case 3010, 2003) for the wild species together with the names that were given by Linnaeus in 1758 to their domesticated derivatives. This list does not include the species that have the same name in the domestic as in the wild species, such as the domestic and wild reindeer, *Rangifer tarandus*.

The formation of breeds

A breed is similar in taxonomic terms to a subspecies in that it is a subpopulation that is reproductively isolated from the parent species, but it is not usually isolated geographically from it. For example, breeds of dogs occur all over the same regions where the wild wolf is still found, but dogs and wolves live apart and only on very rare occasions do they interbreed. Feral breeds of dog, however, such as the dingo of Australia and the New Guinea singing dog, inhabit lands where their ancestor the wolf has never lived. The dingo has been commonly named *Canis dingo*, but apart from this one distinctive breed of feral dog, only the species of domestic animals are given Latin names—not the breeds.

Unconscious selection for the first breeds

The dog may be taken as an example of the means by which the earliest breeds of domestic animals were produced by a combination of natural selection for survival in the new environment of the human social group and unconscious selection for preferred traits.

The domestication of the earliest "dogs," which happened either from the taming of captured wolf cubs or by the intrusion into human hunting camps by scavenging adult wolves, occurred around fifteen thousand years ago according to evidence from archaeological sites in many parts of the world. The tamed or habituated wolves would have had a mutually symbiotic relationship with the humans, and gradually they would have become canine families that were isolated from populations of wild wolves. Survival of puppies would have depended on their response to natural selection from their environment and unconscious selection by their human owners who would have preferred certain canine traits to others. One of the most highly preferred of these traits in the earliest dogs was the mutation from the gray coat color of the wolf to a yellow-gold coat color. This may be assumed

The Arabian horse is tall, light-bodied, fine-limbed, and short coated, well adapted to the deserts of Arabia. © Juniors Bildarchiv/Alamy.

The Shetland pony is short, stocky, heavy-bodied, with a shaggy coat adapted to the cold, wet climate of the Shetland Islands. © willmetts, 2010/Shutterstock.com.

from the fact that the majority of feral dogs around the world have this coat color. These populations are descended from early domesticated dogs that returned to living in the wild hundreds, and sometimes thousands, of years ago, so they were never artificially selected for new breeds. They are the dingo, the New Guinea singing dog, the Carolina dog of the United States, and the Indian and African pariah dogs.

Breeds from artificial selection combined with natural selection

Many of the old, or what may be called native, breeds of livestock, all over the world, have evolved with a combination of natural and artificial selection to be well adapted to the climate and environment of their region as well as being fully adapted for their human use. Take for example the breeds of domestic horses whose homeland may be the deserts of Arabia or the cold wet climate of the Shetland Islands, north of Scotland. The Arab horse is tall, light-bodied, fine-limbed, and short coated, whereas the Shetland pony is short, stocky, heavy-bodied, and shaggy coated, and yet they belong to the same domestic species, *Equus caballus*. Both breeds have been artificially selected for carrying their human owners and their belongings, but their size and bodily shape have also evolved in response to the two rules that relate environmental temperatures to size of body (Bergmann's rule) and length of the extremities (Allen's rule). These rules state that in the hotter southern parts of the world, mammals have lighter bodies and finer, longer limbs than in the cold north, where the same species will tend to be heavier bodied with short legs and small ears.

Another example of a domestic species that follows this adaptation to environment as well as to human use is the sheep. In the Middle East sheep breeds are long-legged, lop-eared, and hairy coated, whereas in the British Isles their bodies are barrel-shaped, their legs are short, and they have heavy fleeces that have been artificially selected to produce wool as a commercial product.

It is not only bodily shape that has responded to environment in the formation of breeds; an example of adapted physiology can be seen in the Boran breed of cattle that has been herded by the

An example of adapted physiology can be seen in the Boran breed of cattle. These cattle have a rare adaptation to their semidesert life in that they need to drink only once every three days. © Michael Freeman/Corbis.

native peoples of Ethiopia and other dry regions of eastern Africa for hundreds of years. These cattle have a rare adaptation to their semidesert life in that they need to drink only once every three days.

Breeds from artificial selection on its own

Those species of domestic animal that have been artificially selected with little or no influence from natural selection have produced breeds that differ most substantially from the wild form, and some of these breeds require constant human aid in order to survive and produce viable young. The domestic dog is the supreme example of this, and many of the four hundred or so breeds in the world have been produced by artificial selection alone. Nevertheless, despite abnormalities that have been produced by artificial selection, such as the huge head and short muzzle of the bulldog, which prevents the natural birth of puppies so they all have to be born by Caesarean section, the basic physiology of the wolf remains in every breed. The digestive system is the wolf's, the gestation period is sixty-three days, every puppy is born with a weight of approximately 400 grams, and all dogs can be infested with the same parasites as the wolf.

The extreme alteration in body shape found in breeds of dog is possible only because their ancestor, the wolf, has a high

degree of genetic variation. This may be a consequence of the wolf's adaptation to many different habitats and climates in its wide distribution over the northern hemisphere. Other examples of wild species with wide distributions that have been artificially selected for many very divergent domestic breeds are the wild boar, ancestor of the domestic pig, and the gibel carp, ancestor of the goldfish (see Table 1). The wild rock pigeon, ancestor of the domestic pigeon, achieved biological fame from being the species that fascinated Darwin with its innumerable exotic descendents of domestic breeds, from racing pigeons to fantails to tumblers. It was because these breeds differ so dramatically from the dull-gray ancestral species that Darwin became convinced of the truth and importance of his principle of the three types of selection—methodical, unconscious, and natural—and their relevance for his theory of evolution.

Artificial selection in plants and genetic engineering

Darwin's three types of selection, unconscious, methodical, and natural, have also been fundamental in cultivation of plants and the development of agricultural systems all over the world. The earliest hunter-gatherers were practicing unconscious selection with the collection of wild plants for multiple purposes and the burning of landscapes in order to stimulate regrowth. Later, as settlement began to replace hunting-gathering in the Neolithic period, there was more active selection with cultivation of cereal crops. This has led through artificial selection to the vast numbers of food plants that sustain the ever-increasing human populations in the modern world, as well as to the great numbers of cultivated garden plants. As with artificial selection of domestic animals, however, resulting plants may be less viable than their progenitors, and many will not survive without careful nurturing by the farmer or gardener.

Darwin had no knowledge of genetics or the laws of inheritance, but scientists now know that selection processes are the routes by which genetic composition is altered, resulting over the course of millennia in the multiplicity of domestic animals and plants inhabiting the human world today.

Within the past decades a new scientific technology for artificial selection has been developed, known as genetic engineering. This new scientific technology, which involves the direct manipulation in a laboratory of the genes of a plant or animal, differs radically from traditional breeding practices and has ever-widening applications in agriculture and medicine. Genetic engineering, however, is a highly specialized and controversial procedure that is very unlikely to replace the selective processes that, as Darwin so famously realized, mimic evolutionary processes and keep the breeds of domestic animals and cultivated plants in step with the changing needs and fashions of human societies.

Bulldogs have been bred for their huge head and short muzzle. This selective breeding prevents the natural birth of puppies so they all have to be born by Caesarean section. © Annette Shaff/Shutterstock.com.

Resources

Books

Clutton-Brock, Juliet. 1992. *Horse Power: A History of the Horse and Donkey in Human Societies*. London: Natural History Museum Publication; Cambridge, MA: Harvard University Press.

Clutton-Brock, Juliet. 1995. "Origins of the Dog: Domestication and Early History." In *The Domestic Dog: Its Evolution, Behaviour, and Interactions with People*, ed. James Serpell. Cambridge, UK: Cambridge University Press.

Clutton-Brock, Juliet. 1999. *A Natural History of Domesticated Mammals*. 2nd edition. Cambridge, UK: Cambridge University Press; London: Natural History Museum.

Corbet, G. B. 1997. "The Species in Mammals." In *Species: The Units of Biodiversity*, ed. M. F. Claridge, H. A. Dawah, and M. R. Wilson. London: Chapman and Hall.

Darwin, Charles. 1890. *The Variation of Animals and Plants under Domestication*. 2nd edition. 2 vols. London: John Murray.

Harris, David R., and Gordon C. Hillman, eds. 1989. *Foraging and Farming: The Evolution of Plant Exploitation*. London: Unwin Hyman.

Harris, David R., and Gordon C. Hillman, eds. 1989. *Foraging and Farming: The Evolution of Plant Exploitation*. London: Unwin Hyman.

Periodicals

Gentry, Anthea, Juliet Clutton-Brock, and Colin P. Groves. 2004. "The Naming of Wild Animal Species and Their Domestic Derivatives." *Journal of Archaeological Science* 31(5): 645–651.

International Commission on Zoological Nomenclature Opinion 2027 (Case 3010). 2003. Usage of 17 Specific Names Based on Wild Species Which Are Pre-Dated by or Are Contemporary with Those Based on Domestic Animals (Lepidoptera, Osteichthyes, Mammalia): Conserved. *Bulletin of Zoological Nomenclature* 60(1, March): 81–84.

Mayr, Ernst. 1966. *Animal Species and Evolution*. Cambridge, MA: Harvard University Press, Belknap Press.

Mayr, Ernst. 1966. *Animal Species and Evolution*. Cambridge, MA: Harvard University Press, Belknap Press.

Pedrosa, Susana, Metehan Uzun, Juan-José Arranz, et al. 2005. "Evidence of Three Maternal Lineages in Near Eastern Sheep Supporting Multiple Domestication Events." *Proceedings of the Royal Society* B 272(1577): 2211–2217.

Juliet Clutton-Brock

Convergent evolution and ecological equivalence

Convergent evolution is the process by which unrelated or distantly related organisms evolve similar body forms, coloration, and adaptations. Natural selection can result in evolutionary convergence under several different circumstances, including mimicry, convergent evolution of eyes and intelligence, convergent evolution of molecules, ecological equivalence, live bearing in squamates, thorny ant-eating lizards, colorful tails, running across water, convergence in sandy deserts, rock mimicry, and community convergence.

Mimicry

Species can converge where they occur together in sympatry, as in mimicry complexes among insects, especially butterflies (coral snakes and their mimics constitute another well-known example). Mimicry evolves after one species, the "model," has become aposematic (warningly colored) because it is toxic or poisonous and therefore protected (Wickler 1968). Two distinct kinds of mimicry are recognized, Batesian and Müllerian. In Batesian mimicry, the mimic is palatable or unprotected, but gains from being mistaken for the model, which is unpalatable or protected. In Müllerian mimicry, two protected model species converge because of the advantage of being mistaken for each other.

Mimicry is an interesting consequence of warning coloration that nicely demonstrates the power of natural selection. An organism that commonly occurs in a community along with a poisonous or distasteful species can benefit from a resemblance to the warningly colored species, even though the "mimic" itself is nonpoisonous and/or quite palatable. Because predators that have experienced contacts with the model species have learned to avoid it, they mistake the mimic species for the model and avoid it as well. Such false warning coloration is termed Batesian mimicry after its discoverer, the English naturalist Henry Walter Bates (1825–1892).

Many species of harmless snakes mimic poisonous snakes (Greene and McDiarmid 1981). In Central America, some harmless snakes are so similar to poisonous coral snakes that only an expert can distinguish the mimic from the "model." A few experts have even died as a result of superficial misidentifications. Similarly, certain harmless flies and clearwing moths mimic bees and wasps, and palatable species of butterflies mimic distasteful species.

Batesian mimicry is disadvantageous to the model species because some predators will encounter palatable or harmless mimics early on and thereby take longer to learn to avoid the model. The greater the proportion of mimics to models, the longer is the time required for predator learning and the greater the number of model casualties. In fact, if mimics became more abundant than models, predators might not learn to avoid the prey item at all but might actively search out model and mimic alike. For this reason, Batesian mimics are usually much less abundant than their models; in addition, mimics of this sort are frequently polymorphic (often only females are mimics) and mimic several different model species.

Perhaps one of the most well-known examples of Batesian mimicry, the viceroy butterfly (*Limenitis archippus*) (top) appears very similar to the noxious tasting monarch butterfly (*Danaus plexippus*) (bottom). © Grant Heilman Photography/Alamy.

Discovered by the German zoologist Fritz Müller (1821–1897), Müllerian mimicry by contrast occurs when two species, both distasteful or dangerous, mimic one another. Both bees and wasps, for example, are usually banded with bright yellows and blacks. Because potential predators encounter several species of Müllerian mimics more frequently than just a single species, they learn to avoid them faster, and the relationship is actually beneficial to both prey species. The resemblance need not be as precise as it must be under Batesian mimicry because neither species actually deceives the predator; rather, each only reminds the predator of its dangerous or distasteful properties. Müllerian mimicry is beneficial to all parties, including the predator; because such mimics merely remind predators that they are distasteful or protected, they can be equally abundant and, in contrast to Batesian mimicry, they are rarely polymorphic.

Convergent evolution of eyes and intelligence

Other examples of convergence involve the development of similar characteristics to solve similar problems of adaptation.

Müllerian mimicry occurs when two species, both distasteful or dangerous, mimic one another. Both bees and wasps, for example, are usually banded with bright yellows and blacks. © Juehua Yin, 2010/Shutterstock.com.

For example, anatomically remarkably similar eyes have evolved independently in cephalopod mollusks (cuttlefish, squid, and octopi) and vertebrates, although the positioning of light-sensitive cells in the retina is different—photoreceptors face the pupil in cephalopods but are on the back side of the retina in vertebrates. (Eyes in these mollusks are actually better designed than those of vertebrates!) Intelligence has arisen multiple times, most notably in cephalopods and monitor lizards, as well as in birds and mammals (especially in cetaceans and primates).

Arid regions of South Africa support a wide variety of euphorbeaceous plants, some of which are strikingly close to American cacti phenotypically. They are leafless stem succulents, protected by sharp spines, presumably adaptations to reduce water loss and predation in arid environments. Similarly, evergreen sclerophyll woody shrubs have evolved convergently under Mediterranean climates in several different geographical regions with winter rain and prolonged summer droughts (Mooney and Dunn 1970). Spines, used in defense, have evolved in several different mammalian lineages, including New World and African porcupines, the tenrecs of Madagascar and the Comoro Islands, African and European hedgehogs, and Australian and New Guinean echidnas (marsupials).

Convergence sometimes occurs under unusual conditions where selective forces for the achievement of a particular mode of existence are particularly strong. Presumably in response to thick-skinned prey, two fossil saber-tooth carnivores, the South American marsupial "cat" *Thylacosmilus* and the North American placental saber-toothed tiger *Smilodon*, evolved long, knifelike canine teeth independently (though these two species were not contemporary). Many other marsupial mammals have undergone convergent evolution with placental mammals, including wombats (woodchucks), numbats (anteaters), quolls (cats), and thylacines (wolves) (Geist 1978). Geist also emphasizes that "mice" and "rats" have evolved repeatedly among many different mammalian lineages.

Still another example of convergent evolution is seen in the similar shape and coloration of fish and cetaceans, both of which have adapted to the marine environment by developing a fusiform body (one tapering at both ends) and neutral buoyancy. They are also countershaded, with a light underbelly and a darker upper surface, which makes them less visible from both below and above. Countershading is actually the rule among both arthropods and vertebrates, however, so it is presumably an ancestral trait that has been retained throughout the evolution of both groups. Countershading is as important for predators as it is for prey species, as it makes both harder to detect.

Convergent evolution of molecules

Even molecules can evolve convergently, especially when parasites mimic molecular messages that signal "self" to immune responses of hosts, which allows the parasite to elude its host's defenses. Molecular convergence could also take place when a particular metabolic function requires similar or identical molecular structure (Doolittle 1994). Some gene circuits and gene networks appear to have undergone convergent evolution by single-gene duplications in higher

The countershading on a killer whale (*Orcinus orca*) is important for the predator, as it makes is harder to detect in the water. © Brandon D. Cole/Corbis.

eukaryotes (Amoutzias et al. 2004; Conant and Wagner 2003). Convergence in DNA nucleotide sequences would lead to erroneous phylogenetic conclusions, which would be problematical for molecular systematic studies.

Ecological equivalence

Evolutionary convergence involving unrelated organisms living in similar environments but in different places (known as allopatry) can also occur in another way. This usually takes place in relatively simple communities in which biotic interactions are highly predictable and the resulting number of different ways of exploiting the environment is limited. Similar environments pose similar challenges to survival and reproduction, and those traits that enhance Darwinian fitness are selected for in each environment. Such organisms that fill similar ecological roles in different, independently evolved biotas are termed "ecological equivalents" (Grinnell 1924; Hubbell 2006).

Examples are legion. For instance, wings and winglike structures have evolved independently several times, in insects, reptiles (pterosaurs and birds) and in mammals (bats) (Dryden 2007). Flight first evolved in insects about 330 million years ago (mya), second in pterosaurs (about 225 mya),

later in birds (about 150 mya), and still later in bats (50–60 mya). Some frogs, lizards, and mammals have also evolved the ability to glide, presumably a precursor to flight. In order to land safely, such hang gliders must time their stall precisely at the right moment and place.

For many years, avian systematists classified Old World and New World vultures as close relatives, both thought to be allied to raptors (hawks and owls). DNA evidence indicates, however, that, although Old World vultures are indeed related to raptors, New World vultures are not and are instead descendents of common ancestors to storks and cranes. Morphological convergence was strong enough to actually mislead students of bird classification. Interestingly, a behavioral trait was conserved in the evolution of New World vultures: When heat stressed, storks defecate/urinate on their own legs to dissipate excess heat. New World vultures do this, whereas Old World vultures do not.

A brown bird of some African prairies and grasslands, the African yellow-throated longclaw (*Macronix croceus*), a motacillid, has a yellow breast with a black chevron "V." This motacillid looks and acts so much like an American meadowlark (*Sturnella magna*), an icterid, that a competent bird watcher might mistake them for the same species, yet they belong to different avian families. Another example is the North American

A brown bird of some African prairies and grasslands, the African yellow-throated longclaw (*Macronix croceus*), a motacillid, has a yellow breast with a black chevron "V." David Hosking/Photo Researchers, Inc.

Little Auk and the Magellan Diving Petrel, two superficially very similar aquatic birdsbelong to different avian orders.

Flightless birds such as the emu, ostrich, and rhea fill very similar ecological niches on different continents. Once thought to be convergent, DNA evidence now suggests that these ratites share a common ancestry. If ratites have indeed evolved from a common ancestor on Gondwana (a southern supercontinent ca. 190 M.Y.B.P.), they would not represent evolutionary convergence but instead are an example of a shared (and conserved) ancestral flightless state.

Live bearing in squamates

Live bearing, or viviparity, has evolved over 100 times among squamate reptiles (lizards and snakes) (Blackburn 1992), usually in response to cold climates. The probable mechanism behind the evolution of viviparity is that, by holding her eggs, a gravid female can both protect them from predators and, by basking, warm them, which would increase the rate of development. Eventually, such selective forces favoring egg retention could lead to eggs hatching within a mother and live birth (Huey 1977). This has happened even in geckos, all of which lay eggs except for one genus in New Caledonia and several related cold temperate New Zealand forms that bear their young alive. In some skinks and xantusiid lizards, embryos attach to their mother's oviducts and grow,

gaining nutrients during development via placental arrangements reminiscent of those in mammals.

Thorny ant-eating lizards

Many convergent evolutionary responses of lizards to arid environments are evident between continents. For example, Australian and North American deserts both support a cryptically colored and thornily armored ant-specialized species: The thorny devil, *Moloch horridus*, an agamid, exploits this ecological role in Australia, while its counterpart, the desert horned lizard (*Phrynosoma*), an iguanid, occupies it in North America. No Kalahari lizard has adopted such a lifestyle. Interestingly, morphometric analysis demonstrates that the thorny devil and the desert horned lizard are actually anatomically closer to one another than either species is to another member of its own lizard fauna, which are much more closely related (Pianka 1993).

Colorful tails

Colorful blue, red, and yellow tails have evolved repeatedly among distantly related lizards in many families (agamids,

An eastern meadowlark (*Sturnella magna*) sits atop a pole. This icterid is very similar to the African yellow-throated longclaw (*Macronix croceus*). Jim Zipp/Photo Researchers, Inc.

anguids, gymnopthalmids, lacertids, skinks, and teiids), presumably a ploy to attract a predator's attention away from the head to the tail, which can be broken off and regenerated should a predator attack it.

Running across Water

The New World iguanid *Basiliscus*, sometimes called "Jesus lizards" because they can run across the surface of water, have undergone convergent evolution with the Old World agamid *Hydrosaurus*. Both *Basiliscus* and *Hydrosaurus* have enlarged rectangular, platelike, fringed scales on their toes, which allow these big lizards to run across water using surface tension for support.

Convergence in sandy deserts

Open sandy deserts pose severe problems for their inhabitants: (1) Windblown sands are always loose and provide little traction; (2) surface temperatures at midday rise to lethal levels; and (3) open sandy areas offer little food, shade, or shelter for evading predators. Even so, natural selection over eons of time has enabled lizards to cope fairly well with such sandy desert conditions. Subterranean lizards simply bypass most problems by staying underground, and they actually

benefit from the loose sand because underground locomotion is facilitated. Burrowing is also made easier by evolution of a pointed, shovel-shaped head and a countersunk lower jaw, as well as by small appendages and muscular bodies and short tails. Such a reduced-limb adaptive suite associated with burrowing habits has evolved repeatedly among squamate reptiles in both lizards and snakes (Wiens and Slingluff 2001).

During the hours shortly after sunrise, but before sand temperatures climb too high, diurnal lizards scurry about above ground in sandy desert habitats. Sand-specialized lizards provide one of the most striking examples of convergent evolution and ecological equivalence. Representatives of many different families of lizards scattered throughout the world's deserts have found a similar solution for getting better traction on loose sand: Enlarged scales on their toes, or lamellae, have evolved independently in six different families of lizards: skinks, lacertids, iguanids, agamids, gerrhosaurids, and geckos (Luke 1986). A skink, *Scincus*, appropriately dubbed the "sandfish," literally swims through sandy seas in search of insect food in the Sahara and other eastern deserts. These sandy desert regions also support lacertid lizards (*Acanthodactylus*) with fringed toes and shovel noses. Far away in the Southern Hemisphere, on the windblown dunes of the Namib Desert of southwestern Africa, an independent lineage of lacertids, *Meroles* (formerly *Aporosaura*) *anchietae*, has evolved into a similar life-form. In North America, this body form has been

The New World iguanid *Basiliscus*, sometimes called "Jesus lizards" because they can run across the surface of water, have undergone convergent evolution with the Old World agamid Hydrosaurus. © Joe McDonald/Corbis.

Sandfish (*Scincus scincus*) escape from enemies by running along the surface and then suddenly diving into loose sand and swimming a short distance. © Chris Mattison/Alamy.

adopted by members of the iguanid genus *Uma*, which usually forage by waiting in the open and eat a fairly diverse diet of various insects, such as sand roaches, beetle larvae, and other burrowing arthropods. They also listen intently for moving insects buried in the sand and dig them up. Sometimes they dash, dig, and paw through a patch of sand and then watch the disturbed area for movements.

All of these lizards have flattened, duckbill-like, shovel-nosed snouts, which enable them to make remarkable "dives" into the sand even while running at full speed. The lizards then wriggle along under the surface, sometimes for over a meter. One must see such a sand-diving feat to appreciate fully its effectiveness as a disappearing act. Some Namib Desert lizards have discovered still another solution to gain traction on powdery sands: froglike webbing between the toes as seen in the geckos *Kaokogecko* and *Palmatogecko*.

Rock mimicry and varanid lizards

Other lizards that have undergone convergent evolution include rock mimics such as the North American round-tailed horned lizard *Phrynosoma modestum*, an iguanid, and the Australian agamid *Tympanocryptis cephalus*. New World teiids (*Tupinambis*) have converged on Old World varanids (*Varanus*):

Members of both genera are large predatory lizards with forked tongues used as edge detectors to find scent trails and track down their prey (other vertebrates).

Community convergence

Sometimes, roughly similar ecological systems support relatively few conspicuous ecological equivalents but instead are composed largely of distinctly different plant and animal types. For instance, although bird species diversities of temperate forests in eastern North America and eastern Australia are similar (Recher 1969), many avian niches appear to be fundamentally different on the two continents. Honeyeaters and parrots are conspicuous in Australia, whereas hummingbirds and woodpeckers are entirely absent. Apparently, different combinations of the various avian ecological activities are possible; thus, an Australian honeyeater might combine aspects of the food and place niches exploited in North America by both warblers and hummingbirds. An analogy can be made by comparing the "total avian niche space" to a deck of cards. This niche space can be exploited in a limited number of ways, and each bird population or species has its own ways of doing things, or its own "hand of cards," determined in part by what other species in the community are doing.

Resources

Books

Blackburn, Daniel G. 1992. "Convergent Evolution of Viviparity, Matrotrophy, and Specializations for Fetal Nutrition in Reptiles and Other Vertebrates." *American Zoologist* 32(2): 313–321.

Conant, Gavin C., and Andreas Wagner. 2003. "Convergent Evolution of Gene Circuits." *Nature Genetics* 34(3): 264–266.

Doolittle, Russell F. 1994. "Convergent Evolution: The Need to Be Explicit." *Trends in Biochemical Sciences* 19(1): 15–18.

Pianka, Eric R. 1993. "The Many Dimensions of a Lizard's Ecological Niche." In *Lacertids of the Mediterranean Basin*, ed. E. D. Valakos, W. Bohme, V. Pérez-Mellado, and P. Maragou. Athens: Hellenic Zoological Society.

Pianka, Eric R. 2000. *Evolutionary Ecology*. 6th edition. San Francisco: Benjamin-Cummings.

Wickler, Wolfgang. 1968. *Mimicry in Plants and Animals*, trans. R. D. Martin. New York: McGraw-Hill.

Periodicals

Amoutzias, Gregory D., David L. Robertson, Stephen G. Oliver, and Erich Bornberg-Bauer. 2004. "Convergent Evolution of Gene Networks by Single-Gene Duplications in Higher Eukaryotes." *EMBO Reports* 5(3): 274–279.

Geist, Valerius. 1978. *Life Strategies, Human Evolution, Environmental Design: Toward a Biological Theory of Health*. New York: Springer-Verlag.

Greene, Harry W., and Roy W. McDiarmid. 1981. "Coral Snake Mimicry: Does It Occur?" *Science* 213(4513): 1207–1212.

Grinnell, Joseph. 1924. "Geography and Evolution." *Ecology* 5(3): 225–229.

Hubbell, Stephen P. 2006. "Neutral Theory and the Evolution of Ecological Equivalence." *Ecology* 87(6): 1387–1398.

Huey, Raymond B. 1977. "Egg Retention in Some High-Altitude Anolis Lizards." *Copeia* 1977(2): 373–375.

Luke, Claudia. 1986. "Convergent Evolution of Lizard Toe Fringes." *Biological Journal of the Linnean Society* 27(1): 1–16.

Recher, Harry F. 1969. "Bird Species Diversity and Habitat Diversity in Australia and North America." *American Naturalist* 103(929): 75–80.

Wiens, John J., and Jamie L. Slingluff. 2001. "How Lizards Turn into Snakes: A Phylogenetic Analysis of Body-Form Evolution in Anguid Lizards." *Evolution* 55(11): 2303–2318.

Other

Dryden, Richard. "Evolution of Flight." Available from http://www.nurseminerva.co.uk/adapt/evolutio.htm.

Eric R. Pianka

· · · · ·

Evolution of the animal eye

The pervasiveness, intricacy, and complexity of animal eyes has long attracted the attention of biologists. Light sensitivity is present in at least two-thirds of all animal phyla, and conspicuous, image-forming eyes are present in approximately 95 percent of all described animal species, dominated by arthropods, mollusks, and vertebrates. In particular, eye intricacy and complexity has demanded the attention of evolutionists, resulting in documentation of a rich interplay between history, structure, and function that produced all animal eyes and that serves as a model for the evolution of complexity by natural processes.

What are eyes?

Eyes are the organs that detect light in many animals and some single-celled organisms such as certain dinoflagellates. Eyes extract important environmental messages by detecting the spatial orientation of light that has been emitted or reflected by nearby objects. Therefore, eyes are crucial for survival and communication in many animals. At minimum, an eye consists of a photoreceptive region adjacent to a pigmented region. The photoreceptive region senses light by converting light energy into a nerve impulse. The pigmented region serves to shield one side of the photoreceptor, allowing for the direction of light to be determined. Although potentially simple, the form and function of animal eyes ranges considerably, from the familiar complex eyes of vertebrates and insects to more exotic forms that use mirrors rather than lenses.

Natural selection and eye evolution

In 1802 William Paley (1743–1805) argued that because we know complex purposeful objects such as watches have intelligent designers, then a structure as complex and purposeful as an eye must also have been designed (Paley 1802). Philosophically, this is an argument by analogy that tenuously rests on the assumption that the complexity of a watch is the same as the complexity of an eye, an assumption that cannot be demonstrated. Despite its philosophical bankruptcy, the design argument is frequently resurrected in the form of intelligent design. More importantly, evolutionary biology long ago proposed an adequate mechanism for the origin of complex eyes by natural processes. In 1859 Charles Darwin (1809–1882) argued that descent with modification,

especially driven by natural selection, provides adequate explanation for the origin of eye complexity and purposefulness without reference to intelligent design. In a section titled "Difficulties on Theory," he indicated that even complex eyes could evolve by natural selection, via a gradual increase in eye complexity, starting from a simple light sensitive nerve.

Evolutionary biologists essentially universally accept Darwin's hypothesis that natural selection drove the eye complexity that is now observed, because the requirements for natural selection are present. First, natural selection requires heritable variation, which is known to exist in eye form. In humans, this includes eye color, eye length, chamber depth, and curvature of the cornea. In insects, heritable variation has been quantified in eye length, size, and color. Another requirement of natural selection is that heritable variations have differential effects on organismal survival. For instance, many New World monkeys lack the genes necessary for color vision. Females in a few species, however, can carry the alleles enabling trichromatic vision. As a consequence of this genetic variation among individuals, females with color perception forage for certain (colored) fruit more effectively than their dichromat counter-parts.

A third requirement for the ability of natural selection to produce eyes is that simple and complex forms must be connected by a graduated series of changes, each being functional and useful. Darwin originally argued for useful intermediates by describing eyes of varying complexity in different animals. More numerous studies of anatomy now allow a detailed understanding of extensive variation of the eyes of closely related animals, including gastropods. Research on cephalochordates (e.g., lancelets), urochordates (e.g., tunicates), hagfish, lampreys, and other vertebrates has suggested probable evolutionary stages in vertebrate ancestors from simple to complex eyes. These "morphological series" mainly support the requirement that eyes of intermediate complexity are functional and useful because the animals that possess them live with such eyes.

In addition to these requirements, natural selection must act quickly enough to produce complex eyes. This assumption has been tested using a conceptual model involving a linear series of eyes, arranged from simple to complex. Changes between each eye were quantified as a percent change in morphological shape. Making conservative assumptions about the rate of morphological change and population sizes allows the conclusion that eyes can evolve from simple photoreceptive spots to complex lens-eyes in only about 500,000 generations, which is consistent with the requirement that natural selection is able to act quickly enough, on a geological timescale, to evolve complex eyes. Given the existence of heritable variation, differential fitness, series of increasingly complex eyes, and sufficient time to evolve, evolutionists infer that natural selection has produced the amazing array of disparate eyes seen in animals today.

Disparate eye morphologies

At the level of overall morphology, eyes may be classed into one of two major structural categories: single-chambered or multichambered eyes, each with multiple subcategories. Each of the structural categories is present in a number of distantly related species. The origin of similar eye morphologies through convergent evolution is often related to functional demands: Many species share similar visual constraints because of their habitat, lifestyle, and ecology.

Pit eyes

In single-chambered eyes, all photoreceptors lie along the interior of a single cup-shaped chamber. The simplest example of such an eye is the pit eye found in some cnidarians, planarians, annelid worms, crustaceans, gastropods, cephalopods (squid and octopus) and bivalve mollusks, echinoderms (sea urchins and sea stars), and early chordates. Pit eyes can form only crude, shadow-based images.

Camera-type eyes

More complex single-chamber eyes improve image resolution over pit eyes, mainly by relying on lenses to refract light onto the retinal layer (e.g., in some jellyfish, vertebrates, spiders, cephalopods, gastropods, and insect larvae) or by using concave mirrors to reflect light onto the retinal layer (e.g., in some scallops and crustaceans). In "camera-type," single-chambered eyes, such as those of vertebrates and cephalopods, a spherical lens focuses light on a retina at the back of the eye. Eyes of many nocturnal vertebrates make use of reflective tissue, called a tapetum, which lies behind the retina to reflect photons back through the photoreceptors, maximizing the number of photons captured in dim light conditions and resulting in eye shine in such animals as cats, deer, and raccoons.

Especially complex lenses, called graded refractive index lenses, have evolved separately in cephalopods and aquatic vertebrates. Lenses of aquatic animals face higher demands than lenses of terrestrial animals. Because lens cells are composed mainly of water, the terrestrial lens refracts the path of light as it enters the lens from the air. Aquatic lenses must be much more powerful because they cannot take advantage of this air-to-water refractive transition. Yet a more powerful lens requires increased curvature, which often results in greater image aberration in lenses of a given size. Aquatic lenses have resolved this dilemma through the evolution of the graded refractive index lens. Arranged like an onion, the core of the lens bends light very significantly (i.e., it has a high refractive index), while the layers outside the core have

Raccoon eyes make use of reflective tissue, called a tapetum, which lies behind the retina to reflect photons back through the photoreceptors, maximizing the number of photons captured in dim light conditions. © Mark L Stephenson/Corbis.

progressively lower refractive indices so as to bend light less and less. The rings of the "onion" thus form a graded series from high refraction in the middle to low refraction on the outside to enable high refractive power with little aberration.

The evolutionary history of the complex graded refractive index lenses of cephalopods has been studied in detail. The proteins making up cephalopod lenses are very similar to a digestive enzyme found in their livers, indicating that a liver protein duplicated during evolution, and new copies of the protein evolved a new function as the main structural component of lenses. Lens proteins, generally termed crystallins, are commonly "co-opted" in this manner for use in many animal eyes from proteins with other bodily functions. Such cases provide a clear example of how the components that make up eyes are not universally shared. Cephalopod crystalline proteins diversified by duplication, and the duplicates vary in electrostatic charge, which differentially affects the refractive index of the lens. Expressing slightly different crystallins in a gradient across the lens allows for a graded refractive index.

In contrast to the hard cephalopod lens, lenses of vertebrate eyes are malleable. The process of adjusting the lens to bring an object into focus is called accommodation. Whereas animals with hard lenses contract the eye's ciliary muscles to move the entire lens forward or away as needed for accommodation, vertebrate eyes rely on the ciliary muscles to manipulate the thickness of the soft lens to adjust the focus for objects at various distances.

Compound eyes

Multichambered eyes differ from single-chambered eyes in that individual photoreceptors are located in separate pigment cups. A simple compound eye results when such structures form aggregations, as seen in a few bivalves and annelids. The familiar compound eyes found in arthropods, however, exhibit greater optical complexity. Arthropod compound eyes are comprised of many individual units called ommatidia. At the ommatidium's surface is the facet, containing a corneal lens (a multisided transparent cuticle). Beneath the facet is a crystalline cone composed of a group of four or so cells that secrete the lens crystals. A group of photoreceptive cells, commonly eight or nine, lie beneath each facet and cone.

Compared to a single-chambered eye of comparable size, compound eyes have much shorter focal lengths and, consequently, provide poorer spatial resolution at a given distance. This means that, for most insects and crustaceans, objects generally must be within a few millimeters to be sufficiently resolved. Even bees, considered to have highly resolved vision among insects, must be sixty times closer to an object than would a human to achieve comparable resolution. Increasing the density of ommatidia improves resolution. Dragonflies, for instance, bear upward of 30,000 ommatidia, enabling them to spot small prey at a distance of several meters. In addition, some insects have evolved specialized regions of wide, flattened ommatidia called fovea to cope with such shortcomings in resolution. Despite such nearsightedness, compound eyes in arthropods allow extreme close-up detail and provide a wide viewing angle with sensitivity to quick motion. Different ommatidia are excited as an object moves across the field of vision, meaning that moving objects receive a greater response than stationary objects. The overall structure of arthropod compound eyes falls into two categories: superposition and apposition.

Superposition compound eyes. Superposition eyes possess a gap between the lens and the photoreceptor cell layer. Consequently, each photoreceptor cluster receives the light refracted through as many as 2,000 lenses. Alternative forms of superposition eyes employ "mirrors" to reflect light onto photoreceptors. Namely, some crustaceans incorporate reflective tissue within the eye's facets to serve as corner mirrors, rerouting the path of incoming light to produce a single, erect image on the photoreceptive cell layer. By allowing each photoreceptor to receive light from multiple lenses or mirrors, superposition eyes are extremely sensitive to light. This light-maximizing morphology appears to have evolved independently from apposition-type eyes in multiple arthropod lineages and is commonly found in nocturnal insects and in crustaceans, which mainly function in low-light conditions.

Apposition compound eyes. In apposition eyes, each ommatidium is isolated from its neighbor by a layer of pigment cells. This structure allows for greater resolution at the cost of less light sensitivity when compared to superposition eyes. As such, apposition eyes are predominant among diurnal arthropods such as crabs and butterflies, as well as among some polychaete annelids. Some apposition eyes exhibit labile screening pigments, giving rise temporarily to a superposition-like eye. For example, in crayfish and in praying mantises, the pigment that normally separates the ommatidia will shift its concentration to the base of the unit, permitting light entering through the lens of an ommatidium to reach photoreceptors in adjacent ommatidia. This mechanism enables greater light sensitivity in dim settings at the expense of image resolution.

Photoreceptor cells

Central to all disparate morphological types of animal eyes are photoreceptor cells: nerve cells bearing specializations for the detection of light. The evolution of photoreceptor cell types has been a matter of discussion for decades. In the 1960s Richard M. Eakin proposed the existence of two classes of photoreceptors, ciliary and rhabdomeric. The cell types were first defined in terms of their morphology and physiology. Morphologically, ciliary cells employ extensive hairlike cilia, whereas rhabdomeric cells rely on tubelike microvilli. Physiologically, ciliary cells respond to light by interrupting a signaling nerve, whereas rhabdomeric respond by initiating a nerve signal. Eakin further suggested that the main photoreceptor types could define major animal groups. For example, animals whose eyes primarily use ciliary photoreceptors include vertebrates and other chordates, echinoderms, and bryozoans ("moss animals"). Animals that primarily use rhabdomeric photoreceptor cells include arthropods and annelids.

Photoreceptor cell morphology is not diagnostic of major animal groups, because many animals have both ciliary and rhabdomeric photoreceptor cells. For example, scallop eyes have dual adjacent retinas with different cell types. Some annelids use rhabdomeric cells in their eyes, but ciliary photoreceptors develop in their brains. Furthermore, molecular evidence suggests that vertebrates also have both cell types: While the principal vertebrate photoreceptors are ciliary, adjacent cells called retinal ganglion cells express the molecular signature of rhabdomeric photoreceptor cells. These retinal ganglion cells serve the eye by integrating

Dragonflies bear upward of 30,000 ommatidia, enabling them to spot small prey at a distance of several meters. © Fritz Rauschenbach/Corbis.

Because of different visual needs, vertebrates vary in the proportion of rod and cone cells, with some species lacking one type altogether. The tokay gecko (*Gekko gecko*) retinas have only rod cells. John Mitchell/Photo Researchers, Inc.

information from the rods and cones. The presence of both cell types in several divergent animal groups indicates that the two cell types evolved very early in the history of animals, persisted during the evolution of many taxa, but were lost in some major groups.

The ciliary photoreceptors of vertebrate retinas are comprised of two classes of cells, called rods and cones. Rods are specialized for scotopic, or dim-light, vision, whereas cones are specialized for photopic, or bright-light, vision. Different cones are further specialized to respond to different wavelengths of light, allowing color vision in many vertebrates. The evolutionary origins of rods and cones allowed vertebrate animals to overcome a trade-off between sensitivity and acuity. Rods are highly sensitive to light but have lower spatial resolution, slower response, and higher noise. Cones are much less sensitive to light (about one-tenth as sensitive as rods in humans) but have high resolution, faster response, and lower noise. Interestingly, because of different visual needs, vertebrates vary in the proportion of rod and cone cells, with some species lacking one type altogether. For example, skate and Tokay gecko retinas have only rod cells. In general, and as one might expect, nocturnal animals have higher ratios and diurnal animals have lower ratios of rods to cones.

The biochemical basis of vision

The molecular biology underlying vision has several highly conserved features, even among animals displaying dramatically different types of eyes. All photoreceptors carry a suite of signaling molecules that comprise a multistep "phototransduction pathway" that converts a photon signal into the nerve impulse that is relayed to the brain or ganglion. Vision begins when light-reactive chemicals called chromophores, such as one Vitamin A derivative known as retinal, isomerize after absorbing a photon of light. (Specifically, a double bond on the retinal's eleventh carbon is energized by light from a range of wavelengths. Once this bond has absorbed energy from a photon, the retinal is transformed from an 11-cis to an all-trans isomer.) Chromophores are bound to abundant proteins, called opsins, which are embedded in the membranes of the cilia or microvilli of photoreceptor cells. The light-induced isomerization of the chromophore triggers a conformational change in a region of the opsin molecule contacting the cytoplasm of the photoreceptor cell. This change to the opsin's shape reveals key enzymatic binding sites to facilitate the initiation of phototransduction and is therefore a fundamental and highly conserved aspect of animal vision.

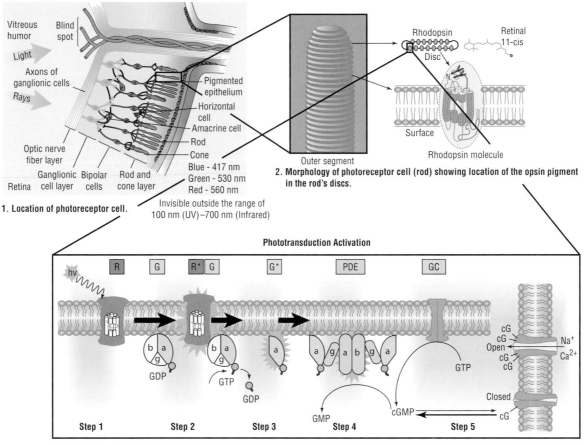

1. Location of photoreceptor cell.

2. Morphology of photoreceptor cell (rod) showing location of the opsin pigment in the rod's discs.

3. Ciliary phototransduction occurring within discs of rod cell.

The molecular pathway used for vision. Reproduced by permission of Gale, a part of Cengage Learning.

All animal eyes examined to date use opsin for photo-transduction, and this gene can be traced to a common ancestral gene present since early in the history of animals. The most distant relatives of humans known to possess opsin are cnidarians, including jellyfish, hydra, and corals. There is no evidence of opsins in sponges, the most distant animal relatives of humans. Opsins are also absent from the genomes of the closest relatives of animals, including the choanoflagellate *Monosiga* and several fungi. Animal opsins belong to a class of proteins called G-protein-coupled receptors (GPCRs) that bond to signaling molecules called G proteins. Non-opsin GPCRs are present in sponges and nonanimals. Therefore, opsins likely originated within animals when a GPCR protein became light sensitive by gaining the ability to bind a chromophore.

After initiation of opsin, subsequent steps in phototrans-duction cascades involve molecules commonly used in other G-protein signaling pathways, which are fundamental for taste, olfaction, and hormone-mediated responses. The origins of several of these molecules predate animals, indicating that the evolution of animal photoreception has its origins in simpler and more ancient processes of transmitting signals from outside to inside cells.

Visual physiology

While natural selection has fine-tuned the cellular machinery and morphological structure of eyes so as to improve their sensitivity and resolution, eyes have evolved to perceive additional properties of light. In particular, many animals have independently evolved color vision through photoreceptors sensitive to differing wavelengths of light. Similarly, many eyes have evolved to take advantage of polarized light, using photoreceptors aligned to receive light from opposing electric fields.

Color vision

Specific wavelengths of light (known as colors) are produced when white light (all wavelengths) interacts with matter, and the individual wavelengths react differently following absorption, scatter, reflection, or refraction. Color vision is possible when different photoreceptors respond to specific and different wavelengths of light. In vertebrate eyes, different cone cells fulfill this role. Vertebrates possess up to four distinct cone types, each characterized by a different opsin protein that responds most strongly to a particular wavelength of light. Most animals with color vision possess three such opsins, although a few arthropods exhibit tremendous diversity

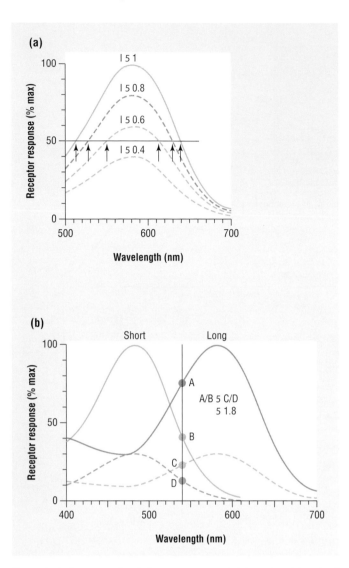

Figure 1. At least two visual pigments are needed for color vision. (top) With only one pigment the response of a receptor does not distinguish between intensity and wavelength. A 50 percent response could have been produced by any of the arrowed combinations. (bottom) With two different visual pigments the ration of stimulation (A/B or C/D) is specific to a particular wavelength, and unaffected by intensity level. Reproduced by permission of Gale, a part of Cengage Learning.

of opsin sensitivities. For example, physiological evidence indicates that the mantis shrimp reigns with about fifteen opsins finely tuned to distinct wavelengths throughout the human-visible and ultraviolet (UV) spectrum. Lacking color vision entirely, monochromatic animals (including cephalopods and many marine mammals) possess only one opsin and cannot differentiate among wavelengths.

For successful wavelength discrimination, all of the opsins expressed within a given type of cone cell must respond most strongly to the same wavelength. Wavelength specificity in the opsin photopigment can result from either the presence of an altered chromophore or from slight changes in the amino acid composition of opsin that "tune" its sensitivity toward a particular wavelength. In contrast to cone cells, all

rod cells contain opsin that responds maximally to the same wavelength and thus report only the presence or absence of visible light.

Opsin sensitivity alone does not give rise directly to color vision. Color perception is foremost a neurological process in which the brain receives photoreceptor input (containing information about the relative amounts of each wavelength) and integrates these inputs to assign a color. The perception of a particular hue is the calculated sum of precise inputs from several wavelength-specific photoreceptors, taking into account the relative amount of signal from each. For example, humans have three cone opsins corresponding to yellow-green (wavelength, 564 nanometers), green (534 nanometers), and blue (420 nanometers). For an object emitting a wavelength of 500 nanometers, the opsins from blue-sensitive and green-sensitive cones will be equally excited. Non-photoreceptor cells in the retina (monopolar cells) receive the inputs and pool the relative responses of each cone type before passing the message on to the optic lobe, where a blue-green color is perceived. By integrating responses from multiple cones, neural processes can infer information about the presence of wavelengths for which no single opsin is attuned.

Color and UV vision functions analogously among many arthropods. Photoreceptors within an ommatidium may belong to separate visual subsystems depending on their function. Of the eight or nine photoreceptors in an ommatidium, roughly one to three bear opsins attuned to blue, green, or UV wavelengths. The remaining cells, analogous to vertebrate rods, are involved in motion detection and dim-light vision.

Although most mammals lack opsins sensitive to ultraviolet wavelengths, many animals including mice, fish, birds, cnidarians, insects, and crustaceans possess such visual pigments and rely extensively on UV discrimination during foraging or migration. For example, UV cues can alter foraging behavior in birds preying on insects that appear cryptic in the visible color spectrum. UV vision is widespread among pollinating insects, which rely on UV-pigmented patches on petals to recognize flowers. Finally, some animals use UV vision when assessing potential mates.

Because slight changes to amino acids in opsin can significantly alter the wavelength of light it responds to, these spectrally tuned opsins have arisen repeatedly throughout evolution. The ancestor of vertebrates likely possessed five opsin genes, probably tuned to specific wavelengths. Throughout the vertebrate radiation some of these genes have been lost numerous times. Most mammals, for example, have lost two opsins during their evolution and have dichromatic vision at present. Some vertebrate groups, however, have independently regained color discrimination as a result of a duplication of the opsin gene. Following duplication, mutations that change specific amino acids responsible for the pigment's spectral sensitivity alter one of the opsin duplicates. Depending on the animal's given ecology and light needs (for foraging, hunting, mate recognition, etc.), natural selection may favor new copies of opsin that allow for improved discrimination of colors. For instance, a single amino acid change in opsin has independently occurred in many bird lineages facilitating UV discrimination. Because

Animals such as mantis shrimp (*Odontodactylus scyllarus*) send signals with polarized light for intraspecific communication. Georgette Douwma/Photo Researchers, Inc.

opsin's spectral sensitivity can shift with relatively few mutations, color vision has evolved (and reevolved) frequently throughout both vertebrate and arthropod taxa. As a result, color vision may be considered another example of convergence in eye evolution.

Polarized vision

Polarized vision occurs when photoreceptors differentiate among the planes of polarization in incoming light. Pure sunlight is unpolarized because it is made up of photons whose electric fields include every direction. Sunlight, however, is scattered by air particles such that the electric fields of the scattered light share the same plane, creating light that is polarized. Similarly, when light is reflected off water or nonmetallic surfaces, the electric vectors of the reflected light are all polarized parallel to the reflective surface. With the exception of some fish and amphibians, most vertebrate eyes interpret such light as glare because their ciliary photoreceptors are not as structurally suited for detecting polarized light as the rhabdomeric cells of invertebrate eyes.

Capitalizing on this distinction, many arthropods and cephalopods have evolved a system in which input is processed from photoreceptors responding to either of two planes of polarization. Much like with wavelength discrimination in color vision, the neural processing of input from cells sensitive to different planes can provide a tremendous amount of additional spatial/visual information. Several invertebrates exploit the ability to discriminate polarized light. Bees and ants are able to use the polarized light resulting from the scattered sunlight for navigation. Aquatic species have also profited in spotting prey items because of the improved image contrast that polarized vision provides underwater. In addition to predation, animals such as mantis shrimp and cuttlefish send signals with polarized light for intraspecific communication.

In eyes of polarized-insensitive vertebrates, the opsins embedded within photoreceptor membranes can be found in any orientation for any cell. Because only some of the opsins in any cell will be orientated to photons of a particular electric field, single cells are not known to distinguish among the various planes from which light may be arriving and will initiate phototransduction regardless of the light's electric vector. In many invertebrate eyes, however, the microvillous structure of the photoreceptors facilitates polarized vision. This is because the majority of any given cell's opsins wind up arranged along the cell's microvillar axis so that their chromophores are parallel to a photon's electric vector. As a consequence of this geometric structure, each photoreceptor is able to distinguish light polarized in a given plane.

How many times did eyes evolve?

Although the number of times eyes evolved is a commonly debated question, there can be no direct numerical answer because some components of eyes are shared among vastly different animals, whereas other components are shared by only some animals. This pattern indicates that animal eyes are complex patchworks of components that have associated and disassociated during evolution in different animal lineages. Although some components probably share a continuous heritage as part of all or at least most animal eyes, other components are relative newcomers to eyes only in certain animals.

Despite the vast disparity in the form of different eyes, many molecular components are now known to be conserved across phyla. Opsins, the visual pigments used in all animal eyes ever examined, can be traced to a single common ancestral gene, underscoring the common ancestry and unity of all life. Additionally, very similar genes are involved in the development of vastly different eyes, such as those of vertebrates and flies. A vertebrate gene (mouse *PAX6*) can rescue the function of a mutation in a homologous fly gene (*eyeless*), indicating that some aspects of function of *PAX6* homologs are conserved between these distantly related species. Several other genes are also now known to have conserved function in the development of eyes. Most of these genes are transcription factors, which regulate the activation of other genes.

Yet some components of eyes are not universal. Lens proteins vary tremendously both within and between different animal lineages, and they are often closely related or identical to proteins used for other functions outside the lens. Another example of nonuniversal components causes the difference between ciliary and rhabdomeric phototransduction cascades. Although both cascades begin with opsin and a G protein, the crucial response at the end of phototransduction signaling pathways differ. Ciliary G proteins signal through an intermediary, PDE, to *close* CNG ion channels. The closing of this channel hyperpolarizes the cell, interrupting an otherwise continuous nervous signal. In contrast, rhabdomeric G proteins signal through PLC to *open* TRP ion channels, which depolarizes the cell and initiates a nervous signal.

Just as origins of eyes cannot easily be distilled into discrete evolutionary events, neither can losses of eyes. Animals whose evolution includes a photic transition because of habitat shift (shallow to deep sea) or a lifestyle shift (pelagic to burrowing) often morphologically lack eyes, while close relatives inhabiting light environments continue to bear eyes. Nevertheless, loss of eye morphology does not mean all components of eyes are lost immediately. In fact, work with blind cavefish has

Adult cavefish eyes probably were lost as a "trade-off" for enhancing features that are highly beneficial in dark environments. © blickwinkel/ Alamy.

revealed that functional eyes can develop in the offspring of crosses of blind parents from evolutionarily distinct cave-dwelling populations. In this case, each cave-dwelling adult lacked eyes because of mutations in different genes. When crossed, offspring received at least one functional copy of all necessary genes, indicating that many genetic components for vision survived, even after the loss of vision itself. During evolution, adult cavefish eyes probably were lost as a "trade-off" for enhancing features that are highly beneficial in dark environments. An expanded jaw is presumably an adaptation for improved foraging that has arisen at the expense of eyes. Cavefish eyes begin to develop in very young fish but are lost by adulthood because of cell death. A similar adult phenotype is found among cave-dwelling amphipods, which have evolved larger sensory antennae, encroaching on the space used by eyes in surface-dwelling species.

Summary

A survey of the animal kingdom reveals tremendous diversity among eye morphologies. These disparate forms represent the numerous "solutions" evolution has found to the challenges of detecting and using light. Yet because light perception presents the same challenges for many taxa, even some distantly related animals have independently evolved similar eye features such as lenses, color and polarization vision, and single- and multichambered eyes. Although such general features may have been lost and gained often during the course of evolution, the underlying cellular and genetic components, such as opsin and *PAX6* and other developmental genes, have been conserved in all animals examined to date. As such, eyes provide a detailed and rich case study in how complex and intricate systems can arise by natural processes such as natural selection, through the dynamic interplay of historical, physical, and functional constraints.

Resources

Books

Darwin, Charles. 1859. *On the Origin of Species by Means of Natural Selection; or, The Preservation of Favoured Races in the Struggle for Life*. London: John Murray.

Horváth, Gábor, and Dezsö Varjú. 2004. *Polarized Light in Animal Vision: Polarization Patterns in Nature*. Berlin: Springer.

Land, Michael F., and Dan-Eric Nilsson. 2002. *Animal Eyes*. New York: Oxford University Press.

Paley, William. 1802. *Natural Theology: Evidences Of the Existence and Attributes of the Deity Collected from the Appearances Of Nature*. Philadelphia: Printed for John Morgan by H. Maxwell.

Salvini-Plawen, Luitfried, and Ernst Mayr. 1977. "On the Evolution of Photoreceptors and Eyes." In *Evolutionary Biology*, Vol. 10, ed. Max K. Hecht, William Campbell Steere, and Bruce Wallace, pp. 207–263. New York: Plenum Press.

Warrant, Eric, and Dan-Eric Nilsson, eds. 2006. *Invertebrate Vision*. Cambridge, UK: Cambridge University Press.

Periodicals

Arendt, Detlev. 2003. "Evolution of Eyes and Photoreceptor Cell Types." *International Journal of Developmental Biology* 47(7–8): 563–571.

Briscoe, Adriana D., and Lars Chittka. 2001. "The Evolution of Color Vision in Insects." *Annual Review of Entomology* 46: 471–510.

Eakin, Richard M. 1979. "Evolutionary Significance of Photoreceptors: In Retrospect." *American Zoologist* 19(2): 647–653.

Fernald, Russell D. 2006. "Casting a Genetic Light on the Evolution of Eyes." *Science* 313(5795): 1914–1918.

Gehring, Walter J., and Kazuho Ikeo. 1999. "*Pax 6*: Mastering Eye Morphogenesis and Eye Evolution." *Trends in Genetics* 15 (9): 371–377.

Jacobs, Gerald H., and Mickey P. Rowe. 2004. "Evolution of Vertebrate Colour Vision." *Clinical and Experimental Optometry* 87(4–5): 206–216.

Lamb, Trevor D., Shaun P. Collin, and Edward N. Pugh Jr. 2007. "Evolution of the Vertebrate Eye: Opsins, Photoreceptors, Retina, and Eye Cup." *Nature Reviews Neuroscience* 8 (12): 960–976.

Nilsson, Dan E., and Susanne Pelger. 1994. "A Pessimistic Estimate of the Time Required for an Eye to Evolve." *Proceedings: Biological Sciences* 256(1345): 53–58.

Plachetzki, David C., and Todd H. Oakley. 2007. "Key Transitions during the Evolution of Animal Phototransduction: Novelty, 'Tree-Thinking,' Co-option, and Co-duplication." *Integrative and Comparative Biology* 47(5): 759–769.

M. S. Pankey
T. H. Oakley

Evolution of flight

When animals evolve the ability to fly, they substantially improve their capacity to obtain resources and exploit their environment. Flight allows animals to reach new habitats; travel faster, farther, and more economically; descend quickly and safely from great heights; and escape from nonflying predators. Flight undoubtedly played a role in the great success and diversity of flying animals. For example, scientists have named more species of birds than any other group of terrestrial vertebrates; insects, the vast majority of which fly, have more known species than all the rest of the animals combined.

Flight means moving through the air by using a wing or winglike surface to produce *lift*, an upward force. Lift is proportional to the area of the wing surface, so bigger wings tend to be more effective. Animals can fly by either *gliding*—gravity-assisted flight without power—or *flapping*—powered flight not dependent on gravity. Many animals have evolved gliding, but powered flight has arisen only a handful of times in evolutionary history (Alexander 2002).

A gliding animal uses gravity to maintain flight. Much like a bicycle coasting downhill, a glider can use accumulated momentum to fly level (or even climb) briefly, but the only way to continue a glide is to descend relative to the air. In still air, a glider may be able to travel a substantial distance, choose a landing spot, and land at a safe speed, but it cannot sustain level flight. One index of a glider's aerodynamic effectiveness is its *glide angle*, which is the angle of the glider's path below the horizontal—the lower, the more effective.

Biologists have traditionally classified animals with glide angles of less than 45 degrees as "gliding" and those that descend at greater than 45 degrees as "parachuting," but this distinction is arbitrary and misleading. First, many gliding animals readily glide at angles both greater than and less than 45 degrees, as needed. Second, any animal that directs its descent significantly away from the vertical is, aerodynamically speaking, gliding, regardless of the glide angle. In contrast, a parachuting animal uses structures or behaviors to increase its drag and slow its fall, but does not direct its descent away from the vertical, aside from being carried by wind. In reality, the two processes overlap, and steep glides can be difficult to distinguish from parachuting.

Wings produce lift whenever they move forward through the air, but without a forward push wings can only glide, using gravity to keep moving. Animals achieve powered flight by flapping their wings to produce *thrust*. Thrust is a forward force that offsets the aerodynamic drag on an animal's body. Thrust keeps the animal moving forward through the air and allows the animal to maintain level flight or climb. Flapping flight is thus synonymous with powered flight in animals. Although many animal lineages have evolved gliding, only insects, pterosaurs, birds, and bats have evolved powered flight.

Flapping flight is a highly specialized ability. Powered flyers require a nervous system that can manage both the complex movements of flapping and the control task of steering in three dimensions. They also have specialized anatomy in addition to wings. Bats, for example, have modified hip and leg joints that allow the legs to sprawl out to the side and adjust the wing shape, and birds have and some pterosaurs had lightweight beaks instead of comparatively heavy bony teeth. Birds also have a chest muscle, the supracoracoideus, located below the wing. Its tendon wraps around a pulley-like structure in the shoulder so that it pulls up, rather than down, on the wing, increasing the distance the wing can be elevated. Flapping flight also requires more power than walking. Powered flyers thus devote a large fraction of their body mass—typically more than 20 percent in birds—to flight muscles.

Flight places a premium on lightweight, yet strong, structures. Air-filled (*pneumatic*) bird bones are a familiar expression of this principle. The lighter a flyer, for a given wing area, the slower it can fly. Larger wing area can also reduce flight speed, but only as long as it does not greatly increase overall weight—weight per unit wing area is called *wing loading*, and any flyer's flight speed is directly proportional to wing loading. This relationship between weight and wing area is particularly critical for gliders, which cannot use flapping to lower their flight speed at touchdown. Thus, large wing area and lightweight wing structures are both beneficial.

Gliding animals

A surprising variety of animals have evolved the ability to glide. At least nine mammal lineages (including over sixty species) have separately evolved gliding, including flying squirrels (Pteromyinae), sugar gliders and related Australian

Bats, such as the heart-nosed bat (*Cardioderma cor*), have hip and leg joints modified so the leg can sprawl out to the side and adjust the wing shape. Image by Merlin Tuttle.

marsupial gliders (Petauridae), and colugos or "flying lemurs" (not true lemurs, but members of their own order, Dermoptera). These animals all have a wing membrane or *patagium* attached along the front and hind limbs and lateral body margin. When these mammalian gliders stretch their limbs out into the typical spread-eagled flight posture, the patagium forms a nearly square wing (Jackson 2000). Gliding mammals are universally *arboreal* (tree-dwelling), and gliding probably began with some animals developing enlarged arms to stabilize or steer leaps from branch to branch. Such stabilizing would also have helped during accidental falls from branches or during jumps to escape from predators. Flaps or webs of skin would have evolved to increase aerodynamic surface area, improving glide angles and extending the length of leaps. Pressure to further reduce glide angle and increase lengths of glides would cause the skin flaps to expand to form the broad patagium of modern gliders.

The gliding ability of extant gliding mammals almost certainly evolved to its present level as an adaptation to improve locomotion through forest canopies, while also reducing the risk of predation while walking on the ground. Gliding from the crown of one tree to the trunk of a neighboring tree is much faster than climbing down, running across the intervening ground, and

climbing back up. It is also faster than running along a likely circuitous route along branches from one tree to another. With this faster locomotion, a glider can travel faster, search a wider area, or spend more time at a location with abundant food. Moreover, this glide-down, climb-up locomotion requires less metabolic energy, under certain conditions, than running through the canopy to travel among the same trees. If a squirrel-sized animal can maintain a glide angle of approximately 20 degrees or less, then gliding and climbing uses less energy—fewer calories—than running through the foliage (Scholey 1986a). Living gliders such as flying squirrels and sugar gliders are reasonably maneuverable. They can change directions in midflight, make flights with two or three sharp turns, or even make U-turns and return to their starting tree (Wells-Gosling 1985). Gliding does have drawbacks, however. All gliding mammals are nocturnal (or largely so), in spite of the fact that they depend on vision to judge flight speed and to locate landing spots. Gliding animals are more visible as they leave foliage to glide across open areas, and their flight paths are fairly predictable, so they are at great risk of attack by predatory birds. Bird predation is thus a major factor selecting for night gliding.

Various species of lizards, frogs, snakes, and fish have also evolved gliding. Lizards in the Southeast Asian genus *Draco*

Sugar gliders (*Petaurus breviceps*) can change directions in midflight, make flights with two or three sharp turns, or even make U-turns and return to their starting tree. ANT Photo Library.

(Agamidae) are commonly called "flying dragons" because of their highly developed gliding. Rather than having a patagium stretched between their arms and legs, flying squirrel-style, these lizards have a patagium running along their flanks supported by modified ribs. These ribs can spread out the semicircular patagium, somewhat like an umbrella, or fold it flat against the lizard's flanks when not in use. Because *Draco*'s arms and legs are completely free of the wing surface, and because its patagium folds up unobtrusively when the lizard is not gliding, these animals are quite agile when clambering about on tree trunks and branches. The patagium and flexible, springy ribs are controlled by a complex set of muscles that apparently can change the wing's shape for steering. These lizards are agile enough gliders that territorial males and the intruding males they chase often perform acrobatic, zigzag gliding pursuits (Colbert 1967). Although *Draco* evolved gliding more recently in evolutionary history, at least three extinct, lizard-like reptiles evolved a very similar, rib-supported wing. The oldest are the kuehneosaurs such as *Icarosaurus*, which had very light, hollow bones and a rib-supported wing larger in proportion to its body than that of *Draco*. *Icarosaurus* lived in the late Triassic, more than 200 million years ago (Colbert 1970).

Gliding geckos (genus *Ptychozoon*) also have lateral flaps of skin, as well as webbed hands and feet, to provide gliding surfaces. The lateral flaps of gliding geckos are held out by the pressure of flowing air with no supporting bones, and their glide angles are not as shallow as those of *Draco* (Young, Lee, and Daley 2002). These geckos seem to be just as maneuverable as *Draco*, however, probably because their enlarged, webbed hands and feet give them powerful steering abilities.

Most frogs also have webbed feet, so they are already equipped with potential gliding surfaces. Various species of arboreal frogs have, indeed, evolved gliding, including many species of tree frogs (Hylidae and Rhacophoridae).

Snakes may seem the least likely of animals to evolve gliding, but the Southeast Asian genus *Chrysopelea* has done just that. Commonly called "flying snakes," they routinely fling themselves off of tree branches, spread their ribs to flatten out their bodies, and descend at glide angles in the neighborhood of 30 degrees (Socha 2002). *Chrysopelea* can keep its body in a nearly horizontal plane while gliding, which is an amazing control achievement in its own right, given the elongated body form of a snake. It also undulates during its glide, rather like an eel swimming, but whether this aids the glide or is a vestigial expression of locomotor behavior from some other context is not known. Gliding lizards, frogs, and snakes are all arboreal, and almost surely evolved gliding for the same reasons as flying squirrels—escape from arboreal predators and avoidance of those on the ground, or improved locomotion through forest canopies, or both.

Due to high flight speeds, one important challenge all arboreal gliders face is landing. Any wing must move through the air fast enough to avoid a stall. If the wing moves too slowly, the smooth flow of air over the wing is disrupted, greatly reducing lift and increasing drag, and causing the animal to fall out of the air. (Even small gliders, such as small flying squirrels [Thorington and Heaney 1981] and gliding geckos [Heyer and Pongsapipatana 1970] glide at nearly 20 miles per hour.) Gliders usually fall steeply to build up speed at the beginning of a flight before leveling out for the more horizontal gliding phase. Typical gliding speeds are too fast for a safe landing, so most gliders end with a short, sharp pull-up just before reaching the intended perch (Colbert 1967; Wells-Gosling 1985). This brief climb allows gravity to slow the flight speed to or below stalling speed while the animal's inertia carries it upward to the perch. The animal alights with its wing stalled, and at a speed much lower than its minimum glide speed. Judging when to pull up for landing is just one of several sophisticated mental abilities any glider needs, which also include an ability to estimate distances accurately, to decide whether potential landing sites are within range, and to control the gliding flight so as to arrive at the perch with just enough airspeed for landing.

Flying fish (Exocoetidae) may be the only gliding animals that did not evolve their aerial ability by leaping or falling from trees. Flying fish leap through the surface of the ocean, spread their enlarged, winglike pectoral fins, and glide considerable distances. Since the nineteenth century, biologists had argued about whether flying fish actually flap their wings and fly under power. In the early 1940s, the marine biologist Charles M. Breder Jr. combined forces with the great photographic genius Harold E. Edgerton to make stroboscopic photos of flying fish in flight. Their photos demonstrated conclusively that flying fish do not flap, but they do have a sort of powered phase in which they beat their tails in the water briefly after emerging. The lower lobe of a flying fish's tail is longer than the upper lobe, so that the fish can actually keep the lower lobe of its tail in the water after

Flying lizards (*Draco*) glide from tree to tree using patagia, wing-like structures formed as they extend their ribs outward. Illustration by Bruce Worden.

breaking through the surface. The fish beats its tail furiously for a second or so to accelerate to a higher airspeed, and then climbs up away from the surface. Flying fish can reach as high as 8 meters above the surface before beginning their inevitable glide back down, and reach distances of 50 meters or more. At the end of the glide, they can either drop back into the water or dip their tail into the water to accelerate for another glide. Some halfbeaks, close relatives of flying fish, have slightly enlarged pectoral fins and make brief, high-speed, squirming dashes over the water surface. Flying fish probably arose when their pectoral fins enlarged beyond those of halfbeaks to form more effective wings. Flying fish gain at least two benefits from gliding in air. First, because drag forces are much lower in air than in water, and because the fish is not actively producing thrust during most of its glide, a fish may use less energy by interspersing glides with bouts of swimming to travel long distances. Second, because gliding is so much faster than swimming, aerial glides can be effective for escaping from predators. Both factors may have aided the evolution of gliding, but predator escape seems to be the main reason modern flying fish glide (Davenport 1994).

Powered flight

Flapping (powered) flight evolved four times in the evolutionary history of animals. Insects evolved powered flight first,

with well-developed wing fossils appearing as far back as the middle Carboniferous, approximately 320 million years ago. Pterosaurs, the first vertebrates to evolve flapping flight, appear in the fossil record approximately 220 million years ago, and went extinct about 155 million years later. *Archaeopteryx*, the earliest and most primitive known bird, lived about 160 million years ago. Bats, the last animals to evolve powered flight, are known from Eocene fossils that are about 53 million years old, and probably actually arose near the time that pterosaurs went extinct.

Insects, the first powered flyers.

Insects evolved flight during a substantial gap in their fossil record. The oldest fossils of insect-like creatures are of wingless bristletails (Archaeognatha) and springtails (Collembola) from the Devonian (395 million years ago). Archaeognathan and collembolan species are still alive today and are still entirely wingless. The next oldest known insect fossils are from 70 million years later, in the Carboniferous. These later fossils consist of a variety of species with fully formed wings, presumably capable of powered flight (Grimaldi and Engel 2005).

Speculation about structures that might have given rise to wings, and the selective pressures that might have favored flight, has gone on for well over a century. Two main theories

The southern flying squirrel (*Glaucomys volans*) can glide at nearly 20 miles per hour. Image by Nicholas Bergkessel Jr.

proponents of the gill theory have adopted it as the beginnings of tracheal gills (Kukalová-Peck 1987). The other theory, the "paranotal lobe" theory, starts with immobile, lateral extensions of the thorax (the *notum* is the exoskeletal plate atop an insect's thorax, so *paranotal* means "alongside the notum"). These extensions evolved for some nonflight function, perhaps as a mate attraction display or for thermoregulation (solar warming). They may have stabilized and extended falls and leaps, and so they became larger to provide longer glides. At some point, an articulation (movable joint) evolved, most likely to allow steering movements. With an articulation, steering movements could evolve into flapping and hence powered flight.

Both of these theories have problems. Although many modern insect species have aquatic juvenile stages, these aquatic stages evolved well after insects evolved flight. The evidence indicates that the earliest insects were terrestrial and would have had no need for gills. Moreover, surface tension would have caused problems for small animals the size of insects trying to move through the air–water interface, and especially for spreading small protowings. As for the paranotal lobe theory, many scientists consider the evolution of an articulation for an immobile structure improbable. Without fossils from that 70-million-year gap, this debate will be difficult to resolve (Grimaldi and Engel 2005).

The ancestors of flying insects faced different selection pressures from those faced by incipient vertebrate flyers because of size differences. Because insects are so small and light, they cannot be injured by falling. Their small size gives them a high surface area for their weight. Aerodynamic drag is proportional to surface area, so they have high drag and low *terminal velocities*. The terminal velocity is the speed at which the drag from falling through the air just equals the body's weight, and it represents the maximum falling speed; this maximum speed is much lower for an insect than even for a small vertebrate such as a mouse. Moreover, insects' low mass gives them low inertia, meaning that collisions are gentler than for heavier bodies. Except for the very largest ones, insects simply do not fall fast enough or hit hard enough to be injured by falls from high places. Thus, a factor that is important in the evolution of gliding (and probably powered flight) among vertebrates—slowing descent speed—is essentially irrelevant in insects because they are so small.

In the absence of fossil evidence, a number of researchers have used biomechanical principles to explore plausible scenarios for the evolution of insect flight. For example, many modern insects bask in the sun to warm their muscles, which confers a number of advantages. Warm muscles contract faster than cool ones, so basking animals can move faster—handy for catching prey or escaping predators. Wings of modern insects are effective solar heat collectors, and basking insects can often start flying at cooler ambient temperatures than insects that do not bask. Experiments using physical models of insects made of epoxy with plastic film, foil, or cardboard wings, under a sunlamp and in a wind tunnel, showed that small protowings could make effective solar collectors while still too small to have much aerodynamic effect. As they enlarge to become more effective for warming,

emerged early on. The "gill" or "pleural appendage" theory suggests that wings evolved from primitively mobile (articulated) structures similar to tracheal gills such as those of modern mayfly nymphs (juveniles). Although gills evolved for gas exchange, many aquatic nymphs use these appendages for paddles, either to drive water currents for gill ventilation or for swimming locomotion. In evolving flight, these appendages would have been used initially to steer during leaps or falls (in air), and selection for extending leaps would have led to enlargement of the thoracic appendages into wings. Because these appendages were movable, steering movements would have evolved into flapping movements over time. Improvements in flapping would lead to powered flight. A variation of this theory, the "exite" theory, suggests that the earliest insects had legs with side branches, or *exites*, like those of modern crustaceans. The most basal segment of the leg had an exite at its joint with the second segment, but the first segment eventually flattened out and became part of the exoskeleton of the lateral body wall. When that happened, the exite moved up on the side of the body, while the rest of the leg stayed low, and the exite evolved into a wing. The original function of this exite is unknown, although

Flying fish (Exocoetidae) can reach as high as 8 meters above the surface before beginning their inevitable glide back down, and reach distances of 50 meters or more. © Anthony Pierce/Alamy.

they also become big enough to be effective for gliding (Kingsolver and Koehl 1985). If insects evolved thoracic surfaces for solar warming (with articulations, to allow them to be aimed at the sun or folded out of the way), those surfaces could have been the basis for the evolution of wings. Another study tested a series of gill-like structures running down the side of the body, based loosely on aquatic mayfly nymphs. The models, of cardboard tubes with a series of small balsa winglets on each side or a single stubby wing with the same total area as the sum of the areas of the small plates, were dropped to simulate falls from trees. These models showed that even very small winglets can greatly improved stability during leaps and falls, and in addition can produce surprisingly long glides (Wootton and Ellington 1991).

Another theory draws on a behavior used by modern insects. Stoneflies (Plecoptera) have aquatic juvenile stages, and many adults stay close to water. Although winged, the adults of some species do not fly but flap their wings to produce thrust to skim along with their feet skating on the water surface. The "surface-skimming" theory suggests that the ancestors of flying insects used some thoracic structure, most likely gills, as passive sails in wind to move about on the water surface. Over time, steering movements evolved into flapping to give thrust for faster and better-controlled locomotion on the water surface. As flapping strengthened, the insects were able to start flying up from the water surface, and eventually to evolve fully powered flight (Marden and Kramer 1994). Although intriguing, this theory has some drawbacks. As noted earlier, the ancestors of flying insects were probably terrestrial, not aquatic. Moreover, no living

animal has evolved flapping first for thrust production on the surface; the only animals that use wing flapping for thrust are those whose ancestors used flapping for flight.

Birds.

Most research on the evolution of bird flight has centered on *Archaeopteryx*, whose pigeon-sized, 160-million-year-old fossils were first discovered in the mid-1800s. Clearly a transitional form, *Archaeopteryx* had the teeth, separate fingers with claws, and long tail of a reptile, but the feathers, including well-developed flight feathers, of a bird. Rather than settling disputes about bird evolution, *Archaeopteryx* seems to raise more questions than it answers. Modern arguments about *Archaeopteryx* have centered on three questions: Determining its ancestry, determining whether it glided or flew under power, and if the latter, what scenario led to flapping flight. Paleontologists agree that *Archaeopteryx* is closely related to maniraptoran dinosaurs (such as *Deinonychus* and *Velociraptor*), which may have settled the first question. *Archaeopteryx* is, however, several tens of millions of years older than these putative ancestors, the maniraptorans, causing some researchers to reserve judgment on its ancestry.

Some fossil *Archaeopteryx* preserve finely detailed feather impressions, so paleontologists know that the wings of *Archaeopteryx* were very well-developed. They look similar in general shape, feather arrangement, and area to short-winged modern birds such as woodpeckers. Even so, researchers throughout the twentieth century contended that *Archaeopteryx* was a glider and could not fly under power. This view was based on the idea that *Archaeopteryx* was reptilian and *ectothermic* ("cold-blooded") and therefore would not have been able to sustain the metabolic power output needed for powered flight. Others argued that *Archaeopteryx* did not seem to have large enough attachment sites for the powerful flight muscles of a modern bird, so it may have flapped feebly to extend its glides but not enough to sustain level flight (Rayner 1988). In the 1980s and 1990s, some researchers argued that *Archaeopteryx* was a fully powered flyer, either because it was endothermic (warm-blooded) with the high-speed metabolism of a bird (Bakker and Galton 1974), or because it had muscles capable of bursts of high power output similar to that of some lizards and snakes (Ruben 1991). Analyses that combined estimates of muscle performance with detailed aerodynamic simulations suggest that *Archaeopteryx* may have required an initial drop from an elevated perch to get airborne, probably could have maintained level flight, but may not have been able to take off from a standstill on level ground (Chatterjee and Templin 2003).

Controversy still surrounds the question of how *Archaeopteryx* evolved powered flight. The competing theories have been dubbed the *cursorial* and the *arboreal* models, and they actually date back to the late 1800s. The cursorial theory starts with a small, fast-running dinosaur with fairly long arms. These long arms help the animal steer during leaps for prey such as flying insects, leading to further enlargement of the arms. These enlarged arms extended leaps into glides, and steering movements evolved into flapping and hence powered flight. A variation on this hypothesis (the "flapping-first cursorial" theory) has the animal evolving small, flapping wings solely to produce thrust for faster running with no gliding, and with increased wing size and flapping power eventually leading to flight (Burgers and Chiappe 1999). This theory was largely discounted until the 1970s, when Yale paleontologist John H. Ostrom noticed a strong resemblance between *Archaeopteryx* and small theropod dinosaurs (Ostrom 1974). Because these dinosaurs were bipedal runners, he revived the cursorial scenario, which has been popular among many paleontologists ever since. In contrast, the arboreal theory starts with a tree-dwelling protobird that leaps from branch to branch. Its arms elongate to help steer during leaps. Longer arms give more surface area and facilitate some gliding, which leads to further enlargement (more area) to enhance gliding. Finally, steering movements evolve into flapping to achieve powered flight (Rayner 1991).

One problem that has plagued efforts to decide between these two theories is that most scientists studying bird origins and the evolution of flight have conflated arguments about the ancestry of birds with arguments about the evolutionary pathway leading to powered flight. Those favoring a theropod ancestry for birds also tend to favor the cursorial theory, whereas those who disagree with the theropod ancestry tend to favor the arboreal scenario. Also, the researchers who have approached flight evolution from a strictly biomechanical background overwhelmingly favor the arboreal model. Moreover, researchers need to be cautious about linking theories of flight evolution tightly to evidence from *Archaeopteryx*. While it was primitive in many ways, *Archaeopteryx*'s wings were already structurally sophisticated and probably closer to a modern bird's wings than to whatever ancestral form first began to evolve flight.

Aerodynamic and physical analyses in the 1980s demonstrated the physical advantages of the arboreal theory and the disadvantages of the cursorial theory. A gliding stage (arboreal theory) requires little control sophistication and can significantly reduce energetic costs of moving through trees. It takes advantage of a "free ride" provided by gravity. A running animal that leaps and glides (the cursorial theory) actually slows down and gains no energy benefit; in a glide, its primary locomotion muscles—for the legs—cannot be used; and a small animal the size of *Archaeopteryx* cannot run fast enough to reach gliding speed. A leaping runner is always working against gravity. Although a runner could, in principle, evolve small flapping wings solely for thrust—to run faster as in the flapping-first cursorial version—no modern animals do so except those that evolved from powered flyers. This contrasts with the multitude of arboreal animals alive today that have evolved gliding, as well as the obvious benefits of gliding mentioned earlier. In spite of the biomechanical evidence, some researchers still prefer the cursorial model because they think the ancestors of birds must have been cursorial. The discovery of the tiny, feathered, probably arboreal dinosaur *Microraptor*, described in 2000, removes the need to accommodate a cursorial ancestor and further bolsters the arboreal scenario (Xu, Zhou, and Wang 2000).

Some newer suggestions have been proposed to overcome objections to the two prominent theories. For example, the "pouncing proavis" scenario is based on discoveries of feathered dinosaurs and suggests a sequence including intermediate forms based on these dinosaurs. It proposes that birds evolved from predators that climbed to elevated sites to pounce on prey, with arms evolving aerodynamic surfaces to better steer during leaps. Arm feathers evolved to improve steering, then to lengthen pounces into glides, and from there, steering movements evolved into flapping (Garner, Taylor, and Thomas 1999). Although drawing on parts of both the arboreal and cursorial models, the pouncing proavis theory envisions an animal descending through the air, so it is, in essence, a variation on the arboreal concept. Another suggestion, the "wing-assisted incline running" theory, is based on the observation that quail chicks can flap their wings and run up steep inclines long before their wings have developed enough to actually fly. In this theory, the ancestors of birds evolved small aerodynamic surfaces on their arms, and flapped their arms to push themselves against a steep surface for climbing (Dial 2003). Like the surface-skimming scenario for insect flight, the wing-assisted incline running theory is based on a behavior of a somewhat specialized modern animal, and also like the surface-skimming scenario, it depends on evolving flapping for thrust before evolving flight. The latter feature makes it, at heart, a variation on the flapping-first cursorial model. Neither of these two theories takes into account the characteristics of *Microraptor*, which still seem to fit the arboreal theory best.

Pterosaurs and bats.

The extinct flying reptiles known as pterosaurs, or sometimes pterodactyls, first appear in the fossil record with complete, apparently fully functional wings. These wings are quite different from those of birds, being formed of a patagium supported by the arm and one enormously elongated finger, and attached along the animal's flanks and legs. Just as for insects, paleontologists have no direct fossil evidence for flight evolution in pterosaurs. Scientists proposed the same two theories as for birds, the arboreal and cursorial scenarios. The arboreal theory was widely accepted into the 1980s, largely because of its mechanical simplicity. A suggestion that pterosaurs share a cursorial ancestry with dinosaurs revived interest in a possible cursorial evolution of flight in pterosaurs in the 1980s (Padian 1985). Other paleontologists were not convinced of this close relationship between dinosaurs and pterosaurs (Bennett 1996). Moreover, the pterosaur patagium was attached along the leg to the knee or ankle (depending on the species) (Chatterjee and Templin 2004), favoring an arboreal origin and making a bipedally running ancestry seem most improbable (Bennett 1997).

The situation for bats is much the same as for pterosaurs. The oldest known bat fossil, approximately 53 million years old, is already a recognizable, fully winged bat (Simmons 1995). In the case of bats, however, researchers generally agree that bats' ancestors were arboreal (Scholey 1986b; Norberg 1990), and the arboreal theory of flight evolution in bats is widely accepted. In fact, Charles Darwin (1809–1882) first proposed the arboreal theory for bats in the mid-1800s, and researchers later adopted it for other vertebrate groups. Moreover, the modern colugo, or flying lemur, provides an excellent model for the appearance and life habits of the

Unlike other mammalian gliders, colugos (*Cynocephalus volans*) also have webbed fingers, so that the hand functions as an accessory flight surface. Image by Tim Laman.

ancestors of bats. Colugos are arboreal gliders from Southeast Asia, and they glide on a patagium stretched between front and hind limbs. Unlike other mammalian gliders, colugos also have webbed fingers, so that the hand functions as an accessory flight surface. The hand's location at the tip of the wing places it ideally to function in steering. If the ancestral protobat had a similar wing structure, enlargement of the hand would have increased the wing's surface area and improved its mobility, enhancing steering and lengthening glides. Continued elongation of the fingers, and evolution of steering movements into flapping, would produce the wing structure seen in modern bats and complete the transformation from arboreal leaper to powered flyer. Although experts disagree on whether or not colugos are the closest relatives of bats (Teeling et al. 2000), the living example of the colugo supports the arboreal origin of bat flight by showing what a protobat may have looked like and how it lived.

A unifying concept: Directed aerial descent

A study published in 2005 provides the basis for considering patterns of flight evolution across animal groups. Stephen P. Yanoviak, then a graduate student at the University of Oklahoma, was studying ants in tropical trees, when he noticed

that if he knocked an ant off a branch, most of the time it fell in a curve that led it back to the tree trunk (Yanoviak, Dudley, and Kaspari 2005). If these forest canopy-dwelling ants fall all the way to the forest floor, they are at risk from unfamiliar predators and of being unable to find the tree inhabited by their colony. Although these ants have no apparent aerodynamic specializations, they can control their fall, like human skydivers, well enough to land back on their home tree trunk more than 80 percent of the time. Such directed aerial descent is now known from a variety of wingless arthropods living in rain forest canopies (Dudley et al. 2007).

The same concept also applies to vertebrates. At least two types of lizards with no obvious aerodynamic specializations—including the green anole, *Anolis carolinensis*, common in the southeastern United States—often leap from tree branches and can control their descent (Oliver 1951; Arnold 2002). Snakes in the genera *Trimeresurus* and *Chrysopelea* nicely illustrate the evolutionary stages of this process. Most snakes, when falling from trees, simply tumble out of control. When *Trimeresurus* falls, it assumes a stable, upright posture and lands right side up, although it does not appear to glide very far. *Chrysopelea* not only remains upright and stable, it also glides at an angle well under 45 degrees (Heyer and Pongsapipatana 1970; Socha 2002).

Yanoviak and colleagues thus suggest that a first step toward the evolution of gliding is the behavioral innovation to assume appropriate body and leg posture during falling, for control and to assure landing right side up. Even a wingless body can achieve a significant glide angle if it falls fast enough. This aerodynamic phenomenon favors arboreal animals that can drop far enough to build up enough speed to achieve modest gliding even without obvious wing surfaces. Once this control behavior has evolved, selection for longer or flatter glides would lead to the evolution of winglike gliding surfaces, which in turn would provide the raw material for the evolution of flapping flight. This theory emphasizes that living in an arboreal environment provides any animal species, regardless of ancestry, an opportunity to benefit from flight over a wide range of sophistication, from barely directed falling to fully developed powered flight (Dudley et al. 2007). Future studies of flight evolution will be incomplete unless they seriously consider the relevance of directed aerial descent.

Resources

Books

Alexander, David E. 2002. *Nature's Flyers: Birds, Insects, and the Biomechanics of Flight*. Baltimore: Johns Hopkins University Press.

Grimaldi, David, and Michael S. Engel. 2005. *Evolution of the Insects*. Cambridge, UK: Cambridge University Press.

Rayner, J. M. V. 1991. "Avian Flight and the Problem of *Archaeopteryx*." In *Biomechanics in Evolution*, ed. J. M. V. Rayner and R. J. Wootton. Cambridge, UK: Cambridge University Press.

Scholey, Keith. 1986a. "The Climbing and Gliding Locomotion of the Giant Red Flying Squirrel, *Petaurista petaurista* (Sciuridae)." In *Bat Flight/Fledermausflug*, ed. Werner Nachtigall. Stuttgart, Germany: Gustave Fischer.

Scholey, Keith. 1986b. "The Evolution of Flight in Bats." In *Bat Flight/Fledermausflug*, ed. Werner Nachtigall. Stuttgart, Germany: Gustave Fischer.

Wells-Gosling, Nancy. 1985. *Flying Squirrels: Gliders in the Dark*. Washington, DC: Smithsonian Institution Press.

Wootton, R. J., and C. P. Ellington. 1991. "Biomechanics and the Origin of Insect Flight." In *Biomechanics in Evolution*, ed. J. M. V. Rayner and R. J. Wootton. Cambridge, UK: Cambridge University Press.

Periodicals

Arnold, E. N. 2002. "*Holaspis*, a Lizard That Glided by Accident: Mosaics of Cooption and Adaptation in a Tropical Forest Lacertid (Reptilia, Lacertidae)." *Bulletin of the Natural History Museum, Zoology Series* 68(2): 155–163.

Bakker, Robert T., and Peter M. Galton. 1974. "Dinosaur Monophyly and a New Class of Vertebrates." *Nature* 248 (5444): 168–172.

Bennett, S. Christopher. 1996. "The Phylogenetic Position of the Pterosauria within the Archosauromorpha." *Zoological Journal of the Linnean Society* 118(3): 261–308.

Bennett, S. Christopher. 1997. "The Arboreal Leaping Theory of the Origin of Pterosaur Flight." *Historical Biology* 12: 265–290.

Burgers, Phillip, and Luis M. Chiappe. 1999. "The Wing of *Archaeopteryx* as a Primary Thrust Generator." *Nature* 399 (6731): 60–62.

Chatterjee, Sankar, and R. Jack Templin. 2003. "The Flight of *Archaeopteryx*." *Naturwissenschaften* 90(1): 27–32.

Chatterjee, Sankar, and R. Jack Templin. 2004. *Posture, Locomotion, and Paleoecology of Pterosaurs*. Special Paper 376. Boulder, CO: Geological Society of America.

Colbert, Edwin Harris. 1967. "Adaptations for Gliding in the Lizard *Draco*." *American Museum Novitates*, no. 2283: 1–20.

Colbert, Edwin Harris. 1970. "The Triassic Gliding Reptile *Icarosaurus*." *Bulletin of the American Museum of Natural History* 143(2): 85–142.

Davenport, John. 1994. "How and Why Do Flying Fish Fly?" *Reviews in Fish Biology and Fisheries* 4(2): 184–214.

Dial, Kenneth P. 2003. "Wing-Assisted Incline Running and the Evolution of Flight." *Science* 299(5605): 402–404.

Dudley, Robert, Greg Byrnes, Stephen P. Yanoviak, et al. 2007. "Gliding and the Functional Origins of Flight: Biomechanical Novelty or Necessity?" *Annual Review of Ecology, Evolution, and Systematics* 38: 179–201.

Garner, Joseph P., Graham K. Taylor, and Adrian L. R. Thomas. 1999. "On the Origins of Birds: The Sequence of Character Acquisition in the Evolution of Avian Flight." *Proceedings of the Royal Society* B 266(1425): 1259–1266.

Heyer, W. Ronald, and Sukhum Pongsapipatana. 1970. "Gliding Speeds of *Ptychozoon Lionatum* (Reptilia: Gekkonidae) and *Chrysopelea Ornata* (Reptilia: Colubridae)." *Herpetologica* 26 (3):317–319.

Jackson, Stephen M. 2000. "Glide Angle in the Genus *Petaurus* and a Review of Gliding in Mammals." *Mammal Review* 30(1): 9–30.

Kingsolver, Joel G., and M. A. R. Koehl. 1985. "Aerodynamics, Thermoregulation, and the Evolution of Insect Wings: Differential Scaling and Evolutionary Change." *Evolution* 39 (3): 488–504.

Kukalová-Peck, Jarmila. 1987. "New Carboniferous Diplura, Monura, and Thysanura, the Hexapod Ground Plan, and the Role of Thoracic Lobes in the Origin of Wings (Insecta)." *Canadian Journal of Zoology* 65(10): 2327–2345.

Marden, James H., and Melissa G. Kramer. 1994. "Surface-Skimming Stoneflies: A Possible Intermediate Stage in Insect Flight Evolution." *Science* 266(5184): 427–430.

Norberg, Ulla M. 1990. *Vertebrate Flight: Mechanics, Physiology, Morphology, Ecology, and Evolution*. Berlin: Springer-Verlag.

Oliver, James A. 1951. "'Gliding' in Amphibians and Reptiles, with a Remark on an Arboreal Adaptation in the Lizard, *Anolis carolinensis carolinensis* Voigt." *American Naturalist* 85(822): 171–176.

Ostrom, John H. 1974. "*Archaeopteryx* and the Origin of Flight." *Quarterly Review of Biology* 49(1): 27–47.

Padian, Kevin. 1985. "The Origins and Aerodynamics of Flight in Extinct Vertebrates." *Palaeontology* 28(3): 413–433.

Rayner, J. M. V. 1988. "The Evolution of Vertebrate Flight." *Biological Journal of the Linnean Society* 34(3): 269–287.

Ruben, John. 1991. "Reptilian Physiology and the Flight Capacity of *Archaeopteryx*." *Evolution* 45(1): 1–17.

Simmons, Nancy B. 1995. "Bat Relationships and the Origin of Flight." *Symposium of the Zoological Society of London* 67: 27–43.

Socha, John J. 2002. "Kinematics: Gliding Flight in the Paradise Tree Snake." *Nature* 418(6898): 603–604.

Teeling, Emma C., Mark Scally, Diana J. Kao, et al. 2000. "Molecular Evidence Regarding the Origin of Echolocation and Flight in Bats." *Nature* 403(6766): 188–192.

Thorington, Richard W., Jr., and Lawrence R. Heaney. 1981. "Body Proportions and Gliding Adaptations of Flying Squirrels (Petauristinae)." *Journal of Mammology* 62:101–114.

Xu, Xing, Zhonghe Zhou, and Xiaolin Wang. 2000. "The Smallest Known Non-avian Theropod Dinosaur." *Nature* 408 (6813): 705–708.

Yanoviak, Stephen P., Robert Dudley, and Michael Kaspari. 2005. "Directed Aerial Descent in Canopy Ants." *Nature* 433 (7026): 624–626.

Young, Bruce A., Cynthia E. Lee, and Kylle M. Daley. 2002. "On a Flap and a Foot: Aerial Locomotion in the 'Flying' Gecko, *Ptychozoon kuhli*." *Journal of Herpetology* 36(3): 412–418.

David E. Alexander

Evolution of limblessness

Early on in life, many people learn that lizards have four limbs whereas snakes have none. This dichotomy not only is inaccurate but also hides an exciting story of repeated evolution that is only now beginning to be understood. In fact, snakes represent only one of many natural evolutionary experiments in lizard limblessness. A similar story is also played out, though to a much smaller extent, in amphibians. The repeated evolution of snakelike tetrapods is one of the most striking examples of parallel evolution in animals. This entry discusses the evolution of limblessness in both reptiles and amphibians, with an emphasis on the living reptiles.

Reptiles

Based on current evidence (Wiens, Brandley, and Reeder 2006), an elongate, limb-reduced, snakelike morphology has evolved at least twenty-five times in squamates (the group containing lizards and snakes), with snakes representing only one such origin. These origins are scattered across the evolutionary tree of squamates, but they seem especially frequent in certain families. In particular, the skinks (Scincidae) contain at least half of all known origins of snakelike squamates. But many more origins within the skink family will likely be revealed as the branches of their evolutionary tree are fully resolved, given that many genera contain a range of body forms (from fully limbed to limbless) and may include multiple origins of snakelike morphology as yet unknown.

Two types of snakelike lizards

These multiple origins of snakelike morphology are superficially similar in having reduced limbs and an elongate body form, but many are surprisingly different in their ecology and morphology. This multitude of snakelike lineages can be divided into two ecomorphs (a type of morphology or body form that is associated with a particular ecology or habitat type). One ecomorph is the short-tailed burrower (burrowing morph hereafter), which has evolved at least twenty times. In these species, the body length (the distance from the tip of the snout to the beginning of the tail, or vent) is elongate, and the tail is (on average) only about half of the body length. This contrasts with more typical lizards, in which the tail is around 1.5 times the body length (Wiens, Brandley, and Reeder 2006). Species of the burrowing morph spend much of their time underground, and include the worm

lizards (amphisbaenians), the California legless lizard (*Anniella pulchra*), and many different lineages of skinks. They also include the snakes, but this requires some explanation (see below). Species with the burrowing morph also tend to have reduced eyes, no external ear openings, and well-ossified skulls, which are used in tunneling.

The other snakelike ecomorph is the long-tailed surface dweller (long-tailed morph hereafter), which has evolved only five times. In these species, the body is also somewhat elongate, but the tail is remarkably so, up to 2.3 times the body length (on average). In at least some groups, the overall elongation is quite similar in both ecomorphs, but is made up by these very different proportions of tail versus body (e.g., Wiens and Slingluff 2001). Many of these species occur in dense vegetation, and they are often referred to as grass swimmers. Members of this ecomorph include the anguid glass lizards (genus *Ophisaurus*), some African plated lizards (cordylids [*Chamaesaura*] and gerrhosaurids [*Tetradactylus*]), and most of the Australian snake lizards (pygopods), which are a specialized group of geckos (Gekkonidae).

Why do surface dwellers evolve unusually long tails and burrowers evolve unusually short tails? The reasons for this dichotomy are still very uncertain. However, tails of many long-tailed surface dwellers can break easily and regrow. This breakability has earned the nickname of glass lizards for the most widespread genus of the long-tailed morph (*Ophisaurus*). The ability to easily lose and regrow parts of the tail is an antipredator adaptation that is widespread in surface-dwelling lizards (Pough et al. 2004). In the long-tailed morph, the retention of a breakable tail may represent a response to higher predation on the surface relative to underground (Wiens and Slingluff 2001). Furthermore, extreme tail elongation has been noted in other grass-dwelling lizards, even those with fully developed limbs (e.g., some *Anolis* and *Takydromus*).

Conversely, one might ask why the length of the tail should be reduced in burrowers. One hypothesis suggests that because many species of the burrowing morph have relatively small body size, reducing the size of the tail may allow them to increase the size of the body cavity, making more room for developing eggs in gravid females (Wiens and Slingluff 2001). In summary, although hypotheses have been proposed to explain differences in body proportions between these ecomorphs, no studies have tested these hypotheses so far.

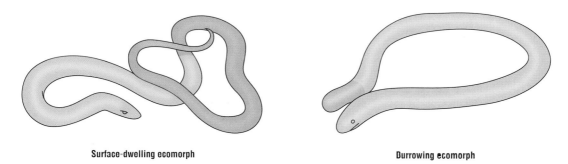

Surface-dwelling ecomorph   Burrowing ecomorph

Snake-like lineages can be divided into two ecomorphs (a type of morphology or body form that is associated with a particular ecology or habitat type). Reproduced by permission of Gale, a part of Cengage Learning.

Why so many origins?

Natural selection is a major driving force for evolutionary change, and most likely was important in the origin of each limb-reduced ecomorph. Direct evidence for this conclusion is limited, however, and many questions remain unanswered. The importance of natural selection is indirectly supported by the repeated association found between the morphology and ecology of the limb-reduced species (Wiens, Brandley, and Reeder 2006). In general, repeated evolution of the same morphology in similar habitats or (in this case) microhabitats is an important line of evidence supporting the role of natural selection in the evolution of that morphology. Furthermore, ecological niches for both of these morphs are present in every major continental region (i.e., Africa, Asia, Australia, Europe, North America, and South America), and it is illustrative that both morphs occur in each region.

Species of the burrowing morph, such as the sharp-tailed worm lizard (*Trogonophis wiegmanni*), spend much of their time underground. Tom McHugh/Photo Researchers, Inc.

Why should natural selection favor evolution of these ecomorphs? Limb-reduced species are likely able to use microhabitats not regularly used by lizards with a normal (fully limbed) body plan. Most lizards that are active underground are limb-reduced (with a few exceptions, such as the limbed, sand-swimming skink [*Scincus scincus*]). Limbs would probably only impede locomotion while burrowing or moving through narrow tunnels underground (Gans 1975). Although the advantages of snakelike morphology for the long-tailed morph are less clear, reduced limbs and an elongate body may facilitate movement through dense vegetation (Gans 1975). Hence, members of this ecomorph are often called grass swimmers. Unfortunately, detailed functional studies are still lacking that demonstrate the advantages of a limb-reduced, elongate body form for locomotion in dense vegetation or underground. In summary, the morphology of these ecomorphs may allow them to better exploit microhabitats not used by other lizard species. In general, competition for finite resources (such as food) is thought to favor evolution of novel traits that allow species to better use resources not used by their close relatives (Schluter 2000), and competition is thought to drive the evolution of ecomorphs in other lizards (e.g., *Anolis* in the West Indies; see Losos et al. 1998).

Natural selection alone does not provide a direct answer for why morphs evolve so frequently. To explain this pattern, the geography of relevant species must also be considered (Wiens, Brandley, and Reeder 2006). With just a few exceptions (see below), the origin of each ecomorph is confined to only a single continent or part of a continent. Furthermore, many continents have multiple origins for an ecomorph, particularly for the burrowing morph (which has evolved at least twenty times across squamates, compared to only five times for the long-tailed morph; Wiens, Brandley, and Reeder 2006). Many of these origins are geographically isolated from each other within a continent. For example, the burrowing morph arose twice in North America—one found only in coastal California and adjacent Mexico (the California legless lizard [*Anniella pulchra*]), the other only in Florida (the Florida sand skink [*Plestiodon reynoldsi*]).

Why should the burrowing morph evolve so frequently? The burrowing morph often seems to evolve in association

A Japanese five-lined skink (*Eumeces japonicus*) with dropped tail, a defensive behavior. The ability to easily lose and regrow parts of the tail is an antipredator adaptation that is widespread in surface-dwelling lizards. © Ryu Uchiyama/ Nature Production/Minden Pictures.

with sandy habitats and patches of these sandy habitats are often geographically isolated from each other. As one striking example, some species of the burrowing genus *Calyptommatus* are found only in sand dunes near the São Francisco River in Brazil (Rodrigues 1991). The basic idea is that species of the normal morph are widespread around the world, but when they encounter an isolated area of microhabitats that can be used by the burrowing ecomorph (e.g., sand dunes), this ecomorph will often evolve from an ancestor of the normal morphology to fill and exploit this empty niche.

In contrast, the long-tailed morph seems to disperse far more readily within and between continents. For example, the genus *Ophisaurus* ranges through North America, Mexico, Europe, and Asia. Intriguingly, the long-tailed morph has not arisen again in any of the regions where *Ophisaurus* occurs. In regions where *Ophisaurus* does not occur (South America, sub-Saharan Africa, and Australia), there have been one or more independent origins of the long-tailed morph, including the anguid genus *Ophiodes* in South America, the plated lizards *Chamaesaura* and *Tetradactylus* in Africa, and pygopodid geckos in Australia (Wiens, Brandley, and Reeder 2006). Thus, the geographic isolation of each origin may also be important because the presence of an ecomorph in a region may prevent this same ecomorph from evolving again in that same region (Wiens, Brandley, and Reeder 2006). Given this, the fewer origins of the long-tailed morph may be explained (at least in part) by the seemingly greater dispersal ability of species with this morphology, relative to the many narrowly distributed species of the burrowing morph.

Multiple origins of the ecomorphs may also be facilitated by their relatively recent origins. Almost all origins of these ecomorphs are less than 100 million years old (Wiens, Brandley, and Reeder 2006). Given scientists' knowledge of the timing of continental drift, these morphs must have originated after the major breakup of continents, an event that doubtless contributed to their geographic isolation.

In addition to *Ophisaurus*, three other lineages of limb-reduced squamates are relatively widespread and occur on multiple continents. Surprisingly, these species are members of the burrowing morph. These are the blind lizards (dibamids), worm lizards (amphisbaenians), and snakes. Dibamids are a primarily Asian group that also includes a species in Mexico (*Anelytropsis*) that is sometimes recognized as a distinct family. Amphisbaenians are a widespread group of burrowing reptiles found in warm or tropical areas of the New World, as well as parts of Africa, Europe, and the Middle East. Within the amphisbaenians are a number of remarkable adaptations associated with their burrowing lifestyle, including a novel element associated with the ear that attaches to the lower jaw (the extracolumella), interlocking skull elements, and a diversity of bizarre head shapes associated with different modes of digging (Kearney 2003; Kearney and Stuart 2004). Intriguingly, these three widespread lineages (amphisbaenians, dibamids, and snakes) also differ from the other limb-reduced lineages of the burrowing morph in being substantially older (all three appear to be more than 100 million years old; Wiens, Brandley, and Reeder 2006).

Snakes: A limb-reduced success story

Snakes are the best known of the limb-reduced squamates. The reason for this seems obvious. Snakes have been wildly successful, making up about 3,000 of the roughly 8,000 species of squamates (Pough et al. 2004).

Why might snakes be so successful relative to other groups of limb-reduced squamates? Snakes are older than many other groups with reduced limbs (more than 120 million years old), but this is not the only factor, as amphisbaenians are of similar age but have far fewer species. Instead, the success of snakes may be more likely explained by their ability to use resources not used by most other squamates. Snakes differ from other squamates in several respects, and one of the most important may be in the remarkable flexibility of their jaws and skulls that characterizes most species (Pough et al. 2004). This flexibility seemingly allows them to open their jaws to eat larger prey than would be possible for most other lizards to eat (at least without dramatically increasing their body mass). Thus, whereas most lizard species feed on insects and other small invertebrates, snakes typically feed on other vertebrates, including birds, mammals, fish, amphibians, and other reptiles, including other snakes (Pough et al. 2004). Along with their use of larger prey items, snakes have evolved novel mechanisms (such as constriction and venom) to subdue or kill oversized prey before swallowing them, in contrast to lizards, which typically swallow their prey alive.

Paradoxically, although the morphology of snakes places most species within the burrowing ecomorph (i.e., relatively elongate bodies and a relatively short tail), most snake species are not burrowers (Pough et al. 2004). Why do so many species have the burrowing morphology but not the lifestyle?

The snake-like form of a slow worm (*Anguis fragilis*) on parched earth. © Mike Buxton; Papilio/Corbis.

The phylogeny within snakes may offer a clue. The earliest branch in snake phylogeny is a group of burrowers (the scolecophidians, or blind snakes), and many other early branches are burrowers as well (such as aniliids, uropeltids, and xenopeltids; Pough et al. 2004). Many snakes appear to have become secondarily surface dwelling, and have maintained their relatively short tails through this transition (Wiens, Brandley, and Reeder 2006). Many other aspects of snake morphology may be a remnant of their underground origin, such as the loss of external ear openings. Despite their subterranean ancestry, snakes have diversified to occupy a range of habitats and microhabitats that is at least as wide as that of all other lizards combined, including arboreal species in tropical rain forests (one of which even glides), many surface-dwelling terrestrial species, a large number of freshwater species, and the most diverse group of living marine reptiles, the highly venomous sea snakes (Pough et al. 2004).

Some authors have suggested that snakes evolved limblessness in association with aquatic habitat use rather than a subterranean lifestyle (e.g., Lee, Bell, and Caldwell 1999). However, the hypothesis of a terrestrial, subterranean origin for snakes was solidified by the reported discovery in 2006 of a fossil snake (*Najash rionegrina*), a species that is clearly the earliest lineage within snakes and that occurred in a terrestrial (and possibly subterranean) environment (Apesteguía and Zaher 2006). This study also found that all of the earliest branching lineages within snakes are terrestrial if not burrowing.

Who loses their limbs?

Can any squamate lose its limbs? Considering the origins of limb-reduced body plans in the context of the evolutionary history of squamates suggests that almost any group of lizards can potentially evolve one or both morphs. For example, the burrowing morph has evolved in families of lizards distributed throughout the squamate tree, including amphisbaenians, anguids, dibamids, geckos, gymnophthalmids, and skinks. Although the long-tailed morph has not evolved as often, its distribution seems equally widespread on the tree, occurring in anguids, geckos, and plated lizards (cordylids and gerrhosaurids). The only major exception to this widespread pattern (or lack of pattern) appears to be the iguanians (agamids, chameleons, iguanas, and relatives), a group of over 1,400 species in which no instances of snakelike limb reduction and body elongation have been reported (Pough et al. 2004). Although some other lizard families lack limb-reduced species (e.g., Lacertidae, Teiidae, Xantusiidae, Varanidae), all of them are closely related to families that do have them. Iguanians

The skinks (*Scincidae*) contain at least half of all known origins of snakelike squamates. Anthony Bannister/Gallo Images/Photo Researchers, Inc.

differ from other lizards in several fundamental aspects of their behavior and morphology. Some of these may prove to be directly or indirectly associated with their failure to evolve these ecomorphs. Some unusual iguanian traits include ambush foraging (versus active foraging in most other lizards), visual prey detection (versus chemical detection), and use of the tongue (versus the jaws) in prey capture (Pough et al. 2004). Lingual prey capture may be particularly difficult or impossible for limbless lizards, which may have difficulty raising their heads well above the ground.

Evolution of limb-reduced body form occurs widely in squamates, but some families seem to be more favored for this transition than others. At least half of the origins of the burrowing morph occur in one group, the skinks. What predisposed skinks to evolve this ecomorph so often? Many skink species seem to be more cryptic (e.g., found under leaf litter and other cover) than many other lizards, and so may be more inclined to evolve burrowing behavior than lizards that are more active on the surface. Another group of lizards in which the burrowing morph evolves repeatedly, the South American gymnophthalmids, also seems to have a tendency toward cryptic behavior and microhabitats (Pough et al. 2004). These trends are suggestive but require further study.

Another interesting pattern is the rarity of transitions between ecomorphs. While it might seem that the easiest way for a given ecomorph to arise in a region is for one limbless ecomorph to originate from another, this appears to have only happened once (the burrowing pygopod genus *Aprasia* evolved from long-tailed surface dwellers; Wiens, Brandley, and Reeder 2006). Intriguingly, it seems to be easier to evolve the dramatic changes in morphology associated with going from lizardlike to snakelike body form than it is to make the shift in ecology between surface dwelling and burrowing. This raises the question of how this major change in morphology actually happens.

How does it happen?

How does the transition from lizardlike to snakelike morphology actually occur? This can be thought of in at least two ways. First, what morphological changes occur? Second, what genetic and developmental changes underlie these changes in morphology?

Analyses of a large database of phylogeny and morphology across squamates (Wiens, Brandley, and Reeder 2006; Brandley, Huelsenbeck, and Wiens 2008) suggest that the evolution of snakelike morphology typically involves three correlated changes (i.e., changes occurring more or less at the same time). First is the elongation of the body length, such that body length increases relative to the width of the body or length of the head. This seems to be associated with an increase in the number of vertebrae that lie between the head and the beginning of the tail. Second is the reduction in limb size relative to other body parts. Third is the gradual loss of digits, going from five fingers and five toes down to one or none on each limb. These trends seem to be very consistent across squamates, despite differences in ecology between the two ecomorphs, the large number of diverse squamate clades, and the huge timescale involved (over 180 million years).

Analyses of morphology show that loss of digits from both limbs is usually correlated, as is reduction in limb size (Brandley, Huelsenbeck, and Wiens 2008). However, digits and limbs are lost from the front limbs rather more often than from the hind limbs. Many taxa retain vestigial hind limbs but have lost all vestiges of the forelimbs, including many snakes. There are also a few exceptions, including at least one (*Biporus*) that has well-developed forelimbs with five digits but lacks hind limbs entirely.

Genetic and developmental bases for these evolutionary changes in morphology remain poorly understood. Different genes or suites of genes are likely involved in different aspects of this transformation (i.e., body elongation, limb-size reduction, digit loss, limb loss). Based on developmental studies in other vertebrates, one set of genes is likely responsible for regulating number of vertebrae and thus may determine the patterns of body elongation (S. Carroll, Grenier, and Weatherbee 2005). Studies by Michael D. Shapiro and colleagues (Shapiro, Hanken, and Rosenthal 2003) on limb-reduced Australian skinks suggest that changes in the duration of expression of the *SHH* (*Sonic Hedgehog*) gene in the developing limb bud are responsible for digit loss (i.e., shortened duration of expression leads to fewer digits). As is common for many developmental genes, *SHH* seems to play different roles at different points in time during limb development.

Going, going … still here?

The transition from a lizardlike morphology to a snakelike morphology is one of the most dramatic transitions in animal evolution. Yet, phylogenetic studies suggest that it has occurred dozens of different times over the evolutionary history of squamates. Furthermore, the presence of fully limbed (i.e., with five digits) and fully limbless species within some genera (e.g., the skink genera *Brachymeles* and *Lerista*) suggests that this transition may happen quite quickly. But how long exactly?

Matthew C. Brandley and colleagues attempted to quantify how long this transition might take (Brandley, Huelsenbeck, and Wiens 2008). In this study, the evolution of body form was reconstructed on an evolutionary tree in which lengths of the branches reflect ages of different species (using a type of modified molecular clock method). Based on these results, the transition from fully limbed to limbless may take roughly 15 to 25 million years. However, more detailed sampling within genera (particularly in the skinks) will most likely show that this change takes place more quickly.

This analysis of the timing of body-form evolution showed another intriguing result. Many authors have noted that

several species seem intermediate in their morphology, between having well-developed limbs and no limbs at all. For example, rather than having all five digits or losing their limbs entirely, they may retain only two to four digits. Almost all authors studying limb-reduced lizards have assumed that such morphologies represent an intermediate stage in the process of transforming from lizardlike to snakelike. But analysis of the timing of body-form evolution shows that these intermediate morphologies are retained for relatively long periods of time (roughly 10 to 60 million years, but most typically around 25 million years), at least as long as it takes for the entire transition to take place in other taxa. Perhaps these species have these intermediate morphologies not because there has not been enough time to become fully limbless, but rather because selection specifically maintains these intermediate morphologies. One interesting implication of this result is that the morphological transformation from fully limbed to fully limbless might actually be quite different from what is inferred based on studying these seemingly intermediate species.

The observation that many species retain vestigial limbs for long periods of time suggests that they serve some purpose. For example, observational studies of some skinks suggest that reduced limbs aid in balance (Bruno and Maugeri 1976; Orsini and Cheylan 1981), and laboratory studies show that even very tiny limbs are used to aid locomotion on some surfaces (Gans and Fusari 1994). In some boas and pythons, tiny hind limbs are retained only in males and seem to be used in courtship and mating (Murphy et al. 1978).

Return of the lost limb

Finally, phylogeny-based studies of snakes and lizards have revealed another surprising result. Several studies have found evidence suggesting limbs and digits that have been lost over evolutionary timescales can be regained. An analysis of snake phylogeny has found that two fossil genera (*Haasiophis* and *Pachyrachis*) with relatively well-developed hind limbs (including four digits) represent relatively advanced snakes, and that these well-developed limbs evolved from ancestors with either vestigial limbs or no limbs at all (Tchernov et al. 2000). Analyses of amphisbaenian phylogeny suggest that *Bipes*, a genus with well-developed forelimbs, is actually a relatively recent group nested among the limbless amphisbaenians (Kearney and Stuart 2004), and that the forelimbs may have reemerged after being lost (Brandley, Huelsenbeck, and Wiens 2008). Analyses of gymnophthalmid and scincid lizards suggest that lost digits may have reevolved repeatedly in these groups also (Kohlsdorf and Wagner 2006; Brandley, Huelsenbeck, and Wiens 2008). These results imply a remarkable flexibility in the genetic systems that control the loss of digits and limbs, and that limbs and other structures that are lost in evolutionary time are not necessarily lost forever. Instead, at least some species that have lost limbs and digits seem to retain some or all of the genetic machinery to develop limbs, even if limbs are not actually expressed in the adult phenotype.

Another limb-reduced ecomorph?

Modern reptiles with limb reduction fall into two ecomorphs, but additional ecomorphs of limb-reduced squamates may have existed in the past. For example, a recently discovered fossil from the Upper Cretaceous (approximately 95 million years old) suggests that there have also been aquatic species with reduced limbs. In *Adriosaurus microbrachis* both sets of limbs are very small, and the forelimb is reduced to only a single upper arm bone. Another group of aquatic marine squamates from the Upper Cretaceous, the dolichosaurs had limbs that were very small in proportion to their overall body size (Caldwell 2000). The specific function of limb reduction in these animals is uncertain, although reduction of some limbs is common in aquatic tetrapods (e.g., manatees, whales) and may aid in locomotion.

Amphibians

Evolution of limb-reduced elongate body forms is not confined to reptiles. Modern amphibians seem to have evolved from an ancestor that resembled a modern salamander (and by extension, a typical lizard), with four limbs and a tail. But a limb-reduced morph evolved quickly thereafter. Caecilians are a widespread and ancient group of elongate, limbless, short-tailed, burrowing amphibians. Caecilians consist of about 175 currently recognized species and are widespread in wet tropical regions in Asia, Africa, and Latin America (Pough et al. 2004). Many species resemble earthworms, which they also feed on. A few caecilian species are also aquatic. Caecilians are widely considered to be the sister group to the clade formed by the frogs and toads (anurans) and salamanders (caudates).

Elongate body form and limb reduction have also evolved repeatedly within salamanders (Wiens and Hoverman 2008). The family Amphiumidae consists of one genus with three species, each with elongate, eel-like body form, highly reduced limbs, and a variable number of digits among species (one, two, or three digits per limb). The family Sirenidae contains two genera, both with elongate body form and no hind limbs. Forelimbs are well-developed, however, with three (*Pseudobranchus*) or four (*Siren*) digits. Both amphiumids and sirenids are aquatic species that also are burrowers (Pough et al. 2004). They also have relatively short tails, and thus are consistent with the short-tailed burrowing ecomorph. As in squamates and caecilians, body elongation appears to be achieved through an increase in the number of presacral vertebrae.

Do the multiple origins of the short-tailed burrower ecomorph in amphibians show the same biogeographic pattern as in reptiles, which had separate origins in different geographic regions? The answer is both yes and no. As might be expected, both amphiumids and sirenids occur outside the range of the tropical caecilians, in temperate North America (Pough et al. 2004). Contrary to expectations, however, ranges of amphiumids and sirenids overlap broadly in the southeastern United States, to which both groups are largely confined. Furthermore, the limb-reduced ecomorph has failed to evolve in salamanders in other temperate regions where salamanders occur but caecilians do not, such as in temperate Asia, Europe, and western North America (Wiens and Hoverman 2008). It is unclear why only the southeastern United States seems favorable to the evolution of this ecomorph, and why two lineages should have evolved this ecomorph in apparent sympatry.

Outside of amphiumids and sirenids, relationships between limb reduction, digit loss, and body elongation are not as clear in other salamanders (Wiens and Hoverman 2008). For example, several other salamanders with elongate body form and relatively small limbs do not lose digits (*Lineatriton* and *Oedipina*) or lose only the fifth toe (*Batrachoseps*). Conversely, many salamanders have lost the fifth toe without body elongation and reduction in relative limb size (e.g., *Hemidactylium*). This sporadic loss of the fifth toe also occurs in some species of lizards that otherwise show no tendency toward limb reduction and body elongation. Remarkably, although body elongation in salamanders typically seems to be achieved through an increase in number of vertebrae, in the Mexican salamander *Lineatriton* elongation is achieved through lengthening of the vertebrae instead (Parra-Olea and Wake 2001).

Evolution of limblessness not only has a geographic component (i.e., multiple origins on different continents) and a taxonomic component (i.e., origins of similar ecomorphs in different groups of reptiles and amphibians), but it may have a temporal component as well. In the Paleozoic era, an entire lineage of limbless amphibians (the aistopodans), not closely related to any modern amphibians, arose and went extinct. Some members of this lineage had more than 200 vertebrae and lacked all limbs and limb girdles, and also had a skull that bore some similarities to that of modern snakes (R. Carroll 1988). Aistopodans lived from roughly 350 million to 250 million years ago and preceded the origin of caecilians, limb-reduced salamanders, and snakelike squamates.

Resources

Books

Bruno, S., and S. Maugeri. 1976. *Rettili d'Italia: Tartarughe e sauri* [Reptiles of Italy: Turtles and lizards]. Florence, Italy: Martello.

Carroll, Robert L. 1988. *Vertebrate Paleontology and Evolution.* New York: Freeman.

Carroll, Sean B., Jennifer K. Grenier, and Scott D. Weatherbee. 2005. *From DNA to Diversity: Molecular Genetics and the Evolution of Animal Design.* 2nd edition. Malden, MA: Blackwell.

Orsini, Jean-Paul G., and Marc Cheylan. 1981. "*Chalcides chalcides* (Linnaeus 1758)—Erzschleiche." In *Handbuch der Reptilien und Amphibien Europas*, Vol. 1, ed. Wolfgang Böhme. Wiesbaden, West Germany: Akademische Verlagsgesellschaft.

Pough, F. Harvey, Robin M. Andrews, John E. Cadle, et al. 2004. *Herpetology.* 3rd edition. Upper Saddle River, NJ: Prentice Hall.

Schluter, Dolph. 2000. *The Ecology of Adaptive Radiation.* Oxford: Oxford University Press.

Periodicals

Apesteguía, Sebastián, and Hussam Zaher. 2006. "A Cretaceous Terrestrial Snake with Robust Hindlimbs and a Sacrum." *Nature* 440(7087): 1037–1040.

Brandley, Matthew C., John P. Huelsenbeck, and John J. Wiens. 2008. "Rates and Patterns in the Evolution of Snake-Like Body Form in Squamate Reptiles: Evidence for Repeated Re-evolution of Lost Digits and Long-Term Persistence of Intermediate Body Forms." *Evolution* 62(8): 2042–2064.

Caldwell, Michael L. 2000. "On the Aquatic Squamate *Dolichosaurus longicollis* Owen, 1850 (Cenomanian, Upper Cretaceous), and the Evolution of Elongate Necks in Squamates." *Journal of Vertebrate Paleontology* 20(4): 720–735.

Gans, Carl. 1975. "Tetrapod Limblessness: Evolution and Functional Corollaries." *American Zoologist* 15(2): 455–467.

Gans, Carl, and Margaret Fusari. 1994. "Locomotor Analysis of Surface Propulsion by Three Species of Reduced-Limbed Fossorial Lizards (*Lerista*: Scincidae) from Western Australia." *Journal of Morphology* 222(3): 309–326.

Kearney, Maureen. 2003. "Sytematics of the Amphisbaenia (Lepidosauria: Squamata) Based on Morphological Evidence from Recent and Fossil Forms." *Herpetological Monographs* 17 (1): 1–74.

Kearney, Maureen, and Bryan L. Stuart. 2004. "Repeated Evolution of Limblessness and Digging Heads in Worm Lizards Revealed by DNA from Old Bones." *Proceedings of the Royal Society* B 271(1549): 1677–1683.

Kohlsdorf, Tiana, and Günter P. Wagner. 2006. "Evidence for the Reversibility of Digit Loss: A Phylogenetic Study of Limb Evolution in *Bachia* (Gymnophthalmidae: Squamata)." *Evolution* 60(9): 1896–1912.

Lee, Michael S. Y., Graham L. Bell Jr., and Michael W. Caldwell. 1999. "The Origin of Snake Feeding." *Nature* 400(6745): 655–659.

Losos, Jonathan B., Todd R. Jackman, Allan Larson, et al. 1998. "Contingency and Determinism in Replicated Adaptive Radiations of Island Lizards." *Science* 279(5359): 2115–2118.

Murphy, James B., David G. Barker, and Bern W. Tryon. 1978. "Miscellaneous Notes on the Reproductive Biology of Reptiles. 11. Eleven Species of the Family Boidae, Genera *Candoia, Corallus, Epicrates* and *Python*." *Journal of Herpetology* 12:385–390.

Parra-Olea, Gabriela, and David B. Wake. 2001. "Extreme Morphological and Ecological Homoplasy in Tropical Salamanders." *Proceedings of the National Academy of Sciences of the United States of America* 98(14): 7888–7891.

Rodrigues, Miguel Trefaut. 1991. "Herpetofauna das dunas interiores do rio São Francisco, Bahia, Brasil," Pt. 1: "Introdução à área e descrição de um novo gênero de microteiídeos (*Calyptommatus*) com notas sobre sua ecologia, distribuição e especiação (Sauria, Teiidae)." *Papéis Avulsos de Zoologia* (São Paulo) 37(19): 285–320.

Shapiro, Michael D., James Hanken, and Nadia Rosenthal. 2003. "Developmental Basis of Evolutionary Digit Loss in the Australian Lizard *Hemiergis*." *Journal of Experimental Zoology* B 297(1): 48–56.

Tchernov, Eitan, Olivier Rieppel, Hussam Zaher, et al. 2000. "A Fossil Snake with Limbs." *Science* 287(5460): 2010–2012.

Wiens, John J., and Jason T. Hoverman. 2008. "Digit Reduction, Body Size, and Paedomorphosis in Salamanders." *Evolution and Development* 10(4): 449–463.

Wiens, John J., and Jamie L. Slingluff. 2001. "How Lizards Turn into Snakes: A Phylogenetic Analysis of Body-Form Evolution in Anguid Lizards." *Evolution* 55(11): 2303–2318.

Wiens, John J., Matthew C. Brandley, and Tod W. Reeder. 2006. "Why Does a Trait Evolve Multiple Times within a Clade? Repeated Evolution of Snakelike Body Form in Squamate Reptiles." *Evolution* 60(1): 123–141.

John J. Wiens
Matthew C. Brandley

Canid evolution: From wolf to dog

The origins of the dog have been hotly debated at least since the early nineteenth century, and it is easy to understand why: There are now more than 350 different dog breeds recognized worldwide ranging in appearance from tiny Chihuahuas and "teacup" Yorkshire terriers to giant breeds such as the Great Dane and Saint Bernard. The domestic dog is in fact far more variable in size, shape, and behavior than any other living mammal, so much so that the English naturalist Charles Darwin (1809–1882) doubted that it would ever be possible to determine with certainty the dog's true ancestry. He, like several later experts, favored the view that the dog was probably the outcome of mixed descent from two or more wild species, such as wolves and jackals. Several other authorities have attempted to trace the origins of the domestic dog to some hypothetical wild dog of Asia that is now extinct, but convincing evidence to support this claim is limited. Nowadays, based on a growing body of anatomical, genetic, and behavioral evidence, most experts believe that the dog originated exclusively from a single species: the gray wolf, *Canis lupus*.

Early comparative studies of behavior strongly support the wolf's claim to be the sole ancestor of the domestic dog. Back in the 1960s, the American ethologist John Paul Scott prepared a detailed ethogram, or catalog of species-specific behaviors, consisting of some ninety different behavior patterns for the dog, of which all but nineteen were also observed in wolves. The missing behaviors tended to be minor activities that probably do occur in wolves but had not, at that time, been recorded. Conversely, all the behavior patterns recorded from wolves but not from dogs occurred in specialized hunting contexts that do not arise under domestic conditions. In contrast, the displays and vocalizations of jackals are rather distinctive, and quite different from those of either dogs or wolves. It appears also that all of the dog breeds that have been adequately studied have common ancestry. Scott and John L. Fuller's important work from 1965 on the genetics and social behavior of six different dog breeds found no example, in any breed, of the total absence of any of the typical canid behavior patterns. More recent anatomical and molecular evidence has confirmed that wolves, dogs, and dingoes are all more closely related to each other than they are to any other member of the family Canidae.

Time and place of domestication

Domestication is a process rather than an event, so it is difficult to pinpoint the precise time and location of the earliest wolf domestication. Apparent associations between the fossil remains of wolves and humans have been unearthed from sites as old as 500,000 years before the present (BP). There is no evidence, however, that these wolves were

A gray wolf (*Canis lupus*) in the snow. Tom Brakefield/Bruce Coleman, Inc.

domesticated or even tamed. The oldest known archaeological remains of probable domestic wolf-dogs consist of a pair of skulls excavated from the Upper Paleolithic site of Eliseye-vichi in western Russia close to the Ukrainian border. The site has been dated at between 13,000 and 17,000 years BP. Both skulls resemble those of Siberian huskies in general shape, although they are larger and broader and have considerably shorter muzzles. A fragment of a dog's lower jaw has also been excavated from the Upper Paleolithic site of Bonn-Oberkassel in Germany dating from around 14,000 years BP. From around 12,000 years BP a whole series of skeletal remains of apparent domestic wolf-dogs have also been found in human dwelling and grave sites in Israel, Iraq, and other parts of the Middle East. All of these animals were morphologically similar to wolves, except that they tended to be slightly smaller and sometimes had overcrowded or compacted teeth. Tooth crowding in early domestic dogs is thought to arise from the fact that selection for small teeth tends to lag behind selection for the smaller and weaker jaws associated with a domestic diet. Overall, the archaeological evidence therefore points to an eastern European or western Asian origin for the domestic dog around 15,000 to 17,000 years BP.

Anatomical and molecular evidence generally supports the date but not the location of dog domestication derived from archaeology. For example, an important feature of dogs' jaws that distinguishes them from those of other canids is the "turned-back" apex of the coronoid process on the rear of the jawbone. The only wild canid that shares this feature is the Chinese wolf (*Canis lupus chanco*) from eastern rather than western Asia. Similarly, a recent analysis of variation in mitochondrial DNA (mtDNA) from a large sample of dogs from all over the world points to a single East Asian origin for the domestic dog, probably around 15,000 years BP. (An older study that derived a date of domestication of 135,000 years BP based on mtDNA evidence is now considered highly controversial and should probably be discounted in the absence of further corroboration.) In the light of current evidence, it therefore appears likely that the domestic dog was derived from a single population of East Asian wolves approximately 15,000 years BP, and that these early domestic wolf-dogs spread very rapidly throughout the rest of Asia and Europe. Domestic dogs were introduced to the New World by Asian colonists around 10,000 years BP, and probably about 5,000 years BP reached Australia, where their feral descendants gave rise to the dingo.

From wolf to dog: How and why?

There are several competing accounts of the process by which wild wolves were transformed into domestic dogs, but it should be emphasized that all of them are little better than fables. The truth is that the archaeological record provides few clues as to the motivations and mechanisms underlying wolf domestication, and this leads inevitably to a great deal of unsupported speculation.

One of the most popular theories proposes that wolves began associating with humans, in much the same way that jackals associate with lions, as opportunistic, commensal scavengers. Gradually, as a result of human tolerance, and

recognition of the dog's potentially useful role as a garbage collector, watchdog, and occasional item of food, this casual association evolved, or so the theory maintains, into a sort of permanent symbiosis. Although favored by a number of authorities, the scavenging hypothesis does not explain how these satellite wolves became fully assimilated into human families and social groups. It suggests a possible mechanism for bringing about a closer *spatial* association between wild wolves and humans, but it fails to account for the development of the kinds of close *social* partnerships so typical of historical and contemporary dog–human interactions. In addition, most studies of prehistoric demography suggest that human populations were very small and highly dispersed during the Upper Paleolithic (Bocquet-Appel et al. 2005), so they may have been insufficient to provide a viable ecological niche for a permanent population of scavenging wolves.

Another theory, now less popular than it used to be, also involves the gradual evolution of a symbiotic or mutualistic relationship between wolves and humans but is based on hunting rather than scavenging. Many recent hunter-foragers undoubtedly use dogs for hunting, and it appears that their hunting success is improved as a result, at least when hunting certain types of prey. There may also be some significance to the fact that the domestication of the dog toward the end of the Paleolithic more or less coincided with the invention of the bow and arrow—the implication being that hunting dogs would have been useful to Paleolithic hunters for trailing wounded game that might otherwise have been lost using this new technology. Unfortunately, the earliest concrete evidence of dogs being used for hunting derives from Neolithic and post-Neolithic rock art, several thousand years after domestication. Furthermore, as with the scavenging hypothesis, it is still necessary to explain the means by which this postulated hunting symbiosis between wild wolves and humans first evolved into a true cooperative partnership.

Gray wolves (*Canis lupus*) exhibit mated pair courtship. Tom Brakefield. Bruce Coleman, Inc.

The idea that wolves, like sheep, goats, or chickens, were originally domesticated as food items receives relatively little support in the literature. Archaeological evidence of dog eating, in the form of cut marks on bones, for example, has been found at a number of late Paleolithic and Neolithic sites in Europe, and the practice is still widespread in certain areas of the world, particularly West Africa, Central America, Korea, China, Southeast Asia, and the Philippines. Dog sacrifice and consumption is (or was) also commonly practiced by many American Indian and Inuit groups. Nevertheless, in global terms, dog eating remains a minority activity, and in most cultures it is viewed as either repugnant or, at best, an extreme starvation measure. At present, there is little evidence to indicate whether or not Paleolithic peoples killed and ate wolves or wolf pups on occasion. And even if they did, one would still need to explain why they would have elected to keep and breed domestic wolves for consumption several thousands of years before they domesticated more palatable and economically viable species such as sheep and goats.

Finally, an alternative theory proposes that the critical step leading to the domestication of wolves (and perhaps other domestic species) toward the end of Paleolithic was the practice of deliberately capturing, adopting, and keeping young wild animals as pets. Pet keeping of this kind is extraordinarily widespread among living and recent hunter-gatherer societies, and it seems reasonable to postulate that Paleolithic hunting and foraging cultures engaged in similar practices. According to this hypothesis, the domestication of wolves would not have been possible until tame wolves had already become an integral feature of Paleolithic village life—a process that would have depended on the active adoption, hand rearing, and socialization of wolf pups by people. Critics of this theory point out that adult wolves are large and potentially dangerous predators that would have posed an unacceptable risk, especially to children living in these prehistoric communities. Yet, this applies equally to wolf-dogs serving as either scavengers or hunting partners, and may also underestimate the socializing effects of handling by humans during a wolf pup's critical early weeks of life. Numerous recent accounts of hand-reared wolves have found that, although they are somewhat nervous and predatory toward other animals, these wolves are remarkably nonaggressive toward people.

Apart from their casual contribution to community hygiene and vigilance, these early pet wolves would not have needed to serve any obvious economic role in order to be valued by their owners. The relationship would have been based on companionship rather than service, just as it is between modern pets and their owners. Eventually, of course, dogs acquired important economic and practical functions, such as hunting and guarding, but this probably occurred sometime after the process of domestication was already complete.

Archaeological evidence offers some support for the pet keeping theory of wolf domestication. Some of the oldest skeletal remains of domestic wolf-dogs from central Europe, the Near East, and North America appear to have been deliberately buried, either in separate graves or with humans.

The Bonn-Oberkassel wolf-dog, for example, was buried in the double grave of a man and woman, while the remains of a puppy from Ein Mallaha in Israel was buried roughly 12,000 years BP with an elderly person whose left hand was apparently positioned so that it rested on the dog's flank. At Koster in Illinois, some of the earliest remains of dogs in the New World dating from about 8,500 years BP were evidently buried intact in individual graves. Some authorities have interpreted such finds as evidence of affectionate rather than purely utilitarian relationships between Paleolithic people and their dogs.

Whichever account of dog domestication one favors, one thing seems reasonably certain: The process probably required the existence of relatively sedentary human communities in which pups could be born and reared in a physically stable location, at least until they were old enough to travel under their own steam. Unlike many mammals, wolf pups are born helpless and blind and must remain in a stationary den or nest until they are old enough to follow adult pack members, usually at around ten to twelve weeks of age. Judging from the archaeological record, sedentary or semisedentary human communities did not appear until the closing stages of the last ice age around 15,000 years BP. Prior to this time, most human populations were highly nomadic, moving from place to place as local food resources became exhausted.

Changes associated with domestication

The central mystery of dog evolution, and the one that led to much of the early confusion and speculation about canine origins, is how humans managed to generate animals as diverse in size, shape, and function as domestic dogs in the space of only 15,000 years. The traditional explanation for this astonishing transformation is that our ancestors favored and selectively bred from individuals displaying unusual or promising physical or behavioral traits in order to attain some particular desired goal, such as the capacity to herd or guard sheep, chase gazelles, bark at strangers, and so on. For this argument to work, however, early dog owners must have had reasonably clear ideas about the kinds of dogs they wanted and plenty of variation in canine appearance and behavior from which to choose. In reality, there is little evidence that early dog breeders ever selected for the gradual accumulation of some desired characteristic in dogs, and it is difficult to imagine how they could have done so because the necessary variation in phenotype did not exist in the wild ancestor. How, by a process of gradual trait-by-trait selection, could our forebears have produced the jowly face and lop ears of the bloodhound, the tightly curled tail of the spitz breeds, or the piebald coat colors of many domestic breeds, when these features are never seen in wolves? The argument simply does not make sense.

To some extent, the same is true when the behavior of dogs and wolves is compared. Consider, for example, hunting behavior. Hunting wolves display highly characteristic and stereotyped predatory sequences involving combinations of trailing, stalking and chasing, biting the living prey, dissecting the carcass, and often carrying or caching the dissected flesh before swallowing it. A wolf deficient in any

Wolves, like dogs, are pack animals. Tom Brakefield/Bruce Coleman, Inc.

of one these components of the predatory sequence would not survive long in the wild, and so natural selection has ensured that there is little individual variation in wolf hunting behavior. In domestic dogs, however, some of these components of predatory behavior are clearly retained and emphasized, whereas others have been de-emphasized or lost altogether. Dogs such as bloodhounds and beagles, for instance, excel at scent trailing; border collies stalk and chase sheep but are inhibited from biting them; greyhound-type breeds delight in chasing and catching prey animals but appear to lose interest once this goal is attained; gun dogs such as Labradors and spaniels enjoy retrieving and carrying things, while livestock-guarding breeds such as the Maremma, ideally, show no predatory behavior at all. How would our ancestors have been able to select for all of these different behavioral specializations when the ancestral wolf was apparently so invariant in behavior?

Some authorities have argued that all or most of these characteristics that distinguish dogs from wild canids such as wolves are the inadvertent outcome of early and intense selection for just a single trait—*tameness*—and evidence for how this might have happened comes from a somewhat unexpected source. In the 1960s the Soviet fur industry developed a problem with farmed silver foxes: The animals were so nervous and intractable that they were becoming unmanageable. A group of Soviet geneticists were therefore delegated the task of breeding a strain of docile and tractable foxes, and they proceeded to do so by subjecting foxes to temperament tests and then selectively breeding from the tamest and most docile individuals. To avoid the problem of inbreeding, these tamer strains were out-crossed to foxes from other farms, and, within about fifteen to twenty generations, some very unusual things started happening. Not only did they succeed in breeding tamer foxes that

Hunting wolves display highly characteristic and stereotyped predatory sequences involving combinations of trailing, stalking and chasing, biting the living prey, dissecting the carcass, and often carrying or caching the dissected flesh before swallowing it. Michael S. Quinton/National Geographic Stock.

behaved toward people much as dogs do, they also began to generate an extraordinary amount of variability: Some foxes actually began to look more like dogs than foxes; some developed piebald coats; others had drooping ears or curled tails; some developed diestrous reproductive cycles (wolf females have only a single estrous each year, in contrast to dogs that normally have two); and many were as playful and friendly as puppies.

It seems reasonable to argue that precisely the same process may have been applied unconsciously to the earliest domestic wolves—that is, intense selection for tameness accompanied by the rapid evolution of increasingly doglike appearance and behavior. Wolves can certainly be tamed and are said to make affectionate pets, but they also tend to be hyperreactive, nervous, or uncontrollable in unfamiliar situations, and prone to displaying full-blown predatory responses toward other animals. Merely by driving away or killing their most intractable pets, and perhaps selectively feeding and caring for the tamest ones, our Paleolithic ancestors may have unwittingly performed the same experiment as the Soviet geneticists—thereby gradually suppressing the wild-type traits and creating a new and highly variable population of domestic wolf-dogs.

The processes responsible for the changes associated with selection for tameness are still quite poorly understood, but it is likely that at least some of them result from the retention of juvenile or "neotenous" features into adulthood. This idea certainly fits with a lot of what scientists know about the behavior of dogs—for example, compared with wolves, dogs are perpetual puppies who never properly grow up—and it also provides an evolutionary route to the emergence of the different working breeds. For example, mammalian infants and adults are rather like different species in that each is adapted and specialized to quite distinct modes of life. Connecting these two stages is a juvenile or adolescent phase during which the animal's behavior and physiology is reorganized from the infant to the adult pattern. In other words, a sort of metamorphosis takes place. This metamorphosis is not as extreme as the one that transforms, say, a tadpole into a frog, but it still represents a fundamental reorganization when one compares neonatal and adult styles of behavior. According to some experts, early and intense selection for tameness arrested the domestic dog developmentally during this metamorphic stage and, as a result, inadvertently generated animals exhibiting a range of incomplete or disorganized patterns of adult behavior that could be subsequently co-opted for the performance of particular specialized tasks.

Two young gray wolves (*Canis lupus*) showing two color phases in same litter. Tom McHugh/Photo Researchers, Inc.

The process is analogous in some ways to deconstructing a piece of equipment such as a bow and arrow. Once disassembled, the constituent parts—the bow, the bowstring, the arrowhead, and so on—can each be used for other specialized functions, none of which could be performed by the fully assembled bow and arrow. Likewise, by interrupting the development of the normal adult hunting sequence in wolves, early domestication could have produced dogs displaying a range of incomplete or partial sequences of hunting behavior that could then be exploited for a variety of novel purposes that wolves were incapable of performing.

The social skills of dogs

One of the first things that happens when animals are domesticated is that they tend to become physically smaller than their wild ancestors (Hemmer 1990). The dog is no exception, and the decrease in size is particularly noticeable in relation to cranial dimensions. When matched for body size, the skulls and brains of dogs are about 20 percent smaller than those of wolves.

Perhaps in keeping with their smaller brains, dogs appear also to be less good at independent problem-solving compared with wolves. When researchers have attempted to compare wolves and dogs for their performance on various problem-solving and observational-learning tasks, the wolves tend to outperform the dogs by a significant margin. Differences in motivation cannot be excluded as a possible explanation for this difference. Anecdotal evidence suggests, however, that wolves are simply better at solving these types of problems. The psychologist Harry Frank, who was responsible for much of the early research in this area, gives a striking example involving the observational learning abilities of a malamute, a malamute–wolf hybrid, and a purebred wolf. The latch on the door connecting the indoor and outdoor runs of Frank's research facility was relatively complex to operate. The malamute never learned to operate the latch despite observing people opening and closing it for six years. The malamute–wolf hybrid learned the task using his muzzle after watching people for only two weeks, and the seven-month-old wolf

learned the task after watching the hybrid only once. Also, the wolf used her paws rather than her muzzle to perform the trick (Frank and Frank 1985). This difference between dogs and wolves makes sense. Existence under domestication greatly reduces the need for independent problem-solving abilities because domestic animals are not required to fend entirely for themselves. They are to some extent insulated from the more complex problems of daily living by their human caretakers. Wolves need ingenuity and insight in order to be successful cooperative hunters in an otherwise unforgiving world. Dogs, at least to some degree, can rely on humans to solve these kinds of problems for them.

It would be a mistake, however, to conclude from this that dogs are necessarily less intelligent than wolves. Rather it appears that they have a different kind of intelligence. Dogs, for example, are highly trainable and wolves, typically, are not. Frank spent six months attempting to train his seven-month-old wolf to sit in response to a verbal command, a task that most dogs can perform after only a few trials. She never learned the trick. Yet this was the same wolf that exhibited one-trial observational learning when it came to opening a complicated door latch (Frank and Frank 1985). So wolves are skilled observational learners and relatively insightful when it comes to solving complex problems, yet they find it extremely difficult, if not impossible, to do what most dogs manage to do with relative ease.

More recently, groups of researchers in the United States, Germany, and Hungary have explored these differences between dogs and wolves in more detail and have discovered that domestic dogs, even from an early age, are far more sensitive to human verbal and nonverbal signals than hand-reared wolves are, and are far better at using these cues to help them solve problems. Using directed patterns of gaze and distinctive vocalizations, dogs also excel at recruiting human assistance to solve problems that they cannot solve themselves. The results of these studies suggest that during the process of domestication, dogs were selected for a unique set of social-cognitive skills that enabled them to communicate and cooperate with humans to an unprecedented degree (Hare et al. 2002). The problem-solving capacities of wolves are limited to situations in which there is an obvious functional connection between an event or an action and its outcome, presumably because this type of intelligence serves them well as wild predators. Dogs, in contrast, tend to take their cues from humans and are able to learn behavioral responses that have no obvious functional connection with their outcomes, such as sitting down in response to a verbal command in return for a pat on the head or a morsel of food. This kind of intelligence would serve little purpose in the life of a wild wolf, but in the context of a domestic existence in partnership with humans it has clear survival advantages.

Humans have been exploiting and enhancing these unique canine attributes throughout history. The best working dogs nowadays are prepared to work exceptionally hard in return for nothing more than human social acceptance and praise. Such dogs are characteristically eager to please, and they exhibit a tremendously strong social orientation toward people compared with hand-reared wolves. Significantly, a

The best working dogs nowadays are prepared to work exceptionally hard in return for nothing more than human social acceptance and praise. © Dale C. Spartas/Corbis.

2005 study found that the descendants of the Russian silver foxes that were originally selected for tameness display a similar degree of sensitivity to human social cues (Hare et al. 2005). This strongly reinforces the idea that human social orientation and trainability in dogs were unforeseen consequences of early selection for a docile temperament.

Conclusions

The best available evidence suggests that dogs were domesticated from wolves in East Asia approximately 15,000 years BP. Although the process by which wolves became dogs has been the subject of considerable speculation, the true story of how or why this extraordinary relationship first developed will probably never be known. The process of domestication transformed the wolf in a variety of ways. It altered the animal's appearance to a remarkable extent and had a dramatic impact on its mental characteristics. Above all, it created variation in morphology and behavior that was not present in the ancestral wolf—variation that later provided the raw material for the development of all the subsequent specialized working dog breeds.

It is likely that many of these changes in morphology and behavior were initiated by early selection for "tameness." By selecting for increasingly tame individuals, our ancestors inadvertently produced a strain of domestic wolves in which a variety of juvenile or neotenous traits were retained into adulthood. Coordinated patterns of "instinctive" behavior characteristic of adult wolves became fragmented by this process, and humans later exploited these "fragments" to generate highly specialized working breeds. A reduced need for independent problem-solving skills and a greatly enhanced ability to detect and respond to social cues provided by humans also emerged as a result of the domestication process. Trainability—the capacity to learn novel behavioral responses to arbitrary human cues—may have been another accidental outcome of early selection for puppylike characteristics. The animal that emerged from the other side of this process of early selection for tameness was the true ancestor of all modern dog breeds. It neither looked like a wolf nor, strictly speaking, did it behave like one. Instead, it was a new and uniquely variable, "all-purpose" creature whose future working talents were still to be realized and exploited by humans.

Resources

Books

Belyaev, D. K. 1979. "Destabilizing Selection as a Factor in Domestication." *Journal of Heredity* 70(5): 301–308.

Bocquet-Appel, Jean-Pierre, Pierre-Yves Demars, Lorette Noiret, and Dmitry Dobrowsky. 2005. "Estimates of Upper Palaeolithic Meta-population Size in Europe from Archaeological Data." *Journal of Archaeological Science* 32(11): 1656–1668.

Hemmer, Helmut. 1990. *Domestication: The Decline of Environmental Appreciation*, trans. Neil Beckhaus. 2nd edition. Cambridge, UK: Cambridge University Press.

Lorenz, Konrad. 1954. *Man Meets Dog*, trans. Marjorie Kerr Wilson. London: Methuen.

Olsen, Stanley J. 1985. *Origins of the Domestic Dog: The Fossil Record*. Tucson: University of Arizona Press.

Sauer, Carl O. 1969. *Agricultural Origins and Dispersals*. 2nd edition. Cambridge, MA: MIT Press.

Scott, John Paul, and John L. Fuller. 1965. *Genetics and the Social Behavior of the Dog*. Chicago: University of Chicago Press.

Serpell, James A. 1989. "Pet-Keeping and Animal Domestication: A Reappraisal." In *The Walking Larder: Patterns of Domestication, Pastoralism, and Predation*, ed. Juliet Clutton-Brock. London: Unwin Hyman.

Serpell, James A. 1995. "From Paragon to Pariah: Some Reflections on Human Attitudes to Dogs." In *The Domestic Dog: Its Evolution, Behaviour, and Interactions with People*, ed. James A. Serpell. Cambridge, UK: Cambridge University Press.

Serpell, James A., ed. 1995. *The Domestic Dog: Its Evolution, Behaviour, and Interactions with People*. Cambridge, UK: Cambridge University Press.

Zeuner, Frederick E. 1963. *A History of Domesticated Animals*. London: Hutchinson.

Periodicals

Clutton-Brock, Juliet. 1995. "Origins of the Dog: Domestication and Early History." In *The Domestic Dog: Its Evolution, Behaviour, and Interactions with People*, ed. James A. Serpell. Cambridge, UK: Cambridge University Press.

Coppinger, Raymond, and Lorna Coppinger. 2001. *Dogs: A Startling New Understanding of Canine Origin, Behavior, and Evolution*. New York: Scribner.

Coppinger, Raymond, and Richard Schneider. 1995. "Evolution of Working Dogs." In *The Domestic Dog: Its Evolution,*

Behaviour, and Interactions with People, ed. James A. Serpell. Cambridge, UK: Cambridge University Press.

Darwin, Charles. 1868. *The Variation of Animals and Plants under Domestication*, Vol. 1. London: John Murray.

Davis, Simon J. M., and François R. Valla. 1978. "Evidence for Domestication of the Dog 12,000 Years Ago in the Natufian of Israel." *Nature* 276(5688): 608–610.

Fentress, John C. 1967. "Observations on the Behavioral Development of a Hand-Reared Male Timber Wolf." *American Zoologist* 7(2): 339–351.

Frank, Harry, and Martha Gialdini Frank. 1982. "On the Effects of Domestication on Canine Social Development and Behavior." *Applied Animal Ethology* 8(6): 507–525.

Frank, Harry, and Martha Gialdini Frank. 1985. "Comparative Manipulation-Test Performance in Ten-Week-Old Wolves (*Canis lupus*) and Alaskan Malamutes (*Canis familiaris*): A Piagetian Interpretation. *Journal of Comparative Psychology* 99 (3): 266–274.

Hare, Brian, Michelle Brown, Christina Williamson, and Michael Tomasello. 2002. "The Domestication of Social Cognition in Dogs." *Science* 298(5598): 1634–1636.

Hare, Brian, Irene Plyusnina, Natalie Ignacio, et al. 2005. "Social Cognitive Evolution in Captive Foxes Is a Correlated By-Product of Experimental Domestication." *Current Biology* 15 (3): 226–230.

Manwell, Clyde, and C. M. Ann Baker. 1983. "Origin of the Dog: From Wolf or Wild *Canis familiaris*?" *Speculations in Science and Technology* 6(3): 213–224.

Miklósi, Ádám, Enikö Kubinyi, József Topál, et al. 2003. "A Simple Reason for a Big Difference: Wolves Do Not Look Back at Humans, but Dogs Do." *Current Biology* 13(9): 763–766.

Miklósi, Ádám, Réka Polgárdi, József Topál, and Vilmos Csányi. 2000. "Intentional Behaviour in Dog–Human Communication: An Experimental Analysis of 'Showing' Behaviour in the Dog." *Animal Cognition* 3(3): 159–166.

Morey, Darcy F. 1992. "Size, Shape, and Development in the Evolution of the Domestic Dog." *Journal of Archaeological Science* 19(2): 181–204.

Morey, Darcy F., and Michael D. Wiant. 1992. "Early Holocene Domestic Dog Burials from the North American Midwest." *Current Anthropology* 33(2): 224–229.

Parker, Heidi G., Lisa V. Kim, Nathan B. Sutter, et al. 2004. "Genetic Structure of the Purebred Domestic Dog." *Science* 304(5674): 1160–1164.

Reed, Charles A. 1959. "Animal Domestication in the Prehistoric Near East." *Science* 130(3389): 1629–1639.

Sablin, Mikhail V., and Gennady A. Khlopachev. 2002. "The Earliest Ice Age Dogs: Evidence from Eliseevichi 1." *Current Anthropology* 43(5): 795–799.

Savolainen, Peter, Thomas Leitner, Alan N. Wilton, et al. 2004. "A Detailed Picture of the Origin of the Australian Dingo, Obtained from the Study of Mitochondrial DNA." *Proceedings of the National Academy of Sciences* 101(33): 12387–12390.

Savolainen, Peter, Ya-ping Zhang, Jing Luo, et al. 2002. "Genetic Evidence for an East Asian Origin of Domestic Dogs." *Science* 298(5598): 1610–1613.

Scott, John Paul. 1967. "The Evolution of Social Behavior in Dogs and Wolves." *American Zoologist* 7(2): 373–381.

Scott, John Paul. 1968. "Evolution and Domestication of the Dog." *Evolutionary Biology* 2: 243–275.

Spady, Tyrone C., and Elaine A. Ostrander. 2008. "Canine Behavioral Genetics: Pointing Out the Phenotypes and Herding Up the Genes." *American Journal of Human Genetics* 82(1): 10–18.

Tchernov, Eitan, and François F. Valla. 1997. "Two New Dogs, and Other Natufian Dogs, from the Southern Levant." *Journal of Archaeological Science* 24(1): 65–95.

Wayne, Robert K., and Stephen J. O'Brien. 1987. "Allozyme Divergence within the Canidae." *Systematic Zoology* 36(4): 339–355.

James A. Serpell

Dinosaurs

Dinosaurs represent one of the most successful groups of vertebrates in the history of Earth. First appearing about 228 million years ago (mya) at the dawn of the Late Triassic epoch (228–199.6 mya), they dominated the medium to large-bodied terrestrial niches throughout the Jurassic (199.6–145.5 mya) and Cretaceous (145.5–65.5 mya) periods. During this interval, nearly all terrestrial animals larger than 10 kilograms in mass were dinosaurs. This group consisted of numerous diverse forms, including armored, plated, horned, dome-headed, duck-billed, crested, and long-necked herbivores and an array of bipedal carnivores. Famously the dinosaurs include the largest terrestrial animals that have every lived, 100-ton giants that have been exceeded in size only by the living fin and blue whales. Yet Dinosauria (from the Greek for "fearfully great lizards") also included numerous small-bodied species as well, including recently discovered forms about the size of a modern robin. Spectacular new fossils from the Early Cretaceous epoch of China reveal that many of the smaller carnivorous dinosaurs were covered with feathers rather than scales; along with an abundance of skeletal information, this confirms that living birds are the direct descendants of dinosaurs. Indeed, under modern classification schemes, Aves (birds) are a subgroup within the larger Dinosauria, just as bats are a subgroup within Mammalia. Discoveries since the 1980s have established that many allegedly "avian" (bird) traits—including skeletal features such as wishbones, soft-tissue features such as the flow-through lung, and behaviors such as brooding on nests—were originally dinosaurian traits retained into modern times.

The success and diversity of the dinosaurs is only part of a larger phenomenon. Bracketed by the Permo-Triassic mass extinction 251 mya, which ended a world dominated by the ancestors of mammals, and the terminal Cretaceous mass extinction 65.5 mya, which ended the reign of the dinosaurs, the Mesozoic era is commonly called the Age of Reptiles. During this interval, reptiles were the dominant forms of medium to large-bodied animals. The marine realm was colonized several times by the reptiles, including marine representatives of surviving terrestrial groups (turtles, tua-taras, lizards, and crocodilians) as well as by the highly specialized ichthyosaurs and plesiosaurs. Reptiles evolved flight twice during this time, first by the pterosaurs—relatives of the dinosaurs—in the Triassic and later by the dinosaurs

themselves in the form of birds. Although dinosaurs were the dominant terrestrial group during the Jurassic and Cretaceous periods, they shared their habitats with a diversity of crocodile-relatives: Representatives of this latter group were the dominant terrestrial animals during the earlier Triassic period. It was in this world of reptiles that the earliest mammals evolved; indeed, most of mammalian history was spent in the shadow of the dinosaurs. Only the great extinction at the end of the Cretaceous period brought the Age of Reptiles to an end and allowed the mammals to occupy the niches once ruled by the dinosaurs.

Fossil record of the dinosaurs

With the exception of living birds, our knowledge of dinosaurs comes strictly from fossils. Fossilized bones and footprints of dinosaurs have been recorded for centuries but were (like most fossils) traditionally attributed to mythological entities such as giants or dragons. Early geologists helped to establish that fossils were simply the remains of parts of once-living organisms buried along with sediment. In the late eighteenth and early nineteenth centuries naturalists had established the principles of comparative anatomy to a degree that allowed them to recognize that many of the animals preserved as fossils were unique extinct forms, distinct from all living species. By the first decade of the 1800s a number of marine (mosasaurs, ichthyosaurs, plesiosaurs) and flying (pterosaurs) reptiles of the Mesozoic era had been discovered and named. Working independently, the English geologists William Buckland (1784–1856) and Gideon Mantell (1790–1852) discovered the first recognized remains of giant Mesozoic terrestrial reptiles in the early 1800s: the carnivorous *Megalosaurus*, the herbivorous *Iguanodon*, and the armored *Hylaeosaurus*. Although none of these were known from complete skeletons, enough was preserved to allow the English paleontologist Richard Owen (1804–1892) to recognize in 1842 that they formed "a distinct tribe or sub-order of Saurian Reptiles, for which I would propose the name of Dinosauria" (Owen 1842, p. 103).

Throughout the mid- and late 1800s many additional discoveries of dinosaurs were made in Europe, India, Madagascar, South America, and, most especially, western North America. These included the first complete dinosaur

Dinosaur exhibit in the museum of natural science in Houston. Image by Richard Cummins.

skeletons, and they revealed that these "fearfully great lizards" had body shapes considerably different from any living reptiles. For example, many dinosaurs were discovered to have been bipedal. During the twentieth century dinosaur fossils were found in nearly any part of the world where Late Triassic, Jurassic, or Cretaceous terrestrial rocks were deposited and exposed. Dinosaur bones and teeth are known from every continent, including Antarctica. While most of these fossil remains are bone fragments or isolated teeth, rare articulated skeletons have allowed paleontologists to confidently reconstruct the anatomy of at least some dinosaur species. Additionally, the fossil record also includes numerous trackways, allowing paleontologists to better understand dinosaur locomotion. Nests and eggs of most major dinosaur groups are now recognized, and occasionally preserved embryonic skeletons or association with a parent have allowed paleontologists to determine the identity of the species that laid those eggs. Skin impressions are known from several groups of dinosaurs, establishing that the herbivorous dinosaurs and the more primitive carnivorous forms had coverings of mosaic rather than overlapping scales, more similar to the pattern on crocodilians and turtles than to lizards and snakes. In the 1990s, however, carbonized remains demonstrated that advanced carnivorous dinosaurs had an integument of feathers or simple feather precursors.

Since the 1980s, technological advancements have opened up important new avenues of dinosaur research. For example, high-quality computed tomography (CT) scans allow access to small, delicate internal structures of dinosaur skulls without damaging the fossils. Such CT scans have revealed great detail of the relative size of the different lobes of the brains of dinosaur species (based on the shape of the brain cavity) and of the natural position of the skull (based on the shape and orientation of the semicircular canals in the inner ear). Microscopic examination of sections of bone polished so that they are transparent—methods developed in the early twentieth century but employed more intensely since the 1980s—have helped to resolve questions concerning the rate and style of dinosaurian growth. Chemical studies of fossils have revealed that some biomolecules survive the process of fossilization and millions of years of burial; new lines of "molecular paleontology" are being developed to use this material to answer questions of dinosaur biology.

Origin and diversity of the dinosaurs

Dinosaurs are one of the major branches of the Archosauria. Archosaurs, which first appeared in the Early Triassic epoch (251–245 mya), differed from typical reptiles in having

Fossil remains of an *Aepyornis maximus* egg. © DK Limited/Corbis.

a semi-improved stance. Rather than walking with legs sprawling out to the sides, archosaurs had legs that were oriented so that the belly was held well above the ground. In this stance archosaurs did not have to resort to the same lateral (side-to-side) undulations that are typical of sprawling vertebrates such as lizards or salamanders. This in turn meant that muscles of the trunk were freed up for use in breathing while the animal was still moving; in sprawling locomotion the same muscles are employed for walking and respiration, so that it is more difficult for the animal to oxygenate its blood while moving. Additional evidence suggests that the pelvic and belly muscles may have acted on the gastralia (the "belly ribs") to provide increased respiration for early archosaurs, bolstering their aerobic capacity.

In the Middle Triassic epoch (245–228 mya) archosaurs are the oldest known representatives of the Dinosauromorpha (dinosaurs and their closest relatives). Dinosaurs were distinguished from other archosaurs by having a parasagittal posture (i.e., their hind limbs were oriented directly underneath the body) and a digitigrade stance (i.e., only the toes were in direct contact with the ground). These adaptations indicate a shift to a striding form of locomotion. Early dinosauromorphs ranged from the 40-centimeter-long *Marasuchus* to the approximately 120-centimeter-long *Pseudolagosuchus* and were relatively minor components of their fauna. Some of these dinosauromorphs were insectivores or

carnivores, but Late Triassic genera such as *Sacisaurus* and *Silesaurus* were herbivores. These latter species were quadrupedal, but *Marasuchus* and some other primitive dinosauromorphs seem to have been bipeds.

Fossils of true dinosaurs are first known from the oldest Late Triassic rocks (228 mya) of Argentina. Because dinosaurs had already diverged into the two main subbranches, Ornithischia and Saurischia, by this time, however, the common ancestor of all dinosaurs likely lived in the later Middle Triassic. (In fact, some footprints from that interval may be tracks of these earlier dinosaurs.) Dinosaurs are deemed specialized relative to their kin by the presence of an open hip socket (one lined only with cartilage rather than bone as in most vertebrates) and a highly modified forelimb. All primitive dinosaurs share an enlarged deltopectoral crest (a muscle attachment surface) on the humerus, and a hand modified for grasping in which digit I (the thumb) is semi-opposable with the other digits, and where digits IV and V are reduced. This transformed forelimb reflects another specialization of early dinosaurs: They were all bipedal. The oldest members of each of the major dinosaur branches are about 1 to 2 meters long, suggesting this is the ancestral length for Dinosauria as a group.

All dinosaurs belong to one of two branches. The Ornithischia ("bird-hipped dinosaurs") is comprised of *Iguanodon* and its allies. Leaf-shaped teeth and a cropping beak (including the characteristic toothless predentary bone in the front of the lower jaw) indicate that ornithischians were herbivores. The oldest and most primitive ornithischian—*Pisanosaurus* of Argentina—may have had the pubis bone in its hip facing in the ancestral forward orientation, but all later ornithischians have a backward-oriented pubis. (This is the "bird-hipped" feature after which the group is named.) This reorientation allowed for increased intestinal volume, an adaptation for digesting greater quantities of plant matter.

Ornithischians are very rare during the Late Triassic, represented only by *Pisanosaurus*, by South African *Eocursor*, and the oldest members of the Heterodontosauridae. These early ornithischians all retained the ancestral large grasping dinosaurian hand; later ornithischian groups had smaller hands, lacking the grasping ability. After the Triassic-Jurassic extinction (199.6 mya) and the disappearance of many other groups of herbivorous archosaurs, the ornithischians evolved into many diverse forms.

The Thyreophora were the armored ornithischians. These had a series of armored osteoderms (bony plates and spikes) in their skin. Primitive thyreophorans were small (between 1.5 and 4 meters) and at least partially bipedal, but later larger forms were habitual quadrupeds because of the weight of their armor. Consequently, their hands and forelimbs became adapted for locomotion. One of the major groups of thyreophorans—the Stegosauria or plated dinosaurs—which first appeared in the Early Jurassic epoch (199.6–175.6 mya), were a major part of the dinosaur fauna in the Middle (175.6–161.2 mya) and Late Jurassic (161.2–145.5 mya) epochs and persisted at low diversity into the Early Cretaceous epoch (145.5–99.6 mya). Stegosaurs reached lengths up to 9 meters and were characterized by pairs of elongated defensive spikes at the end of their tails and by dorsal armor consisting of either tall broad plates or conical spikes. The other major group of thyreophorans was the Ankylosauria, or tank dinosaurs. They are first known from the Middle Jurassic, but do not become major elements of the fauna until the Cretaceous period. Ankylosaurs had a greater percentage of their surface covered in flat plates than stegosaurs, including osteoderms, which fused directly to the bones of the skull. Additionally, some ankylosaurs had extremely large shoulder spines or heavy clubbed tails, which may have served as defensive weapons against attacking predators. Ankylosaurs, which also reached lengths up to 9 meters, are known up to the end of the Late Cretaceous epoch (99.6–65.5 mya).

The remaining bird-hipped dinosaurs, the Neornithischia, were ancestrally small (1- to 2-meter-long), bipedal, fast-running herbivores. From these forms two major branches evolved. The Ornithopoda, or beaked dinosaurs, also included a number of small bipedal genera, but many of these reached much larger size. The largest—including Early Cretaceous *Iguanodon* and the Late Cretaceous duck-billed Hadrosauridae—reached lengths up to 13 meters or more. Their hands were modified as weight supports, and digit V evolved into a grasping finger. The jaws of these advanced ornithopods were highly specialized, with a broad toothless beak and rows of teeth arranged as a dental battery, forming continuous grinding surfaces. Some of the hadrosaurids had nasal passages transformed into expanded crests, which may have served both as visual and sound display organs.

The other neornithischians are the Marginocephalia, or ridge heads. Among these are the bipedal Pachycephalosauria, a group of small dinosaurs, most of which were only 1 to 2 meters long, although a few of the later forms reached 4 meters, with conspicuously thickened skull roofs. Many paleontologists consider these domed skulls to have been used in pushing contests between other members of the same species; others, however, argue that the shape of at least the more-rounded domed forms would have prevented this, and that these may have been only for visual displays. The other marginocephalians were the Ceratopsia. Ancestral ceratopsians were 1- to 2-meter-long bipedal herbivores with deep, powerful snouts. These evolved a frill extending from the back of the skull, increasing the attachment surface for jaw muscles. In later members of the group the frill extended beyond the muscle attachment area, and so may have served as a visual display. The large size of the head of advanced ceratopsians forced them onto all fours. While most groups of ceratopsians were less than 3 meters long, the most specialized forms (the Late Cretaceous Ceratopsidae) were between 4 and 9 meters long. The ceratopsids were characterized by horns over the eyes and on the nose, as well as ornamentation along the edge of the frill. These horns were almost certainly used as weapons, either as defense against attacking predators or in within-species competition.

The other major branch of the dinosaurs was the Saurischia ("lizard-hipped dinosaurs"), *Megalosaurus* and its allies. The most primitive saurischians—such as Late Triassic 1-meter-long *Eoraptor* and 2- to 4-meter-long Herrerasauria—were

Brachiosaurus were a group of herbivorous saurischians and were characterized by elongated necks and proportionately small skulls. © Louie Psihoyos/Corbis.

carnivores, as indicated by their serrated bladelike teeth. The Sauropodomorpha, however, were a group of herbivorous saurischians. Characterized by elongated necks and proportionately small skulls, early sauropodomorphs were 1- to 3-meter-long bipeds. But during the Late Triassic much larger sauropodomorphs evolved. Some paleontologists consider these dinosaurs to form a single group, the Prosauropoda; others, however, interpret some of these bipedal "prosauropods" as being more closely related to the quadrupedal Sauropoda than to other prosauropods. At 4 to 10 meters in length, prosauropods were the first group of large herbivorous dinosaurs; additionally, they were the only diverse group of Triassic dinosaurs. Their success may have stemmed from the fact that they were the only group of large-bodied herbivore capable of reaching high into the trees at that time. The Sauropoda—quadrupedal giant herbivores from 10 to 35 meters in length and 1 to 100 tons in mass—first appeared in the Late Triassic but became more common in the Middle Jurassic after the extinction of their prosauropod relatives. Besides their enormous size, and highly transformed forelimb and hand required to support their massive weight, sauropods were also distinguished by a more precise cropping bite

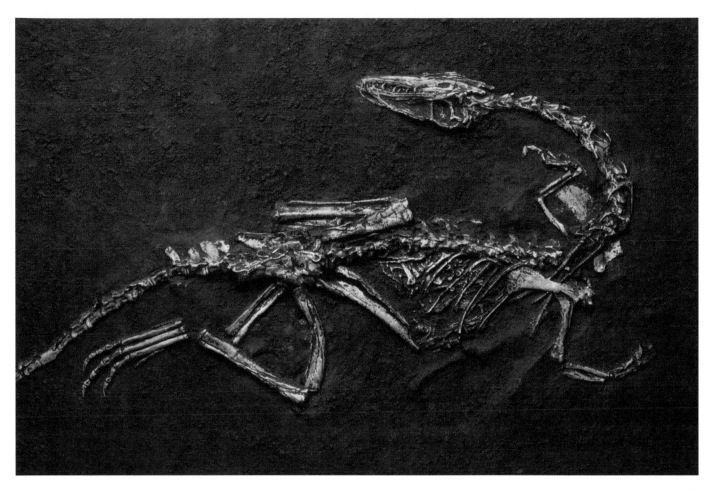

A coelophysis appears to have swallowed its young, but it is hard to know whether these baby coelophysis were its own, or whether it ingested a rival's young. Image by Louie Psihoyos.

with direct tooth-to-tooth contact. Sauropods were the most abundant herbivorous dinosaurs of the Middle and Late Jurassic worldwide, and they continued to be the dominant large herbivores in faunas of the southern continents until the end of the Cretaceous. Within the basic body plan of the sauropods was a great degree of variation, ranging from shorter necked forms specialized for low browsing to various approaches to feeding higher into the trees. The titanosaurs, a primarily Cretaceous group that includes most of the largest dinosaurs, included a number of species that evolved armored osteoderms.

The Theropoda, or carnivorous dinosaurs, remained bipedal throughout their history. Late Triassic theropods such as *Guaibasaurus* and *Coelophysis* were only secondary predators in communities in which larger crocodilian-relatives were predominant; with the extinction of their rivals at the Triassic-Jurassic boundary, theropods became the dominant terrestrial predators for the next 134 million years. Most theropod groups retained a basic set of predatory adaptations: bladelike serrated teeth, lower jaws with a shock-absorbing joint, grasping hands with trenchant claws, and elongate hind limbs for enhanced cursorial (running) ability. Some primitive theropods, such as *Procompsognathus* and *Velocisaurus*, were only 1 meter long or less, whereas

others, such as *Giganotosaurus* and *Spinosaurus*, reached lengths of 14 meters and masses of 6 tons or more.

As far as is known from the fossil record, primitive theropods had a simple scaly integument. In the Coelurosauria, however, there was a simple branching downlike body covering on at least part of the body. Primitive coelurosaurs were 1- to 3-meter-long, swift-moving predators with grasping three-fingered arms. Many specialized variations evolved from these early forms. For example, the Tyrannosauridae (tyrant dinosaurs) became larger (9- to 13-meter-long), more robust species with powerful skulls, thickened teeth, specialized elongated feet, and reduced arms ending in two-fingered hands. The Ornithomimosauria (ostrich dinosaurs) were small-skulled, long-necked forms, with long arms with clasping three-fingered hands and hind limbs resembling those of the tyrannosaurids; in the more advanced ornithomimosaurs the teeth were lost and replaced by a horny beak.

The advanced coelurosaurs, the Maniraptora (hand graspers), had enlarged braincases, highly specialized arms with laterally oriented shoulder joints, and large ossified sterna (breastbones). Additionally, maniraptorans had broad feathers of modern aspect on their arms and tail. The maniraptorans included many diverse forms, such as the beaked Oviraptorosauria; the heavily built, long-necked herbivorous Therizinosauria; and the powerful but short-armed Alvarezsauridae, inferred to be insectivores. The most specialized maniraptorans, the Eumaniraptora, shared elongated feathers along the hind limb, a backward-pointing pubis, and first digits of the foot placed at the base of the foot. The small size (30 to 40 centimeters) of the early eumaniraptorans and their large feather size suggests that at least some were tree-dwelling gliders. From the early eumaniraptorans evolved two major branches. The Deinonychosauria were predominantly carnivorous, and most later forms were too large (from 2-meter-long *Velociraptor* through 6-meter-long *Utahraptor*) to have been tree dwellers or climbers. They represent important small to medium-sized carnivores throughout the Cretaceous.

The other eumaniraptorans are the Avialae ("bird wings"), modern birds and their extinct Mesozoic relatives. (Although a few vocal opponents of the dinosaurian origin of birds remain, they have no compelling counterproposals to offer in place of the vast amount of anatomical evidence demonstrating that birds are part of the eumaniraptoran diversity.) Discoveries during the early twenty-first century

demonstrated that the earliest and most primitive deinonychosaurs (such as *Microraptor*, *Shanag*, and *Mei*) were extremely similar in body size and in anatomy to the earliest and most primitive birds making it difficult to distinguish between these two groups in their early stages. (In fact, some new reconstructions of the interrelationships among these dinosaurs suggest that Late Jurassic *Archaeopteryx*—long considered the oldest and most primitive avialian—may actually be no more closely related to true birds than are the deinonychosaurs.) Biomechanical studies of primitive avialians such as *Archaeopteryx* and Early Cretaceous *Confuciusornis* find that they were incapable of sustained powered flight and may have been no better at flight than the early deinonychosaurs. During the Cretaceous period, however, new lineages of avialians evolved with more sophisticated flight ability. While the earliest birds, like the earliest deinonychosaurs, were comparable in size range from robins to crows, later avialians included a wider range (from sparrow to eagle sized). Cretaceous avialians evolved into a great number of niches and habitats. Several groups became flightless, and one group—the hesperornithines—evolved into wingless foot-propelled swimmers: These were the only

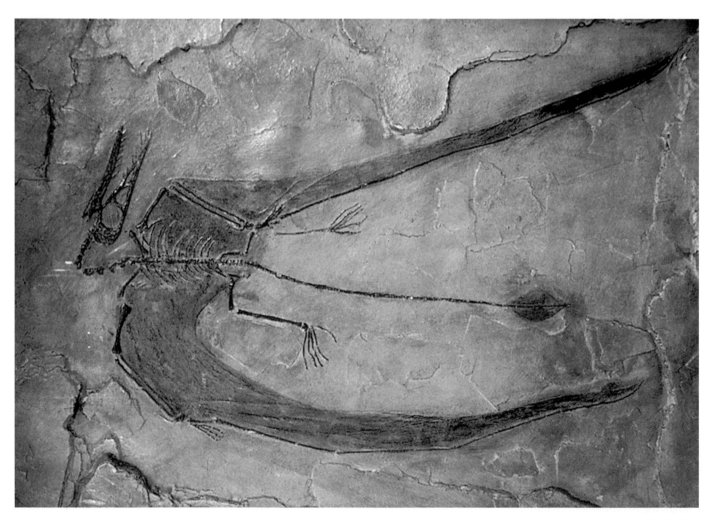

Rhamphorynchus had hollow bones and wings made of skin. Its long, narrow jaw was filled with sharp teeth that pointed outwards. It lived during the Jurassic era. © E.R. Degginger.

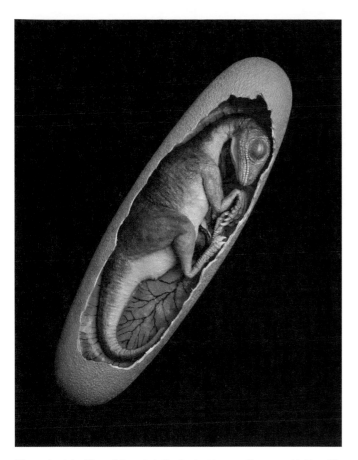

Macroelongatoolithus xixiaensis is the largest known dinosaur egg. Found in the Xixia Basin in China, it is thought to be a therizinosaur. An umbilical cord would have fed the embryo nourishment from the surrounding yolk. Image by Louie Psihoyos.

true "marine dinosaurs" of the Cretaceous period. Nearly all Mesozoic avialians retained teeth, but a few (such as [independently] *Confuciusornis* and the radiation of modern birds) independently lost these. By the end of the Cretaceous period early representatives of Aves proper (the group including all living birds) had evolved. In fact, in latest Cretaceous Antarctica fossils, ducks and duck-billed dinosaurs are found in the same rocks.

In the modern system of classification, only "natural" groups are recognized—that is, groups that include an ancestor and all of its descendants. Dinosauria, for example, is considered the group comprised of the most recent common ancestor of *Iguanodon* and *Megalosaurus* and all descendants of that ancestor. Because living birds are descendants of earlier feathered dinosaurs, Aves is a subgroup of Dinosauria. Recognition that birds are living dinosaurs greatly aids scientific understanding of the biology of the extinct members of this group, but one must be cautious in extrapolating how far down into the dinosaurian family tree any particular "avian" trait goes. In trying to reconstruct dinosaurian biology, one has to also take into account the attributes of crocodilians (the next-closest living relatives of the extinct dinosaur groups) and especially the fossils of the Mesozoic dinosaurs themselves.

Dinosaur biology

Many dinosaurs had anatomical features, including frills, crests, horns, and feathers, that may have served as visual display structures. In addition, because of the limits of the fossil record, paleontologists cannot assess to what degree soft tissue (for example, wattles), sounds, colors, or display behaviors were present in Mesozoic dinosaurs. By inference with living terrestrial vertebrates, however, paleontologists can be relatively secure in concluding that dinosaurs shared at least some of these behaviors with modern animals, even if they cannot tell what behaviors were present in which particular dinosaur species. Statistical analyses have suggested the presence of possible sexual dimorphism in the shapes of certain dinosaurs (ceratopsians and pachycephalosaurs, for instance), but at present it is difficult to find direct evidence of the sex of a given dinosaur fossil. More secure is the evidence of changes due to growth. Because large dinosaurs hatched from eggs only 1 to 4 liters in volume, they went through profound changes in size and often in shape. In fact, many dinosaur species have been named for specimens that later studies suggest were simply the juveniles of previously known species.

Several dinosaur species, including some thyreophorans, ornithopods, ceratopsians, sauropodomorphs, and theropods, have been found in death assemblages consisting of only one species. That these animals died and were buried together is a strong indication that they lived together at least part of the time. Indeed, given that these assemblages also include individuals of multiple growth stages, the weight of the evidence strongly suggests that these dinosaur species lived in herds or packs.

So far as can be ascertained from the fossil record, all dinosaurs hatched from brittle-shelled eggs. Dinosaur nests typically had between 12 and 36 eggs per clutch, fewer than their crocodilian relatives but much greater than most modern bird clutches. Other than the giant eggs of Cenozoic ratite birds (which reached up to 9 liters in the Malagasy elephant bird *Aepyornis*), the largest known dinosaur eggs were only 4 liters. In fact, the eggs of sauropod dinosaurs were rarely more than 1 liter in volume, even though the dinosaurs that hatched from these eggs grew to 50 tons or more.

The growth rate of dinosaurs has long been a matter of speculation. Some estimates, based on comparisons with living tortoises and crocodilians, suggested that it took many decades for large dinosaurs to reach full adult size, and that they had life spans of centuries. Work during the early twenty-first century on the microscopic structure of dinosaur fossils shows, however, that—like many groups of vertebrates—rings known as lines of arrested growth (LAGs) were deposited once a year in their bones. By finding a non-weight-bearing bone (i.e., one that has not undergone much modification because of the stresses of locomotion), one can count the number of LAGs and determine how old the dinosaur was at time of death. These studies have opened up a new understanding of dinosaur biology.

For those dinosaurs species known from multiple individuals, a growth curve can be constructed by comparing

An illustration of a Deinonychus dinosaur. Image by Roy Andersen.

the size of the individual to its age. Typical dinosaurs have been found to have spent a few years at small size, then to have entered a rapid growth phase lasting less than a decade, followed by an adult phase in which growth essentially stopped. Even giant sauropods such as 35-ton *Apatosaurus* reached full body size by age 15 years. Whereas large modern reptiles (such as crocodilians and tortoises) and large modern mammals (such as elephants and rhinoceroses) have life spans of many decades, Mesozoic dinosaurs seem to have died at a relatively young age. For example, no individual of *Tyrannosaurus* yet studied was more than 30 years old at the time of its death.

The rapid growth rate of dinosaurs is inconsistent with models of dinosaur physiology, which consider them to be scaled-up ectothermic ("cold-blooded") reptiles. Calculations of the maximum growth rates of dinosaurs put them within the same range as placental mammals and ground-dwelling birds of the same body size, and higher than ectothermic reptiles or even marsupial mammals. While some might argue that this remarkable speed of growth might be a product of the warmer climates of the Mesozoic, which would have been more favorable to an ectothermy, a useful comparison can be made with the Late Cretaceous giant 3-ton crocodilian *Deinosuchus*. Growth curve analysis of this animal shows that while its growth rate was higher than those of smaller modern crocodilians, it was much slower than dinosaurs of similar body size that lived in the same environment. Additionally, the growth profile of *Deinosuchus* lacks the rapid growth phase seen in dinosaurs.

There are many additional lines of evidence suggesting that the metabolism of Mesozoic dinosaurs was more like their living endothermic ("warm-blooded") bird descendants than their more distant ectothermic crocodilian, lizard, and turtle relatives. For example, living birds breathe using a highly specialized set of air sacs that pump fresh air through their lungs in only one direction. Thus, unlike mammals and most reptiles, the lungs of birds get fresh oxygen both during inhalation and exhalation. The air sacs of birds often invade the vertebrae and long bones, leaving characteristic hollow chambers in these bones. Hollow chambers in precisely the same anatomic positions and of the same structure are found in Mesozoic theropods and in many sauropodomorph dinosaurs, and they are a strong indication that saurischian dinosaurs in general had at least an early version of the flow-through bird lung. The evidence for air-sac-based respiration in ornithischians is less secure, because they lack these particular hollow chambers. Pterosaurs (close relations of the dinosaurs) had similar hollow chambers, however, and it may be that the common ancestor of pterosaurs and dinosaurs had a rudimentary set of air sacs, which became enlarged enough to expand into the bones in pterosaurs and in saurischian dinosaurs (including birds).

Although no fossilized heart of a Mesozoic dinosaur has been recovered (except for a controversial mass of rock in a Late Cretaceous ornithopod), their evolutionary position allows paleontologists to securely infer a four-chambered heart for all of Dinosauria. This is because both crocodilians and birds have four-chambered hearts, and consequently this specialization was almost certainly found in their common ancestor and all of that ancestor's descendants. In four-chambered hearts it is possible to keep the blood that has been freshly oxygenated by the lungs from mixing with deoxygenated blood returning from the body. Thus, higher levels of oxygen can be delivered throughout the body, allowing for greater aerobic activity. Furthermore, the great height of many dinosaurs would require a high arterial pressure in order to keep the brain oxygenated; this would require both an active heart (in order to ensure that the brain maintained oxygen) and separation of the blood going to the lungs (at lower pressure) from that going to the rest of the body (at high pressure).

Compared to the relatively simple teeth and jaw mechanics of herbivorous lizards and turtles, most plant-eating dinosaurs show some form of complex chewing. In the most specialized forms, such as the sauropod *Nigersaurus*, the duck-billed hadrosaurids, and the horned ceratopsids, the teeth form a continuous cropping, grinding, or slicing surface (respectively), allowing them to process plant material more finely and obtain nutrients from it faster. Such wear required rapid replacement of teeth, and microscopic study of these fossils has shown that a given tooth was in use for only 1 to 2 months.

Some paleontologists have observed that the ratio of carnivorous dinosaurs to potential prey is very low (only a few percent), similar to what is seen in modern and fossil mammal populations and much lower than those for fossil communities dominated by mammal precursors and by crocodile relatives. Endothermic animals require much more food per unit time than do ectotherms, the price they pay in order to support a higher metabolic rate. If these fossil predator–prey ratios reflect the actual numbers in the community and are not greatly skewed by preservational or collection biases, they suggest that theropod dinosaurs had at least a high metabolic rate.

All living dinosaurs (birds) are fully warm blooded, and their feathers are used at least in part for insulation. This may have been the original function of the primitive feather precursors found in early coelurosaurs; if so, other groups of dinosaurs were either less endothermic than coelurosaurs or were capable of maintaining a relatively high metabolism with only scaly integument.

Dinosaur contemporaries: Non-dinosaurian amniotes of the Mesozoic

During the Mesozoic era, dinosaurs were the most numerous and most anatomically diverse group of terrestrial amniotes (vertebrates that lay shelled eggs), but they were not the only such group. Other lineages, especially other groups of reptiles, played significant ecological roles during this time.

The closest relatives to the dinosaurs were the pterosaurs, or flying reptiles. Pterosauria and Dinosauromorpha share numerous anatomical features of the skull, vertebrae, and limbs. Like dinosaurs, pterosaurs first appeared as small animals in the Late Triassic epoch. Pterosaurs had greatly elongated arms and an enormously elongated digit IV (equivalent to the "ring finger"

of the human hand). A series of wing membranes (patagia) stretched from the neck to the arm, from the fingertip to the ankle, and from the foot on one side to the other. Microscopic studies of mineralized patagia reveal that they were not simply skin stretched between the limbs, but rather were stiffened internally with slender fibers and contained a sheet of muscle tissue. Thus, pterosaurs could control their wing shapes to great precision. A furlike integument covered pterosaur bodies; this and the evidence for a birdlike air sac system suggests that they were endothermic. Primitive pterosaurs were typically small (wingspans of 30 to 150 centimeters) and may have been relatively helpless on the ground, preferring to climb on the sides of trees and rocks. Fossil trackways and skeletal analysis shows that the later, more advanced group, Pterodactyloidea (the pterodactyls), was capable of more efficient quadrupedal locomotion. Whereas some pterosaurs had wingspans of 30 centimeters or less, others grew to become the largest flying animals of all time, with wingspans reaching 12 meters or more. Pterosaurs were found up until the end of the Cretaceous period.

The Crurotarsi (sometimes called the Pseudosuchia) include the living crocodilians and their extinct kin. Crurotarsans represent the closest relatives to the pterosaur-dinosauromorph group. During the Triassic period, crurotarsans were the dominant terrestrial animal, including diverse types of quadrupedal and bipedal predators (some up to 9 meters long), swift-moving toothless dinosaur-like bipeds, armored herbivorous groups, and long-snouted semiaquatic fish eaters. Although the last were ecologically equivalent to modern crocodilians, they were not their

A Tyrannosaurus devours an Anatosaurus. Image by Roy Andersen.

A dinosaur exhibit at Field Museum. National Geographic.

ancestors. Crocodylomorpha (crocodilians and their closest extinct relatives) were originally small 30- to 100-centimeter-long running quadrupeds with a parasagittal posture and digitigrade stance; in fact, they were very similar anatomically and ecologically to their dinosauromorph contemporaries. At the end of the Triassic period, a mass extinction associated with widespread volcanism and the rifting of the supercontinent Pangaea resulted in the extinction of all crurotarsans except for the crocodylomorphs. It was this event that allowed the dinosaurs to evolve into their dominant role. Despite common belief to the contrary, crocodilians are not unchanged since the early days of the dinosaurs. Mesozoic crocodylomorphs were far more diverse than their contemporary descendants; their ranks included armored blunt-snouted herbivores, long-snouted marine forms with paddles instead of hands and feet and a fishlike tail, deep-snouted terrestrial hunters with bladelike serrated teeth, and more. Crocodilians of modern aspect do not appear until the Early Cretaceous epoch.

More distantly related to the dinosaurs were the euryapsids, the primary group of marine reptiles during the Mesozoic era. Primitive euryapsids spent their time near shore, and they still had fingers and toes and were capable of moving on land. Recent specimens show, however, that even these early forms retained their young in the body until they could function on their own, so that they did lay eggs on land in the fashion of modern sea turtles. This internalization of the egg allowed later groups of euryapsids to evolved entirely pelagic (open-sea) lifestyles. The two most specialized of these groups were the ichthyosaurs and the plesiosaurs, both of

which had limbs that had evolved into paddles. Ichthyosaurs first appeared in the Middle Triassic. Most ichthyosaurs seem to have been fish and cephalopod eaters, with long snouts and very large eyes. Their main organ of propulsion was their tail. Early ichthyosaurs had a relatively eel-like body profile, but later forms (including all Jurassic and Cretaceous species) had a tuna-, mako shark-, or dolphin-like profile with a half-moon-shaped tail. As with those sleek, modern swimmers, ichthyosaurs were almost certainly swift open-sea predators. Triassic species showed a tremendous size range, from 1 to 21 meters (the latter were the largest animals in Earth's history up to that point, only to be overtaken by sauropod dinosaurs in the Jurassic period); Jurassic and Cretaceous ichthyosaurs were more typically 2 to 4 meters long. The last ichthyosaurs died out toward the end of the Early Cretaceous. Plesiosaurs first appeared in the latest Triassic, surviving up until the end of the Cretaceous. They propelled themselves through water primarily by moving their paddlelike limbs. Plesiosaurs ranged from 3 to 15 meters in length. Their body plan varied from small-headed, extraordinarily long-necked forms to huge-skulled, relatively short-necked forms. Their diet seems to have included fish, mollusks, and (especially in the huge-skulled forms) other marine reptiles.

Other marine reptiles of the Mesozoic belonged to more familiar groups still alive today. The oldest sea turtles appeared in the Early Cretaceous epoch. During the Late Cretaceous some reached lengths of about 4 meters, larger than any living turtle. Also during the Late Cretaceous a group of lizards became marine living. These relatives of the varanids (monitor lizards) were the mosasaurs. Like other marine reptiles, mosasaur limbs evolved into paddles. New discoveries showed that mosasaurs retained their young in their bodies until they were ready to swim on their own, allowing mosasaurs to specialize as fully pelagic animals. Mosasaur diets included mollusks, fish, and other marine reptiles.

Not all of dinosaurs' contemporaries among terrestrial amniotes were reptiles, however. The Synapsida is the great branch of amniotes that includes mammals and their extinct relatives. During the Permian period (299–251 mya) synapsids were the dominant group of terrestrial animal. Even after the Permo-Triassic extinction two major groups of synapsids— the herbivorous beaked dicynodonts and the more advanced cynodonts—thrived in the Triassic period. It is among the smaller cynodonts that the first mammals evolved during the Late Triassic period. Although many people think that mammals of the Mesozoic were no more than shrewlike insectivores, recent discoveries shows that there was a large diversity of mammalian contemporaries of the giant dinosaurs. Mesozoic mammals include burrowers, freshwater-swimming fish eaters, tree-dwelling gliders, and other diverse forms. The largest Mesozoic mammals, however, were the size of a small dog, and the majority were in the size range from shrews to rats. Thus, most of mammalian history was in the shadow of the dinosaurs. It was only after the great dinosaur extinction that mammals achieved larger size and their dominance in the terrestrial community.

Terminal Cretaceous extinction

Dinosaurs remained the main group of large-bodied terrestrial animals until the very end of the Cretaceous. While some have proposed that dinosaurs may have been gradually moving toward extinction during the later parts of the Late Cretaceous, more recent statistical analyses of dinosaur diversity do not show this pattern. There are some changes in the makeup of the dinosaur communities of the latest Cretaceous (for example, some groups of ceratopsids disappeared), but similar patterns occurred throughout the Mesozoic without resulting in the extinction of large dinosaurs as a whole.

Considerable evidence demonstrates that an asteroid 10 to 15 kilometers in diameter collided with Earth in the Yucatán Peninsula of Mexico 65.5 million years ago. This impact generated a blast estimated to be equivalent to more than 10^{18} megatons of TNT, generating a crater 180 kilometers in diameter (long since buried but detectable by seismic, magnetic, and gravimetric surveying). This crater, named after the Yucatán town of Chicxulub, attests to one of the greatest natural disasters in Earth's history.

Geologists debate about the relative contribution of different components of the impact on the global extinction event. A powerful atmospheric shockwave, tremendous tsunamis, possibly global forest fires and acid rain, and extreme weather systems have all been predicted based on mathematical modeling and geological evidence. Yet, the most catastrophic effect of the Chicxulub impact may well have been the darkness it would have generated. Ash and dust from the blast (and from the resultant, predicted forest fires) would have remained in higher levels of the atmosphere for weeks, perhaps even months. Cut off from sunlight, plant and algal photosynthesis would have shut down, decreasing the food available for herbivores. With the disappearance of prey, carnivores would have eventually starved as well. Survivorship would favor animals with low metabolic rates, with small body size (and thus small food requirements), or that obtained most of their food from detritus (directly or indirectly, such as bottom feeders in freshwater and marine communities). Animals with elevated metabolic rates, with large body size, and whose food chains did not include detritus feeders would suffer more severely.

This pattern largely holds true for the event at the end of the Cretaceous. Most groups of dinosaurs (all but modern groups of birds), as well as pterosaurs, plesiosaurs, mosasaurs, various groups of mammals, a great variety of mollusks (including the ammonoids—coiled cephalopods that were a major component of the marine community), and even a group of seed plants (the bennettitaleans) became extinct, and many other groups suffered severely. Only a few terrestrial groups survived, but these included the ancestors of all modern land life.

There were other environmental changes occurring during the latest Cretaceous. For example, sea level was undergoing a major drop because of changes in plate tectonic activity, draining the shallow seas that had once covered much of the continents. Additionally, one of the largest episodes of volcanism known, the Deccan Traps volcanism of western India, erupted beginning about 400,000 years prior to the asteroid impact. It is not certain to what degree these changes contributed to the extinction event, or if any of them might have been sufficient to cause an extinction event by themselves.

Nevertheless, it appears that the Chicxulub impact (with possible contribution from other factors) was the primary factor in the end of the Cretaceous period, the Mesozoic era, and the Age of Reptiles. While reptiles would persist, including dinosaurs in the form of Aves, it would be the mammals that would come to be the major players in the ecosystems of the new Cenozoic era.

Resources

Books

Benton, Michael J. 2005. *Vertebrate Palaeontology*. 3rd edition. Malden, MA: Blackwell Science.

Carpenter, Kenneth, ed. 2001. *The Armored Dinosaurs*. Bloomington: Indiana University Press.

Carpenter, Kenneth, ed. 2005. *The Carnivorous Dinosaurs*. Bloomington: Indiana University Press.

Carpenter, Kenneth, ed. 2007. *Horns and Beaks: Ceratopsian and Ornithopod Dinosaurs*. Bloomington: Indiana University Press.

Chiappe, Luis M. 2007. *Glorified Dinosaurs: The Origin and Early Evolution of Birds*. Hoboken, NJ: Wiley.

Currie, Philip J., Eva B. Koppelhus, Martin A. Shugar, and Joanna L. Wright, eds. 2004. *Feathered Dragons: Studies on the Transition from Dinosaurs to Birds*. Bloomington: Indiana University Press.

Currie, Philip J., and Kevin Padian, eds. 1997. *Encyclopedia of Dinosaurs*. San Diego, CA: Academic Press.

Curry Rogers, Kristina, and Jeffery A. Wilson, eds. 2005. *The Sauropods: Evolution and Paleobiology*. Berkeley: University of California Press.

Ellis, Richard. 2003. *Sea Dragons: Predators of the Prehistoric Oceans*. Lawrence: University Press of Kansas.

Farlow, James O., and M. K. Brett-Surman, eds. 1997. *The Complete Dinosaur*. Bloomington: Indiana University Press.

Fraser, Nicholas. 2006. *Dawn of the Dinosaurs: Life in the Triassic*. Bloomington: Indiana University Press.

Luo, Zhe-Xi. 2007. "Transformation and Diversification in Early Mammal Evolution." *Nature* 450(7172): 1011–1019.

Owen, Richard. 1842. "Report on British Fossil Reptiles, Part II." *Report of the Eleventh Meeting of the British Association for the Advancement of Science, Plymouth, England, July 1841*. London: John Murray.

Tidwell, Virginia, and Kenneth Carpenter, eds. 2005. *Thunder-Lizards: The Sauropodomorph Dinosaurs*. Bloomington: Indiana University Press.

Unwin, David M. 2006. *The Pterosaurs: From Deep Time*. New York: Pi Press.

Weishampel, David B., Peter Dodson, and Halszka Osmólska, eds. 2004. *The Dinosauria*. 2nd edition. Berkeley: University of California Press.

Periodicals

Brochu, Christopher A. 2001. "Progress and Future Directions in Archosaur Phylogenetics." *Journal of Paleontology* 75(6): 1185–1201.

Erickson, Gregory M. 2005. "Assessing Dinosaur Growth Patterns: A Microscopic Revolution." *Trends in Ecology and Evolution* 20(12): 677–684.

Thomas R. Holtz Jr.

Megafauna

The fossil record is of great importance in documenting both evolution and extinctions of ancient life-forms. Evolution and extinctions are relatable to global environmental change, an issue of increasing importance to contemporary society. If dominant life-forms of the past have become extinct, is it possible that humans or other extant species could suffer the same fate? Such a question doubtless contributes to the fascination that people have for past episodes of mass extinction, especially those involving large vertebrates. Scarcely 12,000 years ago the Late Pleistocene extinctions removed from the animal communities in much of the world a dominant element of landscape ecology: the megafauna ("large animals"). Both the presence and extinction of megafauna exerted profound influences upon the landscape and its ecosystems. In North America these large mammals included herbivores of several orders—Edentata, Subungulata, Artiodactyla, Perissodactyla, and even Rodentia—as well as the Carnivora (Kurtén and Anderson 1980; Martin and Klein 1984; Guthrie 1984). Although other extinctions had occurred throughout the Cenozoic era, a mass extinction of megafaunal groups such as mastodons, mammoths, giant ground sloths, and other ice-age giants took place over at most a few thousand years, the bulk of it over a few hundred. Given that they disappeared so recently in a geological sense, their extinction has left effects that are still felt today.

The diversity of the North American Pleistocene megafauna indicates that large forms evolved in many distinct lineages. Woolly mammoths (*Mammuthus primigenius*), while iconic as ice-age giants of northern steppe-tundra lands, were not the largest mammoths in North America. Larger species such as the Columbian mammoth (*M. columbi*) ranged as far south as Central America and clearly must not have sported heavy coats of hair. They were part of a western steppe to open mixed forest fauna (the *Camelops* faunal province) that included camels of the genus *Camelops* as well as giant bison (*Bison*), horses (*Equus*), giant short-faced bear (*Arctodus*), and American lion (*Panthera atrox*). Another giant predator was the saber-toothed cat *Smilodon*, so well known from the La Brea tar pits of California. In spruce forest environments (the *Symbos-Cervalces* faunal province) to the east were found mastodon (*Mammut americanum*), the stag-moose *Cervalces*, the "shrub-ox" *Bootherium* (=*symbos*), and the giant beaver (*Castoroides*) (Martin and Neuner 1978).

Mastodon and shrub-oxen were also found in coniferous forests of the west. Giant ground sloths such as *Megalonyx* ranged widely through all zones, likely in riparian woodlands (Kurtén and Anderson 1980). Diverse as these quaternary giants were, they were preceded by an even greater diversity of Tertiary megafauna that included such groups as the uintatheres, brontotheres, chalicotheres, and rhinoceroses of the Eocene, Oligocene, Miocene and Pliocene.

The Tertiary and Quaternary mammalian megaherbivores were not the first giant land animals, for giant herbivores and predators had existed among the dinosaurs as well, before their sudden extinction 65 million years ago. Giant reptiles formerly swam in the seas as well, much as giant whales do today. The term *megafauna* has, however, been applied in the literature most often to large Pleistocene mammals. The Late Pleistocene extinction appears to have been selective, with the larger forms becoming extinct and smaller relatives tending to survive. If that was truly the case, then the distinction between large and small has direct analytical value. This leads to two issues, probably linked: (1) How and why did animals tend toward gigantism in certain lineages, and (2) how and why was the megafauna differentially susceptible to extinction?

The how and why of gigantism

Similar trends toward gigantism between dinosaurs and Quaternary mammals cannot be ascribed to biological affinities, and the two examples are from vastly different geological times. Hence, any parallels between dinosaurian and mammalian gigantism could provide clues as to some of the mechanisms underlying evolution and natural selection. Is there a general tendency among organisms—perhaps a law—involving the selective advantage of larger size? Such a tendency would reflect not the evolutionary innovation of new characters but the process of heterochrony (literally, "different timing"), which involves changes in developmental timing and/or growth rates (Prothero 2004). For example, among the carnivorous theropod dinosaurs, the huge size of *Tyrannosaurus rex* resulted not from a long period of growth but from a greatly accelerated growth rate, even in comparison with close relatives (Erickson et al. 2004). The same was true of the giant herbivorous sauropods, such as "brontosaurus" (*apatosaurus*) (Sander et al. 2004). Celebrated

The term megafauna has been applied in the literature most often to large Pleistocene mammals, such as this Columbian Mammoth. Image by Martin Shields.

vertebrate paleontologist Edward Drinker Cope (1840–1897) believed that size increase (gigantism) was a valid generalization, and "Cope's rule," suggested in several of his publications but never explicitly defined, has been of continuing interest to paleontologists and evolutionary biologists. But is it a general rule?

Cope's rule states that body size in a given group tends to increase over evolutionary time. The logic is that larger size (which need not imply proportionally greater bulk) conferred several advantages, such as in defense against predators, attracting mates, fecundity, mobility, access to food, and greater tolerance of temperature extremes (Wilford 2001; Prothero 2004; Kingsolver and Pfennig 2004; Hone and Benton 2005). Among cold-blooded vertebrates, increased bulk is one way to retain body heat more efficiently, leading to inertial homeothermy (Gillooly, Allen, and Charnov 2006). Nevertheless, large animals have higher nutrient demands in growth and maintenance, so selection is also dependent upon nutrient availability in terms of both overall amount and seasonal variability (Guthrie 1984). Studies seemed to support the applicability of Cope's rule for varied groups from foraminifera to large vertebrates (Newell 1949; Alroy 1998). A critique by David Jablonski (1997), however, suggested that paleontologists and biologists have been biased

by an interest in "bigness" and that the supportive studies were selective. Studies of dinosaurs, for example, were long dominated by the quest for ever-larger fossils, feeding a public hunger for displays of huge extinct animals. Among the horses, multiple lineages did become progressively larger (Kurtén and Anderson 1980; Azzaroli 1992, 1995), but others did the opposite (Guthrie 1984, 2003). There was a considerable range in horse sizes through much of the fossil history, and the analytical bias may reflect the fact that of the modern survivors, horses (*Equus caballus*) are of greater cultural interest in the Western world than are their smaller relatives.

The Mollusca have several lineages with modern giants, including octopus, squid, and bivalve giants (e.g., the giant clam), so it would seem a reasonable place to test Cope's rule. Jablonski's survey of fossil Mollusca showed that closely related lineages can show opposite size trends and that over time the only tendency is an expansion in the overall range of variation—a basic tenet of Charles Darwin's evolutionary model (increased diversity). Of the Cretaceous mollusk lineages studied in detail by Jablonski, 27 to 30 percent showed a net increase in body size over 16 million years, 28 percent showed increases in both extremes of the range, and about the same showed a net decrease. Very few groups showed a decrease in

The Mollusca have several lineages with modern giants, including octopus (*Octopus vulgaris*). Image by David Doubilet.

the overall range of variation. Cope's rule does not apply uniformly here, and "gigantism," then, is one extreme of a wider phenomenon, that of increasing diversity through time. It is also possible that asymmetry toward larger individuals simply reflects structural limits at the *low* end on size change and relatively relaxed constraints on largeness.

Nevertheless, among Cambrian brachiopods multiple lineages increased in size, consistent with Cope's rule, though this appears not to have conferred advantages in longevity of groups. Size-increase trends were scale dependent, noted at the class and order level but with families showing a more random pattern (Novack-Gottshall and Lanier 2008). For Cenozoic deep-sea ostracods, increasing size was widespread among lineages and was correlated with climatic cooling, suggesting that in this case climate change was a driving factor (Hunt and Roy 2006).

For vertebrates, the rule seems broadly applicable, though it is not a "law" in the strict scientific sense (Hone and Benton 2005). Among North American canids, selection for larger size seems to have been nearly pervasive, yet in a broader perspective the fossil record also shows repeated ecological replacement of one type of large carnivore by another. Blaire Van Valkenburgh, Xiaoming Wang, and John Damuth (2004) argued that larger size leads to dietary specialization, which in turn increases vulnerability to extinction. Thus, while Cope's rule may reflect certain adaptive advantages, there is a risky

endgame as well, because extreme specialization is a liability in a context of rapid environmental change.

Defining "megafauna" becomes difficult because not all were "giants" in an absolute sense, though they were so in a relative sense in their own lineages. Thus, for Quaternary mammals, absolute definitions vary from more than 40 kilograms to more than 250 kilograms, depending on the author and the frame of reference, largely based on differential extinction (which introduces a circularity). As compared to other primates, humans certainly rate as megafauna, as do the great apes. If this were a tendency of most or all animal lineages, including invertebrates, a lineage-specific absolute measure or a comparative measure of relative size increase would be the only useful approaches.

Based on their closest modern relatives, most Quaternary megafaunal species appear to have been *K*-strategists (where *K* is carrying capacity of the environment); that is, they had great longevity, few offspring at a time, and hence long generation time and a relatively slow population growth rate. This was balanced by low death rates because size conferred advantages in terms of defense against predators and in decreasing the variety of predators that could take advantage of a given prey species, as the predators became specialized too. Giant herbivores conditioned (were a context for) the evolution of large carnivores, from saber-toothed tigers to wolves (Kurtén and Anderson 1980; Van Valkenburgh, Wang, and Damuth 2004), as well as scavengers such as the hyena-like North American giant short-faced bear, *Arctodus simus* (Matheus 2003). But specialization left all of them vulnerable to rapid environmental change. Large size was linked to mobility and dispersal ability; even the seemingly ungainly Jefferson's ground sloth (*Megalonyx jeffersonii*) was widely distributed from the Atlantic to the Pacific coasts of North America and from the tropics north to Beringia (the vast unglaciated tract of land that stretched from Siberia to Central Alaska and the Yukon when lowered sea level exposed the Bering land bridge), so it must have been a good disperser (McDonald, Harington, and De Iuliis 2000). Dispersal to newly available, nutrient-rich territory in the wake of glacial retreat appears to have been an important factor contributing to gigantism in

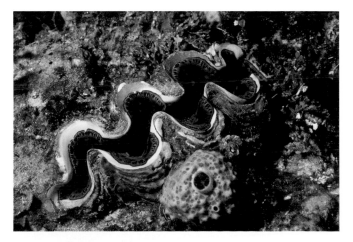

Another Mollusca giant is the giant clam (*Tridacna gigas*). © Hal Beral/ Corbis.

several Quaternary mammal lineages (Geist 1971, 1999; Wilson, Hills, and Shapiro 2008). R. Dale Guthrie (1984) pointed out that monogastric herbivores such as mammoths, camels, and horses were smaller in Beringia than in regions to the south, whereas ruminants were relatively larger, and he attributed this possibly to seasonal availability of nutrients, so dispersal was not the only factor involved.

The Quaternary survivors provide clues as to why there was differential success, because both modern bison (*Bison bison*) and wapiti or elk (*Cervus elaphus*) are highly mobile but also have the ability to rebound rapidly. As grazers and browsers they are ruminants, more efficient at digestion and with broader dietary tolerances than horses, which are monogastric, and which did become extinct in North America. Possible reasons for the success of bison and wapiti despite heavy human predation could also have included selection for shorter generation time (younger maturation), more rapid juvenile growth, and emphasis on dispersal ability (including seasonal migrations)—hence, a move away from *K*-selection to more *r*-type adaptations (where *r* is the reproductive rate) (Wilson, Hills, and Shapiro 2008). Migration confers an advantage in terms of survival through lowered predatory pressure (Fryxell, Greener, and Sinclair 1988), and both wapiti and bison are documented to have had seasonally migrating populations (Jefferson and Goldin 1989; Boyce 1991; Peck 2004).

Two examples show the ability of these species to increase rapidly in numbers when predatory pressures, including hunting by humans, are low. A small band of four or five wapiti moved from the Cascade Mountains to the Energy Research and Development Administration's Hanford Reserve in Washington State in 1973, and by 1986 had increased to eighty-nine with an initial increase close to 30 percent (McCorquodale, Eberhardt, and Eberhardt 1988). In the Mackenzie Bison Sanctuary, in Canada's Northwest Territories, wood bison (*B. bison athabascae*) grew from eighteen founders in 1963 to 1,718 individuals 10 months old or older by 1987 (Gates and Larter 1990).

Susceptibility to extinction

Studies of long-term evolutionary trends, including Cope's rule, are dogged by the fact that ongoing (ambient) selective pressures are rendered irrelevant by rare events such as asteroid impacts, which in a very short time create an entirely new selective regime. The Cretaceous-Tertiary (K-T) extinction 65.1 million years ago, which resulted from the asteroid impact that produced the Chicxulub crater in Mexico, eliminated all non-avian dinosaurs as well as marine reptiles and large invertebrates such as ammonites (Alvarez et al. 1980; Alvarez 1997; Fastovsky and Sheehan 2005). For the vertebrates, the liability imposed by size may have been in lower population

Hippopotamuses (*Hippopotamus amphibius*) are sociable mammals living in groups of up to forty individuals. Image by P. Marazzi.

Hippopotamuses (*Hippopotamus amphibius*) is a modern giant. Image by P. Marazzi.

numbers, long generation time, low recruitment rates, or other factors. Position in the food chain was crucial: If a dust veil inhibited photosynthesis, it would in turn be fatal for obligate herbivores but not for detritus feeders and those (such as insectivores) who fed upon the detritus feeders (Sheehan and Hansen 1986; Robertson et al. 2004). Burrowing and estivation ability may also have been crucial to survival in the face of widespread wildfires and extreme conditions.

The Quaternary megafaunal extinction removed two-thirds of mammalian genera and one-half of the species greater than 44 kilograms in weight (Barnosky 2008). Rapid stepwise climatic change at the close of the Pleistocene was associated with dramatic readjustments of vegetation communities and zones, meaning that there were non-analog communities (communities with no modern direct analog) (Guthrie 1990; FAUNMAP 1994, 1996). These readjustments may also have impacted long-generation megaherbivores, but the sudden appearance of a new predator—humans—with rapidly adaptable cultural strategies would also have been an enormous challenge. Human predation has been offered as one compelling reason for the Quaternary extinctions but remains a topic of considerable debate. Cultural evolution in technology and strategy can occur within the lifetime of a human individual, challenging the ability of prey to adapt: The issue is not the simple arrival of people with technology, but of the speed of their adaptability. The possibility

of substituting trapping and killing technologies would render prey species perennially naive. Long before the era of guns, thrown projectiles of various types were used in conjunction with a wide variety of driving and trapping strategies using natural landscape features as well as constructed traps (Frison 1991). Hence, Paul S. Martin (1967) argued that hunters could have swept rapidly through North America (and elsewhere) in "blitzkrieg" fashion, killing off behaviorally naive species. C. N. Johnson (2002) accepted human agency but argued against the "blitzkrieg" model, offering slow reproductive rate in prey species as an adequate reason in itself coupled with increased predatory pressures. Globally, especially in island environments, a strong correlation has been demonstrated between human arrival and megafaunal extinction (Martin and Steadman 1999; see also papers in Martin and Wright 1967 and Martin and Klein 1984; Wroe et al. 2004). Where humans were delayed in dispersal, as to Tasmania or Arctic Siberia, it can be shown that the megafauna persisted longer (Boeskorov 2005; Diamond 2008; Turney et al. 2008).

Critics of the blitzkrieg hypothesis have pointed to the survival of bison and wapiti as a counter, because both species were clearly hunted intensively by aboriginal peoples. The effects of differential migration ability and ability to rebound rapidly in populations render this argument facile. But there is no reason why extinction cannot have been caused by a

Hippopotamus (*Hippopotamus amphibius*) with mouth agape. Image by Tim Fitzharris.

combination of both climatic and human factors, perhaps even augmented by disease (MacPhee and Marx 1997). Body-size reduction for both bison and horses in Beringia appears relatable to both climatic factors and human predation (Shapiro et al. 2004; Guthrie 2003, 2006). Anthony D. Barnosky (2008) argues that an increase in human biomass intersected with significant climatic change to cause the Quaternary megafaunal extinction and that as human biomass continues to increase, more extinctions will result. But now there is a newly proposed factor, again that of an impact event or an atmospheric explosion, this time of an incoming extraterrestrial object over northern Canada. Evidence for this is debated but includes a "black mat" of organic material, iridium enrichment, magnetic microspherules, carbon spherules, sooty layers, microdiamonds, and fullerenes (Firestone et al. 2007; Haynes 2008).

Megaherbivores had increased needs over those of their smaller relatives in terms of food and water; hence, they significantly influenced the flora and had far-reaching effects throughout their ecosystems, conditioning the landscape for a great diversity of plants and animals. Bison, deer, and sheep are well documented to have influenced establishment of vegetation through importation of seeds on their pelages (hairy coverings) or in passage through their digestive systems (Constible et al. 2005; Myers et al. 2004). By continually browsing trees in established vegetation, megaherbivores could also turn a forest landscape into a mosaic of grasslands, shrubs, and clumps of forest. Elephants in Africa are known to push over trees when they have difficulty in reaching browse and possibly even for social display purposes (Calenge et al. 2002; Midgley, Balfour, and Kerley 2005). Bulky mammoths would probably have challenged modern government regulations for loading on tundra, meaning that their trampling would have helped to change its character. In their strong ability to store nutrients in their roots, grasses also have an advantage over many herbaceous plants in withstanding grazing. It has been argued that there was a former northern grassland community, the "mammoth steppe" or "steppe-tundra," that was structured by grazers and has no precise modern analog (Guthrie 1990; Zimov et al. 1995). Historic bison on the Great Plains inhibited the spread of aspen groves by grazing on suckers (Campbell et al. 1994), yet in tallgrass areas they helped to maintain woody plant abundance because their grazing also tended to reduce fire intensity (Briggs, Knapp, and Brock 2002). In the Arctic, musk ox and caribou may be called on to halt the spread of woody plants in response to global climate change (Post and Pedersen 2008).

In their presence they were influential, but the demise of the megafauna had its own consequences for ecosystems. It stands to reason that the removal of such influential animals would have important consequences in terms of changed fire frequencies, different patterns of floral succession, and changed habitat for other animals. Relict grassland patches ("balds") at high altitudes in the southern Appalachians may have been initiated by megafaunal herbivores in the Late Pleistocene and maintained by bison and wapiti, only to undergo successional change in recent years in their absence (Weigl and Knowles 2006).

These effects could have had a parallel example millions of years earlier. With dinosaur extinction one episode of megafauna ended and large mammals were not to appear for almost 10 million years. Such a factor facilitated the development of the extensive Paleocene closed-canopy angiosperm forests and hence set the table for early primate evolution (Prothero 2006). The extinction of one group of megafauna—the dinosaurs—provided the opportunity for others to come, all responding, in parallel, to shared underlying evolutionary rules.

Resources

Books

Alvarez, Walter. 1997. T. rex *and the Crater of Doom*. Princeton, NJ: Princeton University Press.

Frison, George C. 1991. *Prehistoric Hunters of the High Plains*. 2nd edition. San Diego, CA: Academic Press.

Geist, Valerius. 1999. "Periglacial Ecology, Large Mammals, and Their Significance to Human Biology." In *Ice Age Peoples of North America: Environments, Origins, and Adaptations of the First Americans*, ed. Robson Bonnichsen and Karen L. Turnmire. Corvallis: Oregon State University Press for the Center for the Study of the First Americans.

Guthrie, R. Dale. 1984. "Alaska Megabucks, Megabulls, and Megarams: The Issue of Pleistocene Gigantism." In *Contributions in Quaternary Vertebrate Paleontology: A Volume in Memorial to John E. Guilday*, ed. Hugh H. Genoways and Mary R. Dawson. Pittsburgh, PA: Carnegie Museum of Natural History.

Kurtén, Björn, and Elaine Anderson. 1980. *Pleistocene Mammals of North America*. New York: Columbia University Press.

MacPhee, Ross D., and Preston A. Marx. 1997. "The 40,000-Year Plague: Humans, Hyperdisease, and First-Contact Extinctions." In *Natural Change and Human Impact in Madagascar*, ed. Steven M. Goodman and Bruce D. Patterson. Washington, DC: Smithsonian Institution Press.

Martin, Paul S. 1967. "Prehistoric Overkill." In *Pleistocene Extinctions: The Search for a Cause*, ed. Paul S. Martin and Herbert Edgar Wright Jr. New Haven, CT: Yale University Press.

Martin, Paul S., and Richard G. Klein, eds. 1984. *Quaternary Extinctions: A Prehistoric Revolution*. Tucson: University of Arizona Press.

Martin, Paul S., and David W. Steadman. 1999. "Prehistoric Extinctions on Islands and Continents." In *Extinctions in Near Time: Causes, Contexts, and Consequences*, ed. Ross D. E. MacPhee. New York: Kluwer Academic/Plenum.

Martin, Paul S., and Herbert Edgar Wright Jr., eds. 1967. *Pleistocene Extinctions: The Search for a Cause*. New Haven, CT: Yale University Press.

Matheus, Paul E. 2003. *Locomotor Adaptations and Ecomorphology of Short-Faced Bears (Arctodus simus) in Eastern Beringia*. Whitehorse: Government of the Yukon, Paleontology Program.

Peck, Trevor Richard. 2004. *Bison Ethology and Native Settlement Patterns during the Old Women's Phase on the Northwestern Plains*. Oxford: Archaeopress.

Prothero, Donald R. 2004. *Bringing Fossils to Life: An Introduction to Paleobiology*. 2nd edition. Boston: McGraw-Hill.

Prothero, Donald R. 2006. *After the Dinosaurs: The Age of Mammals*. Bloomington: Indiana University Press.

Wilford, John Noble. 2001. "Horses, Mollusks, and the Evolution of Bigness." In *The New York Times Book of Fossils and Evolution*, ed. Nicholas Wade. Rev. edition. New York: Lyons Press.

Periodicals

Alroy, John. 1998. "Cope's Rule and the Dynamics of Body Mass Evolution in North American Fossil Mammals." *Science* 280 (5364): 731–734.

Alvarez, Luis W., Walter Alvarez, Frank Asaro, and Helen V. Michel. 1980. "Extraterrestrial Cause for the Cretaceous-Tertiary Extinction." *Science* 208(4448): 1095–1108.

Azzaroli, A. 1992. "Ascent and Decline of Monodactyl Equids: A Case for Prehistoric Overkill." *Annales Zoologici Fennici* 28: 151–163.

Azzaroli, A. 1995. "A Synopsis of the Quaternary Species of *Equus* in North America." *Bolletino della Societa Paleontologica Italiana* 34(2): 205–221.

Barnosky, Anthony D. 2008. "Megafauna Biomass Tradeoff as a Driver of Quaternary and Future Extinctions." *Proceedings of the National Academy of Sciences of the United States of America* 105(supp. 1): 11543–11548.

Boeskorov, Gennady G. 2005. "Arctic Siberia: Refuge of the Mammoth Fauna in the Holocene." *Quaternary International* 142–143: 119–123.

Boyce, Mark S. 1991. "Migratory Behavior and Management of Elk (*Cervus elaphus*)." *Applied Animal Behaviour Science* 29(1–4): 239–250.

Briggs, John M., Alan K. Knapp, and Brent L. Brock. 2002. "Expansion of Woody Plants in Tallgrass Prairie: A Fifteen-Year Study of Fire and Fire-Grazing Interactions." *American Midland Naturalist* 147(2): 287–294.

Calenge, Clement, Daniel Maillard, Jean-Michel Giallard, et al. 2002. "Elephant Damage to Trees of Wooded Savanna in Zakouma National Park, Chad." *Journal of Tropical Ecology* 18 (4): 599–614.

Campbell, Celina, Ian D. Campbell, Charles B. Blyth, and John H. McAndrews. 1994. "Bison Extirpation May Have Caused Aspen Expansion in Western Canada." *Ecography* 17(4): 360–362.

Constible, J. M., R. A. Sweitzer, D. H. Van Vuren, et al. 2005. "Dispersal of Non-native Plants by Introduced Bison in an Island Ecosystem." *Biological Invasions* 7(4): 699–709.

Diamond, Jared. 2008. "Paleontology: The Last Giant Kangaroo." *Nature* 454(7206): 835–836.

Erickson, Gregory M., Peter J. Makovicky, Philip J. Currie, et al. 2004. "Gigantism and Comparative Life-History Parameters of Tyrannosaurid Dinosaurs." *Nature* 430(7001): 772–775.

Fastovsky, David E., and Peter M. Sheehan. 2005. "The Extinction of the Dinosaurs in North America." *GSA Today* 15 (3): 4–10.

FAUNMAP Working Group. 1994. *FAUNMAP: A Database Documenting Late Quaternary Distributions of Mammal Species in the United States*. Springfield: Illinois State Museum.

FAUNMAP Working Group. 1996. "Spatial Response of Mammals to Late Quaternary Environmental Fluctuations." *Science* 272(5268): 1601–1606.

Firestone, R. B., A. West, J. P. Kennett, et al. 2007. "Evidence for an Extraterrestrial Impact 12,900 Years Ago that Contributed to the Megafaunal Extinctions and the Younger Dryas Cooling." *Proceedings of the National Academy of Sciences of the United States of America* 104(41): 16016–16021.

Fryxell, John M., John Greener, and A. R. E. Sinclair. 1988. "Why Are Migratory Ungulates So Abundant?" *American Naturalist* 131(6): 781–798.

Gates, C. C., and N. C. Larter. 1990. "Growth and Dispersal of an Erupting Large Herbivore Population in Northern Canada: The Mackenzie Wood Bison (*Bison bison athabascae*)." *Arctic* 43(3): 231–238.

Geist, Valerius. 1971. "The Relation of Social Evolution and Dispersal in Ungulates during the Pleistocene, with Emphasis on the Old World Deer and the Genus *Bison*." *Quaternary Research* 1(3): 285–315.

Grayson, Donald K., and David J. Meltzer. 2003. "A Requiem for North American Overkill." *Journal of Archaeological Science* 30: 585–593.

Guthrie, R. Dale. 1990. *Frozen Fauna of the Mammoth Steppe: The Story of Blue Babe*. Chicago: University of Chicago Press.

Guthrie, R. Dale. 2003. "Rapid Body Size Decline in Alaskan Pleistocene Horses before Extinction." *Nature* 426(6963): 169–171.

Guthrie, R. Dale. 2006. "New Carbon Dates Link Climatic Change with Human Colonization and Pleistocene Extinctions." *Nature* 441(7090): 207–209.

Haynes, C. Vance, Jr. 2008. "Younger Dryas 'Black Mats' and the Rancholabrean Termination in North America." *Proceedings of the National Academy of Sciences of the United States of America* 105(18): 6520–6525.

Hone, David W. E., and Michael J. Benton. 2005. "The Evolution of Large Size: How Does Cope's Rule Work?" *Trends in Ecology and Evolution* 20(1): 4–6.

Hunt, Gene, and Kaustuv Roy. 2006. "Climate Change, Body Size Evolution, and Cope's Rule in Deep-Sea Ostracodes." *Proceedings of the National Academy of Sciences of the United States of America* 103(5): 1347–1352.

Jablonski, David. 1997. "Body-Size Evolution in Cretaceous Molluscs and the Status of Cope's Rule." *Nature* 385(6613): 250–252.

Jefferson, George T., and Judith L. Goldin. 1989. "Seasonal Migration of *Bison antiquus* from Rancho La Brea, California." *Quaternary Research* 31(1): 107–112.

Johnson, C. N. 2002. "Determinants of Loss of Mammal Species during the Late Quaternary 'Megafauna' Extinctions: Life History and Ecology, but Not Body Size." *Proceedings of the Royal Society* B 269(1506): 2221–2227.

Kingsolver, Joel G., and David W. Pfennig. 2004. "Individual-Level Selection as a Cause of Cope's Rule of Phyletic Size Increase." *Evolution* 58(7): 1608–1612.

Martin, Larry D., and A. Michael Neuner. 1978. "The End of the Pleistocene in North America." *Transactions of the Nebraska Academy of Sciences* 6: 117–126.

McCorquodale, Scott M., Lester L. Eberhardt, and Lester. E. Eberhardt. 1988. "Dynamics of a Colonizing Elk Population." *Journal of Wildlife Management* 52(2): 309–313.

McDonald, H. G., C. R. Harington, and G. De Iuliis. 2000. "The Ground Sloth *Megalonyx* from Pleistocene Deposits of the Old Crow Basin, Yukon, Canada." *Arctic* 53(3): 213–220.

Midgley, J. J., D. Balfour, and G. I. Kerley. 2005. "Why Do Elephants Damage Savanna Trees?" *South African Journal of Science* 101(5–6): 213–215.

Myers, Jonathan A., Mark Vellend, Sana Gardescu, and P. L. Marks. 2004. "Seed Dispersal by White-Tailed Deer: Implications for Long-Distance Dispersal, Invasion, and Migration of Plants in Eastern North America." *Oecologia* 139 (1): 35–44.

Newell, Norman D. 1949. "Phyletic Size Increase, an Important Trend Illustrated by Fossil Invertebrates." *Evolution* 3(2): 103–124.

Novack-Gottshall, Philip M., and Michael A. Lanier. 2008. "Scale-Dependence of Cope's Rule in Body Size Evolution of Paleozoic Brachiopods." *Proceedings of the National Academy of Sciences of the United States of America* 105(14): 5430–5434.

Post, Eric, and Christian Pedersen. 2008. "Opposing Plant Community Responses to Warming with and without Herbivores." *Proceedings of the National Academy of Sciences of the United States of America* 105(34): 12353–12358.

Robertson, Douglas S., Malcolm C. McKenna, Owen B. Toon, et al. 2004. "Survival in the First Hours of the Cenozoic." *Geological Society of America Bulletin* 116(5–6): 760–768.

Sander, P. Martin, Nicole Klein, Eric Buffetaut, et al. 2004. "Adaptive Radiation in Sauropod Dinosaurs: Bone Histology Indicates Rapid Evolution of Giant Body Size through Acceleration." *Organisms Diversity and Evolution* 4(3): 165–173.

Shapiro, Beth, Alexei J. Drummond, Andrew Rambaut, et al. 2004. "Rise and Fall of the Beringian Steppe Bison." *Science* 306(5701): 1561–1565.

Sheehan, Peter M., and Thor A. Hansen. 1986. "Detritus Feeding as a Buffer to Extinction at the End of the Cretaceous." *Geology* 14(10): 868–870.

Turney, Chris S. M., Timothy F. Flannery, Richard G. Roberts, et al. 2008. "Late-Surviving Megafauna in Tasmania, Australia, Implicate Human Involvement in Their Extinction." *Proceedings of the National Academy of Sciences of the United States of America* 105(34): 12150–12153.

Van Valkenburgh, Blaire, Xiaoming Wang, and John Damuth. 2004. "Cope's Rule, Hypercarnivory, and Extinction in North American Canids." *Science* 306(5693): 101–104.

Weigl, Peter D., and Travis W. Knowles. 2006. "Megaherbivores and Southern Appalachian Grass Balds." *Growth and Change* 26(3): 365–382.

Wilson, Michael C., Leonard V. Hills, and Beth Shapiro. 2008. "Late Pleistocene Northward-Dispersing *Bison antiquus* from the Bighill Creek Formation, Gallelli Gravel Pit, Alberta, Canada, and the Fate of *Bison occidentalis*." *Canadian Journal of Earth Sciences* 45(7): 827–859.

Wroe, Stephen, Judith Field, Richard Fullagar, and Lars S. Jermin. 2004. "Megafaunal Extinction in the Late Quaternary and the Global Overkill Hypothesis." *Alcheringa* 28: 291–331.

Zimov, Sergei A., Vladimir I. Chuprynin, A. P. Oreshko, et al. 1995. "Steppe-tundra Transition: A Herbivore-driven Biome Shift at the End of the Pleistocene." *American Naturalist* 146 (5): 765–794.

Other

Gillooly, James F., Andrew P. Allen, and Eric L. Charnov. 2006. "Dinosaur Fossils Predict Body Temperatures." *PLoS Biology* 4(8): e248. Available from http://biology.plosjournals.org.

Michael C. Wilson

• • • • •

Non-human primate and human evolution

Modern human beings, *Homo sapiens sapiens*, are a unique African great ape. In many ways, humans are morphologically, genetically, and phylogenetically virtually identical to organisms such as chimpanzees. At the same time, humans are also highly distinctive as bipedal, encephalized, technological, and highly cultural primates that think symbolically, employ complex language, and intensely question their origins and place in nature. These inquiries have uncovered a remarkable human story spanning 7 million years, but it is also clear that human origins are part of a much larger and unprecedented process of primate evolution over the last 50 million years.

Establishing evidence: Approaches in paleoanthropology

The study of human evolution is part of paleoanthropology, one of the principal subfields of physical anthropology. Paleoanthropology is a strongly multidisciplinary field structured by an intensely holistic orientation. Reconstructions of primate and human evolution are based on the integration of multiple and independent lines of evidence. Central perspectives come from comparative anatomy. The study of morphology involves the size, shape, and other observable characteristics of an organism. As has been well established over several centuries of study, the more similar morphology two kinds of organisms share, the more closely related they are to each other. Morphology can also reveal a great deal about behavior (e.g., large eyes correlating to nocturnal behavior; curved finger bones [phalanges] indicative of arboreal locomotion). These principles permit identification—and methodical tracking—of derived traits (ones modified by natural selection from primitive, ancestral states) in bones or teeth that trace changes in evolutionary lineages across time and space (for principles and examples, see Cartmill and Smith 2009; Strait and Grine 2004; Wood 2005). More recently, modern genomics has begun to provide compelling information from ancient and modern DNA as well.

Most evidence of human evolution comes from fossils. Fossilization is exceedingly rare, occurring only when an organism dies and is rapidly buried in a wet environment. Over millennia, iron and silica in the surrounding sediments replace the calcium and phosphates that compose bone and teeth. Bone literally becomes stone. Considering the trillions of animals that have lived, only a tiny fraction became fossils. While immensely informative, one key limitation of the fossil

record is that an individual specimen may not be representative of a species. Secure understandings must be based on many dozens or hundreds of samples. Soft tissue rarely leaves any traces, but indirect evidence of these structures, such as the size and shape of muscle attachments, brain anatomy, or other traces literally "imprinted" on bones in detailed ways, usually abounds (Aiello and Dean 1990; Langdon 2005).

Fossils of primates and hominids occur only in specific geological strata, which greatly aids chances of locating them. For instance, paleoanthropologists in search of fossil primates and human ancestors commonly search late Miocene, Pliocene, and Pleistocene geological beds exposed by erosion and tectonic forces (the East African Rift Valley is a good example) or in caves to which hominid remains were probably transported by ancient predatory cats (e.g., South African sites). When hominids occur in time is another critical baseline variable. Collaborations with geologists define the temporal position of fossil-bearing beds (Brown and Van Couvering 2000; Van Couvering 2000). Paleontologists compare animal remains found alongside primates and hominids to those from known time periods and locations using techniques of biostratigraphy. Physicists contribute radiometric dating techniques, which allow estimations of the number of years that have passed since a specimen died. Carbon dating may be useful for archaeologists, but it has limited application in paleoanthropology: This technique can accurately date organic materials only out to about 45,000 years, and the amount of preserved organic material in older fossils is typically miniscule. Methods that date inorganic materials millions of years old must be used. These include potassium-argon, argon-argon, fission-track, paleomagnetic, electron spin resonance, and thermoluminescence dating (Brown 2000a, 2000b; Schwartz 2006). Destructive removal of samples for dating from one-of-a-kind specimens is usually not an option. Instead, dating materials in intimate association with a key fossil is more feasible, such as potassium-argon or thermoluminescence analysis of sediments surrounding a fossil.

As the English naturalist Charles Darwin (1809–1882) first recognized, environments are the engine of evolution, and understanding the changing nature of ancient ecologies directly addresses the "why" questions regarding the trends of natural selection. Paleoenvironmental and climatological reconstructions are simply vital (Vrba et al. 1995). Study of associated fossils and microfossils by paleontologists and paleobotanists produces volumes of information regarding the nature of flora and fauna living during exactly the same time period to extrapolate detailed environmental parameters. For example, isotopic analysis of long-dead marine protozoans (foraminifera) from seafloor cores are correlated to climate, while ratios of carbon-12/carbon-13 in animal bones and ancient soils (paleosols) can differentiate between the chemical characteristics of an open grassland versus a forest environment.

Primate origins and evolution

Following the mass extinction that closed the Cretaceous period 65 million years ago (mya), mammals began to emerge as the dominant vertebrates and radiated across the land and into the seas. Primates were one of the new mammalians that emerged around 55 mya. Primates share a common set of biological characteristics: They have a ring of bone, called a postorbital bar, which laterally encases forward-facing eyes that provide color stereoscopic vision. They share a generalized, multifunction dentition, consisting of incisors, canines, premolars, and molars, that permits a high degree of dietary flexibility. Primates are very agile. Their brains tend to be larger than proportionally predicted for body size among mammals.

The selective forces that surround primate origins remain a source of debate (Cartmill and Smith 2009; Rasmussen 2002). Three principal hypotheses are entertained. The arboreal hypothesis, proposed in the early 1900s, suggests that origins were related to life in trees; most primates are well adapted for arboreal lifestyles with grasping hands and feet enriched with sensory nerve endings, and brains and eyes that confer precise three-dimensional navigation in forest canopies. The visual predation hypothesis, introduced about seventy years later, holds that the aforementioned traits provided selective advantages in catching small prey such as insects. Many primates today, however, are frugivorous; the angiosperm radiation hypotheses suggests that primate origins and anatomy reflect adaptations to the acquisition and consumption of fruit made available by the radiation of flowering plants (angiosperms). These hypotheses are not mutually exclusive, and each may hold elements explaining why primates evolved anatomies and behaviors that conferred adaptive advantages across a wide range of circumstances.

The first primates emerged in the Cenozoic, but what were the first ancestors and when they arose remain in question. A highly diverse group of probable pre-primates (protoprimates) designated plesiadapiforms spanned western North America, Europe, Asia, and Africa via an adaptive radiation of primitive mammals in the Paleocene. Almost all of these small squirrel-like animals lacked the basic set of cranial, dental, and postcranial characteristics that define all primates. However, one plesiadapid—*Carpolestes simpsoni*, found in the Big Horn Basin of Wyoming—appears to bridge plesiadapids and indisputable fossil prosimians. *Carpolestes* anatomy is compelling: It possessed an opposable big toe, grasping fingers, and nails instead of claws (Bloch and Boyer 2002). *Carpolestes'* possible descendants were the first true primates (euprimates) identified in North America, Europe, Africa, and Asia.

The adaptive radiations of the euprimates—the adapids and omomyids (Covert 2002)—are associated with a rapid climate change about 55 mya, when global temperature rose and triggered growth of widespread tropical forests. Adapids and omomyids are the most common fossil prosimians, with over 200 known species. All possess unambiguous primate morphologies possessed by all their descendants from lemurs to humans: a postorbital bar, forward-facing eyes, grasping hands and feet, generalized dentitions, and larger brains. Among adapids, derived features including long snouts and smaller eyes (indicative of diurnal activity patterns) identify them as ancestors of modern lemurs. Conversely, omomyids' shortened snouts, large eyes, and specialized grasping digits link them closely to the nocturnal and highly arboreal modern tarsiers.

The origins of monkeys and apes are yet to be fully resolved (Covert 2004). Fossil evidence has yet to convincingly link archaic euprimates with later anthropoids. Some hold that adapids were the ancestral group of lemuriforms and monkeys and apes as well. Another school of thought hypothesizes that omomyids are better candidates as anthropoid ancestors. A third possibility involves an Eocene primate found in Shanghuang, China, called *Eosimias* (Gebo et al. 2000). The smallest variety of this extraordinary 42-million-year-old primate was tiny, only about the size of a human thumb, and weighed less than 1 ounce. *Eosimias* morphology reflects a nocturnal arboreal adaptation and an insectivorous diet. The critical link to later monkeys and apes is in the shape of the shortened calcaneus (heel bone) and a monkey-like facial morphology. At 37 mya, the date for *Eosimias* signals that anthropoids likely originally arose in Asia, not Africa. The 2009 discovery of *Darwinius masillae* (Franzen et al. 2009) calls all this into question, however, as it appears to be a transitional adapid ancestor to monkeys and apes living in Europe 47 mya.

By the dawn of the Oligocene, rapid global cooling again triggered the emergence of new habitats and the speciation of new anthropoids, including parapithecids, oligopithecids, and propliopithecids from about 37 to 29 mya. These are particularly well-known groups studied in the fossil-rich site of El Faiyûm, Egypt. One propliopithecid, *Aegyptopithecus*, had a larger brain and body and a more derived dental formula, with only two premolars instead of the ancestral three. This suggests *Aegyptopithecus* was closely related to the first catarrhine, the group of animals that include all Old World monkeys and apes (Begun 2002b).

The origins of New World monkeys remain a vexing question (Fleagle and Tejedor 2002). All early monkey evolution took place in Africa—demonstrated by the close physical and genetic links between Old and New World monkeys. Available evidence suggests the split occurred about 34 mya (Ross and Kay 2004). Parapithecids from El Faiyûm possess three premolars, just like all monkeys of Central and South America today. Similar fossil monkeys are found in South America, such as *Branisella*, which is present in 26-million-year-old deposits in Bolivia. Independent evolution is very improbable, but how they made the Africa–South America voyage needs clarification. One possibility, drawn from studies of modern phenomena, is that parapithecids "rafted" the then-shorter distance to the Americas as they were washed out to sea on large mats of vegetation. Alternatively, they may have

At far left is the shrew-like mammal, Purgatorius (65 million years ago, mya), that is thought to have evolved into a lemur-like animal (second from left, 45 mya). Following this is a succession of four pre-human primates: Aegyptopithecus (35 mya), Proconsul (20 mya), Australopithecus afarensis (4 mya), and Homo erectus (1 mya). Finally, modern humans, Homo sapiens (far right), appeared around 400,000 years ago. Richard Bizley/Photo Researchers, Inc.

migrated via the shallow, island-dotted, and historically much warmer South Atlantic and Antarctic regions.

The Miocene epoch, from about 23 to 5 mya, saw planetary climate again oscillate to warmer conditions. Proconsulids emerged in eastern Africa 22 to 17 mya and were probably descendants of propliopithecids. These were the first apes. *Proconsul* species were larger arboreal quadrupeds with an apelike head and dentition, a divergent big toe, and no tail, but otherwise resembled monkeys. The Miocene is known as the golden age of apes, and proconsulid descendants after 17 mya initiated a remarkable radiation across the Old World. Dryopithecids spread into and thrived in tropical forests in France, Greece, and Hungary. Sivapithecids spread throughout Asia and Indonesia; one of these, *Khoratpithecus*, was probably the direct ancestor to modern orangutans. Another, *Gigantopithecus*, which went extinct only about 500,000 years ago, was the largest primate at about 3 meters (10 feet) tall and 320 kilograms (700 pounds).

Meanwhile, monkeys continued to evolve, and in the latter Miocene, far more monkeys than apes occurred in Africa. Baboon-like *Theropithecus* may have been the size of a modern-day female gorilla. The African *Victoriapithecus* was probably the common ancestor to all modern macaques, baboons, guenons, and colobus monkeys (Benefit and McCrossin 2002). However, the higher primate radiations of the Miocene ended about 7 mya with dramatic alterations

of ecology and climate—the Alps and Himalayas were uplifted, polar caps enlarged, changing ocean currents cooled northerly latitudes, and ape food sources (fruit-bearing trees) were replaced by cooler forests and grasslands. These changes are associated with the extinction of European primates and most Asian apes as well.

Because none of the few late Miocene African apes appear ancestral to gorillas, chimpanzees, and humans, European species such as *Ouranopithecus* may have migrated back to Africa from Europe (Begun 2002a). Fossil and molecular data suggest that gorillas branched off in Africa around 9 to 7 mya, and then around 7 to 6 mya, chimpanzees and human ancestors shared their last common ancestor, which is hypothesized as being very chimp-like. Unfortunately, no chimpanzee fossil record exists, before or after the split, except for 700,000-year-old teeth in Kenya (Wood 2005). This owes partially to an ancient forest environment that was not particularly conducive to fossilization and to the bias of anthropologists who have invested their work immensely into the fossil record for humans but not for chimpanzees.

Natural history and evolutionary anatomy of human ancestors

By the end of the Miocene, a speciation event took place 7 to 6 mya involving the emergence of a bipedal African ape. The

Although the skulls of the human and the chimpanzee look very similar there are some important differences. © DK Limited/Corbis.

advent of bipedalism would have been driven by chance mutations in regulatory and structural genes that govern the manner and timing of how bones and muscles grow. The most fundamental diagnostic anatomical signals of the human lineage emerged among these apes. The foramen magnum, where the brainstem exits the head to become the spinal cord, must be located at the base of the skull to balance the head above the shoulders. A biped's spine requires an S-shaped weight-distributing curvature. The pelvis must flare out at the top to hold the vertically positioned gut. The knee end of the femur is required to form convergent, or valgus, knees (in other words, knock-knees) to maintain effective balance. Relying only on a pair of feet for mobility, bipedalism required a rolling footstep made possible by an arch and an immobile big toe that was brought in line with the rest of the pedal digits.

Bipedalism is probably the oldest anatomical distinction among hominids, but for the first 4 to 5 million years, these bipeds were unique, being neither apelike nor modern human in their locomotion. And, while bipedalism may have been the fundamental defining modification on the basic ape body plan, five other key changes and adaptive transitions transpired to human ancestors that critically contributed to modern human beings: nonhoning chewing (5.5 mya), establishment of material culture (2.5 mya), speech (2 to 1.8 mya), hunting (1 million to 400,000 years ago), and the domestication of food sources (11,000 years ago).

The first bipedal apes

The dawn of the human lineage in the fossil record is marked by *Orrorin tugenensis* and *Sahelanthropus tchadensis*, both discovered in the early twenty-first century. *S. tchadensis* is the earliest, dating back to 7 to 6 mya, and is currently represented by a single cranium found in north-central Africa (Brunet et al. 2002). While somewhat flattened and distorted, this cranium appears to have a foramen magnum at the base of the cranium and dentition that featured nonhoning canines. Honing is seen among other catarrhines but not humans, where canines and premolars interlock to sharpen the canine teeth. *O. tugenensis*, dating to about 6 mya in Kenya, is

represented by some twenty fragments from a few individuals (Senut et al. 2001). Femoral elements demonstrate a long femoral neck associated with a telltale groove for the obturator externis muscle—characteristics seen only in bipeds and never in quadrupedal chimpanzees. While these observations point to *S. tchadensis* and *O. tugenensis* both being obligate bipeds, it remains difficult to fully characterize their anatomy and behavior. An *Orrorin* hand phalanx was curved like an ape finger, which signals this species may have been equally adept in the trees. Their taxonomic status in relation to earlier Miocene apes and later hominids is also unclear, but *Sahelanthropus* especially was probably very close to the common ancestor of apes and humans.

Later early hominids are represented by members of the genus *Ardipithecus* from Ethiopia's Awash region (White, Asfaw, and Suwa 1996). These specimens are more numerous but are fragmentary and incomplete. Still, the earlier *Ar. kadabba* (5.5 to 5.8 mya) appears to bear various ape characteristics, including a small brain, curved finger bones, and a thin layer of enamel on its teeth. An intermediate degree of honing (neither classically ape nor human) is inferred from dental wear patterns. The later *A. ramidus* (4.4 mya) may have

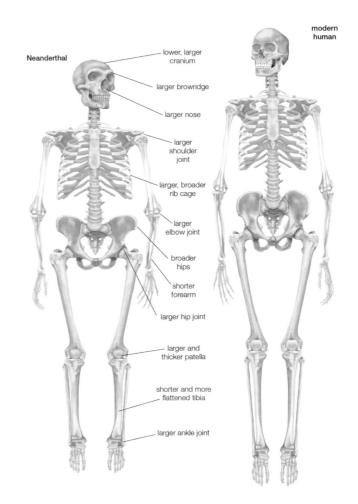

Skeleton of a Neanderthal (*Homo neanderthalensis*) compared with a skeleton of a modern human (*Homo sapiens*). © Universal Images Group Limited/Alamy.

lacked honing. All skeletal anatomy convincingly demonstrates both species were bipeds, probably standing about 1 meter (3.3 feet) tall.

While the anatomical signals of bipedalism are not debated, the reasons for its origins remain unclear (Langdon 2005; Lovejoy 1988). Leading hypotheses surround the selective advantage of freeing the hands for food collection and carrying offspring, greater stature promoting access to more food, increasing line-of-sight for advanced detection of predators, thermoregulation, and long-distance walking. Other ideas, such as Darwin's hunting hypothesis and the aquatic ape concept, have been rejected (hunting occurs only several million years later; swimming leading to bipedalism is ecologically, behaviorally, and biomechanically incorrect). More than likely, there was probably not a single reason for bipedalism to become rooted, but rather, multiple interacting selective factors including those mentioned above. Also, bipedalism was long assumed to have taken root in a grasslands environment 3.5 mya. All paleoecological data associated with *Sahelanthropus*, *Orrorin*, and *Ardipithecus*, however, suggest a more heavily, though perhaps somewhat patchy, forested environment was associated with the origins of bipedal apes.

Australopithecines emerge, diversify, and split

As several hundred fossil specimens now demonstrate, human evolution became more complex following *Ardipithecus* (Aiello and Andrews 2000). The earliest identified member of the genus *Australopithecus*, *A. anamensis*, was found in 4-million-year-old deposits in Kenya (Ward, Leakey, and Walker 1999). Specific primitive traits reveal *A. anamensis* was probably a descendant of *Ardipithecus*. Then, around 3.6 to 3 mya, came the appearance of *A. afarensis*, the best known of all australopithecines because of the discovery of "Lucy" (A.L. 288-1), a 40-percent-complete female *A. afarensis* found in 1974 in Ethiopia's Afar region (Johanson et al. 1982).

Lucy and other members of her species appear as bipedal apes much as the earlier hominids. *A. afarensis* exhibited classic ape morphologies—a small brain (430 cubic centimeters), a funnel-shaped rib cage, curved finger phalanges, small stature (about 1.1 meters), and a U-shaped dental arch. Canines, however, were definitively nonhoning, clearly signaling that the second major event attaining derived human features was definitely in place at this time. The pelvis of *A. afarensis* possessed a broad flare with an anterior warp and a valgus knee. Limb proportions were intermediate between apes and modern humans. *A. afarensis* appears responsible for making the Laetoli footprints in a bed of volcanic ash in Tanzania 3.6 mya. These footprints document the biomechanics of australopithecine bipedalism, revealing the presence of an arch and a nondivergent big toe. Living alongside *A. afarensis* may have been *A. platyops* at 3.5 mya in Kenya. Its flatter face resembles advanced hominids (Leakey et al. 2001), but the skull is heavily distorted and cracked, which makes definitive evaluation difficult.

At 3 mya, Hominidae began to radically change shape as the descendants of *A. afarensis* diverged into two main branches, one of which became an evolutionary dead end.

The first branch, initially populated by *A. garhi*, flourished around 2.5 mya (Asfaw et al. 1999). Cranial, dental, and postcranial fossils indicate a mix of primitive and derived features, including more humanlike limb proportions, signaling lessened dedication to life in the trees than *A. afarensis*. *A. garhi* may have been the first tool user (Semaw 2000) and is a good candidate for the immediate ancestor of *Homo*. Associated with simple stone tools made from one to three strikes of a hammerstone against a core, the Oldowan tradition emerged. Australopithecine hands appear more human- than chimpanzee-like: They possessed the anatomy for precision and power grips to manufacture stone tools. This represents the third major event in human evolution—the dawn of material culture involving the codification of traditions, ideas, and technologies across generations.

The second branch was populated by a diverse group of gracile (*A. africanus*), robust (*A. robustus*), and hyper-robust (*A. aethiopicus*, *A. boisei*) australopithecines in eastern and southern Africa from about 3 to 1 mya (Cartmill and Smith

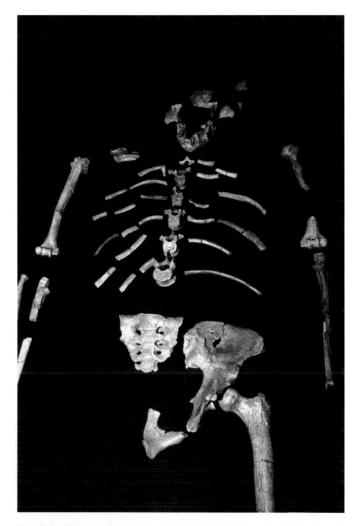

The 3.2-million years old fossilized bones of Lucy, found by Donald Johanson in the Afar region of Ethiopia. The fossils are now housed in the museum in Addis Ababa. © Robert Preston Photography/Alamy.

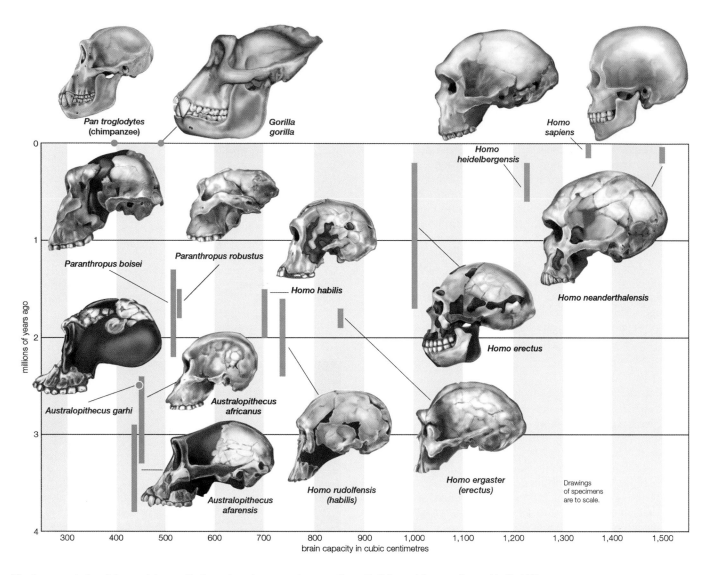

The increase in hominin cranial capacity through various species over time. © Universal Images Group Limited/Alamy.

2009). This group embodies an increasing trend in cranial and dental robusticity (but not for the rest of the body; brain size remained small) when compared to *A. afarensis* and *garhi*. This functionally signifies an increasing dietary specialization involving a diet rich in hard foods such as nuts and tubers, a specialization that corresponded to the shift from forest to grassland. This is particularly clear with the robust and hyper-robust forms. They featured massive molars that acted as crushing and grinding platforms powered by enormous chewing muscles attached to prominent sagittal crests. While this hominid branch flourished for 2 million years, the same narrow dietary specialization that granted them such success resulted in their extinction.

Early genus *Homo*

In the 1960s, a series of 2-million-year-old fossils found in Tanzania's Olduvai Gorge were designated *H. habilis*, meaning "skillful person" (Leakey, Tobias, and Napier 1964). These specimens embodied a jump in brain size—in the 600-cubic-

centimeter range, or at least 50 percent larger than any australopithecine. At this time, the principal trend of human evolution was well underway: encephalization, or the increase of brain size beyond what is proportional for body size. *H. habilis* carried on the Oldowan, and tool cut marks on contemporaneous animal bones suggest the dietary role of meat (obtained from scavenging—not hunting). While the role of meat versus tubers as a key hominid dietary adaptation is actively debated (Bunn 2007; Wrangham et al. 1999), increased access to meat did not hinder protein-hungry encephalization.

As more *H. habilis* specimens emerged, this era of human evolution appeared more complex. In 1972 the cranium of what some argue is a contemporaneous species, *H. rudolfensis*, was found and featured a flatter, advanced face and cranial capacity of 750 cubic centimeters (Leakey 1973). Such a size difference, however, could be a function of sexual dimorphism (the smaller crania represent females), but recent digital morphometric reanalysis suggests brain and face size similar to

H. habilis, thereby questioning the validity of the proposed *h. rudolfensis* taxa. Another key specimen at Olduvai was the highly fragmentary but informative skeleton of an adult individual known as OH-62 (Johanson et al. 1987). Despite a larger brain, OH-62 demonstrates that *H. habilis* retained ape body proportions and short stature (1 meter), similar in size to some australopithecines.

At about 1.8 mya, human evolution took a unique turn as a new taxon, *H. erectus*, emerged in Africa alongside *H. habilis* and robust australopithecines. (While some refer to the African forms as *H. ergaster*, no significant derived character states can be systematically defined to reasonably split Asian and African erectines into two distinct species.) These hominids demonstrate a suite of derived characteristics and behaviors that distinguish them from all previous species. Erectines possessed a larger brain averaging 900 cubic centimeters, a low and long skull with prominent brow ridges, dental reduction, smaller molars, and thinner tooth enamel. The 1.6-million-year-old skeleton of the Nariokotome Boy, an eleven-year-old African erectine (Walker and Leakey 1993), indicates a remarkable change transpired since *H. habilis*. Erectines possessed humanlike body proportions with longer legs than arms, which conferred greater walking and running efficiency. They also grew according to a slower humanlike velocity, but did not experience an adolescent growth spurt. Various clues indicate that the fourth major event in human evolution, speech, probably was in place at this time, though erectine speech was probably not identical to that of modern humans. The study of this skeleton indicates there were probably fewer nerves leading from the spine to the thoracic muscles, and the precise autonomic control of breath necessary for modern speech sounds was probably lacking.

H. erectus was an innovator and developed the Acheulean, a stone tool tradition that produced handheld, roughly triangular bifaces with straight-edged cutting surfaces to signal increasing cultural sophistication, environmental manipulation, and adaptive success. The tradition is well documented in China and Africa, demonstrating that *H. erectus* was the first hominid to control fire, used at the very least for cooking (Weiner et al. 1998).

H. erectus was the first hominid to leave Africa. Evidence of the first, quick migration has been found at Dmanisi in the southwestern Asian republic of Georgia (Gabunia et al. 2001). These fossils appear similar to primitive *H. erectus*, still exhibiting some *H. habilis* morphologies and use of Oldowan choppers. The probable descendants of this group went on to colonize much of Asia and Europe, stopped only by the deep water barriers of the Atlantic and Pacific oceans. They persisted in Asia perhaps as recently as 30,000 years ago. One group, which lived 780,000 years ago in northern Spain, displayed anatomical ties to African erectines but began to diverge into a new morphotype—*H. antecessor*—that exhibited traits seen in later Neandertals (Bermúdez de Castro et al. 2004). They may also have been the first hominids to ritually dispose of their dead as evidenced by the postmortem manipulation, possible ritual cannibalism, and disposal of the dead inside deep caves.

Archaic *Homo sapiens* and Neandertals

The earliest members of *H. sapiens* emerged in Africa around 350,000 years ago during the Middle Pleistocene interglacial as brain size continued to expand and stone tool technologies developed greater complexity. Known as archaic *H. sapiens*, their origins among *H. erectus* is clear. (A debatable use of terminology refers to archaic *H. sapiens* as *H. heidelbergensis*; the type specimen was found in 1907 in Mauer, Germany, and subsequently studied at the University of Heidelberg.) Though possessing prominent brow ridges and thick crania like any erectine, archaic *H. sapiens* exhibited higher foreheads (expanding neocortex) and reduced skeletal robusticity and tooth size (Bräuer 2001). The fossil record amply shows that archaic *H. sapiens* also migrated from Africa and spread throughout the Old World, with well-known specimens originating from South Africa, Spain, England, Greece, India, China, and Indonesia.

The late archaic *H. sapiens* fossil record is remarkable in many ways. Good evidence of the fifth major event in human evolution—active hunting—is evident at this time in the increase of butchered animal bones in the fossil record and the earliest preserved spears around 400,000 years ago. Also, the fossil record includes a very rare hominid speciation event outside of Africa, the one that gave rise to Neandertals.

This form of archaic *H. sapiens* evolved and thrived in glacial Europe and the Middle East from 130,000 to 28,000 years ago. While Neandertals were once envisioned as slow-witted subhumans, more recent research has clarified that they were a very close sister species to modern humans. The Neandertal morphotype is unique among hominids largely because they were shaped by the selective pressures of a cold environment (Churchill 1998; Hubin 1998). Their faces feature massive nasal apertures. Functionally, a larger nose and greater nasal cavity surface area served to preheat cold ambient air before it reached the lungs to avoid reducing the core body temperature. Neandertal bodies were very stocky and relatively shorter than modern humans. This reduced the area of their arms and legs, just as with modern human Inuits, to reduce the amount of heat lost at the body's surface. Neandertal crania were quite long and contained a brain larger than modern humans—and would have generated a notable quantity of heat via its metabolism. Such specialization again may have contributed to their eventual extinction around 28,000 years ago, when climates warmed, their food sources became extinct, and they competed with modern humans (generalists) for resources.

Correlates of Neandertal imagination and creativity are seen in their invention of new technologies (the Mousterian stone tool tradition), complex ritualized burial of their dead, and altruism and compassion potentially reflected in the long-term survival of individuals whose skeletons bore what should have been mortal injuries (Cartmill and Smith 2009). Various lines of evidence indicate Neandertals were efficient hunters. While soft tissues of the vocal tract do not fossilize, Neandertal hyoid bones, located in the throat and vital to producing speech sounds, are identical to those of modern humans. Recovery of exceptionally well-preserved bones from sites in Croatia and Spain has now permitted the once unthinkable complete mapping of the Neandertal mitochondrial genome. A draft version of the 3 billion base-pair-long nuclear genome was completed in early 2009. Among initial findings is that the Neandertal *FOXP2* gene, critically involved with language production among modern humans, is identical to the uniquely derived version shared by all modern humans (Krause et al. 2007). These genomic data indicate not only that Neandertals had very similar potential to produce modern human speech but also that they shared about 99.5 percent of identical DNA sequences with living humans. Yet, Neandertal genetic differences may be outside the range of variation of modern people—implying they were a distinct human species and are not the ancestors of modern humans.

Emergence and radiation of anatomically modern humans

The earliest evidence of *Homo sapiens sapiens*, or anatomically modern humans (hereafter AMHs), is found in Africa between 200,000 and 150,000 years ago. Remains of three individuals were found at Herto, Ethiopia, whose morphologies appear to bridge that of archaic and modern peoples: a long skull and occipital projection mixed with a human-sized brain, vertical foreheads, small brow ridges, and retracted faces (White et al. 2003). Designated *H. sapiens idaltu*, this

potential subspecies is thought to be at the threshold of anatomical modernity. At 130,000 years ago, the descendants of these people led a migrational wave from northeastern Africa that penetrated into the Near East when a wetter climate shaped a more hospitable Levant region. They remained there for several thousand generations but retreated back to Africa when the climate shifted to a cool and dry pattern. Some 30,000 years later, a second wave of migrants left Africa and ultimately established the global distribution of humanity—but the mode of AMH origins and radiations are the subject of contentious debate.

Traditionally, this question has revolved around the out-of-Africa and multiregional models. The former (Stringer 2002) posits that AMHs emerged first in Africa between 200,000 and 80,000 years ago, and they then migrated throughout the Old World to replace all other forms of archaic *H. sapiens*. The multiregional hypothesis (Wolpoff, Hawks, and Caspari 2000) envisions modern humans emerging from earlier archaic populations throughout the Old World due to extensive levels of gene flow between paleopopulations.

Studies of mitochondrial DNA, mapping of other independent markers (i.e., Y chromosome mutation sites), and

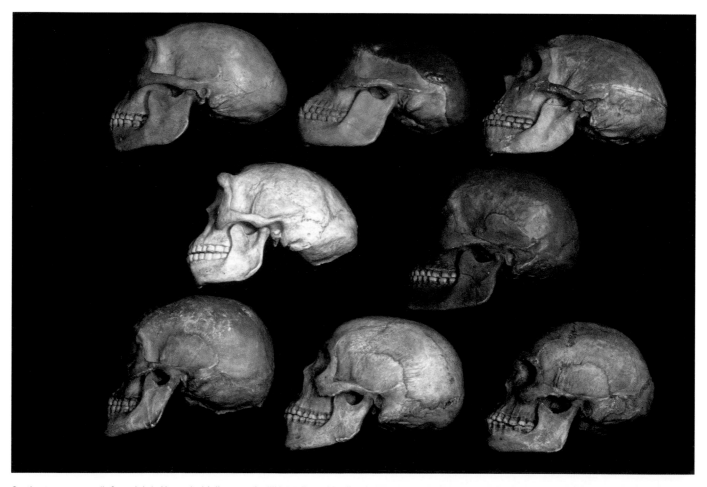

On the top row are (left to right): Homo heidelbergensis ("Heidelberg Man"); Homo erectus ("Java Man"); Homo neanderthalensis ("Neanderthal"). In the middle row are (l to r): Homo erectus pekinensis (Peking Man); and, Homo sapiens ("Cro-Magnon"). On the bottom row are (l to r): Caucasian, African Bushman-Bantu, and Australian Aboriginal skulls. E. R. Degginger/Photo Researchers, Inc.

fossil evidence (i.e., Herto) lends increasing probability to human genetic and anatomical origins in Africa between 130,000 and 80,000 years ago. Also, all modern humans demonstrate a very low degree of genetic diversity—markedly lower than most subspecies—which is consistent with a recent single AMH origin that did not incorporate ancient genes from archaic *Homo*. Yet, evidence of gene flow (consistent with multiregional origins) is suggested from remains of a 24,000-year-old skeleton of a child in Portugal that may demonstrate characteristics, including limb proportions, expected of a Neandertal–human hybrid (Zilhão and Trinkaus 2002). Incisor shoveling, which gives a spatulate shape to the inside of the front teeth, is found among *H. erectus* and later Asian AMHs. Given such evidence, limited interbreeding and gene flow, a "partial replacement" or assimilation model is emerging as another potential scenario.

Yet, one hominid, *H. floresiensis* (called "The Hobbit" by the media), may not have been replaced by AMHs. This remarkable potential species of *Homo* was found to have persisted on the island of Flores, Indonesia, until perhaps 18,000 years ago (Brown et al. 2004). Even more extraordinary is their morphology: about 0.9 meters (3 feet) tall, with a brain the size of an australopithecine, but associated with complex humanlike tools. A fierce debate has ensued. *H. floresiensis* could have shared a common erectine ancestor with AMHs perhaps 2 mya, but became isolated in Indonesia and underwent a process known as "island dwarfism," seen among other island-dwelling mammals when food and predators are both limited. Other paleoanthropologists countered that *H. floresiensis* specimens represented a group of pathological, microcephalic, or pygmy AMHs. The latter views are not well supported by recent detailed studies of endocasts and postcranial anatomy (e.g., Morwood et al. 2005). In particular, aspects of *H. floresiensis* mandibular and wrist anatomy are indistinguishable from *H. habilis* and *erectus*. This implies AMHs shared the planet with another species of *Homo* until the very recent past, and only the catastrophic eruption of a volcano on Flores about 12,000 years ago rendered this species extinct.

AMHs were the first hominids to colonize the entire planet, with their presence firmly established in East Asia and Australia at 50,000 years ago, Europe and Siberia (where Neandertals and AMHs overlapped for tens of thousands of years) at 50,000 years ago, and the Americas between 15,000 and 13,000 years ago. Around 50,000 years ago, however, an unprecedented cultural change began to unfold among AMHs. This process involved the origins of a variety of new technologies, complex burial rituals, multicomponent tools, and art. The so-called dawn of art is best studied in the European archaeological record, though its traces are found throughout the world. Prolific cave art, petroglyphs, "Venus" figurines, and similar expressions are thought to be concrete material reflections of what some have termed the "Inner Eye" and the "Inner I"; in other words, signs of human perceptions and consciousness comparable to that of today's humans (Leakey 1996). This event probably has little to do with evolution of new genes, but is correlated to a global warming trend and greater resource

stability during the Middle Pleistocene when AMH populations reached their highest densities.

The single most impactful change in human history is perhaps the sixth event in human evolution: the shift from foraging to farming. Around 11,000 years ago, the last interglacial period gave way to a period of warming. This had the effect of creating ecological conditions favorable for intensification of harvesting plants, which led to the independent origins of agriculture in Southeast Asia, central Mexico, southern and northern China, the Andes, the eastern United States, and sub-Saharan Africa. While often considered to be a great advance for humanity, agriculture involved major adaptive trade-offs and challenges (Diamond 1987; Larsen 1995) leading to lifestyles seemingly contrary to the directions of the last several million years of evolution. Sedentary lifestyles promoted multiple forms of biological stress among agriculturalists, including an explosive rise of infectious disease, parasitism, anemia, growth disruption, and decreased bone strength. Agricultural diets are less nutritious and varied than those of foragers, and consumption of starchy cultigens promoted a nearly global rise in dental caries (cavities). The first evidence of patterned human-on-human violence and warfare comes following this shift. Humans began a series of negative environmental transformations that continue today. The one factor that may have sustained this evidently maladaptive strategy was the increase in female fertility that agricultural diets sustained, driving population growth and feeding back into ever-increasing food production.

The future

Primate and human evolution resulted from the interplay among millions of random mutations and changing global ecologies over 50 million years. Because future evolution is based on events that have not yet happened, the state of primate and AMH descendants at any point in the future cannot be predicted. But since the dawn of the industrial revolution (when some argue the Holocene ended and the "Anthrocene" began), AMHs have increasingly influenced the nature and tempo of evolutionary change among many forms of life. Modern human economic activities critically endanger the survival of all primates, especially great apes, and humans may be the agents of their extinction within the next 100 to 200 years.

If humans persist, the geographical isolation conferred by the establishment of permanent colonies on other planets could lead to the speciation of new kinds of humans by genetic drift alone. One day, bioengineers may override natural selection as humans artificially shape themselves, molecule by molecule. Or, a postbiological phase may await us, where the tools that we build achieve consciousness, either merging with biological life as recent research in cybernetics has shown is possible, or our machines persisting long after our own extinction. Whatever the speculation, future stages in human evolution will likely be unprecedented—shaped by conscious choices made by the descendant of the first African bipedal apes, choices that will affect them and all life on Earth, whatever forms they shall take.

Resources

Books

Aiello, Leslie, and Christopher Dean. 1990. *An Introduction to Human Evolutionary Anatomy*. London: Academic Press.

Begun, David R. 2002a. "European Hominoids." In *The Primate Fossil Record*, ed. Walter Carl Hartwig. Cambridge, UK: Cambridge University Press.

Begun, David R. 2002b. "The Pliopithecoidea." In *The Primate Fossil Record*, ed. Walter Carl Hartwig. Cambridge, UK: Cambridge University Press.

Benefit, Brenda, and Monte L. McCrossin. 2002. "The Victoriapithecidea, Cercopithecoidea." In *The Primate Fossil Record*, ed. Walter Carl Hartwig. Cambridge, UK: Cambridge University Press.

Bräuer, Günter. 2001. "The KNM-ER 3884 Hominid and the Emergence of Modern Anatomy in Africa." In *Humanity from African Naissance to Coming Millennia*, ed. Phillip V. Tobias, Michael A. Raath, Jacopo Moggi-Cecchi, and Gerald A. Doyle. Florence, Italy: Firenze University Press.

Brown, Frank H. "Geochronometry." 2000a. In *Encyclopedia of Human Evolution and Prehistory*, ed. Eric Delson, Ian Tattersall, John Van Couvering, and Alison S. Brooks. 2nd edition. New York: Garland Publishing.

Brown, Frank H. 2000b. "Paleomagnetism and Human Evolution." In *Encyclopedia of Human Evolution and Prehistory*, ed. Eric Delson, Ian Tattersall, John Van Couvering, and Alison S. Brooks. 2nd edition. New York: Garland Publishing.

Brown, Frank H., and John Van Couvering. 2000. "Stratigraphy Explained." In *Encyclopedia of Human Evolution and Prehistory*, ed. Eric Delson, Ian Tattersall, John Van Couvering, and Alison S. Brooks. 2nd edition. New York: Garland Publishing.

Bunn, Henry T. 2007. "Meat Made Us Human." In *Evolution of the Human Diet: The Known, the Unknown, and the Unknowable*, ed. Peter S. Ungar. Oxford: Oxford University Press.

Cartmill, Matt, and Fred H. Smith. 2009. *The Human Lineage*. Hoboken, NJ: Wiley-Blackwell.

Covert, Herbert H. 2002. "The Earliest Fossil Primates and the Evolution of Prosimians: Introduction." In *The Primate Fossil Record*, ed. Walter Carl Hartwig. Cambridge, UK: Cambridge University Press.

Covert, Herbert H. 2004. "Does Overlap among the Adaptive Radiations of Omomyoids, Adapoids, and Early Anthropoids Cloud Our Understanding of Anthropoid Origins?" In *Anthropoid Origins: New Visions*, ed. Callum F. Ross and Richard F. Kay. New York: Kluwer Academic/Plenum.

Fleagle, John G., and Marcelo F. Tejedor. 2002. "Early Platyrrhines of Southern South America." In *The Primate Fossil Record*, ed. Walter Carl Hartwig. Cambridge, UK: Cambridge University Press.

Hubin, Jean-Jacques. 1998. "Climatic Changes, Paleogeography, and the Evolution of Neandertals." In *Neandertals and Modern Humans in Western Asia*, ed. Takeru Akazawa, Kenichi Aoki, and Ofer Bar-Yosef. New York: Plenum Press.

Langdon, John H. 2005. *The Human Strategy: An Evolutionary Perspective on Human Anatomy*. New York: Oxford University Press.

Leakey, Richard. 1996. *The Origin of Humankind*. New York: Basic Books.

Van Couvering, John. 2000. "Cyclostratigraphy." In *Encyclopedia of Human Evolution and Prehistory*, ed. Eric Delson, Ian Tattersall, John Van Couvering, and Alison S. Brooks. 2nd edition. New York: Garland Publishing.

Vrba, Elisabeth S., George H. Denton, Timothy C. Partridge, and Lloyd H. Burckle, eds. 1995. *Paleoclimate and Evolution, with Emphasis on Human Origins*. New Haven, CT: Yale University Press.

Walker, Alan, and Richard Leakey, eds. 1993. *The Nariokotome* Homo erectus *Skeleton*. Cambridge, MA: Harvard University Press.

White, Tim D., Berhane Asfaw, and Gen Suwa. 1996. "*Ardipithecus ramidus*, a Root Species for Australopithecus." In *The First Humans and Their Cultural Manifestations*, ed. Fiorenzo Facchini. Forlì, Italy: A.B.A.C.O.

Wood, Bernard. 2005. *Human Evolution: A Very Short Introduction*. Oxford: Oxford University Press.

Zilhão, Joa, and Erik Trinkaus. 2002. *Portrait of the Artist as a Child: The Gravettian Human Skeleton from the Abrigo do Lagar Velho and Its Archeological Context*. Lisbon: Instituto Português de Arqueologia.

Periodicals

Aiello, L. C., and P. Andrews. 2000. "The Australopithecines in Review." *Human Evolution* 15(1–2): 17–38.

Asfaw, Berhane, Tim White, Owen Lovejoy, et al. 1999. "*Australopithecus garhi*: A New Species of Early Hominid from Ethiopia." *Science* 284(5414): 629–635.

Bermúdez de Castro, José María, Marcos Martinón-Torres, Eudald Carbonell, et al. 2004. "The Atapuerca Sites and Their Contribution to the Knowledge of Human Evolution in Europe." *Evolutionary Anthropology* 13(1): 25–41.

Bloch, Jonathan I., and Doug M. Boyer. 2002. "Grasping Primate Origins." *Science* 298(5598): 1606–1610.

Brown, Peter, Thomas Sutikna, Michael J. Morwood, et al. 2004. "A New Small-Bodied Hominin from the Late Pleistocene of Flores, Indonesia." *Nature* 431(7012): 1055–1061.

Brunet, Michel, Franck Guy, David Pilbeam, et al. 2002. "A New Hominid from the Upper Miocene of Chad, Central Africa." *Nature* 418(6894): 145–151.

Churchill, Steven Emilio. 1998. "Cold Adaptation, Heterochrony, and Neandertals." *Evolutionary Anthropology* 7(2): 46–60.

Diamond, Jared. 1987. "The Worst Mistake in the History of the Human Race." *Discover* 8(5): 64–66.

Gabunia, Leo, Susan C. Antón, David Lordkipanidze, et al. 2001. "Dmanisi and Dispersal." *Evolutionary Anthropology* 10(5): 158–170.

Gebo, Daniel L., Marian Dagosto, K. Christopher Beard, et al. 2000. "The Oldest Known Anthropoid Postcranial Fossils and the Early Evolution of Higher Primates." *Nature* 404(6775): 276–278.

Johanson, Donald C., C. Owen Lovejoy, William H. Kimbel, et al. 1982. "Morphology of the Pliocene Partial Skeleton (A.L. 288-1) from the Hadar Formation, Ethiopia." *American Journal of Physical Anthropology* 57(4): 403–451.

Johanson, Donald C., Fidelis T. Masao, Gerald G. Eck, et al. 1987. "New Partial Skeleton of *Homo habilis* from Olduvai Gorge, Tanzania." *Nature* 327(6119): 205–209.

Krause, Johannes, Carles Lalueza-Fox, Ludovic Orlando, et al. 2007. "The Derived *FOXP2* Variant of Modern Humans Was Shared with Neandertals." *Current Biology* 17(21): 1908–1912.

Larsen, Clark Spencer. 1995. "Biological Changes in Human Populations with Agriculture." *Annual Review of Anthropology* 24: 185–213.

Leakey, L. S. B., P. V. Tobias, and J. R. Napier. 1964. "A New Species of the Genus *Homo* from Olduvai Gorge." *Nature* 202 (4927): 7–9.

Leakey, Meave G., Fred Spoor, Frank H. Brown, et al. 2001. "New Hominin Genus from Eastern Africa Shows Diverse Middle Pliocene Lineages." *Nature* 410(6827): 433–440.

Leakey, R. E. F. 1973. "Evidence for an Advanced Plio-Pleisto-cene Hominid from East Rudolf, Kenya." *Nature* 242(5398): 447–450.

Oppenheimer, Stephen. 2003. *Out of Eden: The Peopling of the World*. London: Constable and Robinson.

Rasmussen, David Tab. 2002. "The Origin of Primates." In *The Primate Fossil Record*, ed. Walter Carl Hartwig. Cambridge, UK: Cambridge University Press.

Ross, Callum F., and Richard F. Kay. 2004. "Anthropoid Origins: Retrospective and Prospective." In *Anthropoid Origins: New Visions*, ed. Callum F. Ross and Richard F. Kay. New York: Kluwer Academic/Plenum.

Schwartz, H. P. 2006. "Electron Spin Resonance Dating, Fission-Track Dating, Thermoluminescence Dating, and Uranium-Series Dating." In *The Human Evolution Source Book*, ed. Russell L. Ciochon and John G. Fleagle. 2nd edition. Upper Saddle River, NJ: Pearson Prentice Hall.

Lovejoy, C. Owen. 1988. "Evolution of Human Walking." *Scientific American* 259(5): 118–125.

Morwood, Michael J., Peter Brown, Jatmiko, et al. 2005. "Further Evidence for Small-Bodied Hominins from the Late Pleisto-cene of Flores, Indonesia." *Nature* 437(7061): 1012–1017.

Semaw, Sileshi. 2000. "The World's Oldest Stone Artefacts from Gona, Ethiopia: Their Implications for Understanding Stone Technology and Patterns of Human Evolution between 2.6–1.5 Million Years Ago." *Journal of Archaeological Science* 27(12): 1197–1214.

Senut, Brigitte, Martin Pickford, Dominique Gommery, et al. 2001. "First Hominid from the Miocene (Lukeino Formation, Kenya)." *Comptes Rendus de l'Acadmèmie des Sciences*, Series IIA: *Earth and Planetary Science* 322(2): 137–144.

Strait, David S., and Frederick E. Grine. 2004. "Inferring Hominoid and Early Hominid Phylogeny Using Craniodental Characteristics: The Role of Fossil Taxa." *Journal of Human Evolution* 47(6): 399–452.

Stringer, Chris. 2002. "Modern Human Origins: Progress and Prospects." *Philosophical Transactions of the Royal Society of London* B 357(1420): 563–579.

Ward, Carol, Meave Leakey, and Alan Walker. 1999. "The New Hominid Species *Australopithecus anamensis*." *Evolutionary Anthropology* 7(6): 197–205.

Weiner, Steve, Qinqi Xu, Paul Goldberg, et al. 1998. "Evidence for the Use of Fire at Zhoukoudian, China." *Science* 281(5374): 251–253.

White, Tim D., Berhane Asfaw, David DeGusta, et al. 2003. "Pleistocene *Homo sapiens* from Middle Awash, Ethiopia." *Nature* 432(6941): 742–747.

Wolpoff, Milford H., John Hawks, and Rachel Caspari. 2000. "Multiregional, Not Multiple Origins." *American Journal of Physical Anthropology* 112(1): 129–136.

Wrangham, Richard W., James Holland Jones, Greg Laden, et al. 1999. "The Raw and the Stolen: Cooking and the Ecology of Human Origins." *Current Anthropology* 40(5): 567–594.

Other

Franzen, Jens L., Philip D. Gingerich, Jörg Habersetzer, et al. 2009. "Complete Primate Skeleton from the Middle Eocene of Messel in Germany: Morphology and Paleobiology." *PLoS ONE* 4(5): e5723. Available from http://www.plosone.org/article/info:doi/10.1371/journal.pone.0005723.

Haagen D. Klaus

Evolution of the brain and nervous system

This entry is focused on the evolution of complex brains, especially those found in mammals and in primates, the branch of mammalian evolution that led to humans. This focus is partly motivated by challenges posed in describing the great diversity of body plans and nervous system types that have emerged, and the general lack of fossil or comparative evidence on how most nervous systems evolved. In addition, humans are especially curious about their own brains, as they generate all human thoughts, feelings, and actions, and this leads to the question of how the very large and complex brains of humans came to be. Until recently little was known about how human brains evolved, but over the last twenty to thirty years comparative studies of brain organization and function, together with an appreciation of a growing fossil record and an understanding of developmental mechanisms, have led to a greater understanding of the course of brain evolution from early mammals to present-day humans.

What can be learned from the fossil record and from comparative studies of present-day mammals?

Brains are soft tissue and therefore do not fossilize; however, the skulls of mammals do. As the interior of the skull often conforms closely to the shape of the brain, endocasts of the inside of a skull can provide good indications of brain sizes and shapes, proportions of major brain divisions such as the neocortex and olfactory cortex, and even fissure patterns in cortex (Jerison 1973). The fossil record shows that early mammals had small brains for the size of their bodies, which were also small, and that these brains had little neocortex relative to the olfactory bulb, olfactory cortex (piriform cortex), and the rest of the brain (brainstem and cerebellum). In general, mammals with larger brains and especially large amounts of neocortex subsequently evolved, although some present-day mammals have brain sizes and proportions of brain parts very much like those of early mammals. For uncertain reasons, mammals with larger bodies tend to have larger brains, although brains generally increase in size at a lesser rate than bodies. Some mammals, however, have evolved larger brains than expected relative to their body sizes (Jerison 1973). Finally, some things can be deduced about the functional organization of brains of extinct mammals based on studies of endocasts. A brain with a large temporal lobe, for example, would suggest an emphasis on processing visual or auditory information. As fissures in neocortex often form between functionally distinct parts of neocortex, the locations of fissures found in skull endocasts can reveal even more about the functional organization of various mammalian brains.

Comparative studies of the organization of brains of various present-day mammals provide an even more powerful approach toward understanding brain evolution (Butler and Hodos 2005; Striedter 2005). The logic of this approach is that features found widely in current mammals likely evolved early in mammalian evolution, as these traits were more likely to have been retained across the many branches of the mammalian radiation than to have evolved repeatedly. In contrast, traits found in the brains of only a small group of closely related mammals probably evolved recently in immediate ancestors of those mammals. As interpretations of such comparative studies depend greatly on understanding phylogenetic (cladistic) relationships of present-day mammals, we are fortunate that modern techniques of molecular biology have greatly aided reconstructions of how modern mammals evolved from a common ancestor.

Nervous systems of early multicellular animals

Although single-celled animals can respond to external stimuli with movements, they do not have nervous systems. Only multicellular animals have nervous systems. A nervous system consists of an organized, interconnected group of cells, called neurons, specialized for reception of external stimuli or signals from other cells, conduction of excitation from one cell to others, and transmission of excitation to other cells. Some present-day animals, such as jellyfish, have a very simple nervous system consisting of a nerve net of functionally connected nerve cells, but without any central collection of neurons for processing collected information. Such a simple nervous system may represent an ancestral organization or a loss of centralized structures that were no longer needed—the latter sometimes occurring in parasitic or sedentary life-forms. From such simple nervous systems, more complex nervous systems with large concentrations of neurons forming central ganglia or brains in the body and head have evolved independently on numerous occasions. Complex, centralized nervous systems have evolved in mollusks, insects and other arthropods, and vertebrates (Roth and Wullimann 2001; Kaas 2007). As the central nervous systems of arthropods and

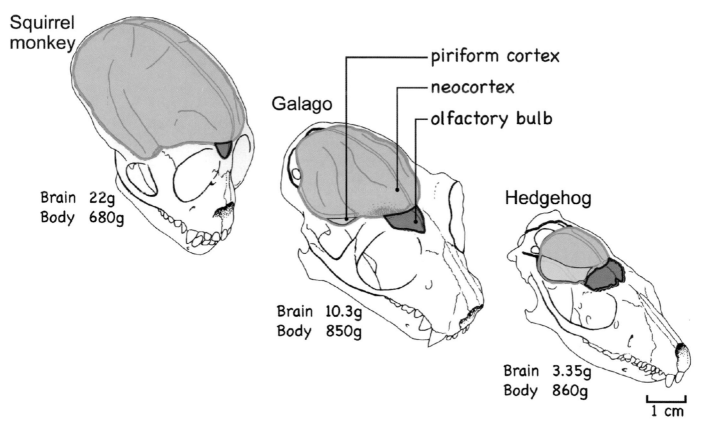

Figure 1. Endocasts of Skulls. Shapes and fissures of brains determined from endocasts of the skulls of a squirrel monkey, prosimian galago, and hedgehog. Modified from Radinsky, L., American Scientist, 1975. Reproduced by permission of Gale, a part of Cengage Learning.

vertebrates have a number of similarities, a long-standing question is whether these two types of nervous systems evolved independently or arose from a common ancestor with a complex, centralized nervous system.

The main difference in nervous systems of arthropods and vertebrates (and other chordates) is that the elongated nervous system (the brain and spinal cord of vertebrates) is located dorsally in vertebrates while the equivalent central nervous system in arthropods is located ventrally (Butler and Hodos 2005). Both have a brain or a brainlike structure in the head, and both have an anteroposterior axis that is subdivided in an anteroposterior direction into anatomical and functional compartments or regions that are under the developmental control of the same genes, known as Hox genes. These genes are expressed along the anteroposterior axis of the body axis in the same order they are found on the chromosome, and they apparently were present in the common ancestor of vertebrates and arthropods where they had a major role in the anteroposterior regionalization of the nervous system (Roth and Wullimann 2001). How then did the nervous systems of these two clades come to differ in being located ventrally or dorsally in the body? One old, but now revived, theory is that the ancient, ventral nervous system of early, bilaterally symmetrical animals was preserved in arthropods, but vertebrates literally "turned over" so that their belly became their back. This is called the dorsoventral inversion hypothesis.

Early stages of evolution of the vertebrate nervous system

Vertebrates belong to a larger clade of chordates, animals with a notochord, a longitudinal, flexible bundle of cells that forms the supporting axis of the body. Nervous systems of other chordates are similar to those of vertebrates but differ enough to make it difficult to surmise exactly how the vertebrate nervous system emerged. The earliest vertebrates were fish. Early fish were jawless, and two groups of jawless fish, the hagfish and lampreys, remain. Other fish include radiations of cartilaginous fishes, ray-finned fishes, and fleshy-finned fishes. Brains and bodies of fish vary greatly, as fish became specialized in various ways (Butler and Hodos 2005). By considering brain features widely shared across clades of fish as primitive characters, and considering those evident in some well-preserved fossil material, the basic organization of early vertebrate brains can be reconstructed (Northcutt 1981).

The earliest vertebrate brains included paired olfactory bulbs at the front (rostral) end of the brain, paired eyes, and optic tracts coursing into the forebrain. The forebrain consisted of a more rostral telencephalon, including the cerebral hemispheres and basal ganglia, and the more caudal diencephalon or thalamus (Butler and Hodos 2005; Kaas 2007; Northcutt 1981). The midbrain is just caudal to the thalamus, followed by the upper brainstem, which is capped

Ray-finned Fish

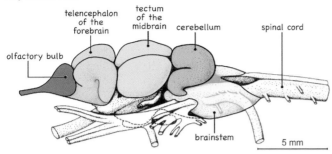

Figure 2. Major divisions of the brain of a ray-finned fish. From Butler and Hodos, 2005. Reproduced by permission of Gale, a part of Cengage Learning.

by a primitive cerebellum, followed by the lower brainstem or medulla, and then the spinal cord, which is part of the central nervous system but not part of the brain. The cranial nerves of the brainstem provide sensory input and motor outputs. These parts of early vertebrate brains have been retained in all vertebrates.

The telencephalon of early vertebrates was likely dominated by olfactory inputs. The telencephalon was small compared to that in most present-day vertebrates, and poorly differentiated into functionally distinct regions that formed along the ventricles, as neurons remained close to where they were generated in development and probably did not migrate laterally to form forebrain structures, as they do to varying extents in present-day vertebrates. The diencephalon included the hypothalamus, which gave rise to the paired retinas, and the pineal apparatus (related to vision). Early vertebrates also had a dorsal thalamus and a ventral thalamus. The midbrain included a sensory "roof," the tectum, with visual and other (lateral line) sensory centers. Parts of this type of brain became variously and sometimes markedly elaborated in the radiations of cartilaginous and bony fishes. Most notably, an elaborate electroreceptive system evolved in some fish for the detection of prey and other objects and for communication.

One line of the fish radiation, the lungfish, gave rise to amphibians over 860 million years ago (mya) (Butler and Hodos 2005). Early amphibians had an aquatic larval stage, as do most modern amphibians. Modern amphibians form a monophyletic group with three major radiations: anurans (frogs and toads), urodeles (newts and salamanders), and caecilians (amphibians that lost their limbs). Modern amphibians are considered to be highly derived, with little resemblance to their early ancestors. Although all of the main divisions of early fish brains have been retained in amphibians, some brain features appear to have undergone a simplification, whereas others appear to be primitive, making it difficult to reconstruct the ancestral amphibian brain. As in fish, the visual midbrain and pretectal regions are responsible for most visual behaviors, and the forebrain (pallium) has a modulatory role. The medial pallium is largely homologous to the mammalian hippocampus, while the dorsal pallium is multisensory and possibly homologous to the limbic cortex of mammals.

Reptiles evolved over 300 mya from an early branch of amphibians that also gave rise to anurans (frogs). Because birds and mammals evolved from reptiles, and all have the extra embryonic amniotic membrane that makes them independent of an aquatic environment, they are collectively called amniotes. Early reptiles gave rise to two major branches, the sauropsids or diapsid reptiles that gave rise to modern reptiles and birds, and the synapsid reptiles (often referred to as early amniotes) that gave rise to mammals, their only surviving descendants (Butler and Hodos 2005). As surviving reptiles are members of the early branch of diapsid reptiles, their brains provide only limited evidence about the evolution of mammalian brains from mammal-like synapsid reptiles approximately 280 mya.

Although a number of different specializations of brain organization occur in reptiles (see figure 3), a few generalizations about sensory systems are possible (Kaas 2007). For the visual system, retinal projections to the hypothalamus, thalamus, pretectum, and tectum of amphibians have been retained. In addition, a projection from a visual nucleus of the dorsal thalamus (the homolog of the dorsal lateral geniculate nucleus in mammals) connects to the dorsomedial pallial cortex, the homolog of the mammalian neocortex. A second visual pathway to the dorsomedial cortex is mediated by a projection from the visual (optic) tectum to a cell group in the dorsal thalamus, and then to the dorsomedial cortex. The visual tectum also projects to a part of the dorsal thalamus that, in turn, projects to a part of the pallium corresponding to part of the amygdala of mammals. The amygdala is very important in motivating behavior, especially in learning what to fear. While the forebrains of both amphibians and reptiles have somatosensory inputs from the dorsal-column and trigeminal nuclei of the brainstem and from the spinal cord, these projections reach the dorsal thalamus in reptiles where they are relayed to the dorsomedial cortex. The auditory systems of reptiles and other tetrapods (four-footed animals) include a peripheral auditory organ (the basilar papilla or cochlea), a cochlear nucleus or nuclei, a brainstem processing center (the superior olive or olivary complex), the auditory midbrain (the inferior colliculus or torus semicircularis), and a thalamic auditory nucleus or nuclei (the medial geniculate complex in mammals). In reptiles, the auditory thalamus projects to the striatum and the dorsal ventricular ridge, the apparent reptilian homolog of part of the mammalian amygdala. In mammals, a major target of the auditory thalamic nuclei is the auditory cortex of the temporal neocortex. Overall, the dorsal thalamus of reptiles has a number of nuclei that appear to be homologous with nuclei or nuclear groups in mammals, although they are typically structurally less distinct.

The existence of homologous brainstem, midbrain, thalamic, and cortical structures in present-day reptiles and mammals suggests that these structures were present in the early amniotes before the divergence of the diapsid line leading to present-day reptiles and birds and the synapsid line leading to present-day mammals. The dorsal cortex of reptiles (see figure 3) is homologous to the neocortex of mammals (see figures 1 and 4), although it is not nearly as laminated or subdivided into areas (Butler and Hodos 2005; Kaas 2007). The functions of the small dorsal cortex in reptiles are uncertain, while the neocortex

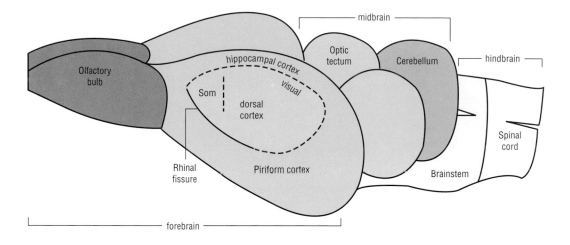

Figure 3. Major divisions of vertebrate brains are shown for a reptile (turtle). The small dorsal cortex in turtles has somatosensory (Som.) and visual regions. The optic tectum is part of the midbrain. Reproduced by permission of Gale, a part of Cengage Learning.

in mammals is important in processing sensory information, sensorimotor control, and decision-making (Allman 1999; Roth and Wullimann 2001). The medial cortex is homologous to the mammalian hippocampus, while the lateral cortex corresponds to the olfactory (piriform) cortex of mammals. Both reptiles and mammals have main and accessory olfactory bulbs, with similar projections to the olfactory cortex and parts of the amygdala.

Finally, birds evolved from archosaurian reptiles (dinosaurs) about 200 mya. Three families of crocodiles represent the other surviving members of the archosaurian radiation. While birds and bird brains vary considerably, birds share many features of brain organization with both reptiles and mammals (Butler and Hodos 2005; Kaas 2007; Roth and Wullimann 2001). Perhaps most importantly, the dorsal pallium of birds, a thickened part of the forebrain with nuclear rather than laminar subdivisions, is the homolog of the reptilian dorsal cortex and the mammalian neocortex. Thus, something like the dorsal cortex of present-day reptiles evolved quite differently into a thick cellular mass in birds and an expansive laminar structure in mammals. In addition, distinct, separate somatosensory and visual regions have been retained but enlarged and elaborated from reptiles in both birds and mammals. Much of the pallial region of birds, now known to be homologous with the neocortex of mammals, was formerly thought to correspond to very large basal ganglia.

Brains of early mammals

The synapsid branch of early amniote (reptilian) evolution gave rise to therapsid mammal-like reptiles, which gave rise to mammals over 280 mya as the only surviving branch of the synapsid radiation. Skeletal remains of advanced therapsids indicate that they had somewhat larger brains than early amniotes. Early mammals likely had hair for insulation as well as sensory hairs similar to the longer sensory vibrissae found in present-day mammals. Casts of the skull interior of early mammals indicate that they had brains that were three or four

times larger than those of therapsids of the same body size (Jerison 1973). Their forebrain included a large olfactory bulb, a very large olfactory (piriform) cortex, and a small dorsal cap of neocortex separated from the olfactory cortex by the rhinal sulcus (marked by a ridge on the inner surface of the skull). Compared to those of most present-day mammals, the brains of early mammals were small relative to body size, and there was little neocortex compared to olfactory cortex (Allman 1999; Kaas 2007; Kaas and Preuss 2008).

Early mammals diverged into three major lines of mammalian evolution. Monotremes, with reptilian features such as egg laying, separated from marsupials some 230 mya. Marsupials, with external pouches for their young, separated from the now dominant placental or eutherian mammals about 160 mya. Present-day mammals include six major clades or superorders and over 4,500 species. They have evolved many different brain specializations and enlargements, especially over the last 65 million years, following the extinction of the dinosaurs. Dinosaurs had brain sizes expected for reptiles of their body sizes, whereas mammals and birds have comparatively larger brains than would be predicted by their body sizes.

While the fossil record has revealed much about the sizes and proportions of major parts of the brains of early mammals, the internal neuroanatomy was not preserved. This anatomy can be reconstructed, however, by considering the common features of brain organization across members of the six major superorders of mammals, because those features were likely retained from a common early mammalian ancestor (Kaas and Preuss 2008). As early mammals had smaller brains with little neocortex, it is especially fruitful to consider the brains of those present-day mammals with smaller brains and little neocortex (see figure 4), because these brains are likely to be less specialized.

The most conspicuous difference between brains of mammals and reptiles is the large, thick, laminated neocortex

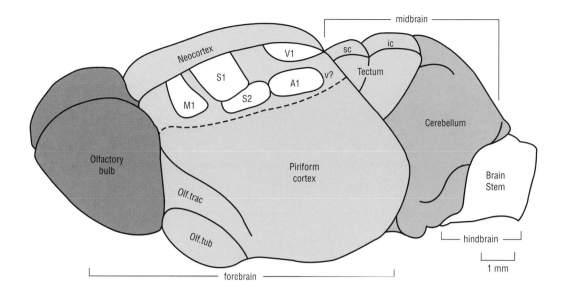

Figure 4. A dorsolateral view of the brain of a small mammal with little neocortex, a Madagascar tenrec. This brain has the proportions and size of the brains of early mammals. Comparative studies indicate that the organization of neocortex has changed little since the early mammal. However, this brain of a placental mammal has a primary motor area (M1), a field lacking in the earliest mammals and present-day monotremes and marsupials. Other cortical areas include primary and secondary visual cortex, V1 and V2, primary and secondary somatosensory cortex, S1 and S2, and an auditory region (Aud) that consists of a primary area (A1) and possibly one or more other auditory fields. Reproduced by permission of Gale, a part of Cengage Learning.

of mammals (Kaas 2007). Reptiles have a thin dorsal cortex largely composed of a single row of pyramidal neurons with inputs from the dorsal thalamus, some intrinsic modulation by smaller neurons, and direct outputs to subcortical structures. Mammals differ by having a thick neocortex of six traditional layers, with each layer distinguished by the types of neurons it contains and by types of connections. Thus, layer 4 receives inputs from the dorsal thalamus or other areas of cortex, layer 6 provides feedback connections to the cortical areas or thalamic nuclei that provide the activating input, layer 5 projects subcortically, and layer 3 projects to other areas of the cortex. Neurons across the layers are interconnected to form computational circuits called modules or columns, and stepwise processing from columns in one cortical area to another adds computational complexity at each step. The cortical areas are subdivisions of the cortex that are specialized via connections and neuronal subtypes to perform specific functions, and they have been called the "organs of the brain."

Overall, the neocortex in all mammals represents a great advance in processing capacity and function over the dorsal cortex of reptiles, but exactly how this complex, laminated, and subdivided structure evolved from something like the dorsal cortex is not known. Comparative studies indicate that the proportionately small neocortex of the first mammals was poorly differentiated into distinct layers, and this cortex was subdivided into as few as perhaps twenty different areas of functional significance (compared to perhaps 200 in the human brain). Functionally distinct areas included primary visual and somatosensory areas, retained from reptilian ancestors, and a primary auditory area that was probably acquired in early mammals or mammal-like reptiles through

the formation of new connections from the auditory nuclei of the dorsal thalamus to the neocortex. Early mammals also had secondary sensory areas, cortex devoted to taste, limbic fields for emotional states, and areas related to memory given that they interconnected with the hippocampus, a structure that evolved from the hippocampal cortex of reptiles. Cortical motor functions of early mammals were indirect, and mediated by somatosensory cortex and cortical connection to the basal ganglia. Separate motor areas of the neocortex emerged with placental mammals.

The organization of the forebrain of early mammals was modified in many ways within the mammalian radiation, even in mammals that retained small brains (Kaas 2007). For example, the brains of bats became specialized for echolocation by altering the organization of the auditory system so that the auditory cortex was large relative to the rest of the cortex and subdivided into a number of areas specialized for specific auditory functions. In addition, a large proportion of auditory neurons became sensitive to the echolocating sound frequencies. As another example, the brain of the unusual duck-billed platypus, a monotreme, became modified so that parts of the somatosensory cortex became sensitive to inputs from newly evolved electroreceptors in the bill. In many mammals, including rats and mice, much of the neocortex became devoted to the tactile receptors activated when they explored the environment with their long facial whiskers or vibrissae. In general, many lines of mammalian evolution evolved larger brains and especially proportionately larger amounts of neocortex (Finlay and Darlington 1995). In general, those with more neocortex added cortical areas, and specialized the functions of these cortical areas (Kaas 2007; Roth and Wullimann 2001). This was especially

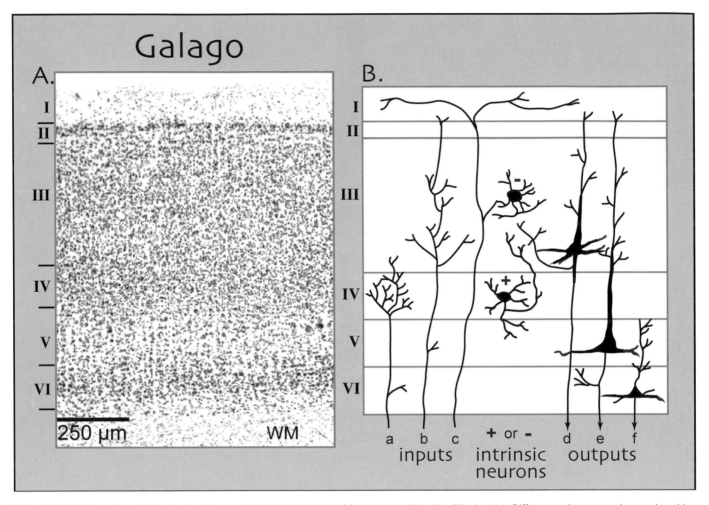

Figure 5. A) A thin section of visual cortex in a prosimian primate stained for neuron cell bodies (Nissl stain). Differences in neuron sizes and packing density allowed early investigators to distinguish six layers. B) A drawing of some of the connections of the six layers. Major activating inputs (a) go to layer IV, while modulating inputs go to layers III and I (b and c). Intrinsic neurons without outputs to other structures can be excitatory (+) or inhibitory (-). Output neurons can project to other cortical areas (d) the thalamus, and brainstem, and spinal cord (e) or provide feedback to sources of input (f).| Reproduced by permission of Gale, a part of Cengage Learning.

evident in primates, where the comparatively huge, complex brain of humans, featuring many cortical areas, ultimately evolved (Kaas and Preuss 2008).

Brain evolution in primates

Early primates diverged from other placental mammals some 70 to 80 mya. The close relatives of present-day primates are tree shrews (Tupaiidae), small squirrel-like mammals; and flying lemurs or colugos (Cynocephalidae), another group of squirrel-like mammals, ones that glide from tree to tree like flying squirrels. Somewhat more distant relatives include rodents and lagomorphs (rabbits). Early primates were small, nocturnal predators living on small vertebrates, insects, leaf buds, and fruit. They lived in the fine branches of the trees in tropical rain forests and depended more on vision than olfaction for finding food and guiding arboreal locomotion (Kaas 2007; Kaas and Preuss 2008). As a result, they evolved forward-facing eyes for increased binocular vision, a short snout to reduce obstruction of vision, and a reduced olfactory

system. Possibly to protect their eyes, they likely used their hands to reach and grasp prey and other food, rather than grasping with the mouth. Thus, visually guided reaching and grasping with the forelimb was emphasized.

Soon after evolution of the first primates, the order divided into three major branches that led to present-day prosimians, tarsiers, and anthropoids. Prosimians further diverged into a number of species in the lemur, loris, and galago lines, but most prosimians retained relatively small bodies and a nocturnal lifestyle. In size and shape, prosimian brains changed little from those of early primates (Kaas 2007). Common ancestors of present-day tarsiers and anthropoids (monkeys, apes, and humans) adapted to a diurnal lifestyle, resulting in such specialization for daytime vision as a fovea in the retina and the loss of the reflecting tapetum behind the retina. Early tarsiers reverted back to a nocturnal life, and they adapted to dim light by acquiring huge eyes to compensate for the loss of a reflecting tapetum. Tarsiers specialized in visually detecting prey in dim light. Present-day tarsiers have a well-developed visual system, with a large proportion of their neocortex devoted to a distinctively laminated primary visual area.

Early ancestors of present-day anthropoid primates were diurnal and depended largely on vision to find food, detect predators, and engage in social interactions. Many became more dependent on fruit and leaves as food, and they formed social groups for predator protection and defending concentrated, dispersed food sources, such as fruits. Only one anthropoid, the owl monkey, reverted to nocturnal life, resulting in large eyes and a degeneration of the fovea (used for detailed vision). Some early African monkeys somehow rafted or island hopped over to South America over 40 mya, forming the radiation of New World monkeys, while Old World monkeys also radiated with one line leading to apes some 30 to 35 mya. One ape line split about 7 mya, leading to present-day chimpanzees and humans (one of apparently many different lines of hominids).

Primate brains vary greatly in size, shape, and organization, and yet they share a number of features that likely have been retained from a common ancestor (Kaas 2007; Kaas and Preuss 2008). Perhaps most importantly, all primates have well-developed visual systems with distinct subsystems from the retina through cortex. The "where and how" subsystem locates objects of interest in the visual scene and uses visual information to guide motor behavior. The "what" subsystem uses detailed and often color vision to identify objects of interest, such as an insect for food or a dangerous snake to avoid. Early primates had visual systems with structures specialized for these two processing streams. The object-identification subsystem required more visual detail, and most of the output neurons of the retina were devoted to this subsystem. In the thalamus, the dorsal lateral geniculate nucleus was divided into thick layers for detailed vision and thin layers for detecting change. This information was relayed to the primary visual cortex, which was large and densely packed with neurons, about twice as many for a given volume as in visual cortex of other mammals. In addition, most of the posterior half of the cortex was devoted to vision and was divided into more than ten visual areas, most of which are unique to primates (see Figure 6). The primary visual cortex distributed information to these areas, which formed interacting networks of areas, with some in a ventral stream of visual information processing for object identification, and others in the dorsal stream providing visual information for guiding reaching and other motor sequences. Ventral stream areas occupied a large and greatly expanded temporal lobe, where they ultimately interacted with memory systems to store information related to object recognition, and with the prefrontal cortex of the frontal lobe where knowledge about objects could access motor areas to guide behavior. Dorsal stream areas projected to a large, expanded posterior portion of the parietal (somatosensory) cortex, where somatosensory, visual, and auditory information was distributed to modular subregions to activate neuronal circuits for relevant motor plans. Major outputs of these sensorimotor modules in the posterior parietal cortex were interconnected with motor areas in the frontal lobe where specifics of behavior actions were programmed. To mediate a range of rapid and skillful forelimb, face, and other body actions, the cortical motor system of early primates had already become greatly elaborated

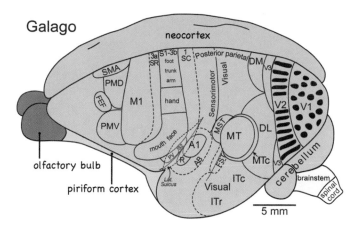

Figure 6. This brain is small relative to most primates, and has only a few fissures. Yet, the neocortex is larger than in early mammals, so that the midbrain, part of the cerebellum, and much of olfactory (piriform) cortex is covered by neocortex. The temporal lobe is greatly expanded, as is posterior parietal cortex, which has sensorimotor functions. Visual cortex includes the primary (V1) and secondary (V2) areas of most mammals, and other areas that are likely specific to primates (the third visual area, V3, the dorsal lateral area, DL, the dorsomedial area, DM, the middle temporal area, MT, the MT crescent, MTc, the middle superior temporal area, MST, the fundal superior temporal area, FST, and several areas of infereo-temporal cortex, IT. Auditory cortex includes a primary area, A1, as well as several other auditory areas. Somatosensory cortex includes a primary area, S1 or area 3b, rostral (Sr) and caudal (Sc) somatosensory belts or areas 3a and 1, a second somatosensory area, S2, and a parietal ventral area, PV. Motor areas of frontal cortex include a large primary area, M1, dorsal and ventral premotor areas, PMD and PMV, a frontal eye field, FEF, and a supplementary motor area, SMA. The lateral sulcus (fissure) has been opened (shaded) to show somatosensory areas S2 and PV. Reproduced by permission of Gale, a part of Cengage Learning.

to include a number of premotor areas in addition to an enlarged and modified primary motor area (see figure 6). The primary motor cortex had an expanded region for mediating finger, hand, and forelimb movements, while dorsal premotor, ventral premotor, supplementary motor, and cingulate motor areas were present and well developed to contribute to the formulation of motor behaviors (Kaas 2007; Kaas and Preuss 2008). These motor fields were directly activated by the sensorimotor modules in the posterior parietal cortex, while also being influenced by object-identification information via connections with the prefrontal cortex. Other changes in brain organization, including those related to this new cortical organization, also occurred, but they were not as dramatic.

Some brains of present-day primates appear to have changed little from the above model, largely based on information from comparative studies. Thus, the brains of present-day prosimians appear to have been modified in only minor ways. The brains of New World and Old World monkeys are more diverse, but some New World monkeys, such as capuchin and spider monkeys, and most Old World monkeys have evolved large brains with proportionately more neocortex that included further specializations and modifications of the visual cortex, modifications and expansions of the somatosensory and auditory cortex, and expansion and modifications of the prefrontal, limbic, and emotional cortical systems for guiding social behavior. Humans benefit from extremely large brains, subdivided into a large but undetermined number of cortical areas, and the emergence of different specializations of the neocortex of the two cerebral hemispheres (Kaas 2007; Kaas and Preuss 2008). Most notably, a large portion of the temporal lobe and part of the frontal lobe of the left cerebral hemisphere became specialized for language, while a large portion of the posterior parietal region of the right hemisphere became specialized for reasoning about spatial relationships. Parts of the temporal lobe in humans, more than in other primates, are devoted to recognizing faces, and can distinguish and recognize hundreds of faces. The enlarged prefrontal region of the frontal lobes of humans allows a heightened ability to read and predict the behavior of others as well as enabling an appreciation of the impact of one's acts on future interactions with others. Human parietal-frontal motor guidance and planning circuits provide an intuitive sense of tool use, and the system's plasticity allows humans to acquire many different motor skills and abilities. Most of these brain changes and new abilities were acquired recently, over the last 2 million years, when the brain sizes of human ancestors were increasing by three times over the brain sizes of the chimpanzee-like ancestors of humans from around 7 mya.

While much is still to be learned, modern noninvasive brain imaging studies are providing an increasing understanding of the organization and functions of human brains, and comparative studies, in conjunction with a more complete fossil record, are leading to a better understanding of how the large, complex human brain evolved.

Resources

Books

Allman, John Morgan. 1999. *Evolving Brains*. New York: Scientific American Library.

Butler, Ann B., and William Hodos. 2005. *Comparative Vertebrate Neuroanatomy: Evolution and Adaptation*. 2nd edition. Hoboken, NJ: Wiley-Interscience.

Jerison, Harry J. 1973. *Evolution of the Brain and Intelligence*. New York: Academic Press.

Kaas, Jon H., ed. 2007. *Evolution of Nervous Systems*. 4 vols. Amsterdam: Elsevier.

Kaas, Jon H., and Todd M. Preuss. 2008. "Human Brain Evolution." In *Fundamental Neuroscience*, ed. Larry R. Squire, Floyd E. Bloom, Nicholas C. Spitzer, et al. 3rd edition. Amsterdam: Elsevier.

Roth, Gerhard, and Mario F. Wullimann, eds. 2001. *Brain Evolution and Cognition*. New York: Wiley.

Striedter, Georg F. 2005. *Principles of Brain Evolution*. Sunderland, MA: Sinauer.

Periodicals

Finlay, Barbara L., and Richard B. Darlington. 1995. "Linked Regularities in the Development and Evolution of Mammalian Brains." *Science* 268(5217): 1578–1584.

Northcutt, R. Glenn. 1981. "Evolution of the Telencephalon in Nonmammals." *Annual Review of Neuroscience* 4: 301–350.

Jon H. Kaas

Evolution of language

Modern research challenges the ancient doctrine that only human beings can comprehend and use human language. Consider the following dialogue between an adult named Roger Fouts and a thirty-month-old infant named Washoe.

Roger: What you want?
Washoe: Tickle.
Roger: Who tickle?
Washoe: Dr. Gardner.
Roger: Dr. Gardner not here.
Washoe: Roger tickle.
Washoe: Tickle.
Roger: Who?
Washoe: You.
Roger: Ask politely.
Washoe: Please you tickle.

Hardly remarkable for conversation between two humans, but Washoe was an infant chimpanzee, Roger an adult human, and these conversations occurred during the first successful attempt to establish two-way communication with a nonhuman being in human sign language. Sign language studies of cross-fostered chimpanzees by R. Allen and Beatrix T. Gardner challenged an ancient claim of human uniqueness. Evidence that members of another species can use a human language has already yielded insight into the nature of language, the process of language acquisition, and the relationship between animal and human intelligence.

Language as problem solving

Traditionally, comparative psychologists separated the role of language in communication from the role of language in problem solving. Early investigators pointed out that common laboratory tasks such as matching, oddity, and conditional discrimination require abilities analogous to fundamental features of language, or at least comprehension of language. In conditional oddity problems, for example, Henry W. Nissen, Josephine Semmes Blum, and Robert A. Blum (1967) presented sets of three objects, such as a blue triangle, a blue circle, and a yellow circle, to laboratory chimpanzees. In a typical experiment, chimpanzees choose the color-odd objects (that is, the yellow circle) when the tray bearing the objects is orange, but the form-odd object (that is, the blue triangle)

when the tray is white. The colors of the trays and the correct choices varied from problem to problem. Nissen and his associates drew analogies between tray colors and linguistic modifiers where "orange" stands for "yours" versus "mine" and "white" stands for "large" versus "small" as a signal for different meanings of the same elements by analogy to modifiers in linguistic combinations such as "my mother" and "your mother" or "cold water" and "hot water."

Following the traditions of Nissen and the American psychologist B. F. Skinner (1904–1990), David Premack (1971) connected problem solving to language even more explicitly. Premack tested a caged chimpanzee, Sarah, with vertical rows of flat plastic tokens that differed in color, size, and shape and that he called "words," together with occasional real objects such as apples and keys. Premack presents a set of elaborate analyses as support for his argument that some eight language functions "considered to be fundamental by linguists," such as competence in the interrogative mode and metalinguistics (the use of language to teach language), are logically equivalent to conditional discriminations that Sarah mastered.

More recently, Sue Savage-Rumbaugh and colleagues (1986) presented similar tasks to a caged bonobo (pygmy chimpanzee), Kanzi, except that Kanzi's language consisted of arbitrary designs on a keyboard that differed in color and shape. Savage-Rumbaugh and her associates sometimes refer to these arbitrary designs as "words" and sometimes as "lexigrams." Sometimes they have included pointing gestures interpreted as "person one" or "person two" according to the direction of Kanzi's pointing.

Language as communication

Sign language studies of cross-fostered chimpanzees depart from studies of "ape language" or "animal language" such as those of Premack, Savage-Rumbaugh, and their associates. With American Sign Language, experimenters compare observations of chimpanzees directly with observations of human children and human adults, with the chimpanzees living in a nearly human environment (Gardner and Gardner 1989). For one study, the researchers' goals were described as follows:

At the outset we were quite sure that Washoe could learn to make various signs in order to obtain food, drink, and other things. For the project to be a success, we felt that something more must be developed. We wanted Washoe not only to ask for objects but to answer questions about them and also to ask us questions. We wanted to develop behavior that could be described as conversation. (Gardner and Gardner 1969, pp. 664–665)

Cross-fostering

Cross-fostering—parents of one genetic stock rearing the young of a different genetic stock—is a traditional tool for studying the interaction between genes (nature) and environment (nurture) in behavioral development. Cross-fosterlings have adopted selected species-specific behaviors of fostering parents, including such characteristic behavior as migration, flight patterns, and feeding habits (Gardner 2007).

Most critical for human–chimpanzee cross-fostering, chimpanzees, like humans, have a comparatively long childhood. Newborn chimpanzees are quite helpless, though not as helpless as human babies. In the Gardners' laboratory, chimpanzees failed to roll over by themselves before four to seven weeks of age, sit up before ten to fifteen weeks, or creep before twelve to fifteen weeks. The change from milk teeth to

adult dentition began at about five years. Under natural conditions in Africa, infant chimpanzees are almost completely dependent on their mothers until they are two or three years old, and weaning begins only at between four and five years of age. Menarche occurs when wild females are ten or eleven, and their first infant is born when they are between twelve and fifteen years old (Goodall 1986). Captive chimpanzees have remained vigorously active, both physically and intellectually, taking tests and solving experimental problems, when their ages have been verified at more than fifty years (Maple and Cone 1981).

Cross-fostering is very different from rearing a chimpanzee in a conventional laboratory staffed by human caretakers. Cross-fostering is also very different from keeping a chimpanzee as a pet. Many people keep pets in their homes and may care for them extremely well, but they hardly treat them like their own children. Providing a nearly human infant environment all day every day for years on end is a daunting laboratory challenge, requiring intensive effort.

All aspects of intellectual growth are intimately related. For young chimpanzees, no less than for human children, familiarity with simple tools such as keys, devices such as lights, and articles of clothing such as shoes are all intimately

Detail of bonobo sign language board that was used with chimpanzees at the Language Research Center, Georgia State University. © Frans Lanting Studio/Alamy.

involved in learning signs or words for keys, lights, shoes, opening, entering, lighting, and lacing. In the Gardners' studies, the human-simulated homes of the chimpanzees Washoe, Moja, Pili, Tatu, and Dar were well-stocked with human objects and activities, and cross-fosterlings had free access to them, or at least as much access as young human children usually have. They ate human-style food at a table, with cups, forks, and spoons. They helped to clear the table and wash the dishes after a meal. They used human toilets (in their own quarters and elsewhere), wiped themselves and flushed the toilet, and even asked to go to the bathroom during boring lessons and chores (Gardner and Gardner 1989).

The daily language of this infant world was American Sign Language (ASL), the naturally occurring language of deaf communities in North America. English, the language of earlier studies, demands vocal apparatus and vocal habits that seem beyond the capability of chimpanzees (see Laitman and Reidenberg 1988 for the evolutionary history of human vocal apparatus). Without conversational give-and-take in a common language, cross-fostering conditions could hardly simulate the environment of a human infant. Whenever a cross-fosterling was present, the sole form of "verbal" communication was in ASL.

Size of vocabulary, appropriate use of sentence constituents, number of utterances, proportion of phrases, and inflection, all grew robustly throughout five years of cross-fostering, and patterns were consistent across chimpanzees. Wherever comparable measurements were available, patterns of growth for cross-fostered chimpanzees paralleled in detail characteristic patterns reported for human infants (Gardner and Gardner 1994, 1998).

Cross-fostered chimpanzees, however, develop more slowly than children learning either signed or spoken languages—more slowly, but without reaching an asymptote while they were under cross-fostering conditions in the Gardners' laboratory. Unlike the caged subjects studied by Premack, Savage-Rumbaugh, and their associates, the chimpanzees in the Gardners' laboratory acquired and used signs in spontaneous conversational interactions with their human foster families the way human children acquire their native languages.

As a naturally occurring human gestural language, the Gardners selected ASL, the most common sign language used in daily communication by many in North America. Because so many human beings, including human parents, use ASL in human communities, this language permits direct

Dr. Sue Savage Rumbaugh worked with chimps to speak using a complex sign language. © Anna Clopet/Corbis.

comparisons between the development of young human signers and cross-fostered chimpanzees.

Modeling

Savage-Rumbaugh and colleagues (1986) reported that Kanzi taught himself to use a computer keyboard by observing his mother, Matata, taking a lesson from Savage-Rumbaugh. Furthermore, these researchers reported that when Matata left her place at the keyboard, Kanzi immediately and without any human intervention proceeded to work the keys by himself. So far, however, neither Savage-Rumbaugh nor any of her associates has published any incident of Kanzi learning to use any keys by observing any human or other bonobo.

Incidents of cross-fostered chimpanzees learning signs by observation have been common. One day, in her tenth month of cross-fostering, Washoe was visiting the Gardners' home and found her way into the bathroom. She climbed up on the counter, looked at the mug full of toothbrushes, and signed TOOTHBRUSH. At that time, the Gardners believed that Washoe understood the sign TOOTHBRUSH, but they had never seen her use one, voluntarily. She hardly had any reason to ask for these toothbrushes because they were well within her reach; and it is very unlikely that she was asking to have her teeth brushed. She was just naming a found object, to her companion or, perhaps, to herself.

Adult-to-adult interest was also critical. In the 1960s many members of Washoe's human foster family were smokers. She must have watched them asking each other for cigarettes and matches over and over again, although she, herself, was not allowed to smoke cigarettes or play with matches. One day, during the thirtieth month of Project Washoe, Naomi (a human nonsmoker) needed to light the stove for cooking, but could not find any matches. Washoe watched the search intently. By way of explanation, Naomi held up an empty box of matches. And Washoe replied: SMOKE. After this first observation, the Gardners discovered that Washoe signed SMOKE to name both cigarettes and matches or their familiar containers (Gardner and Gardner 1974). The second project with Moja, Pili, Tatu, and Dar also included many examples of signs learned by observing human companions.

Washoe often signed to herself in play, particularly in places that afforded her privacy, as when she was high in a tree or alone in her bedroom before going to sleep. Washoe also signed to herself when leafing through magazines and picture books, and she avoided attempts others made to join her in this activity. If someone did try to join her or watched her too closely, she often abandoned the magazine or picked it up and moved away. Records show that Washoe not only named pictures to herself in this situation but also corrected herself. On one occasion, she looked at an advertisement in a magazine and signed THAT FOOD, then looked at her hand closely and changed the phrase to THAT DRINK, which was correct.

Washoe also signed to herself about her own ongoing or impending actions. The Gardners have often seen Washoe

Washoe was the first non-human to acquire a human language and had over 250 "words" in her vocabulary. © HO/Reuters/Corbis.

moving stealthily to a forbidden part of the yard signing QUIET to herself, or running rapidly toward her potty chair while signing HURRY (Gardner and Gardner 1974).

Nearly all of Kanzi's lexigrams refer to foods, drinks, destinations, or services (Savage-Rumbaugh et al. 1986). His response panel is essentially a wish list. In sign language studies, vocabularies were rich and varied (Gardner et al.). Washoe, Moja, Tatu, and Dar also invented combinations of signs to name objects, such as COLD BOX for the refrigerator, DIRTY GOOD for the toilet, METAL HOT for a cigarette lighter, and LISTEN DRINK for Alka-Seltzer.

When Kanzi moved from place to place, his response panel had to be folded and set up again at each location. Washoe, Moja, Pili, Tatu, and Dar usually initiated conversations. They signed to themselves about pictures in books, and they even climbed high up in a tree to sign to themselves in privacy (Gardner and Gardner 1989). Special samples and tests were the only signed interactions in which human beings initiated most of the signing.

A robust phenomenon

Washoe, Moja, Pili, Tatu, and Dar signed to friends and to strangers. They signed to each other and to themselves; to dogs, cats, toys, and tools; and even to trees. Along with their skill with cups and spoons, pencils, and crayons, their signing developed stage for stage much like the speaking and signing of human children (Van Cantfort and Rimpau 1982; Van Cantfort, Gardner, and Gardner 1989). They also used the elementary sorts of sign language inflections that deaf children use to modulate the meaning of signs (Gardner and Gardner 1978; Rimpau, Gardner, and Gardner 1989). Cross-fostered chimpanzees converse among themselves, even when there is no human being present and the conversations must be recorded with remotely controlled cameras. An infant, Loulis, learned more than fifty signs of ASL that he could have learned only from other chimpanzees (Fouts, Hirsch, and Fouts 1982).

When Loulis was ten months old he was adopted by fourteen-year-old Washoe, shortly after she lost her own newborn infant. To show that Washoe could teach signs to an infant without human intervention, Roger S. Fouts introduced a drastic procedure. All human signing was forbidden when Loulis was present. Loulis and Washoe were almost inseparable for the first few years, so Washoe lost almost all her input from human signers. It was a deprivation procedure for Washoe. Later, Moja joined the group in Oklahoma, and still later Tatu and Dar joined the group in Ellensburg, Washington. The signing chimpanzees were allowed to sign to each other; indeed, there was no way to stop them. They became part of Loulis's input.

As Loulis grew older and moved freely by himself from room to room in the laboratory, there were more opportunities for humans to sign to the other chimpanzees when Loulis was not in sight. As expected, however, the rule against signing to Loulis had a generally negative effect on all human signing. Humans hardly signed for five years. It was a deprivation experiment for the cross-fostered chimpanzees.

Washoe, Moja, Tatu, and Dar continued to sign to each other and also attempted to engage human beings in conversation throughout the period of deprivation. Washoe modeled signs for Loulis in ways that could be described only as explicit teaching; and she also molded his hands the way we had molded her's during the teaching process (Fouts, Hirsch, and Fouts 1982; Fouts, Fouts, and Van Cantfort 1989). Loulis learned more than fifty signs from the cross-fostered chimpanzees during the five years in which they were his only models and tutors. Meanwhile, Washoe learned some new signs from Moja, Tatu, and Dar, and the cross-fosterlings signed to each other without any human beings in sight and while their conversations were recorded by remote cameras (Fouts and Fouts 1989).

Semantic range

Vocabulary tests (Gardner and Gardner 1989; Gardner and Gardner 1984) demonstrated that chimpanzees could communicate information under conditions in which the only source of information available to a human observer was the signing of the chimpanzees. Washoe, Moja, Tatu, and Dar named pictures that were out of sight of their human interlocutors. These tests also demonstrated that the signs of the cross-fosterlings referred to generalized natural language categories—that DOG referred to any dog, FLOWER to any flower, and so forth. The chimpanzees accomplished this by naming a varied set of exemplars selected from a large library of photographs. In the tests, each slide appeared once and once only, so that each trial was a first trial (Gardner and Gardner 1989; Gardner and Gardner 1984). That is to say, on each trial the chimpanzees named a picture of an object that they had never seen before.

Cross-fosterlings did well on these tests, but they also made errors. Errors offer little or no information in forced-

Sign language is also used to help children communicate thier needs before they are able to speak. © Huntstock, Inc./Alamy.

choice tests of understanding, as for example, when subjects must choose between a few plastic tokens (Premack 1971) or a few pictures on a testing board (Savage-Rumbaugh et al. 1986). In productive tests, errors contain information because subjects are free to choose their own errors. Signing chimpanzees can produce with their own hands any sign or combination of signs in their vocabularies at any time. Most of their errors on vocabulary tests depended on semantic relationships among the objects in the pictures, or from relationships among the signs. Thus, DOG was a common error for SHOE and SODAPOP (Gardner and Gardner 1984).

Developmental patterns

Gradually and piecemeal, but in an orderly sequence, the language of human toddlers develops into the language of their parents. Cross-fostered chimpanzees developed their sign language gradually along with the rest of their socialization—such as tool use and toilet training—in a nearly human household under conditions approximating those experienced by human infants. The topics of their conversations resemble the topics of human conversations because they had the same things to talk about under nearly the same conditions.

Two studies (Rimpau, Gardner, and Gardner 1989; Chalcraft and Gardner 2005) showed that Dar and Tatu used ASL inflections in conversation to indicate person, place, and object. Two others (Bodamer and Gardner 2002; Jensvold and Gardner 2000) investigated replies of cross-fosterlings to conversational probes of a human interlocutor. When appropriate, Washoe, Moja, Tatu, and Dar incorporated signs from probes of their interlocutors into their own rejoinders. In response to probing questions they clarified and amplified their own previous responses by expanding on signs from probes. Cross-fostered chimpanzees used expansion, reiteration, and incorporation the way human adults and human children use these devices. Their contingent rejoinders maintained the interaction and the topic of the interaction.

Resources

Books

Fouts, Roger S., and Deborah H. Fouts. 1989. "Loulis in Conversation with the Cross-Fostered Chimpanzees." In *Teaching Sign Language to Chimpanzees*, ed. R. Allen Gardner, Beatrix T. Gardner, and Thomas E. Van Cantfort. Albany: State University of New York Press.

Fouts, Roger S., Deborah H. Fouts, and Thomas E. Van Cantfort. 1989. "The Infant Loulis Learns Signs from Cross-Fostered Chimpanzees." In *Teaching Sign Language to Chimpanzees*, ed. R. Allen Gardner, Beatrix T. Gardner, and Thomas E. Van Cantfort. Albany: State University of New York Press.

Fouts, Roger S., Alan D. Hirsch, and Deborah H. Fouts. 1982. "Cultural Transmission of a Human Language in a Chimpanzee Mother–Infant Relationship." In *Studies of Development in Nonhuman Primates*, ed. Hiram E. Fitzgerald, John A. Mullins, and Patricia Gage. New York: Plenum Press.

Gardner, Beatrice T., and R. Allen Gardner. 1974. "Comparing the Early Utterances of Child and Chimpanzee." In *Minnesota Symposium on Child Psychology*, Vol. 8, ed. Anne D. Pick. Minneapolis: University of Minnesota Press.

Gardner, R. Allen, and Beatrix T. Gardner. 1989a. "Cross-Fostered Chimpanzees: I. Testing Vocabulary." In *Understanding Chimpanzees*, ed. Paul G. Heltne and Linda A. Marquardt. Cambridge, MA: Harvard University Press.

Gardner, R. Allen, and Beatrix T. Gardner. 1989b. "A Cross-Fostering Laboratory." In *Teaching Sign Language to Chimpanzees*, ed. R. Allen Gardner, Beatrix T. Gardner, and Thomas E. Van Cantfort. Albany: State University of New York Press.

Gardner, R. Allen, Beatrix T. Gardner, and Thomas E. Van Cantfort, eds. 1989. *Teaching Sign Language to Chimpanzees*. Albany: State University of New York Press.

Goodall, Jane. 1986. *The Chimpanzees of Gombe*. Cambridge, MA: Harvard University Press, Belknap Press.

Premack, David. 1971. "On the Assessment of Language Competence in the Chimpanzee." In *Behavior of Nonhuman Primates*, Vol. 4, ed. Allan Martin Schrier and Fred Stollnitz. New York: Academic Press.

Rimpau, James B., R. Allen Gardner, and Beatrix T. Gardner. 1989. "Expression of Person, Place, and Instrument in ASL Utterances of Children and Chimpanzees." In *Teaching Sign Language to Chimpanzees*, ed. R. Allen Gardner, Beatrix T. Gardner, and Thomas E. Van Cantfort. Albany: State University of New York Press.

Van Cantfort, Thomas E., Beatrix T. Gardner, and R. Allen Gardner. 1989. "Developmental Trends in Replies to Wh-questions by Children and Chimpanzees." In *Teaching Sign Language to Chimpanzees*, ed. R. Allen Gardner, Beatrix T. Gardner, and Thomas E. Van Cantfort. Albany: State University of New York Press.

Periodicals

Bodamer, Mark D., and R. Allen Gardner. 2002. "How Cross-Fostered Chimpanzees (*Pan troglodytes*) Initiate and Maintain Conversations." *Journal of Comparative Psychology* 116(1): 12–26.

Chalcraft, Valerie J., and R. Allen Gardner. 2005. "Cross-Fostered Chimpanzees Modulate Signs of American Sign Language." *Gesture* 5(1–2): 107–131.

Gardner, Beatrix T., and R. Allen Gardner. 1998. "Development of Phrases in the Early Utterances of Children and Cross-Fostered Chimpanzees." *Human Evolution* 13(3–4): 161–188.

Gardner, R. Allen. 2007. "Review of Sign Language Studies of Cross Fostered Chimpanzees." *Journal of the Washington (DC) Academy of Sciences* 93(1): 37–57.

Gardner, R. Allen, and Beatrice T. Gardner. 1969. "Teaching Sign Language to a Chimpanzee." *Science* 165(3894): 664–672.

Gardner, R. Allen, and Beatrix T. Gardner. 1984. "A Vocabulary Test for Chimpanzees (*Pan troglodytes*)." *Journal of Comparative Psychology* 98(4): 381–404.

Gardner, R. Allen, and Beatrice T. Gardner. 1978. "Comparative Psychology and Language Acquisition." *Annals of the New York Academy of Sciences* 309: 37–76.

Jensvold, Mary Lee A., and R. Allen Gardner. 2000. "Interactive Use of Sign Language by Cross-Fostered Chimpanzees (*Pan troglodytes*)." *Journal of Comparative Psychology* 114(4): 335–346.

Laitman, J. T., and J. S. Reidenberg. 1988. "Advances in Understanding the Relationship between the Skull Base and Larynx with Comments on the Origins of Speech." *Human Evolution* 3(1–2): 99–109.

Maple, T. L., and S. G. Cone. 1981. "Aged Apes at the Yerkes Regional Primate Research Center." *Laboratory Primate Newsletter* 20: 10–12.

Nissen, Henry W., Josephine Semmes Blum, and Robert A. Blum. 1949. "Conditional Matching Behavior in Chimpanzee: Implications for the Comparative Study of Intelligence." *Journal of Comparative and Physiological Psychology* 42(5): 339–356.

Savage-Rumbaugh, Sue, Kelly McDonald, Rose A. Sevcik, et al. 1986. "Spontaneous Symbol Acquisition and Communicative Use by Pigmy Chimpanzees (*Pan paniscus*)." *Journal of Experimental Psychology: General* 115(3): 211–235.

Van Cantfort, Thomas E., and James B. Rimpau. 1982. "Sign Language Studies with Children and Chimpanzees." *Sign Language Studies* 34: 15–72.

R. Allen Gardner

Evolutionary medicine and the biology of health

Evolutionary medicine is the study of medical problems using insights from evolutionary biology. The term *Darwinian medicine* carries a restrictive nuance that George C. Williams considered important when he and Randolph M. Nesse wrote the paper "The Dawn of Darwinian Medicine" (1991), which introduced the label to the scientific community. Williams argued that the adjective *Darwinian* emphasizes Darwin's central contribution to evolutionary biology, namely adaptation by natural selection. Darwinian medicine therefore emphasizes considerations of health and disease in the context of human adaptation to past environments and the differences between modern and ancestral environments. Evolutionary adaptation can refer to a process or a characteristic: The process of bacterial adaptation to the presence of antibiotics involves an evolutionary increase in antibiotic resistance, whereas the characteristic of antibiotic resistance is an adaptation that allows the bacterium to survive exposure to antibiotics. Although Williams's point about the central role of adaptation is well taken, in practice the labels evolutionary medicine and Darwinian medicine have been used synonymously, a practice followed in this entry.

A major assumption of evolutionary medicine is that human biology has evolved largely to the environmental conditions that existed prior to the onset of civilization. The logic underlying this assumption is that evolution of humans occurs slowly and human ancestors had vastly longer amounts of time to evolve to the living conditions before the onset of agriculture than to agricultural environments. For *Homo sapiens*, the quarter million or so years of existence prior to agriculture compares with about 10,000 years for those peoples whose ancestry can be traced back to the origins of agriculture. This assumed importance of Paleolithic adaptation has been used particularly to develop hypotheses to explain why humans have diseases that have been associated with diet and levels of activity. These hypotheses propose, for example, that atherosclerosis may result from an agricultural diet. The specific aspects of an agricultural diet that are hypothesized to exacerbate atherosclerosis vary from argument to argument, and include high amounts of saturated fats and cholesterol, low amounts of certain fatty acids (particularly omega-3 fatty acids), or high levels of carbohydrates or low levels of dietary fiber. All the hypotheses, however, suggest that one or more of these dietary differences directly or indirectly contributes to atherosclerosis by contributing to the buildup of fat and cholesterol in the lining of arteries.

Similar hypotheses have been advanced to explain other chronic diseases. Agricultural diets have been hypothesized to contribute to diabetes, for example, because high carbohydrate intake disrupts regulation of blood glucose. The mechanism of this pathogenesis is still hypothetical, but one argument proposes that humans evolved a "thrifty" genotype that encodes for a physiology that functions efficiently when carbohydrate intake is low, but cannot handle the high levels of carbohydrate intake found in modern diets. Similar hypotheses have been proposed for human cancers, incriminating diets rich in calories and poor in fiber.

Although Paleolithic adaptation suggests reasonable hypotheses for diseases such as atherosclerosis and diabetes, these hypotheses need to be critically tested against alternative hypotheses. Indeed, a fundamental challenge for evolutionary medicine is to generate an understanding of the causes of disease that is more balanced and complete than the analyses that characterize the current state of medicine, which tends to foster specialization rather than integration.

Causes of disease

A balanced inquiry into the causes of disease must consider the three major categories of disease causation: inherited genes, nonparasitic environmental factors, and parasitism. Specific candidate causes within each of these general categories need to be identified, and potential interactions between these particular causes—both within and between general categories—need to be evaluated. This process of recognizing and testing the full spectrum of feasible alternative hypotheses is just beginning within evolutionary medicine.

Distinguishing between primary causes and exacerbating causes is critical. Primary causes are essential for the generation of disease, whereas secondary causes are not. For example, the bacterium *Mycobacterium tuberculosis* is the primary cause of human tuberculosis, because people will not develop tuberculosis if they are not infected with this organism. Malnutrition and inherited genetic vulnerabilities to *M. tuberculosis* also contribute to development of tuberculosis, but these are secondary causes because tuberculosis can

A stylized scanning electron microscopic image of *Mycobacterium tuberculosis*. This bacterium causes most cases of tuberculosis. © MedicalRF.com/Alamy.

still occur among people who are well nourished and not particularly genetically susceptible. This distinction has practical importance to medicine because eliminating a secondary cause reduces the damage caused by the disease, but eliminating a primary cause eliminates the disease. Preventing primary causes therefore solves the problem better than preventing secondary causes.

An integrative approach casts the various mixes of primary and secondary causes as alternative hypotheses that are evaluated against the existing evidence. This ongoing process is well illustrated by atherosclerosis. The dietary variables mentioned above fall into the nonparasitic environmental category. The major inherited predisposition to atherosclerosis is an allele (an alternative sequence of a gene) of the apolipoprotein E gene called epsilon 4. The epsilon 4 protein transports fats and cholesterol between cells and confers an increased risk of heart attacks. The epsilon 4 allele is the standard allele for the apolipoprotein E gene among primates. It therefore cannot be considered an inherently defective allele. Rather it seems to have been one that has functioned well in humans throughout their evolutionary history. This raises the hypothesis that epsilon 4 may have functioned well in preagricultural populations but not in agricultural populations. Perhaps it transported too much fat and cholesterol into cells when people began consuming agricultural diets.

Consistent with this hypothesis, people whose ancestries have involved the longest exposure to agricultural diets—those with a distant ancestry in the eastern Mediterranean or China, for example—have low frequencies of the epsilon 4 allele.

This hypothesis is problematical, however. The epsilon 4 allele had already declined to be less prevalent than the other alleles of the apolipoprotein E gene in peoples who had not been exposed to agricultural lifestyles until recent centuries. Whatever the disadvantage of the epsilon 4 allele, that disadvantage must have been occurring long before the onset of agriculture.

A resolution of this paradox is offered by the third category of disease causation: parasitism. Several infectious agents have been found in atherosclerotic lesions. People who have the epsilon 4 allele are especially vulnerable to one of them: *Chlamydia pneumoniae*. This bacterium can get into cells by hitching a ride on the epsilon 4 protein. The paradox can be resolved by focusing on the interaction between infection, genetic predisposition, and noninfectious environmental influences. Specifically, genetic vulnerability to atherosclerosis may result from a vulnerability to *C. pneumoniae*, which has caused the reduction in the frequency of epsilon 4 allele progressively over human prehistory. This reduction may have speeded up in agricultural settings, because *C. pneumoniae* is transmitted by coughing. Therefore, the increased population density associated with agricultural populations likely facilitated the transmission and spread of *C. pneumoniae*. Finally, *C. pneumoniae* facilitates fat and cholesterol deposition in infected cells and may therefore be responsible for early signs of atherosclerosis: fat and cholesterol deposition within the lining of arteries. This explanation is also consistent with results from a large-scale experimental reduction in dietary fat—there was no significant effect on cardiovascular disease. Furthermore, this integration of hypotheses offers an explanation for the exacerbating and protective effects of various dietary components. Garlic, for example, appears to offer some protection against heart attacks, but the most comprehensive studies have not shown any effects of garlic on fat and cholesterol. Garlic has exceptionally powerful antimicrobial effects, however, and may protect against heart attacks by inhibiting *C. pneumoniae* or other infectious causes of atherosclerosis.

This example illustrates how an evolutionary perspective may help draw together the spectrum of relevant evidence, generating a more coherent explanation than collections of risk factors assembled without a conceptual framework based on evolutionary theory. This example also illustrates how a comprehensive evolutionary approach may resolve ambiguities

and contribute to a more coherent understanding of other important diseases, including cancers and severe mental illnesses such as schizophrenia. In each case, coherent understanding arises from the recognition that infectious agents may play an important causal role.

Improved understanding of causes of chronic diseases also affects a more general tenet of evolutionary medicine—application of the evolutionary theory of senescence to human aging. Since the 1950s, evolutionary biologists have embraced a theory of senescence proposing that the greater strength of natural selection during early adulthood would favor alleles that foster survival and reproduction during the early decades of adulthood, with negative effects in later life. The aggregate of these negative effects has then been assumed to be a major cause of human senescence. This effect has been demonstrated in laboratory experiments on short-lived animals such as fruit flies and undoubtedly also applies to humans. What is unclear, however, is the way in which it is applicable to humans. Researchers in evolutionary medicine have proposed that this process explains observed infirmities of age-associated disintegration of health. But as discussed above, atherosclerosis and its attendant effects on heart attack and stroke may be better explained as a slow-burning infectious process

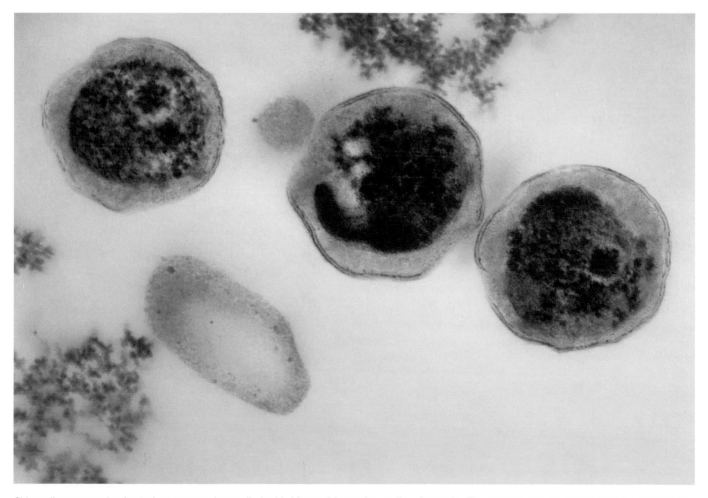

Chlamydia pneumoniae bacterium can get into cells by hitching a ride on the epsilon 4 protein. The paradox can be resolved by focusing on the interaction between infection, genetic predisposition, and noninfectious environmental influences. Dr. Kari Lounatmaa/Photo Researchers, Inc.

Atherosclerosis

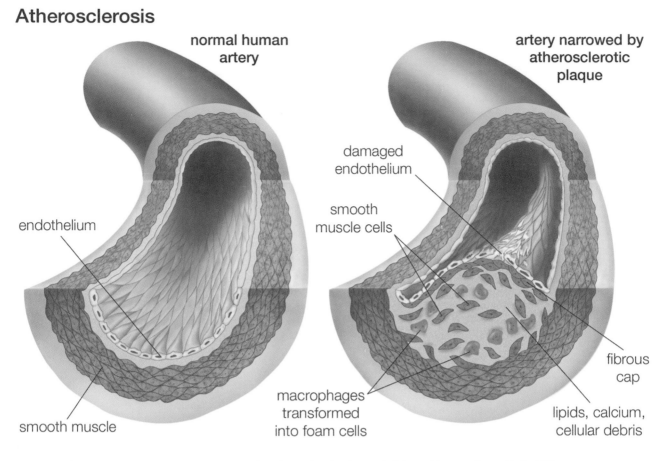

normal human
artery

artery narrowed by
atherosclerotic
plaque

damaged
endothelium

smooth
muscle cells

endothelium

fibrous
cap

smooth muscle

macrophages
transformed
into foam cells

lipids, calcium,
cellular debris

Comparison of a normal artery with an artery narrowed by atherosclerotic plaque. © Universal Images Group Limited/Alamy.

exacerbated by dietary factors and inherited predispositions. Similar arguments are increasingly being supported by research on other age-associated diseases such as cancer, arthritis, diabetes, and Alzheimer's disease. The emerging trend is therefore suggesting that the evolutionary theory of senescence has been overapplied as an explanation of declining health in old age. The evolutionary theory of senescence may still be applicable to age-related illness but at an older age than has generally been thought, perhaps more in the age range of 80 to 110 years than in the range of 50 to 80.

The potential for the three categories of disease causation to act in concert does not mean that all three categories will be acting or interacting in all diseases. Scientists expect, for example, that dangerous new aspects of modern environments, such as pollutants associated with industrial development, may act as the primary causes of modern diseases. In such cases, heritable variation may exist in susceptibility to noninfectious causes of disease agents. Chronic beryllium disease, for example, occurs as a result of excessive exposure to beryllium during mining activities. Susceptibility is associated with a single amino acid substitution in the HLA-DPB1 gene. Interestingly, this gene codes for molecules that present antigens to the immune system. Vulnerability to beryllium poisoning could even involve an interaction with an infectious agent.

Environmental causes of diseases may result from too little or too much exposure to some physical aspect of the environment. For example, too much ultraviolet radiation can contribute to skin cancer by damaging DNA. Too little ultraviolet radiation may contribute to rickets because rickets results from vitamin D deficiency, and ultraviolet radiation facilitates the formation of vitamin D from cholesterol. In both cases, illness results from a mismatch between modern and ancestral environments. Skin cancer is more prevalent in light-skinned people who are living in areas with higher ultraviolet radiation than was common in their ancestral environments. With little pigment in their skin to absorb the ultraviolet radiation, they incur more DNA damage, which can contribute to cancer. Similarly, rickets is common in people who have little contact with sunlight as a result of indoor living, clothing that extensively covers exposed skin, and diets that are deficient in vitamin D.

Vitamin deficiency often arises less circuitously. If a nutrient was ubiquitous in preagricultural diets, natural selection may not favor the development or maintenance of biochemical machinery for synthesizing the nutrient. Vitamin C, for example, is common in fruits, vegetables, and some animal organs such as liver. But when diets do not include such sources, the inability of humans to synthesize vitamin C leads to scurvy, a disease characterized by bleeding gums, hair follicles, and tissue under the fingernails. Most nonhuman animals can synthesize vitamin C, but human ancestors apparently lost this ability because of the ubiquitous presence of vitamin C in their food sources. In some times and places, scurvy has arisen when agricultural diets have excluded such foods (e.g., on long sea voyages).

Manifestations of disease

Physicians divide disease manifestations into subjective symptoms (e.g., feeling feverish) and measurable signs of disease (e.g., body temperature). Research within evolutionary medicine attempts to determine effects of disease manifestations on the evolutionary fitness of the ill individual and any biological agent that causes the disease. If a disease is caused by parasitism, a manifestation could increase the fitness of the host (a "defense"), the fitness of the parasite (an "exploitation"), both, or neither (a "side effect"). Diarrhea, for example, might benefit the person experiencing the diarrhea by facilitating the expulsion of the parasite that initiated the symptom. Alternatively, diarrhea might benefit the parasite by facilitating its transmission to the next host through environmental contamination. Or, diarrhea might result from a regulatory system being out of balance, benefiting neither host nor parasite.

Determining the validity of these possibilities has implications for treatment. If the manifestation is a defense, treatment of the manifestation may be sabotaging a defense and thus hinder recovery. If the manifestation is an exploitation, treatment may be beneficial because it may inhibit the exploitation and thus improve chances for recovery or hinder transmission. The answer generated for one manifestation cannot be generalized to another. Nor can the answer for one manifestation be generalized from one kind of parasitism to another. Each manifestation for each parasite needs to be assessed with controlled experiments that determine whether treatment of the manifestation hurts the host, parasite, both, or neither.

Although this framework is simple and important, little information has been gathered to evaluate the evolutionary function of disease manifestations. Available information indicates, for example, that diarrhea induced by the cholera bacterium *Vibrio cholerae* is an exploitation that facilitates its takeover of the intestinal tract. But the diarrhea induced by the bacterium *Shigella sonnei* appears to be both a defense (because inhibition of the diarrhea prolongs recovery) and an exploitation (because the watery stools facilitate transmission to new hosts). Suppression of diarrhea in cholera therefore helps the infected person. Suppression of diarrhea induced by *Shigella sonnei* harms the infected person but helps control transmission of the parasite. In this latter case, suppression of diarrhea creates an ethical dilemma—it hurts the patient but protects susceptible individuals in the population. One way out of this dilemma is to treat the patient with an antibiotic that compensates for sabotaging the defensive function of diarrhea.

Evolution of virulence

A central issue in evolutionary medicine is the evolution of virulence—that is, why some disease organisms evolve to be

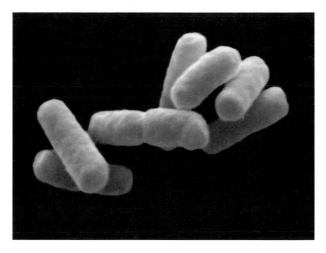

Shigella sonnei causes approximately 10,000 cases of gastroenteritis each year in the United States. Although commonly regarded as waterborne, shigellosis is also a foodborne disease, restricted primarily to higher primates, including humans. Scimat/Photo Researchers, Inc.

extremely harmful, whereas others evolve to be benign or of intermediate harmfulness. Evolutionary biologists developed an interest in this problem because medical dogma during most of the twentieth century clashed with evolutionary principles. This dogma proposed that disease organisms should always evolve to benign coexistence with their hosts, because any parasite that harms its host harms its long-term chances of persistence. Evolutionary principles, however, emphasize that natural selection acts much more strongly on differential success of alternative genetic instructions, than on long-term persistence of a population or species of parasite. Evolutionary principles dictate that the level of exploitation favored by natural selection will depend on the negative and positive effects of the different levels of exploitation on the parasite's fitness. Typically negative effects arise because severely ill hosts do not contact susceptible hosts as frequently as mildly ill hosts. The positive effects of exploitation result from having more parasites within an infected host that can be transmitted to a contacted susceptible host.

The trade-off between these positive and negative effects predicts that parasites will evolve to be highly exploitative, and hence harmful, when they can be transmitted from severely ill hosts to new hosts. Severe human-adapted pathogens conform to this generalization. Mosquito-borne pathogens such as the agents of malaria do not require host mobility for transmission and are more severe than directly transmitted pathogens. Diarrheal pathogens are harmful in proportion to the extent to which they are transmitted by contaminated water, which allows transmission from immobile hosts; this association explains why the agents of cholera, bacillary dysentery, and typhoid are unusually severe. Pathogens of the respiratory tract that are durable in the external environment are less dependent on mobile hosts for transmission than pathogens that die quickly in the external environment. Accordingly, the most severe respiratory tract pathogens, such as the smallpox virus and the tuberculosis bacterium, are more durable in the external environment than are relatively benign respiratory pathogens, such as viruses that cause the common cold.

This theory of virulence provides insight into the threat posed by emerging diseases. Influenza viruses, for example, are in general moderately durable in the external environment and moderately lethal. The influenza viruses that caused the 1918 influenza pandemic, however, were about ten times more lethal than typical influenza. But the extraordinary lethality of the 1918 viruses was first noticed near the western front of World War I, where a unique set of circumstances led to individuals immobilized by illness being transported from one cluster of susceptible hosts to another, again and again, in trenches, tents, hospitals, and trains. Sick people in ships then transported the viruses around the world. Away from the western front, the virulence gradually declined to normal levels within a year. None of the countless trillions of influenza viruses that have infected humans since that time have generated a similarly lethal epidemic, although some nasty variants have arisen in humans. Each year concerns arise over whether a terrible pandemic similar to the one in 1918 could occur. The evolutionary theory of virulence, however, suggests that it will not—unless extremely unusual conditions that permit unrestricted transmission of influenza viruses from immobile hosts once again occur.

This theory provides a basis for evaluating the risk associated with the highly virulent bird flu, caused by the H5N1. Since 1997 when it was apparent that the H5N1 virus could be transmitted directly from birds to humans and cause high lethality in infected humans, flu experts began expressing concerns that a severe H5N1 pandemic was imminent and even inevitable. These flu experts, however, were not often confused mutations and reassortment of viral genes with evolution and therefore were not applying evolutionary principles to an evolutionary problem. In contrast the evolutionary theory presented above indicated that H5N1 would not cause a devastating 1918-type pandemic. The past decade of experience with H5N1 in humans has accorded with this prediction from evolutionary medicine.

Sexually transmitted infectious diseases represent a variation on the theme. Sexual transmission requires that infected hosts be mobile enough for sexual activity and concomitant activities such as courtship. Accordingly, sexually transmitted pathogens of humans tend to be mild during their acute phases: Manifestations in adults are generally restricted to lesions or discharges and are rarely life threatening. Nevertheless, many sexually transmitted infections rank among the most lethal of human infectious diseases if effects are considered over the entire course of infection, because sexually transmitted pathogens cause persistent infections. The most sensible evolutionary explanation for this persistence is that few people in human societies change sexual partners at a rate that would allow for much sexual transmission during the week or so of contagiousness that is typical for an acute infectious disease. For sexually transmitted pathogens, success generally requires persistence within humans for months or years. This persistence requires tricks to avoid the immune system. The gonorrhea bacterium, for example, changes its antigens regularly. The Herpes simplex

virus hides out in neurons. Papillomaviruses cause infected cells to divide indefinitely, allowing the virus to replicate along with the infected cells while staying largely hidden from the immune system. Once pathogens have evolved the ability to persist, they act like wrenches in smooth-running machinery. Eventually they may cause problems, and the bigger the wrench the greater the problem for the infected person. By stimulating infected cells to divide indefinitely, human papillomaviruses, for example, push infected cells toward cancer.

Human populations differ in their potential for sexual transmission, because they differ in such variables as the rates at which new sexual partnerships occur and the kinds of contraception used. Evolutionary theory predicts that sexually transmitted pathogens should evolve to be more exploitative, and hence more damaging, when the potential for sexual transmission is high. Epidemiological comparisons accord with this prediction for all sexually transmitted pathogens that have been tested, including syphilis, the human papillomaviruses that cause cervical cancer, the human immunodeficiency viruses that cause AIDS, and the human herpes viruses that cause genital blisters.

The evolutionary theory of virulence also specifies approaches for "virulence management." Certain public health interventions may tip the competitive balance in favor of mild variants within pathogen populations, and thus favor evolutionary transitions toward reduced virulence. This control may be especially strong and stable because it makes use of mild strains that remain in the population and stimulate immune-mediated protection against more severe pathogens. Specific applications of this idea include interventions that block waterborne transmission of pathogens and the mosquito proofing of houses and hospitals.

One of the most promising applications of virulence management involves vaccine development. When eradication is not feasible, vaccines will be most effective if the pathogens that remain in the wake of the vaccination program are benign. This goal can be accomplished by developing vaccines that target more virulent variants. In practice this strategy specifies that vaccine antigens be molecules that make viable, benign organisms harmful. Mild variants left behind protect unvaccinated individuals against any remaining harmful variants.

The standard vaccine against diphtheria inadvertently accords with this strategy and generated one of the greatest success stories in the history of vaccination. The bacterium that causes diphtheria, *Corynebacterium diphtheriae*, releases a toxin that kills nearby human cells, liberating nutrients for the bacterium. To make the vaccine, this toxin was modified into a harmless toxoid. Vaccination with this toxoid generated antibodies that neutralized the toxin. In a vaccinated individual, the toxin-producing *C. diphtheriae* were therefore wasting about 5 percent of their protein budget producing a toxin that was ineffective. Vaccination therefore put the toxin-producing strains of *C. diphtheriae* at a competitive disadvantage relative to the non-toxin-producing strains. Where toxoid vaccine has been used, *C. diphtheriae* evolved from being largely a toxin-producing bacterium to one that rarely produces toxin. These

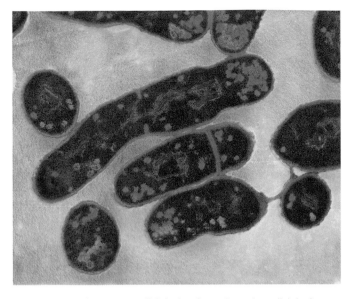

The bacterium that causes diphtheria, *Corynebacterium diphtheriae*, releases a toxin that kills nearby human cells, liberating nutrients for the bacterium. Kwangshin Kim/Photo Researchers, Inc.

mild strains act like free live vaccines, protecting people who did not receive the toxoid vaccine by generating immunity to the harmful strains on the basis of antigens shared by mild and harmful strains.

Genetic diseases and genetic dolutions to new environments

Sickle-cell anemia is a disease characterized by formation of sickle-shaped red blood cells, a low density of red blood cells (anemia), and blockage of blood vessels, resulting in tissue damage and organ failure. During the middle of the twentieth century, geneticists discovered that sickle-cell anemia resulted from a single amino acid substitution in the beta subunit of the hemoglobin molecule. Throughout most of human history, a child who inherited the substituted sickle-cell allele from both parents (i.e., a child homozygous for the sickle-cell allele) would have little chance of surviving into his or her teenage years. Survival has improved substantially in recent decades, but longevity of people with sickle-cell anemia is still cut short by many decades. In spite of the lethal price inflicted by the sickle-cell allele, it typically occurs in more than 5 percent of people who trace their ancestry to certain regions of sub-Saharan Africa, the Mediterranean, and South Asia. Approximately one-third of people living in Nigeria harbor the sickle-cell allele. Such high frequencies are maintained over time because most people who harbor the allele are heterozygous—that is, they have one copy of the sickle-cell allele and one copy of a beta-chain allele that does not have the sickle-cell substitution. Still, one-fourth of children whose parents are heterozygous will inherit two copies of the sickle-cell allele and will consequently develop sickle-cell anemia. If the sickle-cell allele did not confer some compensating advantage, the destructiveness of sickle-cell anemia would cause a steady decline in its prevalence. The

compensating advantage is resistance to one of the most damaging infectious diseases to have attacked humans: falciparum malaria. Malaria mortality is reduced by about 95 percent among people who are heterozygous for the sickle-cell allele.

The sickle-cell allele is thus a quick and dirty evolutionary solution to a grave problem. It would be better for humans if a more refined, less damaging solution could have arisen as a defense against malaria. But natural selection acts on what is available. Once a sickle-cell allele arises by mutation, it will increase in the population so long as the survival it confers against malaria fosters its contribution to the next generation more than the reduced survival from sickle-cell anemia inhibits this contribution. When the prevalence of the sickle-cell allele is low, the number of people with sickle-cell anemia is almost nonexistent. But as the frequency of the sickle-cell allele rises, the number of people with sickle-cell anemia rises disproportionately until the negative effects from sickle-cell anemia balance the positive effects from protection against malaria.

A similar argument applies to some other genetic diseases. In glucose-6-phosphate dehydrogenase (G6PD) deficiency, a mutation in G6PD restricts transfer of electrons to glutathione. Electron-rich glutathione is a critical nutrient for the malaria organism, but it is also used to neutralize toxins. G6PD deficiency thus protects against malaria but makes a person more sensitive to toxins. Accordingly, G6PD deficiency is common in places where falciparum malaria has been common, particularly the Mediterranean and sub-Saharan Africa.

Several other genetic diseases are probably also maintained at elevated frequencies in particular human populations because they provide protection from infectious diseases in particular regions. But the known examples of such situations

Sickle cell anemia is an inherited blood disease. In normal red blood cells hemoglobin carries oxygen from the lungs to organs. The sickle cell is on the left, the healthy cell is on the right. © Dr. Stanley Flegler/ Visuals Unlimited/Corbis.

comprise only a small portion of severe chronic diseases. Alleles that cause disease by directly damaging their hosts will tend to be rare in human populations whenever the damage strongly curtails the transmission of the alleles into future generations. The loss of alleles can be offset by mutations that regenerate these alleles, but mutations alone can maintain damaging alleles only at a very low frequency.

Duchenne muscular dystrophy provides a sense of the upper limit to the commonness of a severe genetic disease that is not associated with a compensating benefit. This disease is characterized by wasting muscles and is almost always lethal prior to sexual maturity. It is attributable to mutations in the gene that codes for dystrophin, a critical component of the internal structure of muscle fibers. Mutations can generate Duchenne muscular dystrophy at an unusually high rate because the dystrophin gene is the longest in the human genome, about 2.5 million base pairs in length. The scope for mutation in this gene is therefore about 10 to 100 times greater than in most other human genes. Yet even Duchenne muscular dystrophy occurs in only about one in 7,000 people.

Known genetic diseases with compensating benefits are characterized by particular patterns of geographic distributions and inheritance. Some chronic diseases of uncertain cause, such as hemochromatosis and Tay-Sachs, do have such patterns and are probably maintained at elevated frequency as a result of some as yet unidentified compensating benefit in the heterozygous state. But most important diseases of uncertain cause—including heart disease, strokes, most cancers, diabetes, schizophrenia, and Alzheimer's disease—do not have such patterns. As illustrated above for atherosclerosis, these diseases will be best explained by integrating knowledge about inherited alleles and environmental risk factors with consideration of infectious processes.

Rate of human evolution

The human genetic endowment is a mishmash of alleles. Some adapted our ancestors to hunting and gathering lifestyles. Others have been spreading to different degrees in different human populations as these populations have adapted to agricultural conditions. One of the best accepted examples is lactose tolerance among adults. Mutations that extended the ability to digest lactose throughout an individual's life span instead of just during childhood were favored in agricultural societies in which milk-producing domestic animals such as cows, goats, and sheep were kept. For such individuals, domestic animals could now be a source of a nutritious, secreted food in addition to meat, fabric, and fertilizer. The spread of this genetic ability suggests that this new food source provided an important competitive advantage.

Studies conducted during the first decade of the twenty-first century indicate that many human alleles have been spreading in human populations in recent millennia, with different alleles spreading in different human populations. Characteristics altered by these alleles are still being worked out, but they seem to be related to three general areas: disease resistance, diet, and neuronal functioning. Disease-resistance alleles may provide compensating benefits associated with genetic diseases as mentioned above, but they also include alleles that encode proteins of the immune system. Dietary alleles may be related to the aspects of agricultural diets that differ from those of hunter-gatherers, such as the availability of milk products as mentioned above. Neuronal alleles presumably are associated with differences between hunter-gatherer lifestyles and agricultural lifestyles in behavior and thought processes most useful for the corresponding environments.

These new discoveries are calling into question the long-held dogma that modern society has slowed down human evolution by removing humans from the selective pressures of nature. To the contrary, the emerging view is that human evolution not only has been continuing but is speeding up. Cultural changes alter the human environment, which in turn alters the human genetic makeup through natural selection. This insight bears on human health because new adaptations are generally not finely tuned. For natural selection to favor the spread of a new allele, the allele has to have only a net positive effect on fitness—it can have negative effects so long as they are more than offset by positive effects. The positive effects may entail improved abilities of certain individuals to function in certain modern settings. The negative effects may often manifest themselves as illnesses in these individuals or their relatives. Over time such alleles can be fine-tuned to reduce the negative effects. Whether this fine-tuning will ever happen for particular alleles is questionable, however, because human environments are changing at such a rapid rate that even the elevated rate of human evolution may not keep up with them. A realistic view of the future, therefore, is that humans will be ever changing but never equilibrating.

Resources

Books

Dieckmann, Ulf, Johan A. J. Metz, Maurice W. Sabelis, and Karl Sigmund, eds. 2002. *Adaptive Dynamics of Infectious Diseases: In Pursuit of Virulence Managements*. Cambridge, UK: Cambridge University Press.

Ewald, Paul W. 1994. *Evolution of Infectious Disease*. Oxford: Oxford University Press.

Ewald, Paul W. 2000. *Plague Time: How Stealth Infections Cause Cancers, Heart Disease, and Other Deadly Ailments*. New York: Free Press.

Frank, Steven A. 2002. *Immunology and Evolution of Infectious Disease*. Princeton, NJ: Princeton University Press.

Nesse, Randolph M., and George C. Williams. 1994. *Evolution and Healing: The New Science of Darwinian Medicine*. London: Weidenfeld and Nicolson.

Stearns, Stephen C., and Jacob C. Koella. 2008. *Evolution in Health and Disease*. 2nd edition. Oxford: Oxford University Press.

Trevathan, Wenda R., E. O. Smith, and James J. McKenna, eds. 1999. *Evolutionary Medicine*. New York: Oxford University Press.

Trevathan, Wenda R., E. O. Smith, and James J. McKenna, eds. 2008. *Evolutionary Medicine and Health: New Perspectives.* New York: Oxford University Press.

Periodicals

Billing, Jennifer, and Paul W. Sherman. 1998. "Antimicrobial Functions of Spices: Why Some Like It Hot." *Quarterly Review of Biology* 73(1): 3–49.

Eaton, S. Boyd, and Melvin Konner. 1985. "Paleolithic Nutrition: A Consideration of Its Nature and Current Implications." *New England Journal of Medicine* 312(5): 283–289.

Ewald, Paul W. 1980. "Evolutionary Biology and the Treatment of Signs and Symptoms of Infectious Disease." *Journal of Theoretical Biology* 86(1): 169–176.

Ewald, Paul W. 1991. "Waterborne Transmission and the Evolution of Virulence among Gastrointestinal Bacteria." *Epidemiology and Infection* 106(1): 83–119.

Ewald, Paul W., and Gregory M. Cochran. 2000. "*Chlamydia pneumoniae* and Cardiovascular Disease: An Evolutionary Perspective on Infectious Causation and Antibiotic Treatment." *Journal of Infectious Diseases* 181(supp. 3): S394–S401.

Gérard, Herve C., E. Fomicheva, Judith A. Whittum-Hudson, and Alan P. Hudson. 2008. "Apolipoprotein E4 enhances attachment of *Chlamydophila* (*Chlamydia*) *pneumoniae* elementary bodies to host cells." *Microbial Pathogenesis* 44: 279–285.

Hawks, John, Eric T. Wang, Gregory M. Cochran, et al. 2007. "Recent Acceleration of Human Adaptive Evolution." *Proceedings of the National Academy of Sciences of the United States of America* 104(52): 20753–20758.

Howard, B. V., L. Van Horn, J. Hsia, et al. 2006. "Low-fat dietary pattern and risk of cardiovascular disease: The Women's Health Initiative Randomized Controlled Dietary Modification Trial." *Journal of the American Medical Association* 295: 655–666.

Ledgerwood, Levi G., Paul W. Ewald, and Gregory M. Cochran. 2003. "Genes, Germs, and Schizophrenia: An Evolutionary Perspective." *Perspectives in Biology and Medicine* 46(3): 317–348.

Neel, James V. 1962. "Diabetes Mellitus: A 'Thrifty' Genotype Rendered Detrimental by Progress?" *American Journal of Human Genetics* 14(4): 353–362.

Williams, George C., and Randolph M. Nesse. 1991. "The Dawn of Darwinian Medicine." *Quarterly Review of Biology* 66(1): 1–22.

Paul W. Ewald

Evolution and biodiversity conservation

Conservation of the world's biological diversity is a topic of increasing concern. Around the globe, biological communities that took millions of years to develop, including tropical rain forests, coral reefs, temperate old-growth forests, and prairies, are being devastated by human actions (R. Primack 2006). Loss of habitat equates to the prominent loss of plant and animal species. As populations of plants and animals shrink, the genetic pool becomes smaller and their ability to adapt is reduced. With the limited gene pool of a small population, organisms are less capable of adjusting to a changing environment. They are also more susceptible to the harmful genetic effects of small populations, and they will lose evolutionary flexibility. Understanding these processes in the context of evolution will help facilitate conservation of these populations.

Conservation biology is an integrated science that has developed in response to many scientists' desire to preserve species and the ecosystems in which they live. The protection of biological diversity is central to conservation biology. Conservation biologists use the term *biological diversity* (or simply *biodiversity*) to mean the complete range of species, biological communities, and ecosystem interactions, and the genetic variation within species.

This definition of biological diversity must be considered on three levels:

1. *Species diversity*: all species on Earth, including single celled organisms such as bacteria and protists as well as the species of the multicellular kingdoms (plants, fungi, and animals). Species diversity represents the entire range of evolutionary and ecological adaptations of species to particular environments.

2. *Genetic diversity*: the genetic variation within species, both among geographically separate populations and among individuals within single populations. Genetic diversity allows species to maintain reproductive vitality, resistance to disease, and the ability to adapt to changing conditions. There may be enough variation among species that separate subspecies or varieties can be distinguished and given distinct names.

3. *Ecosystem diversity*: the different biological communities and their associations with the chemical and physical environment. Ecosystem diversity represents the collective response of species to different environmental conditions.

If species, ecosystems, and populations are adapted to local environmental conditions, why are they facing extinction? Should they not tend to persist in the same place over time? Why are they not able to adapt or evolve to a changing environment? The answers to these questions have a single, simple answer: Massive disturbances caused by people have altered, degraded, and destroyed the landscape on a vast scale, driving species and even whole communities to the point of extinction. The major threats to biological diversity that result from human activity are habitat destruction, habitat fragmentation, habitat degradation (including pollution), global climate change, the overexploitation of species for human use, human–wildlife conflict, the invasion of exotic species, the increased spread of disease, and synergisms among these factors. Most threatened species and ecosystems face at least two or more of these problems, which speed the way to extinction and hinder efforts to protect them (MEA 2005; see Figure1). In most cases these changes are occurring too rapidly for species to undergo evolutionary adaptations. As a consequence, the rate of extinction of existing species is more than one hundred times greater than the evolution of new species (R. Primack 2006). The net result is an overall loss of species, caused by human activities.

These eight threats to biological diversity are all caused by an expanding human population's ever-increasing use of the world's natural resources. The destruction of biological communities has been greatest since 1850. Between that year and 2008, the human population exploded from 1 billion to 6.6 billion. The world population will reach an estimated 10 billion by the year 2050. People use natural resources, such as fuelwood, wild meat, and wild plants, and convert vast amounts of natural habitat to agricultural and residential land. Some degree of resource use is inevitable. Therefore the increase in the human population is partially responsible for the loss of biological diversity. All else being equal, more people means greater human impact and less biodiversity. The increasing use of natural resources by individuals, particularly in developed countries, further damages biological diversity.

Habitat destruction

The primary cause of the loss of biological diversity is not direct human exploitation or malevolence. It is the habitat destruction that inevitably results from the expansion of human populations and human activities. For the remainder

The new Brazilian national highway BR163, criticized by conservationists for its environmental impact, is seen winding around a stretch of the Amazon rain forest near the town of Novo Progresso in Para State. Image by Ricky Rogers.

of the twenty-first century, land-use change will continue to be the main factor affecting biodiversity in terrestrial ecosystems. This will probably be followed by overexploitation, climate change, and the introduction of invasive species (IUCN 2007). Because the major threat to biological diversity is loss of habitat, the most important means of protecting biological diversity is habitat conservation.

In many areas of the world, particularly on islands and in locations where human population density is high, most original habitat has been significantly altered or even destroyed (MEA 2005). Habitat destruction directly affects the size of resident populations. Once a population has been reduced in size, it can often be difficult for it to persist, particularly within a changed landscape. Small populations often have reduced genetic variation and are prone to inbreeding depression that can lead to a further decline in population size.

Habitat fragmentation

In addition to outright destruction, habitats that formerly occupied wide, unbroken areas are now often divided into pieces by roads, fields, towns, and a broad range of other human constructs. Habitat fragmentation is the process whereby a

large, continuous area of habitat is both reduced in area and divided into two or more fragments (see Figure 2). These fragments are often isolated from one another by a highly modified or degraded landscape, and their edges experience an altered set of environmental conditions, referred to as the edge effect. Fragmentation almost always occurs together with a severe reduction in habitat area. It can also occur, however, if the habitat area is minimally reduced when roads, railroads, canals, power lines, fences, or other barriers that prevent animals from moving freely divide it. The island model of biogeography is sometimes applied to such landscapes: The habitat fragments resemble islands in an inhospitable, human-dominated sea.

Habitat fragments differ from the original habitat in three important ways: (1) fragments have a greater amount of edge for the area of habitat (and thus a greater exposure to the edge effect); (2) the center of each habitat fragment is closer to an edge; and (3) a formerly continuous habitat with large populations is divided into pieces with smaller populations. A fragmented habitat can obstruct the movement of individuals, change the size of local populations, and hinder genetic flow within the species. Smaller populations with less genetic diversity are then more likely to go extinct, as described below.

Environmental degradation and pollution

Even when a habitat is unaffected by obvious destruction, alteration, or fragmentation, the ecosystems and species in that habitat can be significantly affected by human activities. The most subtle and universal form of environmental degradation is pollution, commonly caused by pesticides, sewage, fertilizers from agricultural fields, industrial chemicals and wastes, emissions from factories and automobiles, and sediment deposits from eroded hillsides. The general effects of pollution on water quality, air quality, and even the global climate are cause for great concern. Although environmental pollution is sometimes highly visible and dramatic, as in the case of a massive oil spill, it is the subtle, unseen forms of pollution that are probably the most threatening—primarily because they are so dangerous. Chemicals are widely dispersed in the air and water and can harm plants, animals, and people living far from where the chemicals are actually applied. A feature of some chemicals is their ability to persist in the environment for decades after their initial introduction. Populations of aquatic and terrestrial vertebrates directly exposed to pollution may experience immediate declines when in contact with the pollution source and secondary declines from lingering chemicals that can harm their reproductive systems. This can lead to disorders such as sterility and genetic mutations.

Global climate change

Global levels of carbon dioxide (CO_2), methane, and other trace gases have been steadily increasing, primarily as a result of burning fossil fuels—coal, oil, and natural gas (IPCC 2007). Clearing forests to create farmland and burning firewood for heating and cooking, especially in the tropics, also contribute to rising concentrations of CO_2. Carbon dioxide concentration in the atmosphere increased from 290 parts per million in 1900 to 384 parts per million in 2008, and it is projected to double at some point in the latter half of this century.

There is broad scientific agreement by the Intergovernmental Panel on Climate Change (IPCC), a study group of leading scientists organized by the United Nations, that the increased levels of greenhouse gases produced by human activity have affected the world's climate and ecosystems already and that these effects will increase in the future. An extensive review of the evidence supports the conclusion that global surface temperatures have increased by 0.6°C during the last century (IPCC 2007), and ocean water temperatures have also increased by an average of 0.06°C over the last fifty years.

Global climate change has the potential to drastically change biological communities and alter the ranges of many

Bridges (or sometimes tunnels) are constructed to enable terrestrial animals to cross man-made barriers, such as roads. Image by Alan Sirulnikoff.

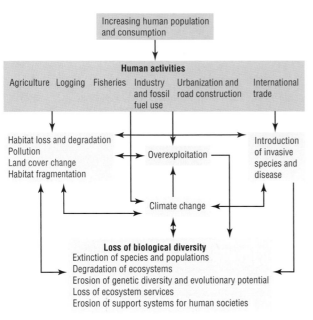

Increasing human population and consumption

Human activities

Agriculture Logging Fisheries Industry and fossil fuel use Urbanization and road construction International trade

Habitat loss and degradation
Pollution
Land cover change
Habitat fragmentation

Overexploitation

Introduction of invasive species and disease

Climate change

Loss of biological diversity
Extinction of species and populations
Degradation of ecosystems
Erosion of genetic diversity and evolutionary potential
Loss of ecosystem services
Erosion of support systems for human societies

SOURCE: *Essentials of Conservation Biology*, Fourth Edition, Figure 9.1.

Figure 1. Effect of human activities on biological diversity. Reproduced by permission of Gale, a part of Cengage Learning.

species. The speed of this change could overwhelm the natural dispersal abilities of species. There is mounting evidence that this process has already begun, with changes in the distribution of bird, insect, and plant species, and reproduction occurring earlier in the spring (D. Primack et al. 2004; Cleland et al. 2007). Warming conditions in the ocean are already affecting the distribution of species in coastal waters. Because the implications of global climate change are so far-reaching, biological communities, ecosystem functions, and climate need to be carefully monitored over the coming decades. In particular, higher temperatures and changing precipitation patterns could lead to crop failures and tree death over large areas, with enormous social, economic, ecological, and political costs.

It is likely that, as the climate changes, many existing protected areas will no longer preserve the rare and endangered species that currently live there. The conditions in such areas may change so quickly that species will not be able to survive at the same place or evolve fast enough to deal with the new conditions (Miller-Rushing and Primack 2004). New conservation areas need to be established now to protect sites that will be suitable for these species in the future, such as sites with large elevational gradients (Hannah et al. 2007). As conditions warm, species will be able to migrate upslope and remain in the same climate. Potential future migration routes, such as north–south river valleys, need to be identified and established. If species are in danger of going extinct in the wild because of global climate change, the last remaining individuals may have to be kept in captivity and later established in new, suitable locations. However, the amount of space available to do this is small, and the potential is limited (IPCC 2007).

Overexploitation

In the early twenty-first century, overexploitation by humans was estimated to threaten about one-third of the endangered mammals and birds (IUCN 2007). People have always hunted and harvested the food and other resources they need to survive. As long as human populations were small and the methods of collection unsophisticated, people could sustainably harvest and hunt the plants and animals in their environment. As human populations have increased, however, humankind's use of the environment has escalated and the methods of harvesting have become dramatically more efficient. Technological advances mean that even in the developing world, guns are used instead of blowpipes, spears, or arrows for hunting in the tropical rain forests and savannahs. Sometimes the problem here is not so much the presence of guns, but the lack of adequate institutional infrastructure to regulate hunting, as occurs successfully in developed countries such as the United States and Canada. Nevertheless, formerly remote forests are now being subdivided by logging roads, which provide access for hunters. Powerful motorized fishing boats and enormous "factory ships" harvest fish from the world's oceans and sell them on the global market.

This notice, which depicts a wolf, informs the public that the area ahead is closed. This is because it is a wildlife corridor to enable terrestrial animals to cross barriers such as roads. Image by Alan Sirulnikoff.

As many overexploited species become rare, it will no longer be commercially viable to harvest them, and their numbers will have a chance to recover. Unfortunately, populations of many species, such as species of rhinoceros and certain wild cats, already may have been reduced so severely that they will require vigilant conservation efforts to recover. National parks, nature reserves, and other protected areas need to be established to conserve the habitat of such overharvested species. When harvesting can be reduced or stopped by the enforcement of international regulations, such as the Convention on International Trade in Endangered Species of Wild Fauna and Flora (CITES), and comparable national regulations, species may be able to recover. Sea otters, elephants, and certain whale species provide hopeful examples of species that have recovered once overexploitation was stopped. Other species may be so reduced in numbers that genetic problems and alterations of their social structure prevent recovery. Intensive harvesting over many generations can also lead to genetic changes in a species; for example, repeated harvesting can lead to fish populations that breed at a younger age and a smaller size. In effect, humans have become the major drivers of evolution for such exploited species.

Invasive species

Exotic species are species that occur outside their natural ranges. In the past, species remained in certain places because they were unable to cross natural barriers such as vast oceans and mountain ranges. But now people transport them deliberately and accidentally beyond their natural range. The great majority of these exotic species do not become established in the places in which they are introduced because the new environment is not suitable to their needs. Yet, a certain percentage of species do establish themselves in their new homes, and many of these are considered invasive species—that is, they increase in abundance at the expense of native species.

Invasive species can impact their surroundings in a number of ways. They may displace native species through competition for limiting resources, they may prey upon native species to the point of extinction, or they may alter the habitat so that natives are no longer able to live there. Invasive exotic species represent threats to 49 percent of the endangered species in the United States, with particularly severe impacts on bird and plant species (Wilcove et al. 1998). In areas of human settlement, introduced domestic cats may be one of the most serious predators of birds and small mammals.

The isolation of island habitats has allowed the evolution of a unique grouping of endemic species, but it also leaves those species particularly vulnerable to predation by invading species. In many cases, animals introduced onto islands have efficiently preyed upon endemic animal species and have grazed down native plant species to the point of extinction. Such endemic species have not evolved with these mainland animals, and are not able to evolve the new adaptations that they need to survive.

Another special class of invasive species is introduced species that have close relatives in the native biota. When invasive species hybridize with the native species and varieties, unique genotypes may be eliminated from local populations. Such hybrids may be better adapted to the host habitat than either of the original species. In this situation the boundaries between species becomes obscured, a situation sometimes called genetic swamping. This appears to be the fate of native trout species that have come in contact with commercial species. In the U.S. Southwest, the Apache trout (*Oncorhynchus apache*) has had its range reduced by habitat destruction and competition with introduced species. The species has also hybridized extensively with rainbow trout (*O. mykiss*), an introduced sport fish, blurring its identity as a distinct species.

Invasive species are considered to be the most serious threat facing the biota of the U.S. national park system. Whereas the effects of habitat degradation, fragmentation, and pollution potentially can be corrected and reversed in a matter of years or decades as long as the original species are present, well-established exotic species may be impossible to remove from communities. They may have built up such large numbers, becoming so widely dispersed and thoroughly integrated into the community, that eliminating them may be extraordinarily difficult and expensive.

Disease

Another major threat to species and biological communities is the increased transmission of disease. This is a result of human activities (such as habitat destruction, which may increase disease-carrying vectors) and interactions with humans (such as populations of wild animals that acquire diseases from nearby populations of domestic animals and people). The increased mobility of people and the spread of exotic species have contributed to the spread of disease. Infections by disease organisms are common in both wild and captive populations. They can reduce the size and density of vulnerable populations. Disease organisms, such as bacteria, viruses, fungi, and protists, can also have a major impact on the structure of an entire biological community.

Such diseases may be the single greatest threat to some rare species. The decline and loss of numerous frog populations from pristine montane habitats in Australia, North America, and Central America is apparently due in part to the introduction of exotic fungal diseases. The decline of these species has been so rapid and complete that species have been unable to evolve disease resistance (R. Primack 2006). The last population of black-footed ferrets (*Mustela nigripes*) known to exist in the wild was destroyed by plague and canine distemper virus in 1987. One of the main challenges of managing the captive breeding program for black-footed ferrets has been protecting the captives from canine distemper, human viruses, and other disease. This is being done through quarantine measures and division of the captive colony into geographically separate groups. The captive population is now sufficiently large that individuals have been used to establish new populations in the wild.

Diseases can sometimes spread between related species when they are in close contact. For example, certain primates, such as gorillas and chimpanzees, have acquired diseases through contact with humans. Likewise, primate diseases,

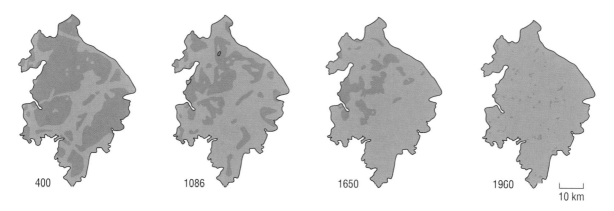

SOURCE: *Essentials of Conservation Biology,* Fourth Edition, Figure 9.11.

Figure 2. The forested areas of Warwickshire, England (shown in green), have been fragmented and reduced in area by paths, roads, agriculture, and human settlements for over seventeen centuries. Reproduced by permission of Gale, a part of Cengage Learning.

such as HIV and Ebola, have moved into the human population (R. Primack 2006).

The problems of small populations

The threats described so far all have the potential to significantly reduce population sizes either directly or indirectly and affect their ability to survive and adapt genetically. Once reduced in numbers, smaller populations carry with them a host of unique issues, most of which lead to further decline. The problems with small populations are so significant that it is often a good idea to prevent species from reaching this condition.

Exceptions notwithstanding, large populations are needed to protect most species. Small populations are subject to rapid decline in numbers and local extinction for three main reasons: (1) loss of genetic variability and related problems of inbreeding depression and genetic drift; (2) demographic fluctuations due to random variations in birth and death rates; and (3) environmental fluctuations caused by variation in predation, competition, disease, and food supply and by natural catastrophes that occur at irregular intervals, such as fires, storms, and droughts.

Genetic variability

Genetic variability is important in allowing populations to adapt to a changing environment. Individuals with certain alleles or combinations of alleles may have just the characteristics needed to survive and reproduce under new conditions (Willi, Van Buskirk, and Hoffmann 2006). Within a population, particular alleles may vary in frequency from common to very rare. In small populations, allele frequencies may change from one generation to the next simply because of chance, depending on which individuals mate and produce offspring. This process is known as genetic drift. It is a separate process from natural selection in which the population changes in response to specific factors in the environment.

When an allele occurs at a low frequency in a small population, it has a high probability of being lost in each generation because of chance. Considering the theoretical case of an isolated population in which there are two alleles per gene, Sewall Wright (1931) proposed a formula to express the proportion of original heterozygosity (H; individuals possessing two different allele forms of the same gene) remaining after one generation for a population of breeding adults (N_e): $H = 1 - 1/[2N_e]$ According to this equation, a population of fifty individuals would have 99 percent of its original heterozygosity after one generation due to the loss of rare alleles, and it would still have 90 percent after ten generations. But a population of ten individuals would have only 95 percent of its original heterozygosity after one generation and only 60 percent after ten generations (see Figure 3).

This formula demonstrates that significant losses of genetic variability can occur in isolated small populations, particularly those on islands and in fragmented landscapes. However, the migration of individuals between populations and the regular mutation of genes tend to increase the amount of genetic variability within the population and balance the effects of genetic drift. The mutation rates found in nature—about 1 in 10,000 to 1 in 1,000,000 per gene per generation—may make up for the random loss of alleles in large populations, but they are insignificant in countering genetic drift in small populations of one hundred individuals or less. Fortunately, however, even a low frequency of movement of individuals between populations minimizes the loss of genetic variability associated with small population size (Wang 2004). If even one or two immigrants arrive each generation in an isolated population of about one hundred individuals, the impact of genetic drift will be greatly reduced. But when a small population becomes isolated because of habitat fragmentation, genetic variation will continue to decline. Such a loss of genetic variation also occurs in captive populations in zoos and aquariums but can to some extent be countered through planned breeding, in which the genetic contributions of founder individuals are equalized to the greatest extent possible.

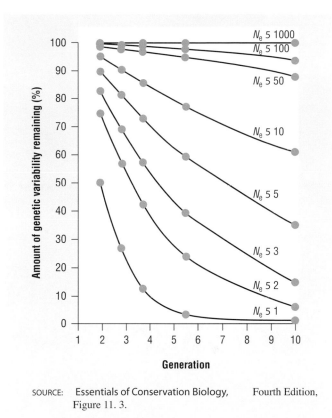

Figure 3. Genetic variability is lost randomly over time through genetic drift. This graph shows the average percentage of genetic variability remaining over ten generations in theoretical populations of various effective population sizes. Reproduced by permission of Gale, a part of Cengage Learning.

Small populations have greater susceptibility to a number of harmful genetic effects such as inbreeding depression and loss of evolutionary flexibility. These factors may contribute to a decline in population size, leading to an even greater loss of genetic variability in subsequent generations and a greater probability of extinction (Frankham 2005).

A variety of mechanisms encourages individuals to mate with unrelated individuals of the same species often leading to healthier offspring (a phenomenon known as hybrid vigor) and prevent inbreeding (mating among close relatives) in most natural populations. In large populations of most animal species, individuals do not normally mate with close relatives. They often disperse from their place of birth or are restrained from mating with relatives by behavioral inhibitions, unique individual odors, or other sensory cues. Mating in animals among parents and their offspring, siblings, and cousins may result in inbreeding depression, a condition that is characterized by higher mortality of offspring, fewer offspring, or offspring that are weak, sterile, or have low mating success (Willi, Van Buskirk, and Hoffmann 2006). The most plausible explanation for inbreeding depression is that it allows for the expression of harmful recessive alleles inherited from both parents. Inbreeding depression can be a severe problem in small captive populations in zoos and domestic breeding

programs. Inbreeding depression results in even fewer individuals in the next generation, which in turn leads to more pronounced inbreeding and further inbreeding depression.

Rare alleles and unusual combinations of alleles that confer no immediate advantages may be uniquely suited for a future set of environmental conditions. Loss of genetic variability in a small population may limit its ability to respond to new conditions and long-term changes in the environment, such as pollution, new diseases, and global climate change (Willi, Van Buskirk, and Hoffmann 2006). A small population is less likely than a large population to possess the genetic variation necessary for adaptation to long-term environmental changes and so will be more likely to go extinct.

Demographic fluctuations

In a stable environment, a population would tend to increase until it reached the carrying capacity of the environment, at which point the average birth rate per individual would equal the average death rate and there would be no net change in population size. Once the population size drops below about fifty individuals, however, individual variation in birth and death rates begins to cause the population size to fluctuate randomly up or down (Jacquemyn et al. 2007). If the population size fluctuates downward in any one year because of a higher than average number of deaths or a lower than average number of births, the resulting smaller population will be even more susceptible to demographic fluctuations in subsequent years. Random size fluctuations upward are eventually bounded by the carrying capacity of the environment, and the population may fluctuate downward again. Consequently, once a population decreases because of habitat destruction and fragmentation, demographic variation, also known as demographic stochasticity, becomes important. Thus, the population has a higher probability of declining more, and even going extinct, because of chance alone (that is, a year with low reproduction and high mortality).

In many animal species, small populations may be unstable because the social structure is unable to function once the population falls below a certain size; this is known as the Allee effect (Angulo et al. 2007). Herds of grazing mammals and flocks of birds may be unable to find food and defend themselves against attack when numbers fall below a certain level. Many animal species that live in widely dispersed populations, such as bears and spiders, may be unable to find mates once the population density drops below a certain point. In this case, the average birth rate will decline, making the population density even lower and worsening the problem. This combination of random fluctuations in demographic characteristics, disruption of social behavior, and decreased population density contributes to instabilities in population size, which often lead to local extinction.

Environmental variation and catastrophes

Random variation in the biological and physical environment, known as environmental stochasticity, can also cause variation in the population size of a species. For example, the

Around the globe, biological communities that took millions of years to develop, including tropical rain forests, coral reefs, temperate old-growth forests, and prairies, are being devastated by human actions. Image by Martin Harvey.

population of an endangered rabbit species might be affected by fluctuations in the abundance of its food plants and the number of its predators. Variation in the physical environment might also strongly influence the rabbit population—rainfall during an average year might encourage plant growth and allow the population to increase, whereas dry years might limit plant growth and cause rabbits to starve.

The most extreme form of environmental stochasticity is natural catastrophes that occur at unpredictable intervals, such as droughts, storms, volcanic eruptions, earthquakes, fires, and even El Niño events. Such catastrophes can kill part of a population or even eliminate an entire population from an area. For a wide range of vertebrates, the frequency of catastrophes in which most of the animals die is around 15 percent per generation (Reed et al. 2003). Even though the probability of a natural catastrophe in any one year is low, over the course of decades and centuries, natural catastrophes have a high likelihood of occurring.

In the light of evolution

This entry has described how genetic changes allow species to adapt to a changing environment and how humans affect

this process. An understanding of evolution is important in order for modern-day conservationists to address the above issues. It aids them in determining conservation priorities based on such factors as the proper identification of species and subspecies, the risks of extinction of different types of species, and the importance of genetic variability in a population. In the context of evolution, this knowledge leads to better-informed decisions within conservation practice and wildlife management.

In some cases, a species is separated into subspecies or varieties. The ability to distinguish between several subspecies has implications for conservation efforts. For example, if a rare subspecies is protected, should a more common subspecies of the same species be protected as well? One way to approach this is to verify how distinct a subspecies is from the other subspecies through advanced genetic analysis techniques using DNA. It is possible that the common subspecies provides evidence of historical evolutionary divergence and may merit conservation attention of its own. It is also possible that the rare subspecies is, in fact, not distinct from the common subspecies and therefore its protection is not of critical importance. Such information leads to more informed decisions on conservation strategies.

A farmer clears brush after cutting trees in this rain forest area in Kalimantan, Indonesia. Image by Charles O'Rear.

Specialist species, those organisms whose environmental requirements are very specific or narrow in scope, are often more vulnerable to extinction. Once their habitat is compromised it can be difficult for them to find alternative sources of food and shelter, for example. This is contrary to the needs of a generalist species that is able to live under a wider range of conditions. If the habitat of a generalist is damaged, it is more likely it will be able to survive in the altered environment or a part of the habitat that is not damaged. When considering the conservation needs of species, it is important to take into account their ability to adapt to changes to their environment. Typically, those species that have generalist requirements are more likely to survive human impacts than those species that have more narrow ecological requirements.

The amount of genetic variability in a population has direct implications for the success of the population. As discussed above, a loss of genetic variability leaves the population more susceptible to the deleterious effects of changes to its habitat. From an evolutionary standpoint, a limited genetic pool also makes it more difficult for the population to evolve. In a population with greater genetic variability it is more likely that a group of individuals possessing the traits necessary for survival exists. Conservation measures that account for genetic variability, through such steps as facilitating movement among isolated populations, or by making it easier for populations to

intermingle across habitats, will experience greater success than those that do not. In some cases in which populations are declining because of low genetic variability and inbreeding depression, introducing new individuals from other populations can restore the genetic health of the population and allow the population to grow again.

Conclusion

Threats to biodiversity affect the ability of species and populations to survive in an altered environment. These threats make species more vulnerable to environmental variation and demographic variation and lead to the loss of genetic variability. As a result, a decline in population size caused by one factor will increase the vulnerability of the population to the other two factors (see Figure 4). The smaller a population becomes, the more vulnerable it is to further demographic variation, environmental variation, and genetic factors that tend to lower reproduction and increase mortality rates. These impacts on reproduction and mortality, in turn, reduce population size even more and drive the population to extinction. This tendency of small populations to decline toward extinction has been likened to a vortex, a whirling mass of gas or liquid spiraling inward—the closer an object gets to the center, the faster it moves toward extinction. At the center

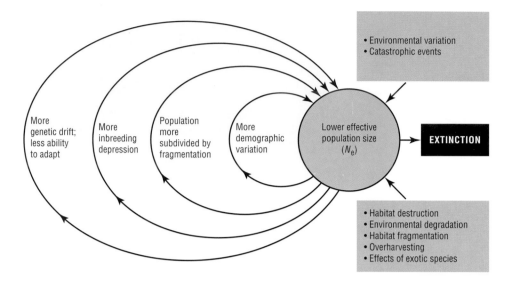

SOURCE: *Essentials of Conservation Biology*, Fourth Edition, Figure 11.16.

Figure 4. Extinction vortices progressively diminish population size, leading to local extinctions of species. Reproduced by permission of Gale, a part of Cengage Learning.

of an extinction vortex is oblivion: the local extinction of the species. Once caught in such a vortex, it is difficult for a species to resist the pull toward extinction.

When a population has declined to a small size, it will probably go extinct unless unusual and highly favorable conditions allow the population size to increase. Such populations require a careful program of population and habitat management to reduce demographic and environmental variation and loss of genetic variability.

Habitat destruction, fragmentation, and degradation, climate change, overexploitation, invasive species, and disease are all cause for immediate concern in the struggle to preserve the world's biodiversity. These anthropogenic factors have led to declines in the populations of many species. With less genetic variability, smaller populations are less able to grow, evolve, and endure further environmental change. Understanding the dynamics of these evolutionary processes enables conservationists to protect wildlife to a greater degree.

Resources

Books

Groom, Martha J., Gary K. Meffe, and C. Ronald Carroll. 2006. *Principles of Conservation Biology.* 3rd edition. Sunderland, MA: Sinauer Associates.

Intergovernmental Panel on Climate Change (IPCC). 2007. *Climate Change 2007: The Physical Science Basis; Contribution of Working Group I to the Fourth Assessment Report of the Intergovernmental Panel on Climate Change.* Edited by Susan Solomon, Dahe Qin, Martin Manning, et al. Cambridge, UK: Cambridge University Press.

Meffe, Gary K., and C. Ronald Carroll. 1997. *Principles of Conservation Biology.* 2nd edition. Sunderland, MA: Sinauer Associates.

Millennium Ecosystem Assessment (MEA). 2005. *Ecosystems and Human Well-Being.* 5 vols. Washington, DC: Island Press.

Primack, Richard B. 2006. *Essentials of Conservation Biology.* 4th edition. Sunderland, MA: Sinauer Associates.

Wilcove, David S., Charles H. McLellan, and Andrew P. Dobson. 1986. "Habitat Fragmentation in the Temperate Zone." In *Conservation Biology: The Science of Scarcity and Diversity*, ed. Michael E. Soulé. Sunderland, MA: Sinauer Associates.

Periodicals

Angulo, Elena, Gary W. Roemer, Ludêk Berec, et al. 2007. "Double Allee Effects and Extinction in the Island Fox." *Conservation Biology* 21(4): 1082–1091.

Cleland, Elsa E., Isabelle Chuine, Annette Menzel, et al. 2007. "Shifting Plant Phenology in Response to Global Change." *Trends in Ecology and Evolution* 22(7): 357–365.

Duncan, Jeffrey R., and Julie L. Lockwood. 2001. "Extinction in a Field of Bullets: A Search for Causes in the Decline of the World's Freshwater Fishes." *Biological Conservation* 102(1): 97–105.

Frankham, Richard. 2005. "Genetics and Extinction." *Biological Conservation* 126(2): 131–140.

Hannah, Lee, Guy Midgely, Sandy Andelman, et al. 2007. "Protected Area Needs in a Changing Climate." *Frontiers in Ecology and the Environment* 5(3): 131–138.

Jacquemyn, Hans, Katrien Vandepitte, Rein Brys, et al. 2007. "Fitness Variation and Genetic Diversity in Small, Remnant Populations of the Food Deceptive Orchid *Orchis purpurea*." *Biological Conservation* 139(1–2): 203–210.

Miller-Rushing, Abraham J., and Richard B. Primack. 2004. "Climate Change and Plant Conservation." *Plant Talk*, no. 35: 34–38.

O'Brien, Stephen J., and Ernst Mayr. 1991. "Bureaucratic Mischief: Recognizing Endangered Species and Subspecies." *Science* 251(4998): 1187–1188.

Primack, Daniel, Carolyn Imbres, Richard B. Primack, et al. 2004. "Herbarium Specimens Demonstrate Earlier Flowering Times in Response to Warming in Boston." *American Journal of Botany* 91(8): 1260–1264.

Reed, David H., Edwin H. Lowe, David A. Briscoe, and Richard Frankham. 2003. "Fitness and Adaptation in a Novel Environment: Effect of Inbreeding, Prior Environment, and Lineage." *Evolution* 57(8): 1822–1828.

Wang, Jinliang. 2004. "Application of the One-Migrant-per-Generation Rule to Conservation and Management." *Conservation Biology* 18(2): 332–343.

Wilcove, David S., David Rothstein, Jason Dubow, et al. 1998. "Quantifying Threats to Imperiled Species in the United States." *BioScience* 48(8): 607–615.

Willi, Yvonne, Josh Van Buskirk, and Ary A. Hoffmann. 2006. "Limits to the Adaptive Potential of Small Populations." *Annual Review of Ecology, Evolution, and Systematics* 37: 433–458.

Wright, Sewall. 1931. "Evolution in Mendelian Populations." *Genetics* 16(2): 97–159.

Other

International Union for Conservation of Nature and Natural Resources (IUCN). 2007. "2007 IUCN Red List of Threatened Species." Available from http://www.iucnredlist.org.

Richard B. Primack
Elizabeth E. Bacon

Further reading

Alcock, John. 2009. *Animal Behavior: An Evolutionary Approach*. 9th ed. Sunderland, MA: Sinauer Associates, Inc.

Andersson, Malte. 1994. *Sexual Selection*. Princeton, NJ: Princeton University Press.

Arnold, Michael L. 1997. *Natural Hybridization and Evolution*. Oxford, UK: Oxford University Press.

Avise, John C. 2000. *Phylogeography: The History and Formation of Species*. Cambridge, MA: Harvard University Press.

Avise, John C. 2004. *Molecular Markers, Natural History, and Evolution*. 2nd ed. Sunderland, MA: Sinauer Associates, Inc.

Bakker, Robert T. 1986. *The Dinosaur Heresies*. New York, NY: William Morrow and Company.

Beebe, Trevor J.C., and Graham Rowe. 2004. *An Introduction to Molecular Ecology*. Oxford, UK: Oxford University Press.

Bell, Michael A., Douglas J. Futuyma, Walter F. Eanes, and Jeffrey S. Levinton. 2010. *Evolution Since Darwin: The First 150 Years*. Sunderland, MA: Sinaeur Associates, Inc.

Benton, Michael J. 2005. *Vertebrate Paleontology*. 3rd ed. Oxford, UK: Blackwell Publishing.

Bowler, Peter J. 2003. *Evolution: The History of an Idea*. 3rd ed. Berkeley, CA: University of California Press.

Bradbury, Jack. W., and Sandra L. Vehrencamp. 1998. *Principles of Animal Communication*. Sunderland, MA: Sinauer Associates, Inc.

Brooks, Daniel R., and Deborah A. McLennan. 2002. *The Nature of Diversity: An Evolutionary Voyage of Discovery*. Chicago, IL: The University of Chicago Press.

Browne, Janet. 1996. *Charles Darwin: A Biography*. Vol. 1 of *Voyaging*. Princeton, NJ: Princeton University Press.

Browne, Janet. 2002. *Charles Darwin: The Power of Place*. Vol. 2 of *A Biography*. Princeton, NJ: Princeton University Press.

Carlquist, Sherwin J. 1974. *Island Biology*. New York, NY: Columbia University Press.

Carroll, Robert L. 1988. *Vertebrate Paleontology and Evolution*. New York, NY: WH Freeman and Company.

Carroll, Sean B., Jennifer Grenier, and Scott Weatherbee. 2005. *From DNA to Diversity: Molecular Genetics and the Evolution of Animal Design*. 2nd ed. Oxford, UK: Blackwell Publishing.

Carroll, Sean B. 2005. *Endless Forms Most Beautiful: The New Science of Evo Devo and the Making of the Animal Kingdom*. New York, NY: Norton.

Carroll, Sean B. 2006. *The Making of the Fittest: DNA and the Ultimate Forensic Record of Evolution*. New York, NY: Norton.

Charnov, Eric L. 1982. *The Theory of Sex Allocation*. Princeton, NJ: Princeton University Press.

Chiappe, Luis M. 2007. *Glorified Dinosaurs: The Origin and Early Evolution of Birds*. Hoboken, NJ: John Wiley & Sons, Inc.

Colbert, Edwin H. 1968. *Men and Dinosaurs*. New York, NY: E.P. Dutton and Company.

Coppinger, Raymond, and Lorna Coppinger. 2001. *Dogs*. New York, NY: Scribner.

Coyne, Jerry A., and H. Allen Orr. 2004. *Speciation*. Sunderland, MA: Sinauer Associates, Inc.

Coyne, Jerry A. 2009. *Why Evolution Is True*. New York, NY: Viking.

Darwin, Charles. 2004. *The Voyage of the Beagle*. Intro. by David Quammen. Washington, D.C.: National Geographic Society. (Orig. pub. 1839 as *Journal of Researches into the Geology and Natural History of the Various Countries Visited by* H.M.S. *Beagle*.)

Darwin, Charles. 1859. *On the Origin of Species by Means of Natural Selection or the Preservation of Favored Races in the Struggle for Life*. London, UK: J. Murray.

Darwin, Charles. 1871. *The Descent of Man, and Selection in Relation to Sex*. London, UK: J. Murray.

Davidson, Eric H. 2001. *Genomic Regulatory Systems: Development and Evolution*. San Diego, CA: Academic Press.

Dawkins, Richard. 1989. *The Selfish Gene*. 2nd ed. Oxford, UK: Oxford University Press.

Dawkins, Richard. 1996. *The Blind Watchmaker: Why the Evidence of Evolution Reveals a Universe Without Design*. With a new introduction. New York, NY: Norton. (Orig. pub. 1986.)

Dawkins, Richard. 1999. *The Extended Phenotype: The Long Reach of the Gene*. With a new afterward by Daniel Dennett. Oxford, UK: Oxford University Press. (Orig. Pub. 1982.)

Desmond, Adrian J. 1975. *The Hot-Blooded Dinosaurs: A Revolution in Paleontology*. New York, NY: The Dial Press/ James Wade.

Desmond, Adrian, and James Moore. 1992. *Darwin: The Life of A Tormented Evolutionist.* New York, NY: Little, Brown, and Company.

Diamond, Jared. 1992. *The Third Chimpanzee: The Evolution and Future of the Human Animal.* New York, NY: HarperCollins Publishers, Inc.

Dingus, Lowell, and Rowe, Timothy 1998. *The Mistaken Extinction: Dinosaur Evolution and the Origin of Birds.* New York, NY: W.H. Freeman and Company.

Dobzhansky, Theodosius. 1951. *Genetics and the Origin of Species.* 3rd ed. New York, NY: Columbia University Press.

Felsenstein, Joseph. 2004. *Inferring Phylogenies.* Sunderland, MA: Sinauer Associates, Inc.

Endler, John A. 1986. *Natural Selection in the Wild.* Princeton, NJ: Princeton University Press.

Endler, John A. 1977. *Geographic Variation, Speciation, and Clines.* Princeton, NJ: Princeton University Press.

Falconer, Douglas S., and Trudy F.C. Mackay. 1996. *Introduction to Quantitative Genetics.* 4th ed. Essex, England: Pearson Education Ltd.

Fisher, Ronald A. 1958. *The Genetical Theory of Natural Selection.* 2nd rev. ed. New York, NY: Dover Publications Inc.

Fortey, Richard. 1999. *Life: A Natural History of the First Four Billion Years of Life on Earth.* New York, NY: Vintage Books.

Fox, Charles W., and Jason B. Wolf, eds. 2006. *Evolutionary Genetics: Concepts and Case Studies.* Oxford, UK: Oxford University Press.

Futuyma, Douglas J. 1998. *Evolutionary Biology.* 3rd ed. Sunderland, MA: Sinauer Associates, Inc.

Futuyma, Douglas J. 2005. *Evolution.* Sunderland, MA: Sinauer Associates, Inc.

Laurie Garrett. 1994. *The Coming Plague: Newly Emerging Diseases in A World Out of Balance.* New York, NY: Penguin Books.

Gavrilets, Sergey. 2004. *Fitness Landscapes and the Origin of Species.* Princeton, NJ: Princeton University Press.

Geist, Valerius. 1998. *Deer of the World.* Mechanicsburg, PA: Stackpole Books.

Gibson, Greg, and Spencer V. Muse. 2009. *A Primer of Genome Science.* 3rd ed. Sunderland, MA: Sinauer Associates, Inc.

Gillespie, John H. 1991. *The Causes of Molecular Evolution.* Oxford, UK: Oxford University Press.

Gillespie, John H. 2004. *Population Genetics: A Concise Guide.* 2nd ed. Baltimore, MD: The Johns Hopkins University Press.

Goldschmidt, Richard. 1940. *The Material Basis of Evolution.* New Haven, CT: Yale University Press.

Gotelli, Nicholas J. 2008. *A Primer of Ecology.* 4th ed. Sunderland, MA: Sinauer Associates, Inc.

Gould, Stephen J. 1977. *Ontogeny and Phylogeny.* Cambridge, MA: The Belknap Press of Harvard University Press.

Gould, Stephen J. 1987. *Time's Arrow: Myth and Metaphor in the Discovery of Geological Time.* Cambridge, MA: Harvard University Press.

Gould, Stephen J. 1989. *Wonderful Life: The Burgess Shale and the Nature of History.* New York, NY: W.W. Norton & Company, Inc.

Gould, Stephen J. 2002. *The Structure of Evolutionary Theory.* Cambridge, MA: The Belknap Press of Harvard University Press.

Gould, Stephen J. 2006. *The Richness of Life: The Essential Stephen Jay Gould.* Paul McGarr, and Steven Rose eds. New York, NY: W.W. Norton & Company, Inc.

Grant, Peter R. 1999. *Ecology and Evolution of Darwin's Finches.* Revised ed. Princeton, NJ: Princeton University Press.

Grant, Rosemary B., and Peter R. Grant. 1989. *Evolutionary Dynamics of a Natural Population: The Large Cactus Finch of the Galápagos.* Chicago, IL: University of Chicago Press.

Grant, Peter R., and B. Rosemary Grant. 2008. *How and Why Species Multiply: The Radiation of Darwin's Finches.* Princeton, NJ: Princeton University Press.

Grant, Verne. 1974. *Plant Speciation.* New York, NY: Columbia University Press.

Graur, Dan, and Wen-Hsiung Li. 2000. *Fundamentals of Molecular Evolution.* 2nd ed. Sunderland, MA: Sinauer Associates, Inc.

Greene, Harry W. 2000. *Snakes: The Evolution of Mystery in Nature.* Berkely, CA: University of California Press.

Hall, Barry G. 2008. *Phylogenetic Trees Made Easy: A How-to Manual.* 3rd ed. Sunderland, MA: Sinauer Associates, Inc.

Harrison, Richard G. 1993. *Hybrid Zones and the Evolutionary Process.* Oxford, UK: Oxford University Press.

Hartl, Daniel L., and Andrew G. Clark. 2007. *Principles of Population Genetics.* 4th ed. Sunderland, MA: Sinauer Associates, Inc.

Harvey, Paul H., and Mark D. Pagel. 1991. *The Comparative Method In Evolutionary Biology.* Oxford, UK: Oxford University Press.

Herrera, Carlos M., and Olle Pellmyr. 2002. *Plant-Animal Interactions: an evolutionary approach.* Oxford, UK: Blackwell Science Ltd.

Hölldobler, Bert, and Edward O. Wilson. 1990. *The Ants.* Cambridge, MA: The Belknap Press of Harvard University Press.

Hölldobler, Bert, and Edward O. Wilson. 1994. *Journey to the Ants: A Story of Scientific Exploration.* Cambridge, MA: The Belknap Press of Harvard University Press.

Hölldobler, Bert, and Edward O. Wilson. 2009. *The Superorganism: The Beauty, Elegance, and Strangeness of Insect Societies.* New York, NY: W.W. Norton & Company, Inc.

Horner, John R., and Dobb, Edwin. 1997. *Dinosaur Lives: Unearthing an Evolutionary Saga.* New York, NY: Harper/Collins.

Horner, John R., and James Gorman. 2009. *How to Build a Dinosaur: Extinction Doesn't Have to be Forever.* New York, NY: Dutton.

Howard, Daniel J., and Stewart H. Berlocher. 1998. *Endless Forms: Species and Speciation.* Oxford, UK: Oxford University Press.

Hughes, Austin L. 1999. *Adaptive Evolution of Genes and Genomes.* Oxford, UK: Oxford University Press.

Jablonka, Eva, and Marion J. Lamb. 2005. *Evolution in Four Dimensions: Genetic, Epigenetic, Behavioral, and Symbolic Variation in the History of Life.* Cambridge, MA: The MIT Press.

Johanson, Donald, and Edey, Maitland. 1981. *Lucy: The Beginnings of Mankind*. New York, NY: Simon and Schuster.

Johanson, Donald, and Blake Edgar. 2006. *From Lucy to Language: Revised, Updated, and Expanded*. New York, NY: Simon & Schuster.

Kielan-Jaworowska, Zofia, Richard L. Cifelli, and Zhe-Xi Luo. 2004. *Mammals From The Age of Dinosaurs: Origins, Evolution, and Structure*. New York, NY: Columbia University Press.

Kimura, Motoo. 1985. *The Neutral Theory of Molecular Evolution*. Cambridge, UK: Cambridge University Press.

Kitcher, Philip. 1987. *Vaulting Ambition: Sociobiology and the Quest for Human Nature*. Cambridge, MA: The MIT Press.

Knoll, Andrew H. 2003. *Life On a Young Planet: The First Three Billion Years of Evolution on Earth*. Princeton, NJ: Princeton University Press.

Kurten, Bjorn. 1971. *The Age of Mammals*. New York, NY: Columbia University Press.

Leaky, Richard. 1994. *The Origin of Humankind*. New York, NY: Basic Books.

Levinton, Jeffrey S. 2001. *Genetics, Paleontology, and Macroevolution*. Cambridge, UK: Cambridge University Press.

Lewin, Roger. 2005. *Human Evolution: An Illustrated Introduction*. 5th ed. Oxford, UK: Blackwell Publishing Ltd.

Lewontin, Richard C. 1974. *The Genetic Basis of Evolutionary Change*. New York, NY: Columbia University Press.

Li, Wen-Hsiung. 2006. *Molecular Evolution*. Sunderland, MA: Sinauer Associates, Inc.

Lomolino, Mark V., Brett R. Riddle, and James H. Brown. 2006. *Biogeography*. 3rd ed. Sunderland, MA: Sinauer Associates, Inc.

Losos, Jonathan B. 2009. *Lizards in an Evolutionary Tree: Ecology and Adaptive Radiation of Anoles*. Berkeley, CA: University of California Press.

Losos, Jonathan B., and Robert E. Ricklefs. eds. 2010. *The Theory of Island Biogeography Revisited*. Princeton, NJ: Princeton University Press.

Lovtrup, Soren. 1974. *Epigenetics: A Treatise on Theoretical Biology*. London: John Wiley and Sons.

Lovtrup, Soren. 1987. *Darwinism: The Refutation of a Myth*. London: Croom Helm.

Lynch, Michael, and Bruce Walsh. 1998. *Genetics and Analysis of Quantitative Traits*. Sunderland, MA: Sinauer Associates, Inc.

MacArthur, Robert H., and Edward O. Wilson. 2001. *The Theory of Island Biogeography*. Preface by Edward O. Wilson. Princeton, NJ: Princeton University Press.

Majerus, Michael E.N. 1998. *Melanism: Evolution in Action*. Oxford, UK: Oxford University Press.

Malthus, Thomas R. 1798. *An Essay on the Principle of Population*. London, UK: J. Johnson.

Margulis, Lynn 1984. *Early Life: How Cells First Evolved*. Boston, MA: Jones and Bartlett Publishers.

Mayr, Ernst. 1942. *Systematics and the Origin of Species*. New York, NY: Columbia University Press.

Mayr, Ernst. 1963. *Animal Species and Evolution*. Cambridge, MA: Belknap Press.

Mayr, Ernst. 2001. *What Evolution is*. New York, NY: Basic Books.

Mindell, David P. 2006. *The Evolving World: Evolution in Everyday Life*. Cambridge, MA: Harvard University Press.

Molles, Manuel C., Jr. 2005. *Ecology: Concepts and Applications*. 3rd ed. New York, NY: McGraw-Hill.

Naskrecki, Piotr. 2005. *The Smaller Majority: The Hidden World of the Animals that Dominate the Tropics*. Cambridge, MA: The Belknap Press of Harvard University Press.

Ohno, Susumu. 1970. *Evolution by Gene Duplication*. Berlin, Germany: Springer-Verlag.

Paige, Roderick D.M., and Edward C. Holmes. 1998. *Molecular Evolution: A Phylogenetic Approach*. Oxford, UK: Blackwell Publishing.

Palumbi, Stephen R. 2001. *The Evolution Explosion: How Humans Cause Rapid Evolutionary Change*. New York, NY: W.W. Norton & Company, Inc.

Pechenik, Jan A. 2005. *Biology of the Invertebrates*. 5th ed. New York, NY: McGraw-Hill.

Pianka, E.R. 2000. *Evolutionary Ecology*. 6th ed. San Francisco, CA: Benjamin-Addison-Wesley-Longman.

Pianka, Eric R., and Laurie J. Vitt. 2003. *Lizards: Windows to the Evolution of Diversity*. Berkeley, CA: University of California Press.

Pigliucci, Massimo. 2002. *Denying Evolution: Creationism, Scientism, and the Nature of Science*. Sunderland, MA: Sinauer Associates, Inc.

Pinna, Giovanni. 1985. *The Illustrated Encyclopedia of Fossils*. New York, NY: Facts on File.

Pough, F. Harvey, Robin M. Andrews, Martha L. Crump, Alan H. Savitsky, and Kentwood D. Wells. 2004. *Herpetology*. 3rd ed. Upper Saddle River, NJ: Pearson Prentice Hall.

Price, Trevor. 2008. *Speciation in Birds*. Greenwood Village, CO: Roberts & Company.

Prothero, Donald R. 2007. *Evolution: What the Fossils Say and Why it Matters*. New York, NY: Columbia University Press.

Provine, William B. 2001. *The Origins of Theoretical Population Genetics*. Chicago, IL: University of Chicago Press.

Quammen, David. 1988. *The Flight of the Iguana: A Sidelong View of Science and Nature*. New York, NY: Touchstone.

Quammen, David. 1996. *The Song of the Dodo: Island Biogeography in an Age of Extinction*. New York, NY: Touchstone.

Quammen, David. 2006. *The Reluctant Mr. Darwin: An Intimate Portrait of Charles Darwin and the Making of His Theory of Evolution*. New York, NY: W.W. Norton & Company.

Rensch, Bernhard. 1960. *Evolution Above the Species Level*. New York, NY: Columbia University Press.

Ricklefs, Robert E. 2008. *The Economy of Nature*. 6th ed. New York, NY: W.H. Freeman and Company.

Ricklefs, Robert E., and Dolph Schluter eds. 1994. *Species Diversity in Ecological Communities: Historical and Geographical Perspectives*. Chicago, IL: University of Chicago Press.

Rickleffs, Robert E., and Gary Miller. 1999. *Ecology*. 4th ed. New York, NY: W.H. Freeman and Company.

Ridley, Mark. 2004. *Evolution*. 6th ed. Oxford, UK: Blackwell Science Ltd.

Ridley, Matt. 1993. *The Red Queen: Sex and the Evolution of Human Nature.* New York, NY: Penguin Books.

Roff, Derek A. 1997. *Evolutionary Quantitative Genetics.* New York, NY: Chapman & Hall.

Roff, Derek A. 2001. *Life History Evolution.* Sunderland, MA: Sinauer Associates, Inc.

Romer, Alfred S. 1966. *Vertebrate Paleontology.* 3rd ed. Chicago, IL: University of Chicago Press.

Rose, Michael R., and George V. Lauder, eds. 1996. *Adaptation.* San Diego, CA: Academic Press.

Roughgarden, Jonathan. 1995. *Anolis Lizards of the Caribbean: Ecology, Evolution, and Plate Tectonics.* Oxford, UK: Oxford University Press.

Roughgarden, Joan. 2004. *Evolution's Rainbow: Diversity, Gender, and Sexuality in Nature and People.* Berkeley, CA: University of California Press.

Sagan, Carl. 1980. *Cosmos.* New York, NY: Ballantine Books.

Sagan, Carl, and Ann Druyan. 1996. *The Demon-Haunted World: Science as a Candle in the Dark.* New York, NY: Ballantine Books.

Sanderson, Michael J., and Larry Hufford. 1996. *Homoplasy: The Recurrence of Similarity in Evolution.* San Diego, CA: Academic Press.

Sax, Dov F., John J. Stachowicz, and Steven D. Gaines. 2005. *Species Invasions: Insights into Ecology, Evolution, and Biogeography.* Sunderland, MA: Sinauer Associates, Inc.

Schluter, Dolph. 2000. *The Ecology of Adaptive Radiation.* Oxford, UK; Oxford University Press.

Scott, Eugenie C. 2004. *Evolution Vs. Creationism: An Introduction.* Berkeley, CA: University of California Press.

Shubin, Neil. 2008. *Your Inner Fish: A Journey into the 3.5-billion-year History of the Human Body.* New York, NY: Pantheon Books.

Simpson, George G. 1944. *Tempo and Mode in Evolution.* New York, NY: Columbia University Press.

Simpson, George G. 1953. *The Major Features of Evolution.* New York, NY: Columbia University Press.

Smith, John Maynard. 1978. *The Evolution of Sex.* Cambridge, UK: Cambridge University Press.

Smith, John Maynard, and Eörs Szathmáry. 1995. *The Major Transitions in Evolution.* Oxford, UK: Oxford University Press.

Sober, Elliott. 1984. *The Nature of Selection: Evolutionary Theory in Philosophical Focus.* Chicago, IL: The University of Chicago Press.

Stanley, Steven M. 2004. *Earth System History.* 2nd ed. New York, NY: W.H. Freeman and Company.

Stearns, Stephen C. 1992. *Evolution of Life Histories.* Oxford, UK: Oxford University Press.

Stebbins, G. Ledyard. 1950. *Variation and Evolution in Plants.* New York, NY: Columbia University Press.

Taylor, P.D., and G.P. Larwood, eds. 1991. *Major Evolutionary Radiations.* Oxford, UK: Oxford University Press.

Templeton, Alan R. 2006. *Population Genetics and Microevolutionary Theory.* Hoboken, NJ: John Wiley & Sons, Inc.

Thompson, John N. 1994. *The Coevolutionary Process.* Chicago, IL: The University of Chicago Press.

Thompson, John N. 2005. *The Geographic Mosaic of Coevolution.* Chicago, IL: The University of Chicago Press.

Trevathan, Wenda R., E.O. Smith, and James J. McKenna, eds. 1999. *Evolutionary Medicine.* Oxford, UK: Oxford University Press.

Valentine, James W. 2004. *On the origin of phyla.* Chicago, IL: The University of Chicago Press.

Waddington, C.H. 1975. *The Evolution of an Evolutionist.* Ithaca, NY: Cornell University Press.

Wakeley, John. 2008. *Coalescent Theory: An Introduction.* Greenwood Village, CO: Roberts & Company.

Wallace, Alfred R. 1869. *The Malay Archipelago.* London, UK: MacMillan & Company.

Wallace, Alfred R. 1881. *Island Life or the Phenomena and Causes of Insular Faunas and Floras.* New York, NY: Harper & Brothers.

Weinberg, Samantha. 2000. *A Fish Caught in Time: The Search for the Coelacanth.* New York, NY: Harper-Collins Publishers Inc.

Weiner, Jonathan. 1994. *The Beak of the Finch: A Story of Evolution in Our Time.* New York, NY: Vintage Books.

Whittaker, Robert J., and José María Fernández-Palacios. 2007. *Island Biogeography: Ecology, Evolution, and Conservation.* 2nd ed. Oxford, UK: Oxford University Press.

Whitlock, Michael C., and Dolph Schluter. 2008. *The Analysis of Biological Data.* Greenwood Village, CO: Roberts & Company.

Wilkins, Adam S. 2002. *The Evolution of Developmental Pathways.* Sunderland, MA: Sinauer Associates, Inc.

Williams, George C. 1966. *Adaptation and Natural Selection: A Critique of Some Current Evolutionary Thought.* Princeton NJ: Princeton University Press.

Wilson, Edward O. 1992. *The Diversity of Life.* Cambridge, MA: Belknap Press of Harvard University Press.

Wilson, Edward O. 1994. *Naturalist.* Washington, D.C.: Island Press.

Wilson, Edward O. 2000. *Sociobiology: The New Synthesis.* 25th anniversary ed. Cambridge, MA: Belknap Press of Harvard University Press.

Winston, Judith E. 1999. *Describing Species: Practical Taxonomic Procedure for Biologists.* New York, NY: Columbia University Press.

Zimmer, Carl. 1998. *At The Water's Edge: Fish with Fingers, Whales with Legs, and How Life Came Ashore but then Went Back to Sea.* New York, NY: Touchstone.

Zimmer, Carl. 2001. *Evolution: Triumph of an Idea.* New York, NY: HarperCollins.

Organizations

American Museum of Natural History
  Central Park West at 79th Street
  New York, NY 10024-5192 USA
  Phone: (212) 769-5100
  http://www.amnh.org

American Society of Naturalists
  http://www.amnat.org

Arizona-Sonora Desert Museum
  2021 North Kinney
  Tucson, AZ 85743 USA
  Phone: (520) 883-2702
  Email: info@desertmuseum.org
  http://www.desertmuseum.org

Center for Biological Diversity
  P.O. Box 710
  Tuscon, AZ 85702-0710
  Phone: (866) 357-3349
  Fax: (520) 623-9797
  Email: center@biologicaldiversity.org
  http://biologicaldiversity.org

Charles Darwin Foundation
  Puerto Ayora, Santa Cruz Island
  Galapagos, Ecuador
  Phone: 593-5-2526-146/147
  Fax: 593-5-2526-146/147, Ext 102
  Email: cdrs@fcdarwin.org.ec
  http://www.darwinfoundation.org

Encyclopedia of Life
  http://www.eol.org

The Field Museum
  1400 S. Lake Shore Drive
  Chicago, IL 60605-2496 USA
  Phone: (312) 922-9410
  http://www.fieldmuseum.org

Harvard Museum of Natural History
  26 Oxford Street
  Cambridge, MA 02138 USA
  Phone: (617) 495-3045
  Email: hmnh@oeb.harvard.edu
  http://www.hmnh.harvard.edu

International Union for the Conservation of Nature
  Rue Mauverney 28
  Gland 1196 Switzerland
  Phone: +41 22-999-0000
  Fax: +41 22-999-0002
  http://www.iucn.org

Institute of Human Origins
  Arizona State University
  Social Sciences Blg. Room 103
  951 South Cady Mall
  Tempe, AZ 85287-4101 USA
  Phone: (480) 727-6580
  Fax: (480) 727-6570
  Email: iho@asu.edu
  http://iho.asu.edu
  http://www.becominghuman.org

Museum of Comparative Zoology
  Harvard University
  26 Oxford Street
  Cambridge, MA 02138 USA
  Phone: (617) 495-2460
  Fax: (617) 495-5667
  http://www.mcz.harvard.edu

Museum of Vertebrate Zoology
  University of California, Berkeley
  3101 Valley Life Sciences Building
  Berkeley, CA 94720-3160 USA
  Phone: (510) 642-3567
  Fax: (510) 643-8238
  http://mvz.berkeley.edu

National Center for Ecological Analysis and Synthesis
  735 State Street, Suite 300
  Santa Barbara, CA 93101 USA
  Phone: (805) 892-2500
  Fax: (805) 892-2510
  Email: nceas@nceas.ucsb.edu
  http://www.nceas.ucsb.edu

National Center for Science Education
  420 40th Street Suite 2
  Oakland, CA 94609-2688
  Phone: (800) 290-6006
  Fax: (510) 601-7204
  Email: info@ncse.com
  http://ncse.com

National Evolutionary Synthesis Center
2024 W. Main Street, Suite A200
Durham, NC 27705-4667 USA
Phone: (919) 668-4551
Fax: (919) 668-9198
Email: info@nescent.org
http://www.nescent.org

National Geographic Society
P.O. Box 98199
Washington, D.C. 20090-8199 USA
Phone: (800) 647-5463
Email: askngs@nationalgeographic.com
http://www.nationalgeogrpahic.com

National Institutes of Health
9000 Rockville Pike
Bethesda, MD 20892 USA
Phone: (301) 496-4000
Email: NIHinfo@od.nih.gov
http://www.nih.gov

National Science Foundation
4201 Wilson Boulevard
Arlington, VA 22230 USA
Phone: (703) 292-5111
Email: info@nsf.gov
http://www.nsf.gov

Natural History Museum
Cromwell Road
London SW7 5BD, UK
Phone: +44 0-20-7942-5000
http://www.nhm.ac.uk

Natural History Museum
University of Kansas
1345 Jayhawk Blvd.
Lawrence, KA 66045-7593 USA
Phone: (785) 864-4450
Email: naturalhistory@ku.edu
http://naturalhistory.ku.edu

Organization for Tropical Studies
North American Office
P.O. Box 90630
Durham, NC 27708-0630
Phone: (919) 684-5774
Fax: (919) 684-5661
Email: ots@duke.edu
http://www.ots.ac.cr

The Royal Society
6-9 Carlton House Terrace
London SW1Y 5AG
Phone: +44 0-20-7451-2500
http://royalsociety.org

School for Field Studies
10 Federal Street
Salem, MA 01970-3876 USA
Phone: (800) 989-4418
Fax: (978) 741-3551
http://www.fieldstudies.org

Smithsonian Institution
P.O. Box 37012 Smithsonian Inst.
Washington, D.C. 20013-7012 USA
Email: info@si.edu
Phone: (202) 633-1000
http://www.mnh.si.edu

Society for the Study of Evolution
Email: sse-manager@evolutionsociety.org
http://www.evolutionsociety.org

Tree of Life Web Project
http://tolweb.org

World Wildlife Fund
P.O. Box 97180
Washington, D.C. 20090-7180
Phone: (202) 293-4800
http://www.worldwildlife.org

Glossary

Abiotic—In ecology, refers to variables of the physical environment, including temperature, rainfall, etc.

Acclimation—The short-term physiological adjustment by an individual to change in a particular environmental factor, such as temperature.

Adaptation—Population level genetic and phenotypic change, due to natural selection, that results in the increased ability of a population to live in a particular environment.

Adaptive landscape—In evolutionary biology, a graphical representation of the relationship between population mean fitness and relative reproductive success.

Adaptive radiation—The diversification of a species in a single lineage into a variety of different adaptive forms, usually differing in resource or habitat use. Often, diversification is over a short period of geological time.

Allele—One of several forms of the same gene, differing by mutation to the DNA sequence.

Allopatric—Pertaining to a population or a species that occupies a geographic region different from that of another population or species.

Allopatric speciation—Speciation that occurs between geographically separated populations. Separation prevents gene flow long enough for reproductive isolation to evolve.

Altruistic—An action that confers benefit to other individuals at a cost to the benefactor.

Anagenesis—Evolutionary change of a feature over time within a single lineage.

Aposematic—Features that advertise toxicity or noxious properties in prey to predators; warning signal.

Asexual reproduction—Reproduction that does not require meiosis, fertilization, or the union of gametes.

Assortative mating—Nonrandom mating on the basis of phenotype; mating with other individuals of similar phenotype.

Batesian mimicry—In this type of mimicry, the mimic is palatable but suffers lower predation due to its resemblance to an unpalatable model.

Biogeography—The study of geographic distribution of organisms.

Biological species concept—Under this concept, a species is a group of individuals that interbreed or could potentially interbreed, and which are reproductively isolated from other such groups.

Biomechanics—The study and application of mechanical principles to living organisms.

Biotic—In ecology, refers to the biological aspect of a community, including competitors, predators, primary producers etc.

Bottleneck—A severe, temporary reduction in population size.

Carrying capacity—The population density that can be sustained by limited resources in the environment.

Character displacement—The phenomenon in which two closely related species exhibit larger difference in a trait value in sympatry compared to allopatry; usually thought to result from evolution to reduce competition for similar, limited resources.

Chromosome—A concentrated, tightly packed complex of DNA and protein.

Cladogenesis—An evolutionary splitting event, usually visualized on a cladogram.

Coevolution—When two or more ecologically interacting species evolve together in response to selection imposed on one by the other(s).

Commensalism—An ecological relationship between species in which one is benefited but the other is little affected.

Community—In ecology, an assemblage of populations of two or more species in the same geographical area.

Competition—An interaction between individuals of the same or different species, whereby resources used by one are made unavailable to others.

Continental drift—The movement of continental plates, floating on the earth's molten core.

Convergent evolution—Independent evolution of similar features in different evolutionary lineages, usually from different developmental pathways.

Crypsis—The state of being camouflaged against detection.

Deme—A local, panmictic population.

Density dependent—In population biology, when a parameter describing a population changes according to population density.

Diploid—A cell or organism possessing two copies of each chromosome.

Directional selection—Selection for the value of a character away from its current mean.

Dispersal—The movement of individual organisms to different localities.

Disruptive (diversifying) selection—Selection that favors two or more phenotypes while disfavoring intermediates.

Ecological equivalence—The idea that different species or clades may play the same ecological role in a community and therefore be interchangeable.

Ecological niche—The range of environmental and biotic variables under which a species or population can exist.

Ecological speciation—Speciation that follows from adaptation to different ecological settings; ecological differentiation amongst incipient species.

Ecological release—The expansion of a population's niche when competition, predation, or parasitism is lessened or removed, often as a result of a species invading a new location.

Ecology—The study of the abundance and distribution of organisms.

Ectotherm—An organism that controls body temperature through external means; dependent on the environment for heat sources.

Endosymbiont—An organism that lives within the body or cells of another organism in a symbiotic relationship.

Endotherm—An organism that controls body temperature through internal metabolism.

Epistasis—When the effect of an interaction between two or more genes is different than the sum of effects for each gene taken separately.

Epigenetics—The study of changes in phenotype or gene expression not caused directly by changes in DNA sequence.

Eukaryote—Any organism whose cells contain complex structures enclosed within membranes, especially the nucleus.

Eutherian mammal—The placental mammals.

Evolution—Change in a population through time; in population genetics, the change in gene frequencies in a population.

Exaptation—The co-option of a trait for a function different than the function for which that trait originally evolved.

Exploitative competition—A form of competition in which individuals of one species use a resource, thereby depleting or making unavailable the resource for another species.

Faunal/Floral exchange—In biogeography, the interchange of species between two previously separated biotic zones that meet because of geological processes such as the formation of an isthmus, the opening of a seaway, or the collision of continents.

Fecundity—The quantity of gametes produced by an individual.

Fitness—A measure of an individual's reproductive success; typically thought of as the total number of alleles or genes contributed to the next or subsequent generations.

Fitness landscape—In evolutionary biology, the graphical representation of the relationship between individual phenotype and reproductive success, where genotype is on the x-axis and reproductive success is on the y-axis.

Fixation—In population genetic terms, when an allele reaches a frequency of 1 in a population.

Food web—A graphical depiction of how energy moves through trophic levels in a community.

Founder effect—A drastic change in allelic frequencies in a founder population compared to the source population due to the small number of individuals that dispersed.

Founder event—The start of a new population from a few individuals through dispersal.

Gamete—A cell that fuses with another cell during fertilization in sexually reproducing organisms; an egg or sperm.

Gene—A section of DNA sequence that serves as the functional unit of heredity. Traditionally thought to encode a protein but this paradigm is breaking down.

Gene flow—The exchange of genetic material between or among populations through migration.

Gene pool—The total collection of allelic variation in a population.

Genetic drift—Random changes in population allele or genotype frequencies.

Genotype—The genetic makeup of an organism in total; also at a single locus.

Gradualism—The evolutionary theory whereby the accumulation over time of many slightly different intermediate forms has led to large phenotypic differences among organisms.

Group selection—The differential rate of origination or extinction of groups (populations, species, demes) based on differences among them.

Handicap principle—In signal theory, a signal that is too costly for a sender to produce if the information in the signal is not true; guarantees an honest signal.

Haplodiploid—A sex determination-mechanism whereby females are diploid and males are haploid. It is thought that eusociality in hymenopterans is associated with haplodiploidy because female workers are more related to their sisters than their offspring on average.

Haploid—A cell or organism possessing one copy of each chromosome.

Hardy-Weinberg equilibrium—Genotype frequencies reached in a population in which mating is random, there is no gene flow, the population is large, there is no selection,

and mutation is minimal. Under these conditions, genotype frequencies do not change from generation to generation.

Heritability—The proportion of phenotypic variation in a population that is attributable to genetic variation among individuals.

Hermaphrodite—An individual that has both male and female reproductive organs.

Heterochrony—A change in the timing of development that leads to evolutionary change in phenotype.

Heterozygosity—An estimate of the proportion of loci at which an individual randomly chosen from a population is heterozygous.

Heterozygote—In dipoids, an organism that possesses two different alleles at any one locus.

Homeotherm—An organism that maintains its body temperature at a constant level, regardless of ambient conditions. Maintenance may be though metabolic or behavioral means.

Homeotic gene—A gene that determines which parts of the body form what body parts during development.

Homozygote—In diploids, an organism that possesses two copies of the same allele at any one locus.

Hybrid—An individual formed by mating of individuals from genetically differentiated populations or species.

Hybrid zone—A region in which individuals of two genetically distinct lineages come into contact and produce at least some offspring.

Inbreeding—Mating between relatives that occurs more frequently than if mates were chosen at random.

Incipient species—Two populations that are in the process of speciating but have not yet gone to completion.

Inclusive fitness—The total contribution of an allele to the next generation; includes the reproductive success of a focal individual and the reproductive success the same allele in other individuals that the focal individual supports.

Interference competition—A form of competition involving direct interactions between individuals, such as fighting, over limited resources.

Invasive species—A species that is highly successful after arriving in a new geographic location, often to the detriment of native species.

Iteroparous—A reproductive strategy in which an individual will reproduce more than once during its lifespan.

Kin selection—Refers to strategies in evolution that favor reproductive success in an individuals relatives, even at a cost to its own survival or reproduction; helps explain altruism.

Law of independent assortment—Mendel's second law; alleles of different genes assort independently of one another during gamete formation.

Law of segregation—Mendel's first law; when any individual produces gametes, the copies of a gene at any one locus separate such that the gamete gets only one copy.

Lineage—A line of descendants from a single common ancestor.

Linkage disequilibrium—The non-random association of alleles at two or more loci.

Logistic growth—In population biology, population growth that reaches a maximum at the carrying capacity of the environment.

Mass extinction—A sharp decrease in the diversity and abundance of life.

Meiosis—The process of gamete formation whereby an organism produces haploid sperm or eggs. The chromosome number is halved.

Meiotic drive—Any process that causes one gametic type to be over- or underrepresented in the gametes formed during meiosis such that that type is over- or underrepresented in the next generation.

Metabolic rate—The rate at which an organism expends energy.

Metapopulation—A set of local, discrete populations among which there may be geneflow, extinction, and colonization.

Migration—The movement of individuals from one geographic location to the next.

Mimicry—The similarity of one species to another, protecting one or both.

Mitosis—The process of cell division.

Molecular clock—In population genetics, the idea that neutral DNA sequences evolves at a constant rate over time.

Monogamy—A reproductive strategy whereby a single male mates with a single female.

Monophyletic—A clade that has decended from a single common ancestor.

Morphology—The physical form and structure of an organism.

Mullerian mimicry—Mimicry in which both the model and the mimic are noxious and the two species share the burden of educating predators.

Mutation—A change in genetic sequence.

Mutation-order speciation—Speciation that arises when different mutations fix in different populations; these mutations are incompatible in hybrids and reproductively isolate the populations.

Mutualism—Species interaction in which both players benefit from the interaction.

Neontology—The study of living (recent) organisms.

Neutral—In population genetics, an allele or mutation that has no effect on fitness.

Non-synonymous mutation—A mutation that changes the amino acid sequence of a protein.

Oscillating selection—Directional selection that changes direction over time.

Oviparity—The reproductive strategy of laying eggs.

Paleontology—The study of extinct organisms, often from fossils.

Parallel evolution—Similar to convergent evolution, the evolution of similar phenotypes from the same underlying genetic pathways.

Parapatric speciation—Speciation in which two populations abut ranges but evolve reproductive isolation, often during adaptation to different environmental conditions along an environmental cline or gradient.

Parasitism—A symbiotic species interaction in which one organism, the parasite, benefits at the expense of the host.

Phenotype—The physical form of an organism (as opposed to genotype).

Photosynthesis—The conversion of sunlight to energy using chloroplasts.

Physiology—The biochemical functioning of a living organism.

Plate tectonics—The large scale motions of the earth's lithosphere.

Pleiotropy—The case where a single gene has multiple phenotypic effects.

Poikilotherm—An organism that conforms in temperature to the surrounding environment.

Polymerase chain reaction (PCR)—A technique for amplifying a single or few copies of DNA that generates thousands to millions of copies of a particular DNA sequence.

Polymorphism—In population biology, the presence of multiple phenotypes within the same species.

Polyploidization—An error during meiosis whereby the ploidy number is not halved. When these gametes take part in fertilization, they may form a polypoid individual that has a greater number of chromosomes than either parent.

Population—A group of organisms within a defined geographic region.

Prokaryote—A single celled organism that does not have a nucleus or tightly bound chromosomes.

Refugia—In biogeography, a part of a species range where a species may persist while going extinct elsewhere.

Reproductive isolation—In speciation, when individuals no longer mate or are physically capable of mating.

Saltationism—The theory that evolution proceeds through discontinuous phenotypic change resulting from mutation

Selfing—The act of mating with oneself.

Semelparous—A reproductive strategy in which an individual will reproduce only once during its life.

Sexual dimorphism—Differences in phenotype between males and females of the same species.

Sexual selection—Arises from individual interactions between sexes or with sexes for access to mates.

Sister species—Two species that descended directly from the same most recent common ancestor.

Species—The fundamental taxonomic assignment for individuals. There are many different species definitions.

Stabilizing selection—Selection against phenotypes that deviate from the optimal value of a character.

Stasis—The absence of evolutionary change over time.

Sympatric—Pertaining to a population or a species that occupies the same geographic region as another population or species.

Sympatric speciation—The evolution of reproductive isolation between two populations that overlap in space.

Synonymous mutation—A mutation that does not change the amino acid sequence of a protein.

Taxonomy—The practice of assigning individuals to categories (e.g. species, genus, family, etc.).

Vicariance—The separation of a continuously distributed ancestral population into separate populations due to the development of a physical or ecological barrier.

Viviparity—The reproductive strategy of giving live birth.

Wild type—The allele, genotype, or phenotype that is most common in a wild population.

Zygote—A single celled individual formed by the fusion of gametes.

Index

205